MW00844881

Handbook of Lipid Research 1

Fatty Acids and Glycerides

Handbook of Lipid Research

Editor: *Donald J. Hanahan*
The University of Texas Health Center at San Antonio
San Antonio, Texas

Volume 1 *Fatty Acids and Glycerides*
Edited by Arnis Kuksis

Volume 2 *The Fat-Soluble Vitamins*
Edited by Hector F. DeLuca

Handbook of Lipid Research 1

Fatty Acids and Glycerides

Edited by
Arnis Kuksis
Banting and Best Department of Medical Research
University of Toronto
Toronto, Ontario, Canada

Plenum Press · New York and London

Library of Congress Cataloging in Publication Data

Main entry under title:

Fatty acids and glycerides.

(Handbook of lipid research; v. 1)
Includes bibliographies and index.
1. Acids, Fatty. 2. Glycerides. I. Kuksis, Arnis. II. Series. [DNLM: 1. Lipids.
QU85 H236] QP751.H33 vol. 1 [QP752.F35] 574.1'9247s
ISBN 0-306-33581-6 [574.1'9247] 77-25277

A Division of Plenum Publishing Corporation
227 West 17th Street, New York, N.Y. 10011

Printed in the United States of America

Contributors

W. Carl Breckenridge, Departments of Clinical Biochemistry and Medicine, University of Toronto, Toronto, Ontario, M5S 2J5, Canada

Dmytro Buchnea, Banting and Best Department of Medical Research, University of Toronto, Toronto, Ontario, M5G 1L6, Canada

Edward A. Emken, Northern Regional Research Center Agricultural Research Service, U.S. Department of Agriculture, Peoria, Illinois 61604

John L. Iverson, Division of Nutrition, Food and Drug Administration, Washington, D.C. 20204

Arnis Kuksis, Banting and Best Department of Medical Research, University of Toronto, Toronto, Ontario, M5G 1L6, Canada

John J. Myher, Banting and Best Department of Medical Research, University of Toronto, Toronto, Ontario, M5G 1L6, Canada

Patrick J. A. O'Doherty, G. F. Strong Laboratory, Department of Medicine, The University of British Columbia, Vancouver, British Columbia, V5Z 1M9, Canada

Alan J. Sheppard, Division of Nutrition, Food and Drug Administration, Washington, D.C. 20204

John L. Weihrauch, Consumer and Food Economics Institute, Agricultural Research Service, U.S. Department of Agriculture, Hyattsville, Maryland 20782

Foreword

The advances in lipid biochemistry over the past 25 to 30 years have been dramatic and exciting. The elucidation of the pathways of fatty acid biosynthesis and oxidation, the delineation of the biogenesis of cholesterol from small-molecular-weight precursors, the structure proof of simple and complex lipids from plants, animals, and microorganisms, are excellent examples of the spectacular advances made during the golden era of lipid biochemistry. The multifaceted discoveries in these diverse areas of study could be attributed to development of highly sophisticated column chromatographic techniques for separation and purification of simple and complex lipids. The advent of thin-layer chromatography as well as gas–liquid chromatography provided an explosive impetus to research developments in this field. Concomitant advances in mass spectrometry allowed an interface with gas–liquid chromatography which spawned even greater insight into the structure of lipids. These eventful days of lipid chemistry nearly 25 years ago led to a relatively quiescent period wherein scientists applied these newly available techniques to investigation of the behavior of isolated (lipid) enzyme systems and to unraveling the intricacies of the metabolic behavior of lipids in the intact cell or whole organisms.

Then, in the early 1960s, a decided change in research emphasis developed with the advent of a simple, reproducible procedure for the isolation of cell membranes. There can be little challenge to the fact that the ready availability of the hemoglobin-free mammalian erythrocyte membrane was an enormous stimulus to investigators interested in the biochemical characteristics of lipids in these structures. In the ensuing years there has been an exponential rise in studies on membrane structure with considerable attention focused initially on the lipids, which was followed later by excellent studies on the membrane proteins. While this has been a provocative era, it had become increasingly apparent that there was developing a generation of investigators whose background in lipid chemistry and biochemistry was limited. This was understandable in part since many of these scientists had come through the halcyon days of RNA–DNA during which such "mundane" topics as lipids were of minor consideration in the graduate sequence in many departments and laboratories. Certainly, molecular biology had made its mark. This trend was of great concern since there was a limited number of books on lipid chemistry available to persons potentially interested in this topic.

Hence, it was decided to undertake a systematically conceived series of books entitled *Handbook of Lipid Research*. The plan was to have each volume in this series edited by an expert (or experts) in the particular area of interest. The topics for consideration in this series would range from fatty acids and glycerides, phospholipids, sphingolipids to the sterols and steroids, lipid vitamins, and then to the more biological facets of lipids. In order to have this undertaking flourish, it was mandatory that experts in the field of lipid chemistry and biochemistry be invited to contribute. In addition, it was our intention to have each of the individual volumes devoted wherever possible to the chemical and/or biochemical aspects of a particular subject. The basic concept was not to achieve an encyclopedic compendium but rather foster a tightly structured reference text of a reasonable length. Thus, each volume would reflect the individual editor's (or editors') thoughts, ideas, and insights into their area of expertise. As a reference source, these volumes should appeal and be helpful to graduate students, postdoctoral students, investigators who wish a "refresher" in a particular area, and novices in related fields. Even the seasoned investigator should find these books of considerable usefulness. These books will make evident the fact that tremendous progress has been made in lipid biochemistry to date and that the exciting years ahead, in which the structure of the membrane will be forthcoming, depend on a basic understanding of the chemical behavior of lipids.

Donald J. Hanahan

San Antonio

Preface

Fatty acids and acylglycerols are the basic constituents of natural fats where they occur in the free form and as components of complex lipids. The universal distribution and industrial importance of these compounds aroused much early interest in their composition and physicochemical properties, but their biochemical significance was largely dismissed as that of an energy source. The introduction of modern methods of separation and determination of structure of fatty acids and acylglycerols over the past two decades has revealed a great complexity in the composition of natural fats and has led to the recognition that, in addition to providing energy, fatty substances serve a number of other metabolic and physiological functions. Although a wide variety of lipid structures may be compatible with the survival of a biological system, there is good reason to believe that the optimal composition and structure of the fatty acids and acylglycerols of animal species and tissues are genetically controlled.

As a result of this change in the emphasis of the biological role of lipids, special methodology, including isotope dilution and enantiomeric probing, has been developed to deal with the more detailed features of the fatty acid and acylglycerol molecules required for metabolic studies. This volume summarizes current methodology and specific findings in the area of fatty acid and acylglycerol separation, determination of structure and chemical synthesis, along with appropriate tabulations of results and a complete documentation of sources. Chapter 1 describes the specific methodology used and the separations obtained for most classes of natural and synthetic fatty acids. It includes an extensive collection of equivalent chain-length values for the fatty acid methyl esters obtained by gas–liquid chromatography with the more common liquid phases. Chapters 7 and 8 tabulate the fatty acid composition of selected dietary fats and animal tissues and provide a critical commentary on the interpretation of these data from the points of collection of representative samples and preparation and quantitation of fatty acids. Chapter 2 describes the synthesis and determination of structure of isotope-labeled fatty acids, while Chapter 5 describes the synthesis and proof of structure of enantiomeric acylglycerols containing both natural and isotope-labeled acids. These chapters contain extensive tabulations of synthetic fatty acids (Chapter 2) and acylglycerols (Chapter 5). Both radioactive- and stable-isotope-labeled fatty acids and acylglycerols have been extensively employed in metabolic studies, as discussed in Chapter 6, along with appropriate listing of specific applications and general routines. Chapters 3 and 4 cover the separation and determination of structure of natural acylglycerols and contain a wealth of tabulated data on the chromatographic behavior of different acylglycerols and on the stereochemical structure of natural triacylglycerols, respectively. The various experimental routines are critically appraised, and specific recommendations for improved analyses are made.

All the contributors have themselves done valuable research in the fatty acid and glycerolipid field, and several of them give personal accounts of their work that would not appear in an ordinary scientific paper. While the discussion has been oriented primarily toward persons engaged in lipid research, the detailed account of the analytical behavior and composition of natural and synthetic fatty acids and acylglycerols provides a unique compendium of knowledge, which should serve as an important reference to workers in related areas, and hasten progress in the understanding of the role of fatty acids and acylglycerols in biochemical and metabolic transformations and in cellular structure.

I express my thanks to the various authors for the excellent coverage and critical review of the material assigned. A special debt of gratitude is owed to Dr. D. J. Hanahan for advice in the selection of the topics and for a constructive criticism of the individual contributions.

Arnis Kuksis

Toronto

Contents

Chapter 1

Separation and Determination of Structure of Fatty Acids

Arnis Kuksis

Chapter 2

Synthesis and Analysis of Stable Isotope- and Radioisotope-Labeled Fatty Acids

Edward A. Emken

Chapter 3

Separation and Determination of the Structure of Acylglycerols and Their Ether Analogues

John J. Myher

Chapter 4

Stereospecific Analysis of Triacylglycerols

W. Carl Breckenridge

Chapter 5

Stereospecific Synthesis of Enantiomeric Acylglycerols

Dmytro Buchnea

Chapter 6

Metabolic Studies with Natural and Synthetic Fatty Acids and Enantiomeric Acylglycerols

Patrick J. A. O'Doherty

Chapter 7

Composition of Selected Dietary Fats, Oils, Margarines, and Butter

Alan J. Sheppard, John L. Iverson, and John L. Weihrauch

Chapter 1

Separation and Determination of Structure of Fatty Acids

Arnis Kuksis

*1.1. Introduction**

The development of precise methods of separation and identification of structure has revealed a great diversity and complexity in the fatty acids prepared from various natural fats. In the past quarter of a century, different positional and geometric isomers of long-recognized and commonly occurring ethylenic fatty acids have been detected and isolated, as have been previously unknown homologues of these acids. In addition, fatty acids with odd numbers of carbon atoms, branched-chain acids, hydroxy and epoxy acids, acetylenic and mixed ethylenic–acetylenic acids both conjugated and unconjugated, and cyclopropanyl and cyclopropenyl acids have been isolated and identified. In many fats the less common acids are present in only minor amounts, while in other fats these acids may represent major components. Since the discovery of gas chromatography, over 500 different fatty acids have been resolved from natural sources or from synthetic products.

* General Abbreviations: TLC, thin-layer chromatography; GLC, gas–liquid chromatography; LLC, liquid–liquid chromatography; GC/MS, gas chromatography/mass spectrometry; 16:0, fatty acids with 16 acyl carbons and zero double bonds; 18:1 $\omega 9$, fatty acid with 18 acyl carbons and 1 double bond located 9 carbons from the methyl terminal; Δ^2–Δ^{18}, fatty acids with double bonds located 2 to 18 carbons away from the carboxyl end of the chain; lc, mt, nk, fatty acids with *cis* and *trans* double bonds or keto groups at carbons number 1, m, and n from the carboxyl end of the chain; saturates, monoenes, dienes, and polyenes, fatty acids with none, one, two, or many double bonds per molecule; *sn*-, stereospecific numbering of glycerol carbons. Abbreviations of trade names of stationary phases: APL and APM, Apiezon L- and M-saturated hydrocarbon greases; BDS, butane-1,4-diol succinate; Carbowax, polyethyleneglycol; CDX, cyclodextrin esters of short-chain acids; CN, cyanoethylsiloxane; DEGA, diethyleneglycol adipate; DEGS, diethyleneglycol succinate; EGP, ethyleneglycol phthalate; EGS, ethyleneglycol succinate; EGSS-X and EGSS-Y, methylsiloxane-ethyleneglycol succinate polymers; EGSP-Z, phenylsiloxane-ethyleneglycol succinate polymer; FFAP, free fatty acid phase; NPGA, neopentylglycol adipate; NPGS, neopentylglycol succinate; PEGA, polyethyleneglycol adipate; PPE, polyphenyl ether; QF-1, fluoroalkylsiloxane; Reoplex 400, polypropyleneglycol adipate; SE-30 and OV-101, methylsiloxane; Silar 10C, 100% cyanopropylsiloxane; TCPE, tetracyanoethylated pentaerythritol; TXP, trixylenyl phosphate; Versamid, polyamide resin; XE-60, cyanoethylsiloxane; XF-1150, General Electric silicone fluid.

Arnis Kuksis • Banting and Best Department of Medical Research, University of Toronto, Toronto, Ontario, M5G 1L6, Canada.

No single analytical method is capable of effectively separating and identifying all of these acids. Usually a combination of several complementary methods is necessary to separate and identify the homologues of any one acid type. Severe limitations on the choice of the methodology are imposed by the sample size and the time required for analysis. Fortunately, the most efficient methods of analysis also require the least sample. It is the purpose of this chapter to survey the more practical methods presently available for complete separation and identification of the vast majority of fatty acids encountered in natural samples of lipids routinely obtained in a research laboratory.

1.2. Preparation and Isolation of Fatty Acids

Prior to separation and identification, it is convenient to obtain natural fatty acids either as free acids or as their simple alkyl esters. This usually involves saponification or transesterification of a fat, but enzymatic hydrolysis may also be used to effect a complete or selective release of fatty acids from natural fats. Great care must be taken during the preparation to avoid losses of short-chain and polar fatty acids, to prevent degradation of and artifact formation from the chemically less stable acids, and to minimize the contamination of all acids from equipment and solvents. Polyunsaturated acids must be protected by inert atmosphere and antioxidants during chemical manipulations and storage (Holman and Rahm, 1966; Schauenstein, 1967; Lundberg and Jarvi, 1970).

1.2.1. Generation of Free Acids

An initial isolation of the fatty acids in the free form from a natural fat serves as the first step in their identification. The free fatty acids are more readily converted into a variety of derivatives necessary for identification than are fatty acid esters.

1.2.1.1. Saponification

A major advantage of saponification is the possibility of removing nonacidic lipids by extraction with organic solvents without resorting to chromatography. However, double-bond migration may take place during saponification, and unconjugated polyunsaturated fatty acids may be converted to conjugated isomers (Markley, 1968). Conjugated acids may undergo polymerization (Hopkins, 1972), and all types of unsaturated acids are subject to autoxidation (Holman and Rahm, 1966). Bottcher *et al.* (1959) have shown that alkali concentrations higher than 0.5 N cause losses of polyunsaturated acids, as do refluxing times exceeding 1 hr.

According to Kates (1972), 15–32 mg of total lipid is refluxed under nitrogen atmosphere with 5 ml of 90% methanolic NaOH (0.3 N) for 1–2 hr. After dilution with aqueous methanol, the nonsaponifiable matter is removed with petroleum ether. The alcoholic phase is acidified with 6 N HCl and reextracted with petroleum to obtain the free fatty acids. If short-chain acids are present, they may be removed by steam distillation after acidification; the residue and the steam distillate are extracted separately (Hilditch and Williams, 1964). This method of hydrolysis does not release fatty acids from amide linkages such as those in sphingolipids. It also retains the long-chain aldehydes in the vinyl ether linkages of plasmalogens, which is a further advantage in the preparation of pure fatty acids.

1.2.1.2. Lipolysis

Various lipases and esterases are known to release fatty acids from the esters of glycerol, cholesterol, and other alcohols (Brockerhoff and Jensen, 1974). Lipolytic hydrolysis of fats occurs with little or no alteration of the structure of the acyl groups, which is a great advantage when dealing with fatty acids of low stability toward heat, acids, and alkalis. There is evidence for different rates of hydrolysis of esters containing acyl groups of different chain length and unsaturation, as well as evidence for different rates with different position of the glycerol molecule.

An effective release of fatty acids from the *sn*-1 and *sn*-3 positions of triacylglycerols may be obtained by means of pancreatic lipase. According to Luddy *et al.* (1964), the triglyceride is dissolved in a small amount of diethyl ether or hexane and added to the buffered enzyme solution. The hydrolysis is conducted at 40°C for 5–10 min, and the reaction is stopped by the addition of alcohol. The enzyme discriminates against acylglycerols with certain fatty acids (Anderson *et al.*, 1967; Kleiman *et al.*, 1970). The pentaenoic and hexaenoic acids which are prominent in fish oils are quite resistant to lipase (Bottino *et al.*, 1967; Brockerhoff, 1971), as are acids with conjugated double bonds (Brockerhoff, 1971). Of the shorter-chain fatty acids, butyric acid is split fastest by the lipase (Brockerhoff, 1971). The lipase from *Rhizopus arrhizus* also attacks the *sn*-1 and *sn*-3 positions of acylglycerols but shows less fatty acid specificity (Fischer, 1973).

A selective release of natural ethylenic fatty acids from any position of a triacylglycerol molecule may be obtained by means of a lipase from the microorganism *Geotrichum candidum*. This lipase preferentially hydrolyzes *cis*-9 18 : 1 and *cis,cis*-9,12 18:2 from triacylglycerols, largely ignoring all other positional isomers of *cis* 18 : 1 as well as *trans*-9 18 : 1 (Jensen *et al.*, 1972). Subsequent studies have shown that the enzyme discriminates also against the *trans,trans* isomer but not between the *cis,trans* and *trans,cis* isomers (Jensen *et al.*, 1973).

Castor bean lipase is another readily prepared lipase which specifically catalyzes the hydrolysis of fatty acids from the *sn*-1 and *sn*-3 positions of triacylglycerols. The enzyme hydrolyzes most readily triacylglycerols of saturated fatty acids having chain lengths between C_4 and C_8, although its natural substrate, castor oil, is also rapidly hydrolyzed (Ory *et al.*, 1969). Castor bean lipase is inhibited by sterculic acid.

An essentially complete release of the common fatty acids from the *sn*-2 position of *sn*-3-phosphatides may be obtained by phospholipase A_2 of snake venoms (Nutter and Privett, 1966). This enzyme also specifically hydrolyzes 1,2-diacyl-*sn*-glycerol-3-phosphophenols, which forms the basis of a stereospecific analysis of triacylgycerols (Brockerhoff, 1971). The reaction is most effectively conducted in the presence of organic solvents such as diethyl ether (Wells and Hanahan, 1969). In general, the saturated fatty acids are liberated faster than the unsaturated ones, and there are differences in the rates of hydrolysis of individual saturated and unsaturated acids (Nutter and Privett, 1966). From synthetic compounds in which the phosphate group is attached to the secondary hydroxyl group of glycerol, the enzyme will release the acyl group at the *sn*-1 position, although at a slower rate (Van Deenen and De Haas, 1963).

Various other lipases and phospholipases have been described (Brockerhoff and Jensen, 1974) which also could be employed for the release of fatty acids from either the *sn*-1 or *sn*-2 positions of glycerolipids.

1.2.1.3. Purification of Free Acids

Since the enzymatic lipolysis usually does not go to completion, the released fatty acids are contaminated with the starting materials and any intermediate transformation products. Likewise, the free fatty acids released from the soaps upon acidification of the saponification mixture may become contaminated with esters as a result of partial reesterification of the acids with the alcohol in the solution. Therefore, the free fatty acids from both lipolysis and saponification must be purified before preparation of derivatives.

For the purpose of gas–liquid chromatography (GLC), sufficient free fatty acid may be purified by thin-layer chromatography (TLC) using a neutral lipid system which contains acetic acid to ensure that all the fatty acids are in the un-ionized form during chromatography. Two systems that give good results since they separate free fatty acids of various chain length from cholesteryl esters, triacylglycerols, partial acylglycerols, free cholesterol, and phospholipids, as well as from fatty acid methyl esters, are hexane–diethyl ether–acetic acid 85 : 15 : 2 (Gloster and Fletcher, 1966) and heptane–isopropyl ether–acetic acid 60 : 40 : 4 (Breckenridge and Kuksis, 1968). The free fatty acids are recovered from the TLC plates by extraction with chloroform–methanol–acetic acid 40 : 20 : 1. The acidity of these solvent systems, however, may interfere with the detection of the shorter-chain fatty acids. This disadvantage does not occur with the solvent system methanol–water 35 : 65, which gives excellent separations of the C_3–C_9 acids in the free form on silanized silica gel (Rodrigues de Miranda and Eikelboom, 1975).

Small amounts of free fatty acids may be recovered from mixtures with large amounts of neutral esters by ion exchange chromatography (Goodridge, 1972).

1.2.2. Preparation of Derivatives

Although free fatty acids of short chain length may be effectively employed for both chromatographic separation and instrumental identification, the longer-chain acids are best handled in the form of simple alkyl esters. The preparation of alkyl esters from free fatty acids and lipids has been reviewed and specific recommendations have been made (Christie, 1972; Sheppard and Iverson, 1975). The preparation of other esters and other derivatives, however, may provide advantages for special separations and identification (Drodz, 1975).

1.2.2.1. Esterification

For preparing methyl esters of heat- and acid-stable fatty acids 2–3% sulfuric acid in anhydrous methanol may be used either at or above reflux temperature in a sealed tube. Dry methanol containing 5% hydrogen chloride is another effective methylating reagent. Since the neutral lipid esters are not appreciably soluble in methanol, an inert solvent such as benzene may be added to increase the solubility of the lipid without affecting the course of reaction. The methanol–hydrogen chloride reagent can be used to methylate free fatty acids in unsaponifiable fat by first allowing the free acids to be absorbed on a strong anion exchange resin and then esterifying the acids directly in the presence of the resin at room temperature (Hornstein *et al.*, 1960). Lankin *et al.* (1974) reported 97–98% yields of fatty acid methyl esters by obtaining calibration curves for each fatty acid to be assessed by this method.

Small quantities of free acids may be methylated for GLC in a few minutes using boron trifluoride–methanol (Morrison and Smith, 1964). The formation of methoxy-substituted acids from conjugated unsaturated fatty acids is a problem (Koritala and Rohwedder, 1972), as is the cleavage of cyclopropane ring (Minnikin, 1972; Vulliet *et al.*, 1974).

Acid catalysis at elevated temperatures has frequently been employed to prepare the higher-molecular-weight alcohol esters of the short-chain fatty acids, which simplifies the handling of these acids because of the lower volatility of these esters (Sheppard and Iverson, 1975). Reaction of methyl iodide with silver salts of fatty acids has been recommended for samples containing a wide range of acids, including those of short chain length. An "on-column" modification of this technique has proved useful for gas chromatography/mass spectrometry (GC/MS) (Johnson and Wong, 1975).

Thenot *et al.* (1972) have prepared methyl, ethyl, *n*-propyl, *n*-butyl, and *t*-butyl esters of long-chain fatty acids using the *N,N*-dimethylformamide dialkylacetals as catalysts. This procedure appears to need further refinement before it can be used routinely.

The *p*-bromophenacyl and *p*-phenylphenacyl esters of C_2–C_{20} acids, the benzyl esters of C_1–C_{20} acids (Hintze *et al.*, 1973), and the 2-naphthacyl esters of long-chain fatty acids (Cooper and Anders, 1974) have been obtained at moderate temperatures by reacting the appropriate alkyl bromides with the free fatty acids in the presence of *N,N*-diisopropylethylamine in dimethylformamide. The latter esters absorb in the UV light and may be used to facilitate the detection of fatty acids during liquid–liquid chromatography (LLC).

Free acids that are sensitive to heat, acids, and alkalis may be readily methylated using diazomethane in the presence of a small amount of methanol (Schlenk and Gellerman, 1960).

Frequently, it is convenient to convert the free fatty acids into simple alcohol esters in the presence of silica gel from the scrapings of the TLC plates on which the fatty acids have been isolated. For this purpose, the dry silica gel is collected in screw-capped (teflon-lined) vials, and enough boron trifluoride (14%) in alcohol is added to cover the powder. The closed vials are heated at 110°C for 5–10 min (Nelson, 1972).

1.2.2.2. Silylation

Trimethylsilyl (TMS) esters of fatty acids are prepared by exposing the dry acid mixture to a suitable silylating reagent, such as hexamethyldisilazane and trimethylchlorosilane or bistrimethylsilylacetamide (Tallent and Kleiman, 1968; Kuksis *et al.*, 1968). These esters are readily formed from the free fatty acids in the presence of other esters of fatty acids (Kuksis *et al.*, 1969, 1976). The TMS esters are highly sensitive to moisture and are best preserved in the silylating reagent. Much more stable are the *t*-butyldimethylsilyl esters, which can be prepared using the corresponding silylating reagent (Phillipou *et al.*, 1975).

The silylating reagents also react with any free hydroxyl groups that may be present in the fatty acids, which provides a further advantage for gas chromatography as well as for mass spectrometry (Tallent and Kleiman, 1968).

Weihrauch *et al.* (1974) have prepared the phenylhydrazine derivatives of keto acids and Langenbeck *et al.* (1975) the quinoxalinols of α-keto acids as an aid to their GLC and mass spectrometric separation and identification. Prior to analysis, the quinoxalinols are converted into the *O*-trimethylsilyl ethers by means of simple or deuterated silylating reagents.

1.2.3. Transesterification of Bound Acids

In most instances, it is convenient to prepare the fatty acid esters of simple alcohols directly by transesterification of the fat. The major disadvantage of the alcoholysis method is that it does not permit a ready separation of the unsaponifiable materials. The methods include both alkaline and acid catalysis.

1.2.3.1. Alkali-Catalyzed Alcoholysis

The alkali-catalyzed reaction can be carried out at or slightly above room temperature, which is of special interest for the alcoholysis of fats containing temperature-sensitive constituents (Luddy *et al.*, 1968). The method recommended by Glass (1971) utilizes a stable reagent consisting of 4% NaOH in a solvent consisting of 60 parts anhydrous methanol and 40 parts reagent-grade benzene. The methanolysis of most lipid esters is accomplished by simply dissolving the lipid in the reagent to the desired concentration (1–10%). The methanolysis of glycerol esters is complete in 1–2 min, while cholesteryl esters may take 15–20 min for complete reaction. The reaction mixture is neutralized by the addition of methanolic HCl. GLC analysis may be done without further treatment of the reaction mixture.

The alkali-catalyzed transesterification may also be performed in the presence of the silica gel scrapings. According to Svennerholm (1968), the P_2O_5-dried scrapings are shaken with 2 ml of 0.1 N sodium methoxide in dry methanol for 1 hr at room temperature. Then 0.2 ml of 1 N acetic acid is added, and the methyl esters are extracted with light petroleum.

1.2.3.2. Acid-Catalyzed Alcoholysis

Alcoholysis with acid catalysts is usually carried out at elevated temperatures for prolonged periods of time and may lead to a partial or complete destruction of certain fatty acids. Usually the methyl esters of fatty acids are prepared by refluxing with 5% hydrochloric acid in methanol. The use of 7.5% and 12% sulfuric acid in methanol is also effective (Christie, 1972; Sheppard and Iverson, 1975).

Acid-catalyzed alcoholysis at or slightly above room temperature may be carried out with methanolic boron trifluoride (Morrison and Smith, 1964). This catalyst can be used in combination with other higher alcohols. However, losses of polyunsaturated fatty acids may occur if the reaction is prolonged (Kleiman *et al.*, 1969), especially if aged reagents are used.

An acid-catalyzed transesterification in the presence of silica gel scrapings from TLC plates may be accomplished with a variety of transmethylating reagents. According to Nelson (1972), about 5 ml of the transmethylating reagent is added to a tube containing the sample and absorbent and the vials are closed with teflon-lined screw caps. The tubes are then heated for a minimum of 2 hr at 90–95°C. After cooling to room temperature, 10 ml of distilled water is added to each tube. The methyl esters are extracted with a minimum of three washes of hexane containing 0.01% butylated hydroxytoluene.

1.2.3.3. Purification of Fatty Acid Esters

Since the fatty acid esters prepared by transesterification contain materials other than the alkyl esters, the sample must be purified, which can best be accomplished by chromatography on a small silicic acid column (Nelson, 1972). The fatty acid methyl esters are recovered by elution with 1% diethyl ether in hexane. The remainder of the lipids in the sample may be recovered by elution with methanol. The fatty acid methyl esters may also be purified by TLC. For this purpose, a plain silica gel H plate is developed with pure benzene and the bands are visualized with water (Nelson, 1972). The methyl esters of both short- and long-chain-length fatty acids are widely separated from hydrocarbons and any butylated hydroxytoluene added as antioxidant, which run ahead, and from cholesterol and more polar compounds, which remain at or near the origin.

1.3. Preliminary Fractionation of Fatty Acids

In addition to chain length and unsaturation homologues, the fatty acid mixture may contain keto, hydroxy, cyclopropane, and epoxy derivatives, as well as dimethylacetals. A preliminary chromatography of the fatty acids serves to ensure the homogeneity and uniformity of the fatty acid class preparation. The more effective chromatographic systems, however, are capable also of effecting a significant resolution among the different molecular species of the fatty acids (chain-length, positional, and geometric isomers), which provides important criteria for the subsequent identification of the acids by GLC and by GC/MS. Of the methods of preliminary resolution of fatty acids, the most important one is argentation TLC, but TLC with borate or on plain silica gel, preparative GLC, LLC, countercurrent distribution, steam distillation, and simple solvent extraction and crystallization with urea may also be employed when necessary.

1.3.1. Conventional TLC

Preliminary fractionation of fatty acid methyl esters by conventional TLC is based largely on differences in the polarity of fatty acids of different chain length and on the presence or absence of any functional groups. In large-scale experiments, the TLC plate may be replaced by an adsorption column (Carroll, 1976).

1.3.1.1. Separations Based on Nature and Number of Functional Groups

Simple esters of fatty acids of long and short chain length may be effectively resolved by TLC on plain silica gel with petroleum ether–diethyl ether 100 : 1. Isolated or conjugated double or triple bonds have little or no effect on mobility relative to the saturated esters (Morris and Nichols, 1970). Likewise, the cyclopropane and cyclopropene fatty esters migrate with the corresponding saturated esters (Christie, 1970). More complex cyclic fatty acids, however, may be resolved from their normal-chain isomers by TLC on plain silica gel (Stoffel and Michaelis, 1976).

An effective separation of mixed methyl esters including epoxy and hydroxy derivatives may be achieved by TLC in hexane–diethyl ether 80 : 20 (Kleiman *et al.*, 1971). Methyl esters of unsubstituted and monoketo and monohydroxy fatty acids may be resolved by TLC in petroleum ether–diethyl ether 60 : 40 or 75 : 25 (Morris *et al.*, 1968). In all instances, the unsubstituted esters migrated the farthest and the hydroxy acid esters the least distance. Excellent separations of the methyl esters of 12-hydroxy-, 9,10-dihydroxy-, 9,10,12-trihydroxy-, and 9,10,12,13-tetrahydroxy stearic acid have been reported by Wood *et al.* (1966) using plain silica gel and chloroform–methanol 92 : 8 as the developing solvent.

Methyl esters of simple and hydroxy fatty acids may be resolved from the dimethylacetals of aldehydes, such as those derived by TLC from plasmalogens after methanolic-HCl treatment, using benzene or toluene as the developing solvent (Morrison and Smith, 1964).

1.3.1.2. Separation of Positional and Geometric Isomers

Detailed investigations in several laboratories have shown that there exist considerable and systematic variations in mobilities of positional and/or geometric isomers of a wide range of substituted fatty acids. Thus Morris and Wharry (1965) were able to show that, of the four *cis*-epoxy C_{18} esters tested, the 6,7 compound was the least mobile of the series, the 9,10 compound was intermediate, and the 12,13-epoxy ester was the most mobile when chromatographed in diethyl ether–light petroleum 1 : 4. Complete separations of the *cis* and *trans* isomers of each compound were obtained, with the *trans* epoxide being the least polar in all instances.

The positional isomers of methyl hydroxystearic acid produce a sinusoidal pattern of spots when chromatographed side by side on a TLC plate in diethyl ether–light petroleum 1 : 1 (Morris and Wharry, 1965). The 18-hydroxystearate, which has a primary hydroxyl group, is more polar than any of the secondary hydroxy isomers. The secondary hydroxy isomers showed increased mobilities from the 17- to the 13- and the 12-hydroxystearates, and then steadily decreasing mobilities down to 5-hydroxystearate, with the 4-, 3-, and 2-hydroxy isomers again being progressively less polar. A similar order of mobilities was observed for the corresponding keto and acetoxy isomers of stearic acid (Morris *et al.*, 1968).

The effect of positional isomerism on TLC mobilities is also seen in the chromatography of vicinal dihydroxy esters. The pattern of migration is the same as that seen for the monohydroxy, acetoxy, and epoxy analogues of these esters. The differences in mobilities between the positional isomers of the *threo* and the *erythro* series of esters observed were sufficient to allow their separation from mixtures of all four samples (6,7-, 9,10-, 12,13-, and 13,14-diols) investigated (Morris and Wharry, 1965).

Geometric isomerism has only a marginal effect on the relative mobilities of *erythro*- and *threo*-dihydroxy esters which may not be sufficient to allow separation of a mixture of a pair of geometrically isomeric esters. Substantial differences in mobilities of *threo*- and *erythro*-dihydroxy isomers are obtained by chromatography on silica gel impregnated with boric acid, sodium borate, or sodium arsenate (Morris, 1966; Wood *et al.,* 1966). Wood *et al.* (1966) have resolved each of the diastereoisomeric methyl esters of the tri- and tetrahydroxystearates (9,10,12-triols and 9,10,12,13-tetraols) on plain silica gel, with the arsenate-treated plates giving somewhat better resolution.

1.3.2. Argentation TLC

Unsaturated fatty acids with both normal and substituted chains are readily resolved by argentation TLC, which retards the migration of the more unsaturated species in relation to that of the more saturated ones. Pure fractions of a uniform number of double bonds may therefore be obtained for each class of fatty esters. In addition, argentation TLC is also capable of considerable resolution of positional and geometric isomers of unsaturated fatty acids of both ethylenic and acetylenic series.

1.3.2.1. Separations Based on Number of Double and Triple Bonds

The methyl esters of normal-chain fatty acids may be resolved according to the number of double bonds by argentation TLC using in succession diethyl ether–light petroleum 5:95 and methanol–diethyl ether 1:9 (Morris, 1966). The first development separates saturates, monoenes, and dienes, and leaves the polyenes at or near the origin. The second development allows the resolution of the polyenes into separate bands. A single-step silver nitrate TLC procedure which separates a methyl ester mixture containing 0–6 double bonds has been described by Dudley and Anderson (1975). For this purpose, a 4% silver nitrate plate is developed twice with a solvent system containing hexane diethyl ether–acetic acid 94:4:2. After the first development, the plate is dried under nitrogen. There is some separation on the basis of chain length within each double bond class (Ackman and Castell, 1966).

Cyclopropanoid fatty acids are recovered with the saturated fatty acids upon argentation TLC (Morris, 1966), but the method is not compatible with the cyclopropenoid acids, which are cleaved quantitatively when deliberately treated with silver nitrate (Christie, 1970). Argentation TLC, however, has proven to be effective for separation of the cyclopentenyl acids from chaulmoogric oil (Mani and Lakshminarayana, 1969).

Unsaturated epoxy esters may be resolved by argentation TLC with benzene–chloroform–diethyl ether 50:50:2 as the developing solvent, which yields separate bands for saturated, monoenoic, and dienoic epoxy esters (Kleiman *et al.,* 1971). Hydroxy fatty esters of different degrees of unsaturation may be resolved by argentation TLC using benzene–chloroform–diethyl ether 50:50:15 as the developing solvent (Kleiman *et al.*, 1971). Furthermore, dihydroxy fatty esters have been separated on the basis of both ethylenic unsaturation and the *threo* or *erythro* configuration of the vicinal glycol groups by double impregnation with silver nitrate and boric acid (Morris, 1966).

An isolated triple bond complexes with $AgNO_3$ about as strongly as does an isolated *cis*-ethylenic double bond, and allows comparable separations between saturated and unsaturated fatty acids (Morris, 1966). Conjugated triple bonds and conjugated ethylenic and triple bonds complex with $AgNO_3$ less readily and become less retarded on argentation TLC. This method therefore has not yet achieved total resolution of mixtures of fatty acid esters containing ethylenic and acetylenic bonds. Nevertheless, argentation TLC has been effective in separating these esters into fractions with and without terminal ethylenic bonds, so that individual components of these fractions can be characterized (Morris, 1966, 1970; Barve *et al.*, 1972).

1.3.2.2. Separation of Positional and Geometric Isomers

Gunstone *et al.* (1967) have examined a complete series of the *cis*- and *trans*-octadecenoates and have shown that their R_f values on the silver nitrate plates conform to a sinusoidal curve similar to that which Morris and Wharry (1965) had observed for the isomeric oxygenated octadecanoates on conventional TLC plates. The *trans* esters (Δ^2–Δ^{16}) formed a very shallow sinusoidal curve with a minimum for the Δ^6 isomer. The *cis* esters lie on a less shallow curve, with lower R_f values than their *trans* isomers except for the Δ^2 ester. Therefore, the separation of individual isomers was easier among the *cis* esters than among the *trans* esters. The most useful separations were possible among the Δ^6, Δ^9, and Δ^{12} isomers.

Christie (1968) has examined the silver ion TLC behavior of the isomeric methylene-interrupted methyl *cis,cis*-octadecadienoates and has found that the migration pattern of the isomers also conforms to a sinusoidal curve. The methyl *cis,cis*-2,5-octadecadienoate had an unusually high R_f value in comparison to other members of the series. The 3,6 to 6,9 isomers could be separated, but the remainder conformed to a gentle curve with a slight but real drop at the 12,15 isomer. A similar pattern was observed for the monoenoic esters (Gunstone *et al.*, 1967), although the curve was more marked and better separations were achieved.

The geometric isomers of methyl 9,12-octadecadienoate and methyl 9,12,15-octadecatrienoate have been resolved by Strocchi and Piretti (1968) by argentation TLC on plates prepared in the presence of ammonia using benzene–light petroleum 30:70 (methyl linoleate isomers) and benzene–light petroleum 40:60 (methyl linolenate isomers) as the developing solvents. The dienoic esters were completely resolved into three classes containing all-*trans*, *trans-cis* and *cis-trans*, and all-*cis* isomers. The trienoic esters were similarly separated into four classes containing all-*trans*, di-*trans*-mono-*cis*, di-*cis*-mono-*trans*, and all-*cis* isomers. In all instances, the molecules with the higher *trans* double-bond content were seen to migrate ahead of those of the lesser *trans* double-bond content. The all-*cis* isomers were retained most strongly by the adsorbent. The results with the double bonds in the 9,12 and 9,12,15 positions suggested that *cis-trans* isomers with double bonds in other positions will also be separated, but this possibility awaits experimental demonstration.

Argentation TLC also separates conjugated dienoic esters into *cis,cis*, *cis-trans* plus *trans-cis*, and *trans,trans* fractions (Emken *et al.*, 1967) and conjugated trienoic acids into all-*cis*, di-*cis*-mono-*trans*, mono-*cis*-di-*trans*, and all-*trans* fractions (Emken *et al.*, 1967; Scholfield *et al.*, 1967). The component mixed *cis-trans* fractions can then be completely resolved by capillary GLC, thus providing the analytical basis for detailed study of alkaline isomerization of polyunsaturated fatty acids, as well as of the intermediate products of hydrogenation of fatty acids.

The unsaturated epoxy esters can be readily differentiated from their saturated analogues by argentation TLC. The position of the double bond in the chain and its relationship to other adsorptive groups have an effect on the degree of such resolution. Thus, although the 9,10- and 12,13-epoxy isomers are clearly separated on plain silica gel, they migrate together on silica gel treated with silver nitrate (Morris and Wharry, 1965).

Hydroxy esters with isolated unsaturated linkages are all retained longer and are clearly separated from their saturated analogues, but the degree of such retention is not always the same (Morris and Wharry, 1965). The differences are due partly to the position of the double bond in the chain, which in the case of the unsubstituted monoenoic esters is sufficient to permit separation of isomers such as 6-, 9-, and 11-monoenes (Bergelson *et al.*, 1964; Morris and Wharry, 1966), and partly to the relative positions of the double bonds and the hydroxyl and carbomethoxyl groups. Conjugated dienols also have been retarded by complexing to some extent. The degree of complexing of such compounds, however, is much less than for an isolated double bond (Morris and Wharry, 1965).

1.3.3. Other Methods

In addition to the readily employed adsorption chromatographic techniques, other chromatographic methods may be employed for a systematic preliminary segregation of the fatty acid mixtures as an aid in separation and identification. Of the latter methods, preparative GLC and LLC are the most effective, but in specific instances the older distillation and crystallization methods can also be utilized to good advantage.

1.3.3.1. Preparative Chromatographic Methods

An effective preliminary separation of fatty acids by molecular weight or carbon number may be obtained on short nonpolar columns by preparative GLC (Smith and White, 1966; Kuksis and Ludwig, 1966). Longer columns prepared with more polar liquid phases allow a separation of many unsaturated fatty acids as well as saturated acids (Organiscian, 1974).

A separation technique which has recently gained increasing attention is liquid–liquid chromatography (LLC) of fatty acids or esters (Scholfield, 1975*a*, *b*). From a commercial hydrogenated vegetable oil, six separate fractions were obtained consisting of octadecatrienoate, octadecadienoate, *cis*-octadecenoate, *trans*-octadecenoate with *cis*-15-octadecenoate, palmitate, and stearate. In this chromatographic system, the effect of adding a double bond was generally greater than that of removing two methylene groups, and the effect increased with the chain length of the ester. Palmitate was separated from oleate, and stearate from eicosenoate. Laurate, myristoleate, and linolenate were found in the same fraction. Geometric isomers of fatty acids in milligram quantities may be separated by reversed-phase LLC on a Bondapak C-18 (Warthen, 1975). The application of reversed-phase LLC to the preparation of highly purified methyl esters of fatty acids has been described by Privett and Nickell (1963) and Privett (1968). Methyl linolenate, arachidonate, eicosapentaenoate, docosapentaenoate, and docosahexaenoate were prepared in high purity.

A comparable older method of fractionation of fatty acids by liquid–liquid partition is countercurrent distribution (Therriault, 1963; Scholfield *et al.*, 1967).

1.3.3.2. Distillation Methods

Distillation methods belong to the gross methods of fractionation which are involved in the initial stages of isolation of fatty acids for complete identification. Fractional distillation when carried out with all-glass apparatus at the highest possible vacuum leads to alteration of polyunsaturated fatty acids containing 4, 5, and 6 double bonds (Privett and Nickell, 1963). Many stable mixtures of fatty acid esters, however, have been effectively fractionated by distillation preliminary to preparative GLC isolation (Pelick *et al.*, 1963).

Steam distillation of short-chain fatty acids is an old method of obtaining mixtures of volatile fatty acids (Hilditch and Williams, 1964), but it has found a new lease on life despite recent advances in chromatography (Perry *et al.*, 1970). Depending on the time allowed for steam distillation, the distillate contains variable amounts of the medium-chain-length C_8–C_{12} fatty acids, which on short runs are largely retained in the aqueous phase.

1.3.3.3. Crystallization

Crystallization also belongs to the gross methods of preliminary fractionation of fatty acids. It is usually preceded by urea adduct formation, which removes most of the straight-chain esters, leaving the branched-chain, cyclic, and polyunsaturated fatty acids in solution (Cason *et al.*, 1953). This method has provided large amounts of various reference materials from marine oil fatty acid esters (Ackman, 1972). The conjugated and polyunsaturated fatty acids can be concentrated by crystallization of the mixed acids from acetone at low temperature (Hopkins, 1972). This crystallization method has been applied to all types of conjugated acids, including dienes, enynes, and hydroxy acids. Polyunsaturated fatty acid esters may also be concentrated by nitromethane (Jangaard, 1965). The esters are dissolved in 1 ml of nitromethane, and the mixture is chilled to –20°C. The more highly unsaturated esters pass into the lower nitromethane-rich layer from which they may be recovered by phase separation.

1.4. Identification of Fatty Acids by Gas–Liquid Chromatography

Gas–liquid chromatography constitutes the most powerful single method of separation of complex mixtures of natural fatty acids, and effective correlations exist between the retention time and structure of an acid (James, 1960; Miwa *et al.*, 1960; Ackman, 1972). The large number of isomeric components potentially present in a fatty acid mixture, however, prevents a positive identification of components except in the simplest mixtures. A preliminary argentation TLC provides a substantial increase in the certainty of tentative identification of most fatty acids (Kuksis, 1971), as does subsequent mass spectrometry (Myher *et al.*, 1974).

1.4.1. Short-Chain Acids

The short-chain fatty acids (C_2–C_6) constitute a frequently overlooked component of natural fatty acid mixtures. This is due to their loss during the extraction and solvent evaporation and to the relatively high temperature of operation of the conventional GLC columns from which they emerge with the solvent front. It is therefore necessary to take special precautions in the preparation and analysis of samples when short-chain fatty acids are suspected to be present. Steam distillation and salt trapping may serve to isolate them (see Section 1.3.3). The short-chain fatty acids may be resolved and identified by GLC as the free acids or as their esters.

1.4.1.1. Separation of Free Acids

The free fatty acids may be separated by GLC on a variety of liquid phases fortified with nonvolatile organic or inorganic acids (Cochrane, 1975), which reduce

Table I. Relative Retention Times of Free Saturated Short-Chain Fatty Acids on Three Stationary Phases and Capillary Columns[a]

	Trimer acid		Tricresyl phosphate		Ucon LB-550-X[b]	
Acid	120°C	140°C	110°C	130°C	125°C	145°C
Acetic	0.138	0.176	0.114	0.176	0.178	0.208
Propionic	0.276	0.317	0.274	0.314	0.314	0.347
Isobutyric	0.405	0.449	0.374	0.416	0.428	0.461
Butyric	0.498	0.542	0.501	0.534	0.542	0.572
Trimethylacetic	0.519	0.555	0.441	0.479	0.521	0.552
3-Methylbutyric	0.697	0.736	0.694	0.725	0.742	0.763
(±)2-Methylbutyric	0.759	0.795	0.717	0.746	0.776	0.796
n-Pentanoic	1.000	1.000	1.000	1.000	1.000	1.000
2,2-Dimethylbutyric	1.080	1.079	0.933	0.943	1.040	1.039
3,3-Dimethylbutyric	1.215	1.212	1.605	1.539	1.542	1.489
(±)2,3-Dimethylbutyric	1.236	1.229	1.154	1.146	1.215	1.200
2-Ethylbutyric	1.286	1.274	1.267	1.241	1.308	1.280
(±)2-Methylpentanoic	1.381	1.360	1.333	1.296	1.340	1.306
(±)3-Methylpentanoic	1.431	1.426	1.449	1.407	1.427	1.386
4-Methylpentanoic	1.524	1.490	1.540	1.480	1.485	1.436
n-Hexanoic	1.855	1.776	1.933	1.812	1.816	1.717

[a] From Hrivnak *et al.* (1972).
[b] Polypropylene glycol with phosphoric acid.

dimerization of the free acid molecules as well as block adsorptive sites on the support and column material. However, ghosting may remain as a recurrent problem, and caution should be used in employing this method. GLC columns made up of porous plastic beads have given some of the best separations of short-chain fatty acids in the free form. Ackman (1972) has claimed, however, that the addition of formic acid vapor to the carrier gas is necessary to achieve adequate separation and recovery of free fatty acids also from plastic beads.

Hrivnak *et al.* (1972) have compared the relative retention times of the C_2–C_6 fatty acids on a steel capillary column using three different liquid phases. The best separation was obtained on a polypropylene glycol Ucon LB-550-X (Carlo Erba, Milan, Italy). Fewer than 100,000 theoretical plates were necessary to obtain baseline separation for all the critical pairs of acids at 125°C. The separations on the trimer acid and on the tricresyl phosphate were not so good because some of the critical pairs of acids were not resolved. Table I gives the relative retention times of all 16 saturated C_2–C_6 fatty acids.

1.4.1.2. Separation of Esters

Esterification of the short-chain fatty acids greatly improves their GLC properties but does not affect their relative order of elution. Either conventional (Tanaka and Yu, 1973) or capillary (Allen and Saxby, 1968) GLC columns may be used for the separation and identification of the simple esters of complex mixtures of the short-chain acids. Table II gives the relative retention times of C_1–C_7 fatty acid methyl esters on polar and nonpolar liquid phases on capillary columns. It is obvious that essentially complete separations of closely related fatty acids may be achieved by an appropriate selection of liquid phases, and tentative identifications of the short-chain fatty acids thereby be obtained. The resolution of the isomeric branched pentanoates is better when they are analyzed as the methyl esters on conventional 20% dioctyl phthalate columns, but complete separations are not obtained even at low temperatures and low gas flows (Tanaka and Yu, 1973).

Ashes and Haken (1974, 1975) have examined in great detail the structure–retention time relationships of saturated and unsaturated short-chain fatty acids in the form of various saturated and unsaturated simple alkyl and isoalkyl esters on a variety of polar and nonpolar liquid phases.

1.4.2. Saturated Normal- and Branched-Chain and Cyclic Acids

The saturated normal and branched medium- and long-chain fatty acid esters are isolated as a group by argentation TLC, along with any cyclic chain derivatives. GLC of free long-chain fatty acids is usually unsatisfactory for the separation of any but the simplest of mixtures (Ottenstein and Supina, 1974; Kuksis, 1977).

1.4.2.1. Normal and Monomethyl-Branched Medium- and Long-Chain Esters

The resolution of normal and monomethyl-branched fatty esters of the same carbon number depends on the location of the branching on the hydrocarbon chain. Abrahamsson *et al.* (1963) have made a detailed study of the methyl esters of

Table II. Relative Retention Times of Short-Chain Fatty Acid Methyl Esters on Polar and Nonpolar Phases[a]

	Nonpolar		Polar	
Fatty acid	SE-30	PPE	TXP	DEGS
Formic	1.51		2.73	0.86
Acetic	2.19	2.39	3.29	1.20
Propionic	3.35	3.06	3.76	1.64
Isobutyric	4.12	3.42	4.10	1.86
n-Butyric	4.67	4.03	4.67	2.44
Trimethylacetic	4.72	3.59	4.24	1.77
(±)2-Methylbutyric	5.50	4.54	5.15	2.80
3-Methylbutyric	5.50	4.67	5.21	3.01
3,3-Dimethylbutyric	6.07	4.95	5.62	3.52
n-Pentanoic	6.08	5.52	6.08	3.88
2,2-Dimethylbutyric	6.24	4.95	5.62	3.20
(±)2,3-Dimethylbutyric	6.55	5.53	6.20	3.88
2-Ethylbutyric	6.70	5.66	6.39	3.88
(±)2-Methylpentanoic	6.86	5.94	6.39	4.16
(±)3-Methylpentanoic	6.95	6.23	6.74	5.01
4-Methylpentanoic	7.09	6.42	6.90	5.20
n-Hexanoic	7.48	7.03	7.40	5.88
(±)2-Methylhexanoic	8.06	7.32	7.66	6.18
(±)3-Methylhexanoic	8.12	7.56	7.93	6.75
(±)4-Methylhexanoic	8.55	8.25	8.41	7.32
5-Methylhexanoic	8.36	8.03	8.20	7.15
n-Heptanoic	8.78	8.60	8.75	7.92
n-Octanoic	10.00	10.00	10.00	10.00

[a] Modified from Allen and Saxby (1968).

positionally isomeric monomethyl-branched fatty acids (mostly 19:0) on both polar and nonpolar liquid phases. Ackman (1972) has recalculated their data as fractional equivalent chain length (ECL) values. These data are reproduced in an abbreviated form in Table III. There are barely detectable separations of authentic 8-, 9-, and 10-methyloctadecanoates. Only the methyl groups in the 2–5 and 12–17 positions of a chain length of 18 showed ECL values significantly different from the values for the 6–11 positions (Ackman, 1972). With shorter chain length, useful separations may be obtained because of an increasing proximity of the methyl substituent to the terminal methyl group. Thus the separation between the 9- and 10-methylhexadecanoates has been reported (MacLean *et al.*, 1968).

The 10-methyl derivatives, of which the most popular is 10-methyl octadecanoic or tuberculostearic acid, are readily distinguishable from *iso* and *anteiso* acids by a shorter retention time.

Table IV gives the ECL values for a variety of *iso*, *anteiso*, and *neo* isomers of monomethyl-branched fatty acids on selected polar and nonpolar liquid phases (Jamieson, 1970). From nonpolar GLC columns the common *iso* and *anteiso* C_{n+1} acids are usually eluted together about halfway between the straight-chain acids C_n and C_{n+1}. From polar columns, the *anteiso* acid C_{n+1} may be seen to emerge at about two-thirds of the distance from C_n to C_{n+1}. The less common *neo* isomer of C_{n+1} is eluted at about one-third the distance from C_n to C_{n+1}. Complete peak resolution is obtained for all isomers on open tubular columns of both nonpolar and polar type (Ackman, 1972).

1.4.2.2. *Multimethyl-Branched Medium- and Long-Chain Esters*

The GLC behavior of fatty esters with two or more methyl branches in the hydrocarbon chain has been shown to abide by the concept of additivity of fractional chain length values. The multimethyl-branched acids may be considered in two separated groups: methylene-interrupted and propylene-interrupted series.

The GLC behavior of the methylene-interrupted multimethyl-branched fatty acids has been studied in great detail by Odham (1967). Chemical synthesis of methyl 2L,4L-dimethylhexanoate and methyl 2D,4L-dimethylhexanoate allowed the demonstration of their GLC separation on a Versamid packed column (Odham, 1967). Furthermore, the diastereoisomeric structures methyl 2L,4D- and 2D,4D-dimethyl-dec-9-enoates could be completely resolved under similar conditions (Odham and Waern, 1968). A structurally similar pair of higher molecular weight, 2L,4D,6D- and 2D,4D,6D-trimethyldodec-11-enoate, could also be completely resolved. The synthesis of other acids of this series (2,4,6-trimethylnonanoates) allowed the demonstration of the following GLC elution order: 2D,4D,6D, 2L,4L,6D, 2L,4D,6D, and 2D,4L,6D (Odham, 1967). Odham (1967) noted that in certain acids with two (2,4-dimethylhexanoates) or three (2,4,6-trimethyloctanoates) asymmetrical centers the shortest retention time was found for that acid with the same

Table III. ECL Values of Methyl Esters of Methyl-Branched Octadecanoic Acids on Selected Polar and Nonpolar Liquid Phases[a]

Position of substitution	Polar		Nonpolar	
	Reoplex	BDS	SE-30	Apiezon L
2	17.91	17.89	18.22	18.15
3	18.13	18.13	18.28	18.29
4	18.41	18.39	18.41	18.41
5	18.30	—	18.29	—
6	18.29	—	18.30	—
7	18.30	—	18.31	—
8	18.28	18.30	18.26	18.32
9	18.30	—	18.26	18.32
10	18.30	18.32	18.29	18.32
11	18.33	—	18.31	—
12	18.34	18.35	18.35	—
13	18.33	—	18.35	—
14	18.44	—	18.45	—
15	—	18.51	—	18.53
16	18.64	18.70	18.69	—
17	18.58	—	18.56	—

[a] Abbreviated from Ackman (1972).

Table IV. ECL Values of Monomethyl-Branched Fatty Acids on Selected Polar and Nonpolar Liquid Phases[a]

	Polar				Nonpolar		
Fatty acid	EGSS-X 170°C	DEGS 160°C	PEGA 197°C	BDS 160°C	Apiezon M 197°C	Apiezon L 197°C	SE-30 200°C
Iso acids							
10:0	—	—	—	—	—	9.59	—
12:0	11.47	11.46	11.54	11.55	—	11.58	11.60
13:0	12.44	12.48	12.54	12.55	12.66	12.64	12.62
14:0	13.48	13.48	13.40	13.55	13.63	13.49	13.63
15:0	—	14.49	—	14.55	14.57	14.64	14.65
16:0	—	15.65	15.49	15.56	15.70	15.68	15.67
17:0	—	16.58	—	16.55	16.57	16.61	16.81
18:0	17.48	17.50	17.49	17.52	17.62	17.62	17.65
20:0	19.47	19.51	19.55	19.51	19.63	19.62	19.65
22:0	—	—	21.63	—	21.62	—	—
24:0	—	—	23.55	—	23.61	—	—
Anteiso acids							
13:0	—	—	—	12.70	—	12.67	—
15:0	—	—	14.65	14.72	14.66	14.70	14.77
16:0	—	—	—	—	—	—	15.81
17:0	—	16.71	16.69	16.70	16.68	16.78	16.83
19:0	18.70	18.70	18.72	18.69	18.70	18.71	18.72
21:0	20.70	20.70	20.79	—	20.69	20.70	20.72
Neo acids							
16:0	—	—	14.96	—	15.14		
18:0	—	—	16.91	—	17.11		
20:0	—	—	19.01	—	19.12		
24:0	—	—	23.08	—	23.12		

[a] Abbreviated from Jamieson (1970).

configuration at all centers. Furthermore, in the trimethyloctanoates the isomer with alternating configuration (e.g., 2D,4L,6D) was the longest retained, and when the configuration was the same at two centers the isomer with opposite configuration at carbon atoms 2 and 4 (e.g., 2L,4D,6D) had a longer retention time than when the opposite configurations were at carbon atoms 4 and 6 (e.g., 2L,4L,6D). An interesting feature observed when the four 2,4,6-trimethylnonanoates were analyzed by capillary GLC was that the retention times increased by three almost exactly equal increments between the four peaks (Odham, 1967). However, the 2,4,6-trimethyloctanoates (Odham, 1967) showed a different behavior. The all-D natural isomer coincided with the first peak eluted from the four racemic peaks of methyl 2,4,6-trimethylundecanoate derivatives of the preen gland acids.

The trimethylene- or propylene-interrupted multimethyl-branched fatty acids possess a carbon skeleton similar to the isoprenoid lipid materials such as carotenoids and squalene, which upon hydrogenation and partial oxidation may give rise to short-chain isoprenoid fatty acids. The great majority of the saturated isoprenoid fatty acids occurring in nature appear to be derived from plant phytol (3,7D,11D,15-tetramethylhexadec-2-en-1-ol) or derivatives thereof (Ackman, 1972).

Table V. ECL Values of Trimethylene-Interrupted Multimethyl-Branched Fatty Acids on Selected Polar and Nonpolar Liquid Phases[a]

Fatty acid		Polar		Nonpolar
		PEGA 197°C	BDS 150°C	Apiezon L 197°C
Dimethyl acids				
3,7-	18:2 (2c,6)	10.68		9.45
3,7-	18:2 (2t,6)	11.41		10.01
Trimethyl acids				
3,7,11-	12:0	12.80	12.87	13.36
3,7,11-	12:3 (2c,6t,10)	15.66		14.00
3,7,11-	12:3 (2t,6t,10)	16.41		14.55
4,8,12-	13:0		14.14	
4D,8D,12-	13:0		14.12	
Tetramethyl acids				
2,6,10,14-	15:0		15.82	
2D,6D,10D,14-	15:0		15.76	
2D,6D,10D,14-	15:0		15.80	
3,7,11,15-	16:0	17.03	17.02	17.48
3L,7D,11D,15-	16:0		—	
3D,7D,11D,15-	16:0		—	
3,7,11,15-	16:4 (2c,6t,10t,14)	20.67	16.98	18.40
3,7,11,15-	16:4 (2t,6t,10t,14)	21.66	17.02	18.76

[a] Abbreviated from Jamieson (1970).

Ackman (1968) has shown that the fractional chain length values derived from literature data for the calculation of ECL values of multi-branched methyl esters yield good agreement with the experimental values obtained for 3,7,11-trimethyldodecanoate, 4,8,12-trimethyltridecanoate, 2,6,10,14-tetramethylpentadecanoate, and 3,7,11,15-tetramethylhexadecanoate on BDS and for 2,14-dimethylpentadecanoate, 3,6-dimethylpentadecanoate, and 3,6,13-trimethyltetradecanoate on CDX acetate and butyrate. Table V gives the ECL values of a number of trimethylene-interrupted multimethyl-branched fatty acids on selected polar and nonpolar liquid phases.

Halobacterium cutirubrum contains an all-D phytanyl skeleton, which, converted to phytanic and pristanic acid forms, has served as the reference material in the GLC of the isoprenoid fatty acids as both methyl and L-menthyl esters. Useful separations of the primary diastereoisomeric acids 3L,7D,11D- and 3D,7D,11D-phytanic or 2L,6D,10D- and 2D,6D,10D-pristanic, as their methyl esters, may be obtained on polar open tubular columns of 40,000–50,000 plates (Ackman and Hansen, 1967). In these resolutions, the DDD peak always appears after the LLD peak. Terminal oxidation of mesopristane (Ackman *et al.*, 1972) produces 2L,6L,10L,14 and 2D,6L,10L,14 diastereoisomers as well as the familiar 2L,6D,10D,14 and 2D,6D,10D,14 diastereoisomers. Pristanic acids of ruminant and marine origin both give two similar peaks as methyl esters analyzed on polar open tubular columns (Ackman and Hansen, 1967), but when converted to L-menthyl esters the marine pristanic acids show the anticipated four diastereoisomers (Ackman *et al.*, 1969). The ruminant pristanic acids show only two components, corresponding to the first and last diastereoisomers of marine origin. These have been shown to be identical by GLC on both polar and nonpolar columns to L-menthyl esters of phytol-derived 2L,6D,10D,14- and 2D,6D,10D,14-pristanic acids, the latter emerging last. The shorter-chain acidic degradation products of phytol include 4,8,12-trimethyltridecanoic acid, which is widely distributed in nature. Chemical oxidation of *meso*-pristane gives equal amounts of two 4,8,12-trimethyltridecanoic acids completely resolved as L-menthyl esters (Ackman, 1972). The first to elute is the 4D,8D,12-trimethyltridecanoate. The next lowest acid degradation product of phytol is 3D,7D,11-trimethyldodecanoate. The L-menthyl esters of their synthetic mixture (prepared from farnesol) can be resolved on a single open tubular BDS column into four components in a triplet peak pattern. The phytol-derived 3D,7D,11 diastereoisomer corresponds to the last peak to emerge from the mixture, and this appears to be also true of the presumed 2,6,10-trimethylundecanoate, a secondary oxidation product resolved in the same GLC run. The 3D,7D,11 and 4D,8D,12 diastereoisomer emerges last for methyl esters (MacLean *et al.*, 1968; Ackman, 1968). The general order of emergence of the isoprenoid derivatives in these GLC analyses is the opposite of that proposed for the fatty acids arising from propionate condensation (Ackman, 1968). In the isoprenoids, the methyl esters of the fatty acid with all asymmetrical centers of the same configuration emerge last. Thus the enantiomeric pair 3L,7L,11- and 3D,7D,11- will follow the pair 3D,7L,11- and 3L,7D,11- in the methyl trimethyldodecanoates. The L-menthyl esters of 4L,8D,12D,16- and 4D,8D,12D,16-tetramethylheptadecanoates show no resolution whatever, whereas there is virtually complete resolution of the methyl esters.

1.4.2.3. Cyclic Esters

Cyclic esters are isolated along with the methyl-branched acids. Fatty acids containing cyclic structures in their molecules have been identified as products of certain bacteria and plants, from which they find their way into the food and bodies of animals and man. The major group of structures of this type is provided by the cyclopropanoid fatty acids. All the natural forms are *cis*, and they have been most widely studied by GLC. The *trans* forms have been prepared synthetically and have been shown to possess slightly shorter retention times than the *cis* forms (Christie and Holman, 1966).

The cyclopropanoid fatty acids have polar GLC properties similar to those of monoethylenic fatty acids. Table VI gives the ECL values of methyl *cis*-methylene-octadecadienoates on several liquid phases. These C_{19} propanoids show what appears to be a smooth asymmetrical curve from 2,3 to 17,18 isomers. There is no obvious interaction of a 2,3- or 4,5-cyclopropane ring with the carboxyl group. Ackman (1972) has pointed out that this smooth curve contradicts the dip seen for 5,6-methyleneoctadecanoates on more polar liquid phases and has claimed that the abnormality must be due to other than a physicochemical reason. The ECL values for selected cyclopropanoid fatty acids of different total carbon number have been given by Ackman (1972). A modifying effect of the chain length was apparent in the ECL values for BDS and APL liquid phases. The difference in ECL values between

Table VI. ECL Values of Methyl cis-Methyleneoctadecanoates on Polar and Nonpolar Phases[a]

	Polar			Nonpolar
Position of substitution	DEGS 180°C	PEGA 190°C	NPGS[b] 190°C	Apiezon L[b] 220°C
2,3	19.64	19.55	19.38	18.98
3,4	19.62	19.50	19.32	18.85
4,5	19.58	19.46	19.30	18.84
5,6	19.51	19.36	19.20	18.76
6,7	19.55	19.37	19.20	18.76
7,8	19.52	19.35	19.18	18.75
8,9	19.51	19.37	19.15	18.73
9,10	19.55	19.37	19.24	18.73
10,11	19.55	19.40	19.25	18.78
11,12	19.58	19.44	19.25	18.78
12,13	19.65	19.48	19.29	18.80
13,14	19.71	19.56	19.41	18.85
14,15	19.82	19.65	19.45	18.96
15,16	19.94	19.80	19.50	19.02
16,17	20.30	20.19	19.80	19.22
17,18	20.54	20.38	20.04	19.38

[a] From Christie *et al.* (1968).
[b] Open tubular.

the methyleneoctadecanoates and octadecenoates (excluding 2,3, 3,4, and 17,18 isomers) averaged 1.06 for NGPS, 1.03 for DEGS, and 1.10 for APL (Christie *et al.*, 1968). The isomers with the functional group located in the terminal position (e.g., 17,18-methyleneoctadecanoates) did not have a relative retention time shorter than that of the isomer in the penultimate position. It is noteworthy that the methyl esters of the 9,10- and 11,12-methyleneoctadecanoic acids, which occur together in nature, are easily separated on certain capillary columns (Christie, 1970).

GLC may not be applicable for the identification of cyclopropene esters which readily rearrange or decompose on the column to give spurious peaks. Wolff and Miwa (1965) have shown that variation in type and content of liquid phase or solid support as well as the size of the sample injected and column temperature all affect the analysis. For analytical purposes, GLC can be combined with a prior chemical modification of the cyclopropenoid components by such means as hydrogenation or reaction with mercaptans or silver nitrate (Christie, 1970).

Other long-chain fatty acids with cyclic structures are provided by the ω-alicyclic fatty acids. Table VII gives the ECL values for a number of the fatty acid esters with cyclic structures at the terminal carbon on selected liquid phases (Sen Gupta and Peters, 1969). When the total carbons (base chain plus terminal group) are subtracted from the observed ECL values for APL, the differences (deltas) are roughly in order of the proportions of carbons in the base chains and terminal groups. On an APL column the 11-cyclohexylundecanoate (C_{17}) has an ECL value of 17.92, while on BDS or NPGS it is 18.4. The 10-cyclopentyldecanoate (C_{15}) has an ECL value of 15.8 and 9-cyclobutylundecanoate (C_{15}) an ECL value of 15.6 on

Table VII. ECL Values of Some Fatty Acid Esters with Cyclic Structures at Terminal Carbon on Selected Liquid Phases

	Nonpolar	Polar	
Ester	Apiezon L	BDS	NPGS
9-Cyclobutylundecanoate[a]	15.6		16.1
10-Cyclopentyldecanoate[a]	15.8		16.5
5-(2-Cyclopentenyl)pentanoate[b]	10.9		12.6
7-(2-Cyclopentenyl)heptanoate[b]	12.9		14.6
9-(2-Cyclopentenyl)nonanoate[b]	14.9		16.6
11-(2-Cyclopentenyl)undecanoate[c]	16.9		18.6
13-(2-Cyclopentenyl)tridecanoate[c]	18.9		20.5
13-(2-Cyclopentenyl)-6-tridecenoate[b]	18.6		21.0
11-Cyclohexylnonanoate[a]	15.9	16.7	
11-Cyclohexylundecanoate[c]	17.9	18.4	
13-Cyclohexyltridecanoate			

[a] Sen Gupta and Peters (1969).
[b] Calculated from data of Zeman and Pokorny (1963).
[c] Ackman (1972).

the APL column. The principal natural acids are 11-(2-cyclopentenyl) undecanoic (hydnocarpic) acid, 13-(2-cyclopentenyl)tridecanoic (chaulmoogric) acid, and 13-(cyclopentenyl)-6-tridecaenoic acid (gorlic) acid. These structures are those of palmitic, stearic, and petroselinic acids. Butterfat fatty acids have been reported to contain 11-cyclohexylundecanoic acid (Schogt and Haverkamp-Begeman, 1965; Hansen, 1967). Oshima and Ariga (1975) have used GLC to identify the 11-cyclohexylundecanoate and 13-cyclohexyltridecanoate in acidophilic thermophilic bacteria, and Glass *et al.* (1975) have identified furanoid fatty acids in fish lipids.

1.4.3. Ethylenic Acids

The ethylenic components of the common unsaturated fatty acids are recovered according to the total number of double bonds by argentation TLC, which allows the separate examination of the mono-, di-, tetra-, penta-, and hexaunsaturated fatty acid classes. However, cross-contamination can and does occur, either because of incomplete resolution or because of the presence of positional and chain-length isomers. Since most GLC columns cannot separate *cis* and *trans* isomers, this resolution must be accomplished by $AgNO_3$ TLC (Emken, 1972; Emken and Dutton, 1974).

1.4.3.1. Monoethylenic Acids

The monoethylenic fatty acids isolated by $AgNO_3$ TLC comprise several chain lengths and two or more positional isomers each. The GLC analysis of the isomeric monoethylenic fatty acids has received extensive study. Data have been tabulated for all or most of the octadecenoates, including *cis* and *trans* isomers, for APL, NPGS, and XE-60 (Gunstone *et al.*, 1967), and for DEGS, CNES, and polyphenyl ether

Table VIII. ECL Values of Normal-Chain Octadecenoates on Selected Polar and Nonpolar Liquid Phases

Position of unsaturation	DEGS[a]	NPGS[a]			Apiezon L		XF-1150[b]		TCPE[b]	
	cis	*cis*	*trans*	XE-60[a]	*cis*	*trans*	*cis*	*trans*	*cis*	*trans*
2	18.32	18.07	19.34	17.93	17.98	18.80				
3	18.74	18.42	18.42	18.23	17.87	17.87				
4	18.42	18.14	18.10	18.02	17.72	17.72				
5	18.34	18.13	18.13	18.11	17.66	17.79				
6	18.45	18.14	18.14	18.04	17.65	17.75				
7	18.45	18.14	18.14	18.01	17.64	17.73				
8	18.45	18.12	18.12	18.02	17.64	17.74				
9	18.49	18.14	18.14	18.07	17.63	17.74	13.42	18.25	18.54	18.39
10	18.53	18.19	18.19	18.11	17.66	17.75				
11	18.58	18.23	18.23	18.15	17.68	17.78				
12	18.62	18.27	18.21	18.18	17.73	17.73	18.57	18.40	18.72	18.52
13	18.76	18.32	18.26	18.24	17.80	17.80				
14	18.86	18.40	18.26	18.32	17.86	17.83				
15	19.01	18.46	18.32	18.37	17.89	17.84	18.79	18.49	18.96	18.65
16	19.38	18.75	18.53	18.63	18.14	18.03				
17	18.91	18.49	18.49	18.34	17.89	17.89				

[a] Gunstone *et al.* (1967).
[b] Scholfield and Dutton (1971).

(Scholfield and Dutton, 1970). Particularly reproducible retention data have been obtained on APL columns, both packed and open tubular (Gunstone *et al.*, 1967; Ackman and Hooper, 1969). Table VIII gives the ECL values of positional isomers of normal-chain octadecenoates. On polar liquid phases, there is a sharp differentiation of the 3,4 isomers from the adjacent isomers, with the minimal values seen for the 2,3 isomers. The *cis*-2,3 isomer is only slightly below an extension of the general curve when the values are plotted. Once the double bond reaches the 6,7 position, there is a smooth curve to the 16,17 position (penultimate isomer), since the influence of the carbonyl group is negligible. There is a drop to the 17,18 which approximates the 15,16 isomer value. On apolar liquid phases (APL), the *cis*-octadecenoates give an almost smooth curve from the 2,3 to the 14,15 position.

The *trans*-octadecenoates differ greatly with respect to the 2,3 isomer, which has an extremely enhanced retention time on both polar and nonpolar liquid phases. It appears that the 3,4 and other adjacent isomers do not differ radically in ECL values from their *cis* analogues, but on APL the data reported give higher (about 0.1) ECL values than the *cis* in the 6,7–14,15 range and lower in the 14,15–16,17 positions. On NPGS, the *trans* ECL values are lower than the *cis* in the 12,13–16,17 range and on DEGS for the whole 6,7–16,17 range. The cyanoalkylmethylsiloxane phase gives the most effective resolution of the *cis-trans* isomers. There is a nominal difference of about 0.3 ECL unit between respective *cis* and *trans* isomers, whereas on polyester phases the nominal difference is only somewhat better than 0.1 ECL unit. The cyanoalkyl siloxanes yield much better resolutions of the *cis* and *trans* isomers (Ottenstein *et al.*, 1976; Golovnya *et al.*, 1976) but have not yet been widely explored.

Table IX. ECL Values of Normal-Chain Methyl Monoalkenoates on Polar and Nonpolar Liquid Phases[a]

	Polar					Nonpolar
Fatty acid	EGSS-X	EGS	CDX	DEGS	BDS	Apiezon L
14:1	14.80			14.70		13.83
14:1 (7c)		14.75	14.71		14.38	13.83
14:1 (9c)		14.75	14.71			
15:1	15.69	15.67		15.58		
16:1 (3t)					16.48	
16:1 (7c)					16.25	
16:1 (9c)	16.74	16.56	16.50	16.61	16.33	15.70
16:1 (10c)					16.36	
16:1 (11c)					16.50	
17:1	17.67	17.57		17.53		
17:1 (9c)		17.55	17.56		17.27	16.73
18:1 (6c)		18.54	18.52	18.57	18.17	17.71
18:1 (6t)		18.47	18.45	18.47		17.75
18:1 (8t)		18.52	18.49	18.50		17.73
18:1 (9c)	18.62	18.50	18.56	18.46	18.20	17.71
18:1 (9t)		18.43	18.47	18.47	18.17	17.76
18:1 (10t)		18.50	18.50	18.51		17.78
18:1 (11c)					18.30	
18:1 (11t)		18.53	18.60	18.58		17.80
18:1 (13c)		18.64	18.60	18.75	18.35	17.75
18:1 (13t)		18.58	18.54	18.65		17.80
18:1 (13)					18.48	
18:1 (17)		18.82	18.75	19.00		17.90
19:1	19.64			19.43		
19:1 (9)					19.18	
19:1 (11)		19.50	19.53	19.60	19.25	18.60
19:1 (12)					19.30	
19:1 (13)					19.39	
20:1 (9)					20.16	
20:1 (11)	20.45	20.38	20.32	20.39	20.19	19.78
20:1 (13)					20.19	
20:1 (14)					20.36	
20:1 (15)					20.43	
22:1 (11)					22.08	
22:1 (13)	22.47	22.30	22.27	22.32	22.14	21.57
22:1 (15)					22.27	
24:1 (15)		24.40	24.20	24.27	24.08	23.67

[a] Abbreviated from Jamieson (1970).

Table IX gives the ECL retention data for the natural *cis* and *trans* monoethylenic fatty acids. These acids are of various chain lengths but have only a limited number of positional isomers. Nevertheless, this separation is of great practical importance since oleic (18:1 c9) is usually accompanied by 10–30% *cis*-vaccenic (18:1 c11) acid (Ackman and Castell, 1966; Brockerhoff and Ackman, 1967). In most animal lipids, there are major proportions of normal even-chain monoethylenic

Table X. ECL Values of Normal-Chain Monomethylene-Interrupted Methyl Octadecadienoates and Octadecatrienoates on Polar and Nonpolar Liquid Phases[a]

Position of unsaturation	Polar									Nonpolar	
	TCPE[b]	EGS	CN[b]	DEGS	XF-1150[b]	EGA	Carbowax	NPGA	NPGS	Apiezon L	PPE[b]
2c,5c		18.65		18.64		18.37	18.25		18.14	17.68	
3c,6c		19.68		19.54		19.36	18.94		18.65	17.62	
4c,7c		19.28		19.05		18.69	18.45		18.36	17.47	
5c,8c		19.40		19.06		18.71	18.38		18.38	17.43	
6c,9c		19.57		19.19		18.81	18.46		18.44	17.46	
7c,10c		19.63		19.23		18.80	18.46		18.46	17.44	
8c,11c		19.65		19.28		18.87	18.51		18.53	17.48	
9c,12c		19.75		19.38		18.95	18.57		18.60	17.50	
10c,13c		19.81		19.46		19.03	18.60		18.70	17.60	
11c,14c		20.00		19.62		19.15	18.75		18.82	17.68	
12c,15c		20.16		19.75		19.28	18.88		18.90	17.78	
13c,16c		20.60		20.25		19.68	19.20		19.27	18.00	
14c,17c		20.18		19.78		19.33	18.95		19.04	17.80	
9c,12t				19.46				18.16		17.59	
9t,12c				19.55				18.23		17.64	
9t,12t				19.35				18.23		17.64	
12c,15c	19.82			19.45[b]						17.83[b]	18.46
12c,15t	19.56			19.30[b]							18.51
12t,15c	19.68			19.38[b]							18.61
12t, 15t	19.34			19.23[b]						17.85[b]	18.55
9c,12c,15c	20.49		20.00	19.91[b]	19.90						18.52
9c,12t,15c			19.80		19.91						18.55
9t,12c,15t	20.15		19.52		19.50						18.79
9t,12t,15t	19.91		19.12	19.75[b]	19.14						18.83

[a] From Jamieson (1970).
[b] Scholfield and Dutton (1971).

fatty acids, as well as lesser amounts of those of odd chain length (Brockerhoff and Ackman, 1967; Jacob and Grimmer, 1968). Elaidic acid (18:1 t9) is frequently encountered in dietary fats as a result of partial industrial hydrogenation of unsaturated fatty acids. The natural *cis* and *trans* fatty acids have been effectively resolved by glass capillary GLC (Jaeger *et al.*, 1976). Monoethylenic fatty acids with unusually long chains (C_{20}–C_{30}) have been reported in lactobacilli (Uchida, 1974).

1.4.3.2. Methylene-Interrupted Polyethylenic Esters

The diethylenic fatty acid methyl esters isolated by $AgNO_3$ TLC also comprise several chain lengths, positional isomers, and two or more geometric isomers. In addition, the methylene-interrupted dienes may be contaminated with conjugated dienes. Gunstone *et al.* (1967) have determined the ECL values for all the isomeric methyl *cis*- and *trans*-octadecadienoates on NPGS and APL open tubular columns. Table X gives the ECL values recorded for the isomeric octadecadienoic acids by Gunstone *et al.* (1967). In general, the GLC resolution of these isomers follows the general patterns established for the methyl-branched, cyclopropanoid, and monoethylenic fatty acids from both polar and nonpolar phases. The greatest differences occur when the center of unsaturation is near a terminal position in the carbon chain. On nonpolar columns, the *cis* isomers elute before the *trans* isomers, and the order is reversed on polar columns. Using the APL and NPGS columns, it is theoretically possible to separate all the isomeric *cis*-octadecadienoates from the corresponding *trans* isomers except for the position 3 and 4 isomers. The presence of more than one olefinic group increases the effects of geometric isomers on retention time, and by using both a polar and a nonpolar column it is possible to analyze the four possible geometric isomers of methyl linoleate.

Table XI. ECL Values of Isomeric Dimethylene-Interrupted Methyl cis,cis-Octadecadienoates on Nonpolar and Polar Liquid Phases[a]

	Nonpolar				Polar				
Isomer	Apiezon L	OV-101	SE-30	XE-60	Carbowax 20M	FFAP	DEGA	DEGS	Silar 10C
2,6	17.69	17.62	17.62	17.87	18.32	18.43	18.56	18.55	18.49
3,7	17.54	17.64	17.68	18.17	18.63	18.62	18.93	19.10	19.15
4,8	17.42	17.55	17.56	17.96	18.37	18.33	18.68	18.82	18.87
5,9	17.39	17.47	17.52	18.05	18.32	18.38	18.66	18.78	18.90
6,10	17.41	17.47	17.51	18.14	18.46	18.44	18.81	18.95	19.06
7,11	17.41	17.51	17.52	18.16	18.46	18.50	18.83	18.97	19.09
8,12	17.47	17.55	17.60	18.23	18.52	18.60	18.90	19.00	19.21
9,13	17.49	17.63	17.63	18.32	18.59	18.67	19.01	19.14	19.30
10,14	17.59	17.71	17.72	18.42	18.72	18.71	19.09	19.20	19.36
11,15	17.68	17.78	17.76	18.50	18.82	18.85	19.18	19.28	19.50
12,16	17.88	17.95	17.99	18.79	19.16	19.18	19.57	19.84	19.89
13,17	17.66	17.72	17.82	18.50	18.82	18.90	19.25	19.49	19.57

[a] From Lam and Lie Ken Jie (1976*a*).

Lam and Lie Ken Jie (1976*a*) have reported the ECL values of all 12 dimethylene-interrupted methyl *cis,cis*-octadecadienoates on five polar and three nonpolar stationary phases. These results are summarized in Table XI. On the three nonpolar stationary phases (Apiezon L, SE-30, and OV-101), the methyl *cis,cis*-octadecadienoates gave ECL values ranging from 17.39 to 17.88. In all instances, the 5,9, 6,10, and 7,11 isomers gave the lowest ECL values, while the 12,16 isomer exhibited the highest value. On the semipolar XE-60 phase, the ECL values were in the range 17.87–18.79. The highest ECL values were obtained on a Silar 10C phase ranging from 18.49 to 19.89. Only a few isomers of the dimethylene-interrupted methyl *cis,cis*-octadecadienoates have been isolated from natural sources. The 5,9- and the 11,15-octadecadienoates have been found in trace amounts in seed oils (Smith, 1970*a*) and animal fats (Hoffmann and Meijboom, 1969), respectively. Murawski *et al.* (1971) have isolated the 8,12 and the 9,13 isomers from human milk, while others have identified these compounds also in partially hydrogenated seed oils (Lam and Lie Ken Jie, 1976*a*).

Table XII gives the ECL values (*cis,cis* and *trans,trans* isomers) of synthetic diethylenic fatty acids with different numbers of methylene groups between double bonds. The table also includes some mixed *cis-trans* isomers. A number of the positional and geometric isomers can be resolved with any one of the liquid phases when capillary columns are employed. Table XIII lists the ECL values for the natural octadecadienoates on a variety of liquid phases. It is seen that in essentially all instances the elution of the longest-retained diunsaturated isomer is complete before the emergence of the saturated member of the next highest even-carbon-number homologue.

The tri- and tetraethylenic fatty acids isolated by $AgNO_3$ TLC contain progressively fewer chain-length and positional isomers. Furthermore, the natural

Table XII. ECL Values of Normal-Chain Polymethylene-Interrupted Methyl Octadecadienoates on Polar and Nonpolar Phases[a]

	Polar		Nonpolar					
	DEGS		XE-60		Apiezon L		PPE[b]	
Position of unsaturation	*cis,cis*	*trans,trans*	*cis,cis*	*trans,trans*	*cis,cis*	*trans,trans*	*cis,cis*	*trans,trans*
5,12	18.92	18.83	17.94	17.92	17.42	17.56	18.04	
6,12	18.98	18.91	17.96	17.87	17.41	17.53	18.10	
7,12	19.04	18.96	17.89	17.96	17.37	17.59	18.05	
8,12	19.04	18.85	18.03	17.87	17.44	17.50	18.15	
6,10	18.97	18.78	17.94	17.79	17.39	17.46	18.06	
6,11	18.92	18.91	17.86	17.98	17.34	17.57	18.01	
7,15	19.21		18.14		17.54			
8,15	19.23		18.15		17.55			
9,15	19.31	19.00	18.17		17.58		18.32	
9c,15t[b]		19.04			17.60			
9t,15c[b]		19.18			17.75			

[a] From Jamieson (1970).
[b] Scholfield and Dutton (1971).

Table XIII. ECL Values of Normal-Chain Methyl Alkadienoates on Polar and Nonpolar Liquid Phases[a]

	Polar						Nonpolar
Fatty acid	EGSS-X	EGS	CDX	DEGS	PEGA	BDS	Apiezon L
15:2 (6,9)		16.42	16.30	16.43			14.60
15:2 (9,12)		16.70	16.50	16.77			14.77
16:2 (6,9)	17.39	17.32	17.25	17.50	16.99	16.73	15.47
16:2 (7,10)		17.32	17.25	17.50	17.02		15.47
16:2 (9,12)	17.77	17.32	17.45	17.47	17.30	16.93	
17:2 (9,12)		18.40	18.32	18.50			16.62
18:2 (6,9)	19.31			19.16	18.85		
18:2 (8,11)		19.43	19.40	19.46	18.96		17.48
18:2 (9,12)	19.45	19.22	19.23	19.30	18.98	18.71	17.53
18:2 (10,13)		19.55	19.37	19.55	19.05		17.57
18:2 (11,14)		19.73	19.47	19.60	19.24		17.62
18:2 (12,15)		19.63	19.50	19.63			17.75
18:2 (5,11)		18.90	18.93	19.03	18.72		17.40
18:2 (9t,11t)		21.00	20.60	20.70			18.68
18:2 (9,15)		19.43	19.33	19.46			17.60
19:2 (8,11)				20.28			18.25
19:2 (9,12)		20.10	20.10				
19:2 (11,14)		20.43		20.43			18.32
20:2 (8,11)	21.14	20.95		21.11	20.82	20.45	
20:2 (11,14)	21.19	21.13	21.13	21.36	21.00	20.75	19.48
20:2 (5,11)		20.80	20.77	21.00			19.32
20:2 (7,13)		20.93	21.00	21.17			19.30
22:2 (13,16)					22.95	22.60	

[a] Abbreviated from Jamieson (1970).

esters are confined largely to the *cis* isomers. Table XIV gives the ECL values for the natural tri- and tetraethylenic fatty acids thus far isolated and identified. The table also includes certain isomerization products of the natural acids which have been thus far isolated and identified or partially synthesized. The natural trienes are of two types, the $\omega3$ and $\omega9$, but the $\omega6$ isomer may also be encountered in certain instances. The trienes of various chain length may be completely resolved on packed columns containing a variety of polar liquid phases. Table XV gives the ECL values of the natural pentaenoic and hexaenoic fatty acid esters. The natural long-chain polyenes have been effectively resolved by glass capillary GLC (Lin and Horning, 1975*a*, *b*).

1.4.3.3. Conjugated Ethylenic Esters

In addition to the usual 1,4-methylene-interrupted (nonconjugated) polyethylenic acids, there occur in nature and in industrial preparations the 1,3-methylene-interrupted (conjugated) polyethylenic acids. Hopkins and Chisholm (1968) and Hopkins (1972) have surveyed the conjugated fatty acids of seed oils and have

Table XIV. ECL Values of Normal-Chain Methyl Alkatrienoates and Alkatetraenoates on Polar and Nonpolar Liquid Phases[a]

	Polar						Nonpolar
Fatty acid	EGSS-X	EGS	CDX	DEGS	PEGA	BDS	Apiezon L
16:3 (4,7,10)					17.50	17.26	
16:3 (6,9,12)	18.42	18.33	18.10	18.52	17.66	17.42	15.47
16:3 (7,10,13)		18.33	18.10	18.52	17.83	17.55	15.44
17:3 (6,9,12)		19.00	18.93	19.23			16.40
18:3 (6,9,12)	20.15	19.78	19.70	20.00	19.45	19.15	17.30
18:3 (9,12,15)	20.50	20.13	20.10	20.40	19.82	19.50	17.51
18:3 (3,9,12)		20.32	20.00	20.33			17.42
18:3 (3,11,14)		19.63	19.60	19.86			17.24
19:3 (8,11,14)		20.85	20.80	21.05			18.25
20:3 (5,8,11)	21.69	21.57	21.53	21.65		20.70	19.15
20:3 (7,10,13)		21.63					19.23
20:3 (8,11,14)	22.06	21.65	21.72	22.13	21.40	21.06	19.23
20:3 (11,14,17)	22.40			22.30		21.45	
20:3 (5,11,14)		21.60	21.50	21.75			19.15
22:3 (7,10,13)		23.17	23.30	23.73			21.15
22:3 (10,13,16)		23.38	23.36	23.94			21.33
16:4 (4,7,10,13)		18.95			18.26	17.85	
16:4	19.30	19.28		19.00	18.36	18.05	
18:4 (6,9,12,15)	21.13	21.15	20.73	21.00	20.27	19.86	17.30
19:4 (5,8,11,14)		21.30	21.27	21.54			18.02
20:4 (5,8,11,14)	22.50	22.25	22.15	22.43	21.76	21.29	19.00
20:4 (8,11,14,17)	23.08	22.68	22.94	22.94	22.21	21.75	
21:4 (7,10,13,16)		23.45	23.27	23.50			20.00
22:4 (7,10,13,16)	24.56	23.85	24.03	24.58	23.71	23.23	20.93
24:4 (9,12,15,18)		25.73	25.87	25.87			22.87

[a] Abbreviated from Jamieson (1970).

Table XV. ECL Values of Normal-Chain Methyl Alkapentaenoates and Alkahexaenoates on Polar and Nonpolar Liquid Phases[a]

	Polar						Nonpolar
Fatty acid	EGSS-X	EGS	CDX	DEGS	PEGA	BDS	Apiezon L
20:5 (5,8,11,14,17)	23.56	22.92	22.88	23.45	22.52	21.97	19.00
21:5	24.56	23.94					
21:5 (4,7,10,13,16)		23.97	23.70	24.05			19.78
22:5 (4,7,10,13,16)	24.95	24.37	24.43	24.97	24.05	23.50	20.87
22:5 (7,10,13,16,19)	25.42	24.93	24.87	25.38	24.51	23.90	21.00
22:6 (4,7,10,13,16,19)	25.91	25.31	25.35	26.03	24.86	24.18	20.73

[a] Abbreviated from Jamieson (1970).

Table XVI. ECL Values of Normal-Chain Conjugated Methyl Octadecadienoates and Octadecatrienoates on Polar and Nonpolar Liquid Phases[a]

	Polar				Nonpolar			
Position of unsaturation	XE-1150[b]	DEGS	BDS	XE-60	NPGA	TCPE[b]	Apiezon L	PPE[b]
6c,8c		20.25		18.58			17.54	
6t,8t		20.59		19.25			18.60	
9c,11c		20.32			18.74	20.87	18.12	19.31
9c,11t	20.10	20.10			19.06	20.60	18.12	19.04
9t,11c	20.15				19.02	20.65	18.20	
9t,11t	20.59	20.70			19.58	21.19	18.61	19.60
10c,12c				18.76			17.67	19.32
10c,12t					19.02		18.20	19.07
10t,12c	20.24	20.30			19.28	20.77	18.25	19.16
10t,12t	20.57	20.61		19.16	19.58	21.16	18.61	19.61
12c,14t		19.68				20.00		19.59
13t,15c		19.82				20.16		19.70
9c,11t,13t			21.97				19.12	
9t,11t,13t			22.40				19.47	
10t,12t,14t							19.47	

[a] Abbreviated from Jamieson (1970).
[b] Scholfield and Dutton (1971).

pointed out the nearly random distribution. The natural conjugated acids are characterized by the presence of the conjugated double-bond system near the middle of the fatty chain, which occasionally may include an acetylenic bond and/or a hydroxyl group.

Table XVI gives the ECL values for a number of conjugated methyl octadecadienoates and octadecatrienoates on different liquid phases. It is seen that the pattern of elution changes little with the position of double bonds in the isomers but that it is affected by double-bond configuration. For esters with the same double-bond position, isomers with *trans* double bonds tend to have lower ECL values than *cis* isomers on polar liquid phases and higher ECL values on nonpolar liquid phases. ECL values of conjugated octadecadienoates are always higher than values for unconjugated ones and increase in the order mono-*trans*, *cis,cis*, and *trans,trans* (Scholfield and Dutton, 1971). The data showing that the *cis*-12,*trans*-14 and *trans*-13, *cis*-15 are eluted before the 9,11 and 10,12 series on polar columns represent an exception to the rule that esters with double bonds near the methyl end of the chain are eluted later than those with double bonds nearer the middle of the chain.

The presence of conjugated ethylenic bonds increases the retention times considerably compared to the methylene-interrupted isomers on both polar and nonpolar columns. The all-*cis* isomers have the shortest and the all-*trans* isomers the longest retention times on both nonpolar and polar columns. It has not been possible with the columns available to separate all the isomers obtained by the alkaline isomerization of linoleic acid since the 9t,11c and 10c,12t isomers have similar retention times on both nonpolar and polar columns as do the 9t,11t and 10t,12t isomers. Lam and Lie Ken Jie (1976*a*,*b*) have published ECL values for a series of methyl *cis,cis*- and *trans,trans*-octadecadienoates. On Apiezon L all the *cis,cis* compounds elute before the corresponding *trans,trans* isomers, but on XE-60 the reverse order of elution occurs, except for the conjugated isomers on DEGS and except for the conjugated isomers and those with the methylene groups between the two double bonds.

1.4.4. Acetylenic Acids

Some natural fatty acids contain either isolated or conjugated acetylenic bonds. These acids occur largely in fungi and in plants, but may be found in animal bodies as a result of their presence in foods. The acetylenic bonds frequently occur along with one or more ethylenic bonds in the same fatty acid molecule and may become conjugated to them. The acetylenic bonds are isomeric with the vicinal double bonds found in allenic acids, which may undergo chemical interconversion. Both acetylenic and allenic acids may occur in the form of hydroxy fatty acids.

1.4.4.1. Simple Acetylenic Esters

The methyl esters of simple long-chain acetylenic acids are recovered by argentation TLC in mixtures with the olefinic fatty acids of comparable degree of overall unsaturation. GLC on both polar and nonpolar liquid phases and conventional columns is effective in their tentative identification provided that proper reference compounds are available. Table XVII gives the ECL values of various synthetic and natural monoacetylenic fatty acids. It is seen that the long-chain acetylenic acids are retained more strongly than the corresponding olefinic esters on both polar and nonpolar columns.

Anderson and Rakoff (1965) found that all the isomeric methyl nonynoates were eluted after methyl nonanoate on an Apiezon L column. Methyl stearolate was eluted with methyl linolenate on a DEGS column and on one PEGA column, but after methyl linolenate on another PEGA column (Jamieson, 1970). Gunstone *et al.* (1967) have shown that of the seven isomeric methyl dodecynoates studied only the 9 and 10 isomers were significantly separated on Apiezon L from methyl dodecanoate with increased retention time. All the isomeric methyl octadecynoates studied, however, except the 2 isomer, were eluted before methyl octadecanoate. Comparable values have been obtained by Scholfield and Dutton (1970). With methyl octadecynoates there is a greater, more regular increase in ECL values as the triple bond moves away from the ester group. The polyphenyl ether phase gave the best resolutions, but recoveries from these capillary columns were not very high.

Table XVIII gives the equivalent chain lengths of all acetylenic and *cis*-ethylenic undecanoic acids on Apiezon L, diethylene glycol succinate polyester, and Silar 10C (Lie Ken Jie and Lam, 1974). The methyl undecynoates gave ECL values ranging from 11.00 to 11.17 on the nonpolar Apiezon L column, except for the 2 isomer (11.61) and the 9 isomer (11.52), for which the values were significantly higher. The 4, 5, and 10 isomers had retention times nearly identical to that of the saturated undecanoic acid ester. On polar stationary phases, the ECL values were found to be in the range 13.38–14.87 on DEGS and 12.99–14.39 on Silar 10C. On

Table XVII. ECL Values of Normal-Chain Methyl Dodecynoates and Octadecynoates on Polar and Nonpolar Liquid Phases

Position of unsaturation	Dodecynoates[a]			Octadecynoates[b]						
	DEGS	NPGS	Apiezon L	CN[c]	EGS	CDX	DEGS	NPGS	Apiezon L	PPE[c]
2								20.47	18.60	
3										
4								19.02	17.91	
5	14.47	13.23	12.05	19.58				19.07	17.87	18.88
6	14.68	13.34	12.00	19.60	20.40	20.03	20.33	19.16	17.84	18.96
7	14.78	13.42	12.03		20.45	20.04	20.35	19.20	17.82	
8	14.88	13.48	12.06	19.74	20.45	20.04	20.40	19.22	17.80	18.99
9	15.13	13.62	12.15	19.78	20.48	20.12	20.44	19.23	18.00	19.63
10	16.00	14.26	12.52	19.86				19.25		19.09
11	15.74	14.00	12.01	19.91	20.60	20.17	20.53	19.28	17.87	19.14
12				20.02			20.65	19.39		19.42

[a] Hofstetter *et al.* (1965).
[b] Gunstone *et al.* (1967).
[c] Scholfield and Dutton (1970).

Table XVIII. Equivalent Chain Lengths of Isomeric Acetylenic and cis-Ethylenic Undecanoic Acids on Nonpolar and Polar Liquid Phases

	Apiezon L[a]		DEGS[a]		Silar 10C[a]		SE-30[b]		FFAP[b]	
Isomer	Acetylenic	Ethylenic	Acetylenic	Ethylenic	Acetylenic	Ethylenic	Acetylenic	Ethylenic	Acetylenic	Ethylenic
2	11.61	10.98	14.49	11.28	14.39	11.03	11.63	10.86	13.61	11.16
3	11.17	10.87	14.12	11.88	13.65	11.70	11.31	10.89	13.31	11.48
4	11.02	10.76	13.38	11.62	12.99	11.44	11.18	10.82	12.67	11.25
5	11.02	10.75	13.43	11.58	13.12	11.48	11.18	10.77	12.67	11.21
6	11.05	10.79	13.64	11.73	13.38	11.73	11.18	10.86	12.80	11.38
7	11.07	10.83	13.72	11.81	13.45	11.77	11.24	10.88	12.86	11.41
8	11.14	10.89	13.99	11.95	13.61	11.86	11.33	10.96	13.10	11.49
9	11.52	11.10	14.87	12.29	14.38	12.19	11.59	11.13	13.77	11.82
10	11.00	10.89	14.60	11.93	14.18	11.88	11.15	10.88	13.42	11.52

[a] Lie Ken Jie and Lam (1974).
[b] Lie Ken Jie (1975*b*).

Table XIX. ECL Values of Isomeric Dimethylene-Interrupted Methyl Octadecadiynoates on Nonpolar and Polar Liquid Phases[a]

	Nonpolar				Polar				
Isomer	Apiezon L	SE-30	OV-101	XE-60	Carbowax 20M	FFAP	DEGA	DEGS	Silar 10C
2,6	18.40	18.61	18.63	20.55		22.09	22.62	23.18	23.92
3,7	18.16	18.35	18.38	19.90			22.14		
4,8	17.90	18.30	18.27	19.52	21.13	21.22	21.59	22.37	22.44
5,9	17.87	18.27	18.25	19.60	21.10	21.19	21.64	22.40	22.62
6,10	17.90	18.26	18.25	19.67	21.19	21.28	21.80	22.59	22.79
7,11	17.93	18.27	18.29	19.80	21.21	21.31	21.87	22.63	22.95
8,12	17.96	18.32	18.32	19.83	21.31	21.36	21.98	22.72	23.05
9,13	18.03	18.33	18.37	19.90	21.39	21.47	22.03	22.80	23.14
10,14	18.05	18.38	18.40	20.01	21.44	21.54	22.13	22.86	23.24
11,15	18.14	18.44	18.48	20.09	21.64	21.79	22.37	23.05	23.43
12,16	18.51	18.82	18.82	20.61	22.28	22.47	23.19	23.98	24.29
13,17	17.96	18.20	18.21	20.51	21.86	22.01	22.81	23.59	23.95

[a] From Lam and Lie Ken Jie (1975).

both phases, the 4 isomer gave the lowest and the 9 isomer the highest ECL value. The ECL values were also markedly increased when the triple bond was close to the carbomethoxy end of the chain, as in the 2 and 3 isomers.

The ethylenic isomers gave ECL values which were smaller than 11.00 on Apiezon L, with the exception of the 9 isomer (11.10). There was little difference in the retention times of these isomers, although it was possible to recognize that the 4 isomer had the lowest ECL value in the series. On polar phases, the ECL values ranged from 11.28 to 12.29 on DEGS and from 11.03 to 12.19 on Silar 10C. Both sets of ECL values gave an almost parallel plot except in the case where the double bond was located between carbons 6 and 7, as the 6 isomer had identical ECL values on DEGS and on Silar 10C.

In subsequent studies, Lie Ken Jie (1975*b*) has examined the GLC behavior of the isomeric methyl undecynoates and *cis*-undecenoates on other polar and nonpolar liquid phases. FFAP, Carbowax 20M, and XE-60 have been found to be more efficient than DEGS in these separations, while OV-101 and SE-30 are comparable to Apiezon L in the separation of isomeric unsaturated fatty esters.

Table XIX gives the ECL values for the complete series of dimethylene-interrupted methyl octadecadiynoic acids on polar and nonpolar liquid phases (Lam and Lie Ken Jie, 1975). The lowest ECL values for these isomers were recorded on the nonpolar Apiezon L phase, with the values ranging from 17.87 to 18.51, while on OV-101 and SE-30 phases these isomers gave almost identical retention behavior and ECL values ranged from 18.25 to 18.82 and from 18.26 to 18.82, respectively. On all three nonpolar phases, the 2,6 and 12,16 isomers exhibited the highest ECL values while the 4,8 to 7,11 isomers gave the lowest values. On the semipolar XE-60 phase, the ECL values were in the range 19.52–20.61, with the 2,6, 12,16, and 13,17 isomers exhibiting nearly identical ECL values. The 4,8 isomer gave the lowest ECL value of all isomers. On the polar phases, the ECL values were lowest on Carbowax, ranging from 21.10 to 22.28, and highest on Silar 10C phase, ranging from 22.44 to 24.29. The 3,7 isomer appeared to decompose when injected onto all polar stationary phases except DEGA. The Silar 10C phase was superior to all the other phases in separating the isomers. Table XX gives the ECL values for several trimethylene-interrupted methyl octadecadiynoates and the corresponding *cis,cis*-octadecadienoates on nonpolar and polar liquid phases (Lie Ken Jie, 1975*a*). On the Apiezon L column, the ECL values of the isomeric methyl octadecadiynoates ranged from 17.89 to 18.44. The 2,7 isomer exhibited the highest ECL value and was readily

Table XX. ECL Values of Methyl Octadecadiynoates and cis,cis-Methyl Octadecadienoates on Nonpolar and Polar Liquid Phases[a]

	Apiezon L		DEGS		Silar 10C	
Isomer	Acetylenic	Ethylenic	Acetylenic	Ethylenic	Acetylenic	Ethylenic
2,7	18.44	17.57	23.48	18.74	24.05	18.73
3,8	18.07	17.43	23.40	19.25	23.12	19.40
4,9	17.89	17.32	22.68	18.98	22.49	19.04
5,10	17.92	17.30	22.89	19.11	22.87	19.34
6,11	17.93	17.29	22.90	19.12	22.86	19.34

[a] From Lie Ken Jie (1975*a*).

separated from the other isomers. On the polar stationary phases, the ECL values ranged from 22.68 to 23.48 on DEGS and from 22.49 to 24.05 on Silar 10C; on both of these phases, the 2,7 isomer gave the highest ECL value and the 4,9 isomer the lowest. Separation of mixtures on Silar 10C was more efficient than that on DEGS. Three distinct peaks were obtained when a mixture of the 2,7, 3,8, and 4,9 isomers was examined, while a mixture of the 4,9 and 5,10 isomers gave twin peaks on Silar 10C.

The isomeric *cis,cis*-methyl octadecadienoates gave ECL values ranging from 17.29 to 17.57 on the nonpolar Apiezon L, whereas on DEGS and Silar 10C they ranged from 18.74 to 19.25, and on Silar 10C from 18.73 to 19.40. Unlike the diynoate isomer, the 2,7 isomer gave the lowest ECL values on both DEGS and Silar 10C; the 4,9 isomer also gave closely similar ECL values on DEGS and Silar 10C. None of these compounds seems to occur in nature except the 4,9 isomer, which has been isolated from *Pinus sibirica* (Lie Ken Jie, 1975*a*). Lemarchal and Munsch (1965) noted the conversion of elaidic acid to the 4,9- and 5,9-octadecadienoic acids in rats fed a linoleic acid-free diet.

Table XXI gives the ECL values for various methyl octadecadiynoates on polar and nonpolar liquid phases. The order of emergence on all the phases studied by Lie Ken Jie (see Jamieson, 1970) was (1) isomers with more than one (2–5) methylene group between the two triple bonds, (2) isomers with one methylene group between the two triple bonds, and (3) conjugated isomers. It was possible to separate mixtures of the different isomer classes but not mixtures of isomers of the same class.

1.4.4.2. Mixed Ethylenic-Acetylenic Esters

In certain seed oils there occur major proportions of more complex acetylenic and hydroxyacetylenic acids containing conjugated systems of acetylenic and

Table XXI. ECL Values of Normal-Chain Methyl Octadecadiynoates on Polar and Nonpolar Phases[a]

Position of unsaturation	Polar		Nonpolar
	DEGS	XE-60	Apiezon L
5,12	22.29	19.28	17.85
6,12	22.35	19.27	17.83
7,12	22.40	19.33	17.89
8,12	22.43	18.39	17.90
9,12	23.39	19.76	18.23
10,12	24.94	21.58	19.60
6,8	25.02	21.57	19.61
6,9	23.21	19.72	18.18
6,10	22.27	19.24	17.85
6,11	22.32	19.29	17.88
7,15	22.72	19.51	17.94
8,15	22.82	19.57	17.95
9,15	22.77	19.49	17.95

[a] From Jamieson (1970).

Table XXII. ECL Values of Mixed Ethylenic-Acetylenic Octadecanoates and Eicosanoates on Polar and Nonpolar Liquid Phases[a]

	Polar			Nonpolar
Fatty acid	EGS	CDX	DEGS	Apiezon L
18:2 (9e,12a)	21.23	20.97	21.48	17.90
18:2 (9a,12a)	23.74	23.75	24.10	18.30
20:2 (7e,13a)	22.75	22.63	23.10	19.50
20:2 (7a,13e)	22.75	22.63	23.10	19.50
20:2 (7a,13a)	24.47	24.20	24.92	19.77

[a] From Hofstetter *et al.* (1965).

ethylenic bonds. Neither $AgNO_3$ TLC nor GLC has yet achieved total resolution of such mixtures, but in combination these methods have given resolutions of many individual components. Thus $AgNO_3$ TLC can effect separations into fractions with and without terminal ethylenic bonds so that the individual components of these fractions can be chromatographed by GLC (Morris, 1966; Morris and Nichols, 1970; Hopkins, 1972).

Table XXII gives the ECL values of various ethylenic-acetylenic acids on some polar and nonpolar columns. It has been concluded (Hofstetter *et al.*, 1965; Jamieson and Reid, 1976) that the contribution of the acetylenic group to the retention time is similar to that of three ethylenic groups.

1.4.4.3. Allenic Acid Esters

Fatty acids with acetylenic bonds are isomeric with acids containing vicinal double bonds of the type –CH=C=CH–, and chemical interconversion by movement of a proton is a possibility. These acids also are recovered by $AgNO_3$ TLC in mixtures of unsaturated fatty acids of comparable number of ethylenic double bonds. The natural allenic acids occurring in fungi have intermediate chain lengths (C_8–C_{13}), and the allenic function is usually conjugated (Brennan *et al.*, 1975). In higher plants, the characteristic allenic acid has a structure of (–)5,6-octadecadienoic acid (Bagby *et al.*, 1965). An analogue, 5,6-*trans*-16-octadecatrienoic acid, has been isolated by Mikolajczak *et al.* (1967). The allenic acids also occur as the hydroxy derivatives. An 8-hydroxy-5,6-octadienoate has been isolated from the seeds of *Sapium*, where it is esterified at the hydroxyl group with *trans*-2, *cis*-4-decadienoic acid (Sprecher *et al.*, 1965).

1.4.5. Oxygenated Acids

There have been numerous identifications made of hydroxy acids from natural sources, particularly from bacteria, from fungi, and from brain lipids. Epoxy acids have been isolated from the seed oil of plants from several families. A few keto acids also have been reported to occur in plants. Furthermore, mixtures of oxygenated acids result from the autoxidation or chemical oxidation of unsaturated fatty acids.

1.4.5.1. Monofunctional Derivatives

Tulloch (1964) has reported ECL values for all the isomeric methyl hydroxy-, acetoxy-, and oxostearates on polar and nonpolar liquid phases, and these values have been reproduced in Table XXIII. With both the hydroxypalmitate and -stearate, the 2 and 3 isomers have much lower ECL values on all the stationary phases used than the other isomers.

Hydrogen bonding is not possible with acetoxy esters, and it was found that on all stationary phases the isomers formed a series of gradually increasing ECL values, although the increases were small for the 6–14 isomers. The isomer with the oxo function on the terminal carbon atom had the highest ECL values on the polar stationary phases.

Tulloch (1964) has devised a scheme for the identification of most of the isomeric hydroxystearates using three stationary phases: EGS, QF-1, and SE-30 and the 12-hydroxy-, acetoxy-, and oxostearates derived from hydrogenated castor oil as standards. If the unknown ester is a hydroxy ester, then only the 2-, 3-, 4-, 5-, 17-, and 18-hydroxystearates can be positively separated from 12-hydroxystearate on any of the stationary phases. Conversion of the hydroxy esters to the acetoxy esters and chromatography on an EGS column then allow separation of the 2, 3, 4, 15, 16, 17, and 18 isomers from 12-acetoxystearate. If the unknown hydroxy ester is converted to the corresponding oxoester, chromatography on the QF-1 column allows separation of the 2, 4, 5, 6, 7, 8, 16, and 17 oxoisomers from each other and from 12-oxostearate. Thus, by using a combination of the three stationary phases and the three types of oxygenated esters, all except the 9, 10, 11, and 13 isomers can be characterized. O'Brien and Rouser (1964) have reported the retention times of all the isomeric methyl hydroxypalmitates on EGS, DEGS, and Apiezon L and of the acetoxypalmitates on EGS. Wood *et al.* (1966) have observed that the TMS ethers are well suited for the GLC of the hydroxy fatty acid methyl esters. Karlsson (1974) has described the resolution and chromatographic analysis of configuration of 2-hydroxy fatty acids.

A few keto acids have been reported to occur in plants (Hitchcock and Nichols, 1971), some of which also have conjugated ethylene bonds (4k-18:3, 4k-18:4, 4k-18:4, 9c11t13t15c; 9k-18:2, 10t12t; 13k-18:2, 9t11t) or cyclopropane and epoxy functions. There have been no systematic studies made of the GLC behavior of these complex structures. Conacher and Gunstone (1969), however, have recorded the ECL values for a number of these and related compounds which they have obtained during chemical transformations of certain natural fatty acids. Table XXIV gives the ECL values for the latter compounds on conventional GLC columns containing either polar or nonpolar liquid phases.

1.4.5.2. Di- and Polyhydroxy Derivatives

Wood *et al.* (1966) have reported the GLC separations of polyhydroxy methyl esters as the TMS ethers. On an Apiezon L open tubular column, partial separations of methyl *threo*- and *erythro*-9,10-hydroxystearate were obtained. Partial separations were also realized for the TMS derivatives of the diastereoisomeric methyl 9,10,12-trihydroxystearates and the methyl 9,10,12,13-tetrahydroxystearates. Analyses of a

Table XXIII. ECL Values of Methyl Hydroxy-, Acetoxy-, and Oxo-octadecanoates on Polar and Nonpolar Liquid Phases[a]

Position of substituent	Hydroxy			Acetoxy			Oxo		
	EGS 224°C	QF-1 202°C	SE-30 220°C	EGS 224°C	QF-1 202°C	SE-30 220°C	EGS 224°C	QF-1 202°C	SE-30 220°C
2	23.55	20.10	19.25	23.50	21.95	20.30	22.55	21.00	18.95
3	24.45	20.80	19.50	23.95	22.35	20.35			
4	27.15	24.55	19.85	24.40	22.45	20.40	24.70	21.95	19.40
5	28.00	25.10	20.10	24.45	22.60	20.45	24.85	22.40	19.50
6	26.15	21.70	19.85	24.50	22.85	20.50	25.20	22.80	19.65
7	26.20	21.75	19.90	24.60	22.95	20.50	25.25	22.90	19.65
8	26.20	21.75	19.95	24.60	23.00	20.50	25.30	23.00	19.65
9	26.20	21.80	20.00	24.65	23.00	20.50	25.30	23.05	19.70
10	26.25	21.80	20.00	24.65	23.05	20.50	25.35	23.05	19.70
11	26.25	21.80	20.00	24.70	23.05	20.55	25.35	23.10	19.70
12	26.25	21.80	20.00	24.70	23.10	20.55	25.40	23.15	19.75
13	26.30	21.80	20.05	24.80	23.15	20.60	25.40	23.15	19.80
14	26.30	21.80	20.05	24.90	23.20	20.65	25.50	23.15	19.85
15	26.35	21.80	20.05	25.10	23.30	20.75	25.60	23.15	19.90
16	26.50	21.90	20.10	25.50	23.60	20.95	25.90	23.36	19.95
17	26.95	22.10	20.10	25.95	23.90	21.15	26.40	23.75	20.00
18		23.05	21.00	27.40	24.90	21.90			

[a] From Tulloch (1964).

Table XXIV. ECL Values of Some Synthetic Oxo Fatty Acids[a]

Fatty acid	Liquid phase	
	DEGS	Apiezon L
9:0 4-keto	15.2	10.3
17:0 9c,10-methylene, 12-keto	25.6	19.2
17:0 9t,10-methylene, 12-keto	24.8	18.8
18:0 9c,10-methylene, 12-keto	26.8	20.2
18:0 9t,10-methylene, 12-keto	25.8	19.8
17:0 10,11-methylene, 12-keto	25.2	19.1
17:0 9c,10-methylene	18.6	17.8
17:0 9t,10-methylene	18.0	17.0
17:0 9c,10-methylene, 12-hydroxy	26.4	
17:0 9t,10-methylene, 12-hydroxy	25.7	
17:0 10,11-methylene, 12-hydroxy	25.7	
18:0 12-keto	24.9	19.4
18:1 9c,12-keto	25.3	19.1
18:1 10t,12-keto	26.8	19.9
18:0 12c,13-epoxy	24.0	19.3
18:1 9c,12c,13-epoxy	24.6	19.1
18:1 12a,9c,10-epoxy	26.0	19.1

[a] From Conacher and Gunstone (1969).

mixture of the TMS ethers of the methyl esters of 12-hydroxystearic, *threo*-9,10-dihydroxystearic, *threo*-9,10-12-trihydroxystearic, and methyl-9,10-*erythro*-12,13-tetrahydroxystearic acids on a DEGS-packed column and on an Apiezon L capillary column gave the expected elution order: mono-, di-, tri-, and tetrahydroxy derivatives.

Wood *et al.* (1966) have discussed the use in gas chromatography of the trifluoroacetyl derivatives of hydroxy acids. Useful separations of isomers were obtained using both EGSS-X- and SE-30-packed columns. A mixture of mono-, di-, tri-, and tetrahydroxystearates as their trifluoroacetyl derivatives gave this unexpected elution order on a packed column of SE-30: tetra-, tri-, di-, and mono-substituted derivatives. Wood (1967) has shown that packed columns can be used for the separation of the isopropylidene derivatives of the dihydroxy compounds derived from oleic and elaidic acids. The two isopropylidene derivatives were easily separated on both EGSS-X- and SE-30-packed columns, the *trans* isomer having the shortest retention time. The four geometric isomers of linoleic acid furnish eight diastereoisomeric tetrahydroxystearic acids whose isopropylidene derivatives gave five peaks on an EGSS-X column. These were, respectively, due to the following isomers: one *threo-threo* isomer; the other *threo-threo* isomer; all the mixed *threo-erythro* isomers; one *erythro-erythro* isomer; and the other *erythro-erythro* isomer. Alkaline permanganate oxidation of ricinoleic and ricinoelaidic acids gives four diastereoisomeric trihydroxy acids. The isopropylidene–trifluoroacetyl derivatives of these acids were separated on an EGSS-X-packed column with the following relative retention times: *threo*-9,10-*erythro*-10,12- (1.00); *threo*-9,10-*erythro*-10,12- (1.12); *erythro*-9,10-*threo*-10,12- (1.25); *erythro*-9,10-*erythro*-10,12- (1.46).

1.5. Identification of Fatty Acids by Gas Chromatography/Mass Spectrometry

In recent years, gas chromatography/mass spectrometry has been widely accepted as one of the most valuable techniques for the identification of fatty acids and their derivatives. However, complete identifications of all compounds are not obtained, and considerable prefractionation and chemical derivatization may be required until all alternatives have been eliminated. Since the most complete GLC resolutions of fatty acids are obtained on polar liquid phases which are not generally compatible with effective mass spectrometry (Myher *et al.*, 1974), it is necessary to prefractionate the fatty acid methyl esters by argentation TLC, leaving the molecular weight resolution to the nonpolar columns which are compatible with mass spectrometric analysis. There exists an extensive literature on both practical aspects of GC/MS and on systematic interpretation of the mass spectra of fatty acids and derivatives (Ryhage and Stenhagen, 1963; Sun and Holman, 1968; McCloskey, 1969, 1970; Odham and Stenhagen-Ställberg, 1972). The following section summarizes and updates the main forms of information provided by GC/MS and points out its advantages and disadvantages for the identification of specific fatty acids and fatty acid derivatives. Murata (1975) has described an application of chemical ionization mass spectrometry to analysis of fatty acids.

1.5.1. Saturated Normal- and Branched-Chain and Cyclic Acids

The saturated fatty esters are recovered as a mixture from argentation TLC and may be examined by GC/MS in this form. A detailed study of the mass spectra of any minor components, however, usually requires a preliminary isolation of enriched fractions by preparative GLC.

1.5.1.1. Saturated Normal-Chain Acids

The mass spectra for homologues of the saturated normal-chain acids are well defined and characteristic of individual members. The relative abundance of the molecular ion (M) increases from methyl pentanoate upward (Rhyhage and Stenhagen, 1960). The identification of M is verified by the acylium ion, $[M - 31]^+$, which is due to loss of the methoxy group. The normal-chain methyl esters produce a base peak at *m/e* 74, which is the ion formed by a γ-hydrogen migration to a double bond, followed by a β cleavage. This is known as the McLafferty rearrangement (McLafferty, 1959). In addition, ions of the series $CH_3OCO(CH_2)_n$ are formed which are found at *m/e* $(59 + 14n)$. Both simple cleavage and rearrangement processes contribute to the formation of hydrocarbon ions, of which the most prominent are *m/e* 69, 83, 97, etc., arising from the saturated series C_nH_{2n+1}. In general, the mass spectra of all normal-chain fatty acid esters from methyl butyrate upward are similar.

The mass spectra of fatty acids which are esterified with alcohols other than methanol can usually be predicted from the behavior of the corresponding methyl ester, although the overall appearance of the spectrum may be somewhat different. The McLafferty rearrangement product for the TMS ester of palmitic acid is *m/e*

132, and members of the TMSOCO $(CH_2)_n$ series are found 58 mass units higher than their methyl ester counterparts. $[M - 29]^+$ and $[M - 43]^+$ are observed, but the acylium ion $[M - OTMS]^+$ is not present. Virtually all mass spectra of the TMS derivatives contain the TMS ion, *m/e* 73, and the rearranged ion, *m/e* 75 (McCloskey, 1970). A characteristic $[M - 15]^+$ ion is obtained by loss of a methyl radical from the TMS function. A prominent $[M - 57]^+$ ion is generated by loss of a butyl radical from the *t*-butyldimethylsilyl esters (Phillipou *et al.*, 1975).

1.5.1.2. Saturated Branched-Chain Acids

Ryhage and Stenhagen (1960, 1963) have discussed in detail the effects of monomethyl, polymethyl, and higher alkyl branching. In general, the presence of branching does not greatly disturb the fragmentation pattern of a saturated acid. For this reason, the interpretation of the spectrum is made in terms of variations of the spectrum of a normal-chain ester. If the position of substitution is C_2, C_3, or C_4, ions of *m/e* 74, 87, $[M - 29]^+$, or $[M - 43]^+$ are shifted in mass since they represent either directly or by difference the atoms adjacent to the ester linkage. Mass 74 contains C_2 but not C_3; therefore, it is shifted to higher masses only on α substitution. Similarly, *m/e* 87 and $[M - 29]^+$ are shifted to higher masses only when C_2 or C_3 is substituted. The same considerations are used to predict the pattern for multiple branching. Thus a 3,4-dimethyl substitution leaves *m/e* 74 unshifted, while *m/e* 87 is shifted to *m/e* 101 (Ryhage and Stenhagen, 1960). Elimination of C_2 and C_3 plus hydrogen involves 43 mass units, while the usual $[M - 43]^+$ is found 28 mass units lower at $[M - 71]^+$. If branching occurs near the center of the chain, the effects are less obvious but are easily discernible over the regular pattern of normal-chain esters. Thus the mass spectrum of methyl 10-methyl stearate appears similar in many respects to that of a normal-chain ester, but several diagnostically important changes are present (McCloskey, 1970). Methyl branching at position 6 leads to an intensive peak at $[M - 76]^+$ (Ryhage and Stenhagen, 1963). The most difficult is the identification of the *iso* branching. The mass spectra of these compounds are very similar to those of normal-chain esters. The terminal isopropyl group can be characterized by a peak of very low intensity at $[M - 65]^+$ (Ryhage and Stenhagen, 1963). However, the ECL values of the GLC peaks can be used to provide the additional assurance necessary for a positive identification of an *iso*, *anteiso*, or *neo* branching in a fatty acid chain, provided that GLC columns of sufficient resolution are employed.

A number of mass spectra of dimethyl- and trimethyl-branched fatty acids have been described by Odham (1967), Odham and Stenhagen-Ställberg (1972), Jacob and Poltz (1975), and Nicolaides *et al.* (1976). The mass spectra of the tetramethyl fatty acids have been discussed by Lough (1973).

Foglia *et al.* (1976) have employed GC/MS methods for the analysis of synthetic α-branched saturated fatty acids. The products were mixtures of isomeric branched-chain acids composed of either α-monosubstituted (dialkylacetic acids) or α,α-disubstituted (trialkylacetic acids) moieties. The mass spectra of the α-monoalkyl derivatives were characterized by prominent McLafferty rearrangement ion peaks, whereas those of the α,α-dialkyl isomers contained the former ions plus an α-cleavage ion. Foglia *et al.* (1976) have reported the major mass peaks for a series of α-branched fatty acids of various chain length of the alkyl groups.

Andersson and Holman (1975) have shown that localization of a methyl branch in a fatty acid molecule by GC/MS is facilitated by using the pyrrolidides rather than the methyl esters.

1.5.1.3. Cyclic Fatty Acids

The geometric and positional isomers of monocyclopropane esters yield mass spectra which are virtually indistinguishable from those of monounsaturated fatty esters of the same chain length (Wood and Reiser, 1965; Christie and Holman, 1966). Measurement of the ratio of the intensity of the parent molecular ion to that of the parent molecular ion minus 32 (the mass of methanol) can distinguish between monoenoic and cyclopropanoid fatty acid methyl esters with the same carbon number. A similar technique can be used to distinguish between isomeric normal, iso, and anteiso fatty acid methyl esters (Campbell and Naworal, 1969). Polycyclopropane esters or cyclopropane esters with other functional groups in the chain, however, yield characteristic fragmentation patterns which either permit isomers to be distinguished or functional groups to be located depending on the complexity of the molecule. A simple and rapid mass spectrometric method of ring location involves a reductive opening of the ring which produces a normal-chain ester and a mixture of the two possible methyl-branched isomers (McCloskey and Law, 1967). When the reaction mixture is analyzed by GC/MS, the branched-chain isomers emerge ahead of the straight-chain compound. The mass spectrum of the branched ester readily identifies the position of branching as described above. Location of the branch points is determined by pairs of characteristic ions which are always 14 mass units apart.

Minnikin (1972) has described a method of ring location in cyclopropane fatty acid esters based on BF_3-catalyzed methoxylation. Mass spectra of the methoxylation products are characterized by intense peaks due to cleavage adjacent to methoxy functions, which allow the position of the ring in the original ester to be easily assigned.

The characterization of cyclopropene esters by mass spectrometry is hindered by the high reactivity of the cyclopropene ring. Therefore, chemical conversion to more stable derivatives is necessary. A reduction to a cyclopropane ester can be employed, followed by the ring-opening techniques described for cyclopropane esters (Hooper and Law, 1968; Minnikin, 1972). Alternatively, the cyclopropene ring can be converted by ozonolysis to a methylene-interrupted diketone, or to a thiol adduct by methanethiol (Raju and Reiser, 1966). The mass spectrum of the diketo ester derived from methyl sterculate exhibits most of the ions observed for keto esters. All four possible simple α cleavages are observed, followed by the corresponding elimination of methanol. Quantitive addition of methanethiol to methyl sterculate produces an unresolved mixture of products in which sulfur is attached to either C_9 or C_{10}. The position of the ring, however, is clearly marked by the appropriate fragments. The mass spectra of methyl sterculate and malvate have been also discussed by Pawlowski (1972).

The mass spectra of the methyl 11-cyclopentylundecanoate and methyl 13-cyclopentyltridecanoate have been determined by Christie *et al.* (1969) and that of the methyl 15-cyclopentylpentadecanoate by Spener and Mangold (1974). In all instances, a base peak at *m/e* 74 (McLafferty rearrangement), prominent peaks at *m/e* 87, 101, 129, 143, 185, 199, 255 [$+(CH_2)_nCOOCH_3$ for $n = 2, 3, 5, 6, 9, 10, 14$], 69 (cyclopentane), 281 $[M - 43]^+$, and the molecular ion at *m/e* 324 with an abundance of 63% were obtained. Spener and Mangold (1974) also determined the mass spectrum of hormelic acid [15-(2-cyclopenten-1-yl)pentadecanoic acid], which they synthesized by chain elongation from chaulmoogric acid [13-(2-cyclopenten-1-yl)tridecanoic acid]. The hormelic acid gave a base peak at *m/e* 67 (cyclopentene), prominent peaks at *m/e* 82, 191, 209 [$CH_2{=}CH(CH_2)_{11}CO$], 241 [$+(CH_2)_{13}COOCH_3$], 290 $[M - 32]^+$, and molecular ion at *m/e* 322 with an abundance of 12%. The GC/MS spectra of the ω-cyclohexyl fatty acids have been determined by Oshima and Ariga (1975). The fragment ions at *m/e* 83, 69, 55, and 41 gave evidence of fragmentation of the cyclohexane ring. The mass spectra of furanoid fatty acids have been reported by Glass *et al.* (1975).

The anthracene-labeled fatty acids prepared by Stoffel and Michaelis (1976) all yielded the same characteristic fragmentation ions as well as molecular ions.

Cyclic structures are also found in fatty acids as a result of internal cyclization. Thus linoleic and linolenic acids yield cyclic monomers upon heating (Artman and Smith, 1972). The cyclic monomer esters resolved by a combination of argentation TLC and GLC gave mass spectra (Perkins and Iwaoka, 1973) similar to those published for pure synthetic aromatic cyclic monomers and for aromatic acids produced from linolenate and linoleate.

1.5.2. Ethylenic and Acetylenic Acids

Since the location and stereochemistry of the double and triple bonds in the monounsaturated esters cannot be determined directly by mass spectrometry, complementary chromatographic separation or chemical derivatization must be employed.

1.5.2.1. Ethylenic Acids

The mass spectra of the ethylenic fatty acid esters are distinctly different from those of their saturated counterparts, but differ relatively little from each other according to degree of unsaturation (McCloskey, 1970). The molecular ions are abundant, and are confirmed in each case by loss of a methoxyl radical $[M - 31]^+$ and by elimination of methanol. $[M - 74]^+$ and $[M - 116]^+$ correspond to loss of the ester moiety plus a rearranged hydrogen by exchange of the C_2–C_3 and C_5–C_{16} bonds, respectively (Odham and Stenhagen-Ställberg, 1972). Examination of all the methyl nonenoates (Groff *et al.*, 1968) and C_{18} monoenes with double bonds in positions 6, 8, 10, 13, and 17, including both isomers of positions 6 and 9, has revealed essentially indistinguishable mass spectra. This is due to double-bond migration at the molecular ion stage, giving a number of common intermediate products from which likewise common fragment ions are produced (Groff *et al.*, 1968). An exception to the similarity of spectra of monoenes is afforded by unsaturation in position 2. The mass spectra of methyl *cis*- and *trans*-2-octadecenoates not only differ from each other but also show a characteristic peak at *m/e* 113, which is the base peak in the *cis* isomer (Ryhage *et al.*, 1961). The mass spectra of short-chain (C_6) monoenes also exhibit different behavior (Rohwedder *et al.*, 1965).

Andersson *et al.* (1975) have shown that pyrrolidides of polyenoic fatty acids give different spectra for each acid and that with enough reference compounds there should be no problem identifying each positional isomer without further chemical derivatization.

The mass spectra of all the isomers of methyl linoleate are generally similar with the exception of the 2,5- compound (Christie and Holman, 1967), which shows the usual peak associated with -2- unsaturation. In that case, an abundant ion at *m/e* 139 was also observed. The available data on trienes show that methyl 6, 9, 12-octadecatrienoate and its isomer methyl 9,12,15-octadecatrienoate yield spectra which show intensity but not mass differences (Holman and Rahm, 1966). In the mass spectrum of methyl linolenate, a preferred ion arises from loss of 56 units, which has been postulated to involve cleavage of C_{15}–C_{16} with hydrogen rearrangement to the neutral species which is lost.

Myher *et al.* (1974) have recorded GC/MS spectra for standard polyunsaturated fatty acids as eluted from polar siloxane columns. It was observed that an abundant ion at *m/e* 108 was characteristic of the $\omega 3$ esters and an ion at *m/e* 150 of the $\omega 6$ esters. Plattner *et al.* (1976) have developed a rapid microprocedure for double-bond location in polyenoic acids through oxymercuration and GC/MS. Only one methyl group is present per molecule, and the mass spectra of the mixture are indicative of all olefinic positions.

Characteristic mass spectra for the benzyl esters of C_1–C_{20} fatty acids have been recorded by Hintze *et al.* (1973), while Cooper and Anders (1974) have obtained the mass spectra of the 2-naphthacyl esters of long-chain acids.

1.5.2.2. Acetylenic Acids

Fatty acid methyl esters with triple bonds yield mass spectra with prominent peaks resulting from cyclic elimination with hydrogen transfer (Bohlmann *et al.*, 1967). Esters with unsaturation at 2, 3, 4, or terminal positions have spectra different from those having the triple bond at the central positions of the molecule. Similar results have been obtained also for the spectra of a series of isomeric methyl nonynoates (Groff *et al.*, 1968). Because of a secondary fragmentation and shift of the triple bond to other positions, it is not possible to determine the exact location of the triple bond. However, the position of the triple bond in the ester molecule can be located after deuteration by the method of pyrolysis, or by any of the other methods applicable to olefinic esters after a partial reduction of the triple bond to a double bond (Sun and Holman, 1968).

The spectra of monoynoic esters show an intense peak at *m/e* 154 (75% RA) in the C_{18} esters and 95% RA in the C_{20} homologue (Pearl *et al.*, 1973). This formation of allenes in the mass spectrometer has been shown by Bohlmann *et al.* (1967) with methyl 6-tetradecynoate (*m/e* 154) and by Sun and Holman (1968) with methyl stearolate (*m/e* 196 and *m/e* 210).

The mass spectrum of 17-octadecen-6-ynoate has been reported by Pearl *et al.* (1973) to be similar to that of methyl tarirate (6-octadecynoate). Ions at *m/e* 154 (75% RA) and *m/e* 140 (8% RA) indicated that the triple bond is in the 6 position. Both the molecular ion (*m/e* 292) and molecular ion minus methoxyl (*m/e* 261) were 2 mass units less than those of methyl tarirate. Pearl *et al.* (1973) also identified the methyl 6-hexadecynoate.

1.5.3. Oxygenated Acids

GC/MS of oxygenated fatty acids is employed for the identification of natural oxygenated acids and synthetic oxygenated acids derived from unsaturated and cyclopropanoid acids.

1.5.3.1. Epoxy Acids

The fatty acid epoxides yield characteristic ions upon fragmentation in the mass spectrometer corresponding to cleavages at either side of the ring if it is in the central position of the hydrocarbon chain. Saturated epoxy esters give mass spectra which are so readily interpreted that epoxidation and mass spectrometry form an established procedure for the location of double bonds (Eglinton *et al.*, 1968; Aplin and Coles, 1967). Thus the spectrum of methyl 9,10-epoxystearate (Ryhage and Stenhagen, 1960) has a base peak at *m/e* 155 arising from cleavage α to the epoxide ring. An α-cleavage on the other side of the functional group produces a much smaller but significant peak at *m/e* 199. However, introduction of a double bond into the molecule changes the spectrum so radically that assignment of the location of the epoxide ring is impossible (Kleiman and Spencer, 1973). Unsaturated epoxy methyl esters produce spectra difficult to interpret. When the epoxide is converted to the methoxy-hydroxy derivative by BF_3-methanol, the spectrum locates the position of the epoxide group. Trimethylsilylation of the hydroxyl group produces a compound that gives a less complicated spectrum which also locates the original epoxy group.

1.5.3.2. Keto Acids

The spectra of the keto esters show six prominent peaks which indicate the position of the keto group (Sun and Holman, 1968). The McLafferty rearrangement involving the carbonyl group in the chain is structurally specific for γ-hydrogens. Therefore, all the major ions are not found in cases in which the carbonyl group is closer than three carbons from either end of the polymethylene chain (Ryhage and Stenhagen, 1960).

The position of the keto group in each of the two isomeric keto esters arising from the isomerization of an epoxide ring can be located by the respective sets of characteristic peaks, and each peak from one set should have 14 mass units difference from the corresponding peak of another set.

Weihrauch *et al.* (1974) have reported the mass spectra of the keto fatty acids isolated from milk fat and of authentic standards. The location of the carbonyl group was made readily as both α and β cleavage occurred on either side of the carbonyl group. The molecular ion which is usually less than 1% of the base peak was readily discernible. Mass spectra of the methyl esters of the ketostearates have been discussed by Ryhage and Stenhagen (1963), who noted large β-cleavage ions from both sides of the keto group.

Kleiman and Spencer (1973) have shown that the mass spectra of the unsaturated keto esters closely resemble those of the saturated keto esters. Thus, in 17-oxo-*cis*-20-hexacosanoate, α cleavage was observed on both sides of the keto group, with ions *m/e* 153 and 297 present. However, the *m/e* 297 ion was much stronger than expected from the behavior of an analogous saturated compound. Also, the β-cleavage ion on the olefin side (*m/e* 312) was more abundant than the corresponding ion on the other side of the keto group (*m/e* 168).

1.5.3.3. Monohydroxy Acids

The mass spectra of the hydroxy fatty acid esters also show readily recognized peaks for the characteristic *m/e* 74 and 87 ions (Ryhage and Stenhagen, 1960). The major ions are associated with cleavages β to the oxygen atoms, resulting in a loss of the methyl end of the chain, which serve to mark the position of substitution in the chain. Thus the 10-hydroxystearate yields *m/e* 210 ($[M-(CH_2)_7CH_3]^+$) as a characteristic ion. The identity of this ion is confirmed by ions 32 mass units lower (*m/e* 169) due to elimination of CH_3OH. In methyl 10-methoxystearate, two methoxy groups are present, and the characteristic ion (*m/e* 215) undergoes two successive losses of 32 mass units. The molecular ion is of low abundance in these esters but can be calculated from the ions representing the other portion of the chain (*m/e* 143 and *m/e* 157 in the 10-hydroxy- and methoxystearates, respectively). The mass spectra of methyl esters containing a hydroxyl group in position 3 are dominated by *m/e* 103 so that M and the upper mass range fragment ions are virtually absent (Ryhage and Stenhagen, 1963; Eglinton *et al.*, 1968). This requires that the chain length of the ester be determined by GLC. If the hydroxyl group is located on the α-carbon atom, cleavage between C_1 and C_2 is facilitated, leading to loss of the carbomethoxy group. Trimethylsilylation of the hydroxyl function results in cleavages on either side of the substituted carbon atom, which yields prominent ions of corresponding *m/e*. In general, all silylated hydroxy methyl esters exhibit $[M-15]^+$ (methyl), $[M-31]^+$ (methoxyl), and $[M-47]^+$ (methyl-methanol) ions. In all these esters, the ions *m/e* 73 $(CH_3)Si^+$ and *m/e* 75 $(CH_3)_2SiOH^+$ are major peaks (Eglinton *et al.*, 1968).

Kleiman and Spencer (1973) have shown that silylation of hydroxyl groups in methyl esters of unsaturated hydroxy fatty acids provides compounds that give mass spectra which can be readily interpreted. In esters that have the TMS group separated from the double bond by one methylene group, the ions caused by α cleavage at the TMS group on the side closest to the olefinic group are much more abundant than those produced from α cleavage on the other side of the TMS group. In esters that have the TMS group and the double bond separated by two methylene groups, α-cleavage ions are approximately equal. When the TMS group and the double bond are allylic, no fragmentation results between them. Cleavage, however, occurs on either side of the system, and those ions resulting from cleavage α to the TMS group are in greatest abundance. In esters that have a conjugated diene or ene-yne system adjacent to a hydroxyl group, large mass spectral peaks are observed that arise from α cleavage at the TMS group and at the other end of the olefinic system. Kleiman and Spencer (1973) have shown that the spectrum of the TMS derivatives of methyl ricinoleate allows positive identification. The base peak is *m/e* 187 from α cleavage at the unsaturated side of the TMS ether group. Cleavage of the molecule on the other side of the oxygenated site provides a much less intense though significant peak at *m/e* 299. These intensities found for the α-cleavage peaks are in contrast to those found from mass spectra of the silylated saturated analogue of methyl ricinoleate (McCloskey, 1970) where both α-cleavage peaks are greater than 50%.

The mass spectrum resulting from GC/MS of the TMS derivative of methyl dimorphecolate (9-hydroxy-*trans*-10,*trans*-12-octadecadienoate) has two major peaks that define the structure of the esters: *m/e* 225 ($[M - (CH_2)_7COOCH_3]^+$, base peak); *m/e* 311 ($[M - C_5H_{11}]^+$), and *m/e* 382 (molecular ion). Two of these ions clearly define the location of the conjugated dienol system, while the large *m/e* 225 peak shows the location of the TMS group itself. The *m/e* 311 peak locates the double bonds in the 10 and 12 positions (13-hydroxy-*cis*-9,*trans*-11-octadecadienoate) (Kleiman and Spencer, 1973). The mass spectrum from the TMS derivative of methyl 7-hydroxy-*trans*-10-heptadecen-8-ynoate was similar to those for the hydroxy-conjugated diene esters. Major peaks arose from α cleavage at the TMS groups (*m/e* 237), and α cleavage at the double bond (*m/e* 281) also marks the location of the functional group. A total of some 20 esters of unsaturated hydroxy fatty acids and their reduction products were analyzed by mass spectrometry as the TMS ethers by Kleiman and Spencer (1973).

Hammarstrom (1974) and Laine (1974) have discussed the mass spectra of several 2-hydroxy and 3-hydroxy fatty acids of both enantiomeric types. In all instances, the mass spectra of diastereoisomeric compounds are mutually indistinguishable. Some of the fragments obtained from the (–)menthyloxycarbonyl derivative of 2-D-hydroxystearic acid methyl ester and of the 2-D-phenylpropionate derivative of 3-D-hydroxystearic acid methyl ester retain the fatty acid part of the molecule, whereas others are formed by elimination of this part of the molecule.

1.5.3.4. Dihydroxy Acids

The mass spectra of the dihydroxy fatty acids which may be obtained by oxidation of methyl esters of monounsaturated fatty acids exhibit intense peaks from one of the α-cleavage ions and subsequent methanol elimination (Ryhage and Stenhagen, 1960). A more characteristic spectrum is obtained following the conversion of the diol into the corresponding TMS ether derivative (Capella and Zorzut, 1968), which yields a more informative fragmentation pattern. The spectrum of 9,10-dihydroxystearate TMS reveals two intense peaks due to α cleavage between carbon atoms 9 and 10. The spectrum of the TMS derivative of methyl ω-undecenoate (Capella and Zorzut, 1968) shows the ion containing the terminal end of the chain to be of low intensity but that of the carboxyl end of the chain to be the second most intense peak in the spectrum. The molecular ion is represented by $[M - 15]^+$ ($-CH_3$) and $[M - 31]^+$ ($-OCH_3$).

Of special interest is the mass spectrometric fragmentation pattern of the *O*-isopropylidene derivatives of fatty acids with vicinal diols, because it allows differentiation between geometric isomers. The vicinal diols produced by stereospecific oxidation of double bonds retain in the 1,3-dioxolone ring system the relationship *cis*-erythro and *trans*-threo, and lead to marked intensity differences between the mass spectra of isomers (McCloskey and McClelland, 1965; Wolff *et al.*, 1966). The mass spectrum of the 1,3-dioxolone derivatives of vicinal dihydroxy esters gives two peaks, which correspond to cleavages of the bonds at either side of the ring. The intensity of the peak corresponding to the fragment of the end chain decreases with increasing chain length of the ester. The intensities of the peaks at $[M - 89]^+$ $[CH_3(CH_2)mCH{-}CH(CH_2)nC{=}O^+]$ in the spectra of the dioxolone derivatives can be used to distinguish between two geometric isomers (McCloskey and McClelland, 1965). In the spectrum of the *erythro* isomer, which is derived from a *cis*-unsaturated ester, the peak $[M - 89]^+$ is at least twice as intense as that in the spectrum of the *threo* isomer derived from the *trans*-unsaturated ester.

1.5.3.5. Polyhydroxy Acids

The characterization of multiply hydroxylated fatty acid esters may be made by extending the methods applied to the mono- and dihydroxylated acids. This approach has been utilized by Niehaus and Ryhage (1967) and Eglinton *et al.* (1968) for the location of multiple double bonds following hydroxylation. In general, the upper mass ions are of very low abundance owing to the numerous fragmentation pathways possible in a polyhydroxylated molecule.

If a polyenoic ester is oxidized to a polyol, permethylation affords a polyether whose mass spectrum (Niehaus and Ryhage, 1967) is similar to that of a monomethoxy ester (Ryhage and Stenhagen, 1960). The spectrum of the hexamethoxy derivative of methyl 6,9,12-octadecatrienoate, for example, yields the most prominent ions by simple cleavages between the vicinal methoxyl groups, followed by successive multiple eliminations of methanol. The ions of the original cleavages can be distinguished from their daughter ions by their occurrence as the highest members of a series of ions differing by 32 mass units. Positions of the methoxyl groups and hence of the double bonds can then be determined by fitting observed mass values of prominent peaks to allowable values, which correspond to possible structures (McCloskey, 1970). No ions are observed beyond *m/e* 363, but the value of M can be indirectly confirmed by matching values obtained from the major simple cleavage ions.

Alternatively, a polyol obtained by oxidation of a polyene can be subjected to silylation. Thus the tetra-TMS derivative of methyl linoleate (McCloskey, 1970) undergoes the principal modes of cleavage between the substituted carbon atoms, as the polymethoxy derivatives. However, some of the ions of the initial cleavage may be absent but instead eliminate trimethylsilanol (90 mass units). This effect is also found in the spectrum of the penta-TMS derivative of methyl 9,10,12,13,18-pentahydroxystearate (Eglinton *et al.*, 1968).

The mass spectra of the trihydroxy fatty esters have been reported by Mikolajczak and Smith (1967). The methyl ester of phloionolate (9,10,18-trihydroxyoctadecanoate) has been reported by Eglinton *et al.* (1968). The spectrum of an unsaturated trihydroxy ester (9,10,18-trihydroxyoctadec-*cis*-12-enoate) has been reported by Kleiman and Spencer (1973). The ions resulting from the cleavage between the carbon atoms containing the TMS groups (*m/e* 259 and 301) were again conspicuous, and two additional peaks (*m/e* 271 and 361) became prominent.

1.5.4. Stable Isotope-Labeled Fatty Acids

Mass spectrometry is essential for determination of the chemically or biologically incorporated heavy isotopes into fatty acids or their derivatives. Both the total content and molecular distribution of the isotope may be determined, but each measurement varies with the nature of the isotope and its mode of incorporation. A special case arises from the increasing use of stable isotope-labeled fatty acids as internal standards in the quantitation of metabolites by GC/MS methods.

1.5.4.1. Chemically Labeled Acids

Chemical labeling with heavy isotopes such as deuterium has been extensively used as an aid in interpreting the mass spectra of model fatty acids and in elucidating certain basic structural features in unknown fatty acids. In both instances, the mass spectra of labeled and unlabeled compounds are compared and shifts in mass fragment ions due to the presence or absence of heavy isotopes are observed and interpreted (Dinh-Nguyen, 1968; McCloskey, 1970). The chemical synthesis of such compounds involving ^{2}H, ^{13}C, and ^{18}O is described by Emken in Chapter 2 of this book.

For the present purpose, reference may be made to the use of deuterium labeling in the location of double bonds in unsaturated fatty acids. The double bond is saturated with deuterium by reacting the ester with deuteriohydrazine and oxygen (Dinh-Nguyen *et al.*, 1961). The mass spectrum of the saturated ester then gives deuterium-containing ions derived from $[(CH_2)nCO_2CH_3]^+$, if the carbon atom which has the isotope is involved in the fragment. For instance, in the spectrum of methyl 6,7-dideuteriooctadecanoate, which is obtained from methyl petroselinate, the peak *m/e* 129 is shifted to 130 and the peak *m/e* 143 to 145, indicating that one deuterium is in position 7 of the compound. The original double bond must therefore have been between carbons 6 and 7 in the ester. The practical application of this method is limited because of the complications caused by hydrogen and skeletal rearrangements (Dinh-Nguyen, 1968; McCloskey, 1970). Furthermore, many of the $[(CH_2)nCO_2CH_3]^+$ peaks in the spectrum are too small to be identified, especially when the isotope content of the ester is low. Sun and Holman (1968) have described an improved version of double-bond location by means of deuterium reduction with deuterated hydrazine. The method is applicable to any position of mono- and polyunsaturated fatty acid esters. The corresponding deutero ester is subjected to pyrolysis in a flow system at 600°C; the pyrolysis products are separated by GLC and then identified by mass spectrometry. The principal products are a homologous series of unsaturated esters having shorter chain length than the original ester, plus a homologous series of olefins. The absence or presence of deuterium in a fragment may be used to positively locate deuterium in the ester, because—under the conditions of pyrolysis—expulsion, rearrangement, and secondary reactions are minimized. The position of the deuterium in the fatty chain is then determined by examining the chain lengths and deuterium content of both series of fragments.

Deuterium-labeled TMS reagents may be employed to interpret the spectra of polyhydroxy fatty acids. The number of TMS functions in a fragment ion can be determined by shifts of 9 mass units per intact TMS group (McCloskey, 1970).

1.5.4.2. Biologically Labeled Acids

Stable isotopes in biological products are conventionally determined by mass spectrometric analyses following chemical degradation. However, if the intact molecule is submitted to GC/MS, the distribution of labeled species as well as the location of the labels may also be determined. A general example is provided by the generation of deuterium-labeled fatty acids during perfusion of an isolated rat liver with deuterated water (Wadke *et al.*, 1973; Kuksis *et al.*, 1975). The incorporation level must be sufficiently high to be measured with significance over natural isotope species. The minimum useful level is usually around 1 mole %, depending on the spectrum and the type of isotope used. Incorporation of a single deuterium atom necessitates subtraction of background due to naturally occurring ^{13}C, while ^{18}O label causes a shift of 2 mass units to a region where much less correction (due mostly to ^{13}C) is usually necessary.

A more specialized example of the technique is offered by the investigation of the biosynthesis of 10-methylstearic from oleic acid (McCloskey, 1970). Tuberculostearic acid was isolated from *Mycobacterium* growth in the presence of d_3-methylmethionine. The mass spectrum of the methyl ester was compared to that of the synthetic compound. Shift of the abundant molecular ion from *m/e* 312 to 314 revealed the incorporation of only two deuterium atoms, while *m/e* 315 (which corresponded to d_3 species) was calculated to arise solely from naturally occurring heavy isotopes. Fully deuterated fatty acids have been isolated and characterized from *Scenedesmus obliquus* cultures by Graff *et al.* (1970).

1.6. Ancillary Methods of Fatty Acid Identification

In many instances, further information about the structure of a fatty acid must be obtained by chemical derivatization, spectrophotometric measurements, and enzymatic transformations (Christie, 1970; Schlenk, 1970; Polgar, 1971).

1.6.1. Chemical Methods

If the acid is a well-known member of a homologous series, its identity can usually be readily confirmed by chemical methods provided that it is isolated in sufficient quantity and in a high state of purity. Markley (1967) has prepared an exhaustive compilation of melting points of characteristic derivatives of fatty acids. If the acid is unknown, not only must its constitution or structure be determined, but also it should be synthesized by one or more specific methods, and the properties of the isolated and synthetic products compared. Appropriate techniques and experimental strategies for this purpose are discussed by Emken in Chapter 2 of this book. The present discussion is restricted solely to minor chemical modifications of the fatty acid structure as an aid in their identification and isolation.

1.6.1.1. Determination of Number of Double Bonds

A convenient laboratory method for the estimation of the number of double bonds in an isolated fatty acid is by quantitative catalytic hydrogenation. It may be performed on minute amounts of material using commercially available microhydrogenation equipment (Brown, 1967). The catalysts commonly employed are platinum oxide or palladium or platinum on a suitable support such as carbon or calcium carbonate. These catalysts are not suitable for partial hydrogenation, since they cause double-bond migration and *cis-trans* isomerization (Dutton, 1968). Partial hydrogenation, however, may be effected by means of hydrazine as the reducing agent (Aylward and Sawistowska, 1962). For the partial reduction of acetylenic bonds to *cis* double bonds, the lead-poisoned palladium catalyst of Lindlar has been found useful (Lie Ken Jie, 1975*b*). Hydrogenation is often used in combination with GLC to resolve ambiguities arising from the analyses of mixtures of methyl esters such as the identification of 20:1 $\omega 9$ and 18:3 $\omega 3$, which overlap on most liquid phases, and for differentiation between unsaturated and branched-chain acids on nonpolar columns.

1.6.1.2. Determination of Position of Double Bonds

Oxidative cleavage followed by analysis of the oxidation products is the most widely used method for determining the position of the double bonds in unsaturated fatty acids. The oxidizing agents commonly employed for this purpose are potassium permanganate and ozone, and the use of these agents in the analysis of unsaturated acids has been discussed by Privett (1966) and Roehm and Privett (1969).

The combined potassium permanganate–sodium metaperiodate reagent of Lemieux and von Rudloff (1955) is satisfactory for monounsaturated but less reliable with di- and triunsaturated fatty acids. Medium- and long-chain monoenoic and dienoic acids derived from oxidation are readily analyzed by GLC as the methyl or higher alcohol esters. Permanganate oxidation is not generally suitable for polyunsaturated fatty acids containing more than three double bonds, because a dibasic acid is obtained from each of the segments of the molecule between the double bonds as well as from the carboxy end of the molecule, and therefore more than one structure may be proposed for the original acid.

Ozonization of fatty acid methyl esters is generally carried out with gaseous ozone in a nonpolar solvent such as pentane at very low temperature (e.g., −70°C). The ozonides of fatty acid methyl esters, once obtained, are relatively stable and may be separated by TLC or solvent extraction techniques (Privett and Nickell, 1963). The fatty acid ozonides are cleaved by hydrolysis to monobasic and dibasic acids as on permanganate oxidation, but cleavage by reduction may be preferred, in which case the products are aldehydes and aldehyde esters. The aldehydes, dialdehydes, and aldehyde esters are readily analyzed by GLC. Reductive ozonolysis gives an unequivocal fragmentation pattern for mono- and diunsaturated fatty acids or esters, but more than one structure can be proposed for acids containing more than two double bonds because more than one dialdehyde will be produced during cleavage. The cleavage and analysis of ozonides may also be carried out by pyrolysis GLC, in which case the cleavage takes place in the inlet of the GLC instrument (Privett, 1966). In order to resolve ambiguities in the assignment of structures, Roehm and Privett (1969) have combined partial hydrogenation with ozonolysis to determine the location of *cis* and *trans* double bonds in polyunsaturated fatty acids. After partial hydrogenation with hydrazine (Roehm and Privett, 1969), the mixture of the monounsaturated isomers was separated by argentation TLC into *cis* and *trans* components. The separated isomers could then be individually cleaved by reductive ozonolysis, and the structure of the original polyunsaturated fatty acid was deduced by identifying the fragments.

1.6.1.3. Determination of Configuration of Double Bonds

A chemical method of investigating the geometry and position of double bonds in mono- and diunsaturated fatty acids involves the complete stereospecific hydroxylation of the compound with alkaline permanganate (Wood, 1967) or osmium tetroxide (McCloskey and McClelland, 1965), followed by conversion of the hydroxy compounds to their corresponding isopropylidene derivatives. These are analyzed by GLC and GL/MS as described in Section 1.5.3.

Chemical modification is also useful in determining the amount of *trans* acid in a mixture of mono- or diunsaturated fatty acids. Emken (1971) has proposed a procedure based on epoxidation of the esters followed by standard GLC. The *cis* and *trans* esters yield the corresponding *cis* and *trans* epoxides on treatment with peracetic acid at room temperature for 3–4 hr. The epoxides are readily separated on polyester columns and the *trans* isomer content is determined. An excess of peracid, however, may lead to a conversion of the *cis*-epoxy acid to the hydroxy-acetoxy acid, while the *trans* epoxide is stable under these conditions. This technique has been applied to the methyl esters of *trans,trans*-, *cis,trans*-, and *cis,cis*-9,12-octadecadienoic acids. Diastereoisomers of the *cis,cis*-diepoxide and *cis,trans*-diepoxide apparently were also separated by GLC, resulting in a more complex chromatogram than those observed with the monoenes.

1.6.1.4. Functional-Group Analysis

Functional groups other than sites of unsaturation also may be detected by chemical modification in conjunction with GLC or GC/MS. The presence of hydroxy acids in a sample of fatty acids can be confirmed by acetylation (Markley, 1967). The advantages of preparing TMS derivatives containing deuterated methyl groups in the GC/MS analysis of hydroxy fatty acids are discussed in Section 1.5.4.

The preparation of the 2,4-dinitrophenylhydrazines from the keto fatty acids has been extensively utilized by Weihrauch *et al.* (1974) for the analysis of the oxo fatty acids of milk lipids. Problems, however, arise because of the presence of the *syn* and *anti* isomers in the diphenylhydrazones, which double the number of isomeric fractions each time a separation is made.

Langenbeck *et al.* (1975) have made a systematic survey of the GLC properties of the TMS-quinoxalinol derivatives of keto acids. These derivatives offer several advantages for GLC of α-keto acids because of their high stability and the absence of stereoisomerism, and because of the presence of specific, common, and abundant fragments in electron-impact mass spectra, which allows the low-level detection of whole groups of keto acids by single-ion monitoring.

The chemical transformations of the cyclopropene (Section 1.5.1.3) and of the epoxy (Section 1.5.3.1) fatty acids for the purposes of chromatographic and mass spectrometric characterization have been described elsewhere.

The method of Rapport and Alonzo (1955) is useful for the microdetermination of ester groups in fatty acids. It depends on the reaction of the ester with an alkaline hydroxylamine solution to yield an hydroxamic acid, which forms strongly colored complexes with ferric ion.

1.6.2. Spectrometric Methods

Specific groups in the fatty acid molecules absorb electromagnetic radiation of particular energy levels, provide valuable information about molecular structure, and can be used quantitatively (O'Connor, 1968). A special case is provided by the rotation of polarized light by fatty acid molecules.

1.6.2.1. Infrared Spectroscopy

All fatty acids show strong absorption in the 2750–3000 cm range because of the large number of CH_2 and CH_3 groups in the molecules (Fischmeister, 1975). The presence of aliphatic ring systems alters the relative intensities of the bands associated with hydrocarbon chains of molecules, as compared to the acyclic systems. Cyclopropane fatty acids give pronounced characteristic infrared absorption bands at 1020 cm^{-1}, which are attributable to in-plane wagging vibrations of the methylene CH_2 group, and at 3050 cm^{-1}, which are due to the stretching frequency of the C—H bonds in the cyclopropane ring (Christie, 1970). There is little difference between the spectra of *cis* and *trans* isomers (Wood and Reiser, 1965). There are alterations in the relative intensities of bands associated with hydrocarbon chains of molecules in the polyisoprenoid acids (Wright *et al.*, 1967).

Unsaturation in fatty acids may be detected by the C—H stretch at or near 3020 cm^{-1}. *Trans* double bonds are associated with C—H deformation at 950–1000 cm^{-1}, which is prominent in the spectrum of methyl elaidate, for example. This characteristic can be used for quantitative determination of *trans* double bonds in fatty acids but the estimation is subject to error if appreciable double-bond conjugation is present (Chapman, 1965). A marked feature of fatty acids containing a hydroxyl group is the strong C—O stretch of the alcohol group at 1045 cm^{-1}. Baumann and Ulshoefer (1968) have made a detailed study of the C—O stretch of ester and hydroxyl groups in fatty acids.

An infrared spectrum may be obtained with as little as 1 μg of material by the use of beam condensers and microcells (Kates, 1972). It has been applied to GLC fractions (Freeman, 1968).

Charts showing the positions of the various absorption bands most commonly encountered in fatty acids, together with structural assignments, have been given by several authors (Chapman, 1965; Freeman, 1968; Fischmeister, 1975). Table XXV is an abbreviated version of the compilation prepared by Kates (1972).

1.6.2.2. Ultraviolet Spectroscopy

Fatty acids containing one double bond in conjugation with the carboxyl group show absorption maximum near 208–210 m and have a lower extinction coefficient than do the conjugated dienes. Several of these acids have been synthesized and studied (Markley, 1967).

Conjugation of unsaturated linkages brings the absorption bands into the easily accessible region. As the conjugation increases, the extinction rises. With three or more conjugated double bonds, the absorption shows three peaks, a main one and

Table XXV. Major Infrared Absorption Bands in Spectra of Fatty Acids in Chloroform or Carbon Tetrachloride or as Oils[a]

Functional group	Absorption mode	Frequency (cm^{-1})	Intensity[b]
$-CH_3$	CH stretching	2962 and 2872 ± 10	s
	CH bending, asymmetrical	1450 ± 20	m
	CH bending, symmetrical	1375 ± 5	m
$-OCH_3$	CH bending	1430	m
$-CH_2$	CH stretching	2926 and 2853 ± 10	s
	CH bending	1465 ± 10	m
	$-(CH_2)_4$ skeletal	750–720	m
	$-CH_2-$ in cyclopropane, CH stretch	3100–3000	m
	$-CH_2-$ in cyclopropane, skeletal	1025–1000	m
$-C(CH_3)_2$ (isopropyl)	CH bending	1385 and 1365 ± 5	m (db)
	skeletal	1170 ± 5	m
$-C(CH_3)_3$ (tert-butyl)	CH bending	1395–1385 (m); 1365	s
	skeletal	1250–1200	s
$-CH$ (tertiary)	CH stretching	2890 ± 10	w
	CH bending	1340 ± 20	w
C=C (aliphatic)	C=C stretching, nonconjugated	1680–1620	v
	C=C stretching, C=O or C=C conjugated	1600 ± 20	v
	CH stretching	3040–2010	m
$-CH=CH$ (*trans*)	CH out-of-plane deformation	970–960	s
	CH in-plane deformation	1310–1295	s–w
$-CH=CH$ (*cis*)	CH out-of-plane deformation	ca 690	m
$-CH=CH_2$ (vinyl)	CH stretching	3095–3075	m
	CH out-of-plane deformation	995–985	s
	CH_2 out-of-plane deformation	915–905	s
	CH_2 in-plane deformation	1420–1410	s
$R_1R_2C=CH_2$	CH stretching	3095–3075	m
	CH_2 out-of-plane deformation	895–885	s
	CH_2 in-plane deformation	1420–1410	s
$R_1R_2C=CHR_3$	CH out-of-plane deformation	840–790	s
$-C(CH_3)=CH-$ (isoprenoid)	CH out-of-plane deformation	835	s
	C=C skeletal	1600 and 1500	v
$R-C\equiv C-R$	$C\equiv C$ stretching	2260–2190	m
$R-C\equiv C-H$	$C\equiv C$ stretching	2140–2100	m
	C–H stretching	3300	m
$-C-OH$	O–H stretching; free OH	3650–3590 (sharp)	v
	O–H stretching; associated OH (hydrogen bonded)	3400–3200 (broad)	v
	C–O stretching	1150–1140 (tertiary)	s
		1120–1100 (secondary)	s
		1075–1010 (primary)	s
$-C-O-C-$ alkyl ether	C–O stretching	1150–1060	vs
$-\overset{H}{C}=\overset{H}{C}-O-C$ (*cis* vinyl ether)	C–O stretching	1270–1230	s
	C=C stretching	1670	m
	CH deformation	732–730	m
$-C-O-C-$ (epoxide)	C–O stretching	1250	s
	cis	830	m
	trans	890	m

Table XXV—Continued

Functional group	Absorption mode	Frequency (cm^{-1})	Intensity[b]
Alkyl peroxides	C—O stretching	890–820	v–w
Acyl peroxides	C=O vibrations	1820–1810 and 1800–1780	s
Aldehyde C=O	C=O vibrations:		
	saturated aldehyde	1740–1720	s
	$\alpha\beta$-unsaturated aldehyde	1705–1680	s
Ketone C=O	C=O stretching:		
	saturated ketone	1725–1705	s
	$\alpha\beta$-unsaturated ketone	1685–1665	s
Carboxyl, RCOOH	C=O stretching:		
	saturated fatty acid	1725–1700	s
	$\alpha\beta$-unsaturated acid	1715–1690	s
	C—O stretching	1320–1210	s
Ester, RCO—OR	C=O stretching: saturated ester	1750–1730	s
	$\alpha\beta$-unsaturated ester	1730–1717	s
	β-keto ester (enol)	1650–1540	s
	C—O stretching:		
	formates	1200–1180	m
	acetates	1250–1230	m
	propionates and higher	1200–1150	m
Anhydride, CO—O—CO	C=O stretching	1850–1800 and 1790–1740	s
	C—O stretching	1170–1050	

[a] Abbreviated from Kates (1972).
[b] Abbreviations: s, strong; m, medium; w, weak; v, variable.

subsidiary maxima on each side. As the number of double bonds increases, the peaks are spread farther apart (Shenstone, 1971). Differences in absorption spectra also occur with the different geometric isomers of the same acid. The all-*trans* form has the highest extinction coefficient, and the *cis* isomers usually have their maxima displaced to longer wavelengths.

With acetylenic acids, selective absorption in the accessible ultraviolet region occurs only when this group is conjugated with a double bond or another triple bond.

Acids containing *cis* and *trans* isomers of conjugated polyene and poly-yne groups have been examined in detail, and a long series of papers listing ultraviolet maxima and molar absorptivities is available (Bohlmann and Hanel, 1969). Conjugated polyene and polyene-yne acids of natural occurrence have been determined spectroscopically (Hopkins, 1972). Table XXVI gives the ultraviolet absorption data for the conjugated chromophores in various synthetic and natural fatty acids (Shenstone, 1971).

1.6.2.3. Nuclear Magnetic Resonance Spectroscopy

High-resolution spectra of simple lipids such as fatty acids can be used to determine patterns of unsaturation, branching, or substitution. The NMR signal

Table XXVI. Ultraviolet Absorption Data for Conjugated Polycenoic Fatty Acids[a]

Chromophore group	X_{max} (nm)	E_{max}
	–(CH=CH)*n*–	
Dienes, $n = 2$		
trans,trans	231	33,000–35,000
cis,trans	232–235	24,600–28,700
Trienes, $n = 3$		
	259	47,000
trans,trans,trans	268	61,000
	279	49,000
cis,cis,cis	262	36,200
trans,trans,cis	271	47,000–48,000
	281–283	37,000–38,000
	265	
cis,cis,trans	275	47,000
	387	
Tetraenes, $n = 4$		
	288	56,400
	301	87,100
	315	77,900
all *trans*	292	50,000–53,000
	305–306	76,000–78,000
	319–321	66,000–69,200
	315	
Pentaene, $n = 5$	328	
	346	
	333	
Hexaene, $n = 6$	353	
	374	
Octaene, $n = 8$	396	157,000
Nonaene, $n - 9$	469	160,000
	–C≡C–	
	227	370
Diyne, $n = 2$	238	344
	253	120
Poly-ynes, $n = 3$–6	200–280	130,000–450,000
	300–390	100,000–200,000
trans-Ene-ynes	229	16,200
	216, 240, 255	
trans-Ene-diyne	269	24,000
	284	
Monoene-one	231–263	6,000–18,000
	300–312	100–120
Diene-one	277–314	7,600–26,300
Monoene-dione	252–270	5,000–11,400
2-Enoic acid	210–215	13,200–11
2,4-Dienoic acid	260	25,800

[a] Abbreviated from Shenstone (1971).

Table XXVII. Proton Resonance Signals for Various Structural Groups[a]

Compound or group[b]	τ (ppm)
$(CH_3)_4$Si	10.00
$-CH_2-$, cyclopropane	9.78
CH_3-terminal, in paraffin chains	9.10–9.12
$(CH_3)_2C-$, isopropyl methyls	8.7–8.8
$-CH_2-$, saturated paraffin chains	8.65–8.80
R_3C-H, saturated	8.35–8.60
$CH_3-C{=}C-$, allylic methyl (e.g., squalene)	8.1–8.4
$CH_3-C{=}O$	7.4–8.1
$-CH_2-CO$ (in fatty acid esters)	7.8–7.9
$-CH_2-CO$, in ketones	7.5–7.8
$-CH_2-C{=}C$, allylic methylene	7.96
$H-C{\equiv}C-$, nonconjugated	7.35–7.55
$H-C{\equiv}C-$, conjugated	6.9–7.2
CH_3-O-, ethers	6.2–6.7
CH_3-O-, aliphatic esters	6.2–6.4
$-CH_2-O-$, aliphatic saturated alcohols or ethers	6.3–6.6
ROH (concentrated, 0.1–0.9 mole fraction)	4.7–7.0
$C{=}CH-O-$, vinyl ethers	4.2 (*cis*), 4.0 (*trans*)
$CH_2{=}C$, nonconjugated	5.0–5.4
$H-C{=}C-$, olefinic, nonconjugated	(4.1) 4.3–4.8 (4.9)
$H-C{=}C-$, cyclic, nonconjugated	4.3–4.8
CH_3 | $H-C{=}C-$, isoprenoid olefinic H (e.g., squalene)	4.88–5.00
$CH_2{=}C$, conjugated	(3.75) 4.3–4.7
$H-C{=}C-$, olefinic, conjugated	(219) 3.5–4.0 (4.5)
RCHO, aliphatic, α, β-unsaturated	0.35–0.50
RCHO, aliphatic, saturated	0.2–0.3 (0.5)
RCO_2H, dimer, in nonpolar solvents	(−3.2)–2.2 to −1.0 (+0.3)

[a] Modified from Hopkins (1965) and Kates (1972).
[b] Assigned proton designated as *H*.

obtained for a particular proton depends on its magnetic environment within the molecule and on its interactions with neighboring magnetic nuclei. Because of these interactions, each chemically distinct proton gives rise to a particular resonance peak. In the spectrum for linoleic acid, for example, the methylene groups α to the carboxyl group and double bonds give peaks at 7.7τ and 8.0τ that are well separated from the main methylene peak at 8.7τ. The diallylic ($-CH{=}CHCH_2CH{=}CH-$) methylene group appears at 7.2τ. The protons of the double bonds give a triplet at 4.7τ, distinct from the other peaks. A wide variety of fatty acids have been investigated by nuclear magnetic resonance (NMR) spectroscopy (Hopkins, 1968; Frost and Barzilay, 1971; Frost and Gunstone, 1975), and peaks representing various special groups have been identified. Table XXVII lists the chemical shifts of protons observed in various groups present in fatty acids (Hopkins, 1965; Kates, 1972). Gunstone and Inglis (1971) and Gunstone and Jacobsberg (1972) have published an extensive listing of references which include useful data on the NMR spectra of specific fatty acids and derivatives. Pawlowski *et al.* (1972) have developed an assay of cyclopropenoid lipids by nuclear magnetic resonance. ^{13}C NMR has been employed by Tulloch and Mazurek (1976) for structural studies with saturated, unsaturated, and oxygenated fatty acids.

1.6.2.4. Polarimetry

Fatty acids containing one or more asymmetrical carbon atoms exhibit optical isomerism, which may be detected by measuring the degree to which a given acid is capable of rotating plane-polarized light. Commercial instruments with automatic digital readout now allow measurements with an accuracy of $\pm 0.001°$ at selected wavelengths (Kates, 1972).

The assignment of absolute configuration to compounds with weak optical activity, however, can be hazardous unless possible solvent and concentration effects are considered. The most reproducible rotations therefore are obtained on pure substances in the absence of solvents.

a. Methyl-Branched Acids. Levene and co-workers (see Markley, 1960) reported data on the rotatory dispersion of a series of configurationally related fatty acids of the general type $RCH(CH_3)(CH_2)nCOOH$, where n varied from 0 to 1 and R from C_2H_5- to $C_{11}H_{21}-$. The rotation of the first member of this series was negative and that of all successive members was increasingly positive, indicating that the contri-

Table XXVIII. Maximum Molecular Rotations of Configurationally Related Methyl-Substituted Fatty Acids without Solvent[a]

Fatty acid	Molecular rotations $[M]^{25}_{5892.6}$
2-Methylbutanoic	−18.0
2-Methylpentanoic	−21.4
2-Methylhexanoic	−24.3
2-Methylnonanoic	−27.3
2-Methyldodecanoic	−27.3
3-Methylpentanoic	−10.4
3-Methylhexanoic	+3.6
3-Methylheptanoic	+6.1
3-Methyloctanoic	+8.1
4-Methylhexanoic	−13.6
4-Methylheptanoic	−6.9
4-Methyloctanoic	−4.1
4-Methylnonanoic	−1.9
5-Methylheptanoic	−11.1
5-Methyloctanoic	−3.7
5-Methylnonanoic	−1.7
5-Methyldecanoic	−0.6
6-Methyloctanoic	−12.2

[a] Markley (1960).

bution of the heavier alkyl radical is positive and that of the group containing the carboxyl is negative. Furthermore, a carboxyl group attached directly to the asymmetrical carbon atom furnishes a higher negative rotation than the corresponding group of the second member. The third member furnishes a higher contribution than the second and lower than the fourth. Table XXVIII gives the molecular rotations of configurationally related methyl-substituted fatty acids of synthetic origin. It appears that in all instances thus far recorded, measurements of optical activity of *anteiso* fatty acids have revealed dextrorotation. These dextrorotatory acids have the L configuration, an assignment which rests on correlation of the common asymmetrical center with those of (−)2L-methylbutanol and L-isoleucine (Smith, 1970*b*). Table XXVIII includes optical rotations of seven *anteiso* acids occurring naturally.

Table XXIX. Molecular Rotations of Methyl-Substituted Fatty Acids

Fatty acid	Molecular rotation $(M)_D$	Fatty acid	Molecular rotation $(M)_D$
Monomethyl-branched acids[a]		Dimethyl-branched acids	
D(−)-3-Me-5:0	−10.03	2D,4D-diMe-10: 1(9)	−36.4
D(+)-3-Me-6:0	+3.16	2L,4D-diMe-10: 1(9)	+34.2
L(−)-3-Me-6:0	−3.18		
D(−)-2-Me-6:0	−27.7		
D(−)-4-Me-6:0	−13.7	Trimethyl-branched acids[b]	
L(−)-3-Me-7:0	−5.77	2D,4D,6D-triMe-12:1 (11)	−42.5
D(+)-3-Me-11:0	+10.15	2L,4D,6D-triMe-12:1 (11)	+19.6
(+)-2-Me-12:0	+28.3	2D,4D,6D-triMe-9:0	−41
(+)-12-Me-14:0	+11.4	2L,4L,6D-triMe-9:0	+3.6
(+)-12-Me-14:0	+4.7	2L,4D,6D-triMe-9:0	+19.2
(+)-14-Me-16:0	+5.0	2D,4L,6D-triMe-9:0	+75.2
D(−)-2-Me-18:0	−12.1	3L,7D,11-triMe-12;0	−16.6
D(−)-3-Me-18:0	+14.5	3L,7L,11-triMe-12:0	−11.4
(+)-10-Me-18:0	+0.2	3D,7L,11-triMe-12:0	+16.6
(−)-10-Me-18:0	−0.15	3D,7D,11-triMe-12:0	+11.4
(−)-15-Me-18:0	−3.01		
(+)-16-Me-18:0	+4.6		
D(+)-3-Me-19:0	+4.5	Tetramethyl-branched acids[c]	
D(+)-3-Me-19:0	+13.1	D-2,4,6,8-tetraMe-10:0	−80.6[d]
D(+)-3-Me-20:0	+13.1	D-2,4,6,8-tetraMe-11:0	−59[d]
L(−)-3-Me-22:0	−13.4	D-2,6,10,14-tetraMe-15:0	−37
D(+)-3-Me-23:0	+15.5	D-3,7,11,15-tetraMe-16:0	+11
D(−)-21-Me-23:0	−13.0	D-4,8,12,16-tetraMe-17:0	−2.4
(+)-3-Me-24:0	+13.2	D-5,9,13,17-tetraMe-18:0	−0.67
D(+)-4-Me-24:0	+1.1		
D(−)-21-Me-24:0	−3.3		
D(+)-5-Me-25:0	+0.4		
L(+)-6-Me-25:0	+1		
L(+)-21-Me-25:0	+2.6		
(+)-2-Me-26:0	+28.0		
(+)-3-Me-27:0	+13.5		

[a] Markley (1960).
[b] Ackman (1972).
[c] Kates *et al.* (1972).
[d] Smith (1970*b*).

Table XXIX includes the molecular rotations of other mono- and polymethyl-branched fatty acids. The monomethyl-substituted acids of the D series are levorotatory if the methyl groups are attached to even-numbered carbons. Furthermore, the 2D isomers have larger molecular rotations than the 4D isomers. Odham (1967) has summarized his findings on the methyl-branched acids, including 2,6-dimethyloctanoic, 4,6-dimethyloctanoic, 2,4-dimethylnonanoic, 2,6-dimethyl-decanoic, 2,8-dimethyldecanoic, and 2,6-dimethylundecanoic acids. The 2D,4D (or 2R,4R) dimethylheptanoate had a molecular rotation of −43.3 and the 2L,4D (or 2S,4R) dimethylheptanoate had a rotation of +29.1.

Table XXIX concludes with the listing of some natural and synthetic fatty acids with four methyl branches. The natural 2,4,6,8-tetramethyldecanoic and 2,4,6,8-tetramethylundecanoic acids possess the all-D configuration. The naturally occurring 3,7,11,15-tetramethylhexadecanoic (phytanic) acid has the 3D,7D,11D or 3R,7R,11R configuration. The naturally occurring 2,6,10,14-tetramethylpentadecanoic acid (pristanic) also has the all-D configuration.

Table XXX. Optical Rotations of Some Di- and Trihydroxyoctadecanoic Acids[a]

Position of hydroxyl groups	Source	Specific rotation $[\alpha]_D$
Threo series		
9,10 (C_{16})	Synthetic	+23.2
		−23.2
9,10	Synthetic	+23.5
		−23.7
9,10	Enzymic	+22.5
9,10	Enzymic	+27.0
9,10	Natural	+25.6
11,12	Synthetic	−21.1
12,13	Synthetic	+23.8
		−23.4
12,13 (9c)	Enzymic	+19.0
	Acetolysis	−18.6
12,13	Synthetic	+23.8
		−23.8
12,13 (9c)	Acetolysis	+18.9
		−19.0
12,13 (9c)	Natural	+20
9,10,18	Natural	+22.3
9,10,18 (12c)	Natural	+18.2
Erythro series		
9,10	Synthetic	+1.6
		−2.1
9,10	Enzymic	+1.4

[a] Modified from Smith (1970*b*).

b. Hydroxy and Epoxy Acids. Smith (1970*b*) has reviewed the stereochemistry of these long-chain compounds and has pointed out some of the interrelationships of their structures and optical rotations. Table XXX gives the optical rotations for a large number of hydroxy and epoxy fatty acids. A pattern emerges which shows that the 3D-hydroxy compounds are levorotatory, while the enantiomeric L isomers are dextrorotatory. The synthetic (+)12-L-hydroxyoctadecanoic acid was shown (Smith, 1970*b*) by optical rotation and melting point behavior to be the enantiomer of (–)12-hydroxyoctadecanoic acid prepared by hydrogenation of ricinoleic acid. This assignment of the D configuration to ricinoleic acid has provided the key to the stereochemistry of related hydroxy acids derived from seed oils. The absolute configuration and optical purity of the 2-hydroxy fatty acids has been correlated by Tatsumi (1974).

The stereochemistry of naturally occurring epoxides is intimately associated with that of the vicinal diols that are derived from these epoxides (Morris and Crouchman, 1972). Under acidic conditions, 1,2-epoxides undergo ring cleavage, with the formation of vic-diols or their derivatives. These cleavages occur stereospecifically with isomerization of one carbon so that *cis* epoxides yield *threo* compounds while *trans* epoxides afford *erythro* isomers (Smith, 1970*b*). The product obtained after acetolysis of (+)vernolic acid gives the same magnitude of optical rotation but of opposite sign to that of the (+)*threo*-diol obtained by enzymatic action. Accordingly, these acids are optical antipodes. These correlations have been further extended and substantiated by Morris and Wharry (1966), by correlating that absolute configuration of 2D,13D0(+)vernolic acid with ricinoleic (12D-hydroxyoctadec-*cis*-12-enoic) acid. Likewise, Powell *et al.* (1967) established that coronaric and *cis*-9,10-epoxyoctadecanoic acids have the 9L, 10L configuration.

Because of internal compensation, *erythro* isomers are much less optically active than those of the *threo* series, and there is more uncertainty in measuring their rotation.

1.6.3. Enzymatic Methods

Certain structural features of the fatty acids may be assessed by enzymatic methods. The number, location, and configuration of functional groups are important, although individual members of a homologous series usually cannot be recognized. The effectiveness of application of this technique in the identification of fatty acids depends on the ease with which the enzymatic transformation products can be recognized or the enzyme activity determined.

1.6.3.1. Lipoxidases

The lipoxidases are highly specific for the peroxidation of unsaturated fatty acids containing *cis*-1,*cis*-4-pentadiene configuration, such as linoleic, linolenic, and arachidonic acids and their esters, but not oleic acid or its esters. The principal product is an optically active *cis-trans* conjugated hydroperoxide (Hitchcock and Nichols, 1971). The initial products of oxidation of linoleic acid are 13-hydroperoxy-*cis*-9,*trans*-11-octadecadienoate and 9-hydroperoxy-*trans*-10,*cis*-12-octadecadienoate. Optical rotatory dispersion spectroscopy showed that the 13-hydro-

Table XXXI. Lipoxygenase Specificity and Rates of Oxidation of Unsaturated Acids[a]

Fatty acid	Oxidation rate relative to 18:2 (9c,12c)
18:2 (9c,12c)	100.0
22:6 (4c,7c,10c,13c,16c,19c)	93.6
20:5 (5c,8c,11c,14c,17c)	91.4
20:2 (11c,14c)	86.9
19:3 (10c,13c,16c)	83.3
18:3 (9c,12c,15c)	79.0
20:4 (5c,8c,11c,14c)	76.8
20:3 (11c,14c,17c)	55.7
21:4 (6c,9c,12c,15c)	44.6
18:3 (6c,9c,12c)	39.3
18:3 (5c,8c,11c)	39.1
18:2 (9c,15c)	32.6
18:4 (6c,9c,12c,15c)	27.6
17:3 (5c,8c,11c)	25.6
17:2 (9c,12c)	19.3
16:3 (6c,9c,12c)	9.7
19:2 (10c,13c)	3.5
18:2 (9c,12a)	3.8
20:4 (5c,8c,11c,14c)	0.0

[a] Modified from Holman *et al.* (1969).

peroxy acid belonged to the L series (Hamberg and Samuelsson, 1967). The *trans* analogues of linoleate (18:2,9c12t; 18:2,9t12t) are not acceptable substrates. Hamberg and Samuelsson (1967) and Holman *et al.* (1969) have tested a large number of the methylene-interrupted *cis,cis* isomers of linoleic acid as substrates for purified soybean lipoxygenase. A comparison of the rate of oxidation of different unsaturated fatty acids showed that the presence of double bonds in the $\omega 6$ (and $\omega 9$) position usually resulted in efficient oxidation.

The various lipoxygenases differ in their positional specificity. Thus peanut lipoxygenase differs from that of flaxseed, which attacks both positions in ratios of 80% at the C-13 and 20% at C-9 (Zimmerman and Vires, 1970) and from that in corn, which attacks predominantly (83%) the C-9 position and to a lesser degree (17%) the C-13 (Gardner and Weisleder, 1970). Table XXXI summarizes the specificity and reactivity of lipoxygenase with various unsaturated fatty acids (Hitchcock and Nichols, 1971).

1.6.3.2. Hydrases

Soluble extracts of a pseudomonad (Niehaus and Schroepfer, 1965) have been shown to convert oleic acid by *trans* addition of water across the double bond (Davis *et al.*, 1969; Morris, 1970). Oleate yields D-10-hydroxystearate and palmitoleate yields 10-hydroxypalmitate (Niehaus *et al.*, 1970); linoleate yields D-10-hydroxy-*cis*-12-octadecenoate (Schroepfer *et al.*, 1970).

Under appropriate conditions, (+)vernolic acid (*cis*-12,13-epoxyoleic) is converted to 12,13-dihydroxyoleic acid in the crushed seeds of *Vernonia anthelmintica* and *Euphorbia lagasca.* The enzymatic hydration is stereospecific at the 12 position of (+)D-12D-13-epoxyoleate with inversion at that position, and the formation of (+)*threo*-L-12, D-13-dihydroxyoleate (Morris and Crouchman, 1969). In spores of plant rusts, *cis*-9,10-epoxystearate is similarly hydrated to (+)*threo*-9,10-dihydroxystearate. In this case, the L-9,L-10 epoxide yields the L-9,D-10 diol (Morris, 1970).

1.6.3.3. Dehydrogenases

The marked substrate and product specificity of the desaturation systems of animal and plant cells also may be used to confirm the identity of certain fatty acids. Howling *et al.* (1968) have shown that the homologous series of acids from myristic to nonadecanoic are converted to the corresponding 9-monoenoic acids when incubated with *Chlorella vulgaris* cells. In addition, myristate, pentadecanoate, and palmitate also yielded the 7-monoenoates. Johnson *et al.* (1969) have similarly shown that hen liver preparations desaturate fatty acids of chain length C_{12}–C_{22} to the corresponding 9-monoenoates, again two optima being observed, one at C_{14} the other at C_{17-18}. The desaturation is highly stereospecific. Results with hydrogen-labeled stearates have shown that the enzyme cleavage involves the D-9 and D-10 hydrogen atoms, bringing about a simultaneous *cis* elimination (Morris, 1970). The stereospecificity of desaturation of oleate to linoleate in *Chlorella* has been shown by Morris (1970) to be the same as that of stearate desaturation. Again, a *cis*

Table XXXII. Specificity of Hen Liver Microsomal Dehydrogenase and Relative Rates of Desaturation of Isomeric Methylstearic (Nonadecanoic) Acids[a]

Fatty acid	Desaturation rate (%) relative to 18:0
18:0	100
2-Me-18:0	74
3-Me-18:0	10
4-Me-18:0	39
5-Me-18:0	0
6-Me-18:0	0
8-Me-18:0	0
9-Me-18:0	0
10-Me-18:0	0
11-Me-18:0	0
12-Me-18:0	0
14-Me-18:0	0
15-Me-18:0	0
16-Me-18:0	21
17-Me-18:0	42
18-Me-18:0	74

[a] Modified from Brett *et al.* (1971).

elimination of hydrogen was shown to take place by means of appropriately labeled substrates.

The high specificity of these transformations apparently depends on a strong nonpolar interaction between the flexible hydrocarbon chains and a lipophilic enzyme surface. These forces are weakened when the chain is substituted. Thus, when a range of 15 isomeric monomethylstearic acids was compared with stearic acid itself as substrates for the 9-desaturase of *C. vulgaris*, the presence of the methyl group in different positions partially or completely prevented desaturation (Brett *et al.*, 1971). The 5-, 6-, 8-, 9-, 10-, 11-, 12-, and 14-methylstearates did not react to any significant extent, while substitutions at the 2, 3, 4, 16, 17, or 18 position caused a fall in desaturation rate to 10–74% of that of stearate. Table XXXII summarizes the positional specificity of the 9-desaturate system of hen liver.

1.6.3.4. Lipases

A lipase elaborated by the microorganism *Geotrichum candidum* has been found to attack natural triacylglycerols, specifically hydrolyzing fatty acids having *cis*-9- and *cis*,*cis*-9,12- unsaturation (Brockerhoff and Jensen, 1974). Jensen *et al.* (1972) have demonstrated that the enzyme hydrolyzed very little of the other 14 positional isomers of oleic acid when *cis*-9,*cis*-12-18 : 2 was present in the triacylglycerol substrates and was hydrolyzed extensively. There was no preference between 9-18:1 and 18:2. The hydrolysis of the isomeric 18:1 acids from synthetic triacylglycerols was generally less than the hydrolysis of the saturated acids. The enzyme also discriminates against the *trans*,*trans* isomer but not between the *cis*,*trans* and *trans*,*cis* isomers of linoleate (Jensen *et al.*, 1973).

1.7. Summary and Conclusions

The complete identification of a given fatty acid ultimately depends on its isolation in pure form and sufficient amounts for characterization by chemical and physical methods. This is impractical for every analysis, and various means have been developed to predict with increasing degree of certainty the exact nature of the material in a given fraction with progressing analysis. With simple mixtures, it is possible to proceed from isolation to final instrumental analyses with a minimum of preliminary fractionation. For many complex mixtures, a preliminary fractionation by complementary chromatographic techniques is required before even a tentative identification can be made.

There are several chromatographic systems which lend readily to a sequential application resulting in progressively simpler mixtures of fatty acids. The specific combination of the systems for an adequate preliminary resolution of the fatty acids depends on the complexity of mixture and the needs of the analysis. The most popular and effective combination of complementary chromatographic systems is provided by argentation TLC followed by GLC of the individual unsaturation classes, which is adequate for identification of most known acids. Routine combinations of TLC and GLC for the tentative identification of fatty acids in natural mixtures have been described by many workers, and this approach is now generally favored over complex calculations of theoretical retention times and fractional chain length values. TLC/GLC is compatible with quantitative recovery of the TLC fractions so that the relative proportions of the various components of the total mixture may be established. Furthermore, the TLC prefractionation frequently yields minor components in high purity which can be recovered and identified more readily than is possible by GLC alone. This method also allows the systematic preparation of fatty acid subfractions which can be meaningfully examined by GC/MS using nonpolar siloxane columns.

Another effective combination of complementary chromatographic techniques is provided by preparative GLC on nonpolar columns, which resolve the fatty acid methyl esters according to molecular weight. The trapped fractions are then rechromatographed on a polar column to determine the individual unsaturated fatty acids in each chain length. In the past, much hope has been expressed for a complementary analytical GLC on polar and nonpolar liquid phases as a means of unambiguous identification of fatty acids. It has been shown that, in simple mixtures of standards, characteristic differences in the elution times are observed which may be used in the form of semilog grids for the identification of unknown fatty acids on the basis of their retention times on a polar and a nonpolar liquid phase. This approach, however, is not justified when dealing with complex mixtures of fatty acids, because it is not possible to recognize which peak is which in the GLC elution patterns, especially when the relative concentrations of the components are also unknown. This problem is effectively overcome and the advantages of the use of the two liquid phases are retained when the peaks from one or the other of the columns are collected. Although capillary columns are more effective for the analysis of close isomers, problems arise from poor reproducibility and low recovery of the higher-molecular-weight components. Complete analyses of fatty acid esters including minor components usually require exhaustive chromatographic fractionation along with some chemical or physiochemical method of simple enrichment of the minor components. Some of the methods of enrichment are highly specific, while others are of a more general nature.

Ultraviolet and infrared spectroscopy can confirm the presence of conjugated double bonds and *trans* double bonds, respectively. Likewise, infrared spectroscopy can usually provide a confirmation of the presence of hydroxy or keto functions in the fatty chain. Further evidence for the tentative structure of the fatty acid derived on the basis of the chromatographic fractionations may be obtained from nuclear magnetic resonance spectroscopy, which can also suggest appropriate further analyses and the strategy of structure determination by chemical means. The chemical determination of the structure of an unknown fatty acid is essentially a problem in structural organic chemistry, which involves a series of traditional and modern techniques. Excellent examples of the confirmation of fatty acid structure by chemical synthesis may be found in many early studies as well as in several of the recent systematic syntheses. In special instances, it may be desirable to demonstrate the metabolic equivalence of the natural and the synthetic products, especially if the latter have been prepared by methods which may have involved inversion of asymmetrical carbons.

GC/MS potentially provides the most powerful combination of analytical methods for the identification of fatty acids. Inadequacies, however, are experienced in the determination of the location of the double bonds and in the determination of their geometric configuration. The method also fails to yield any information about the stereochemical configuration of other functional groups, although this may be overcome by chemical derivatization and the availability of polar liquid phases suitable for GC/MS work. Nevertheless, the identification is complete for most of the well-characterized fatty acids as achieved by routine TLC/GLC or GC/MS techniques in terms of molecular weight and number, and occasionally location of double bonds. Furthermore, the certainty of identification of the fatty acids increases with the knowledge of the source or origin of the material. Therefore, reliable identifications of fatty acids may be obtained routinely by relatively simple chromatographic systems.

Acknowledgments

The studies of the author and his collaborators referred to in this chapter were supported by funds from the Ontario Heart Foundation, Toronto, Ontario, the Medical Research Council of Canada, Ottawa, Ontario, and the Special Dairy Industry Board, Chicago, Illinois.

1.8. References

Abrahamsson, E., Ställberg-Stenhagen, S., and Stenhagen, E., 1963, The higher saturated branched chain acids, in: *Progress in the Chemistry of Fats and Other Lipids*, Vol. 7 (R. T. Holman and T. Malkin, eds.), Part 1, p. 41, Pergamon Press, Oxford.

Ackman, R. G., 1968, Prediction of retention times in the GLC of diastereoisomers of methyl-branched fatty acids, *J. Chromatogr.* **34**:165.

Ackman, R. G., 1972, The analysis of fatty acids and related materials by gas–liquid chromatography, in: *Progress in the Chemistry of Fats and Other Lipids*, Vol. 12 (R. T. Holman, ed.), pp. 165–284, Pergamon Press, Oxford.

Ackman, R. G., and Castell, J. D., 1966, Isomeric monoethylenic fatty acids in herring oil, *Lipids* **1**:341.

Ackman, R. G., and Hansen, R. P., 1967, The occurrence of diastereoisomers of phytanic and pristanic acids and their determination by gas–liquid chromatography, *Lipids* **2**:357.

Ackman, R. G., and Hooper, S. N., 1969, The load effect in open tubular GLC: Relationships among methyl *trans*-11-octadecenoate and *cis*-isomers with similar retention times, *J. Chromatogr. Sci.* **7**:549.

Ackman, R. G., Hooper, S. N., Kates, M., Sen Gupta, A. K., Eglinton, G., and Maclean, I., 1969, Gas–liquid chromatographic separation of diastereoisomers of phytanic acid L-methyl esters, *J. Chromatogr.* **44**:256.

Ackman, R. G., Cox, R. E., Eglinton, G., Hooper, S. N., and Maxwell, J. R., 1972, Stereochemical studies of acyclic isoprenoid compounds. I. Gas chromatographic analysis of stereoisomers of a series of standard acyclic isoprenoid acids, *J. Chromatogr. Sci.* **10**:392.

Allen, G. R., and Saxby, M. J., 1968, Gas chromatography of isomeric fatty acid methyl esters, *J. Chromatogr.* **37**:312.

Anderson, R. D., Bottino, N. R., and Reiser, R., 1967, Pancreatic lipase hydrolysis as a source of diglycerides for the stereospecific analysis of triglycerides, *Lipids* **2**:440.

Anderson, R. E., and Rakoff, H., 1965, Gas-liquid chromatography of the positional isomers of methyl nonynoates, *J. Am. Oil Chem. Soc.* **42**:1102.

Andersson, B. A., and Holman, R. T., 1975, Mass spectrometric localization of methyl branching in fatty acids using acylpyrrolidines, *Lipids* **10:**716.

Andersson, B. A., Christie, W. W., and Holman, R. T., 1975, Mass spectrometric determination of positions of double bonds in polyunsaturated fatty acid pyrrolidides, *Lipids* **10:**215.

Aplin, R. T., and Coles, L., 1967, A simple procedure for localization of ethylenic bonds by mass spectrometry, *Chem. Commun.* **1967:**858.

Artman, N. R., and Smith, D. E., 1972, Systematic isolation and identification of minor components in heated and unheated fat, *J. Am. Oil Chem. Soc.* **49:**318.

Ashes, J. R., and Haken, J. K., 1974, Gas chromatography of homologous esters. VI. Structure-retention increments of aliphatic esters, *J. Chromatogr.* **101:**103.

Ashes, J. R., and Haken, J. K., 1975, Gas chromatography of homologous esters. IX. Structure-retention increments of unsaturated esters, *J. Chromatogr.* **111:**171.

Aylward, F., and Sawistowska, M., 1962, Hydrazine—A reducing agent for olefinic compounds, *Chem. Ind. (London)* **1962:**484.

Bagby, M. O., Smith, C. R., Jr., and Wolff, I. A., 1965, Labellanic acid. A new allenic acid from *Leonotis nepetaefolia* seed oil, *J. Org. Chem.* **30:**4227.

Barve, J. A., Gunstone, F. D., Jacobsberg, F. R., and Winlow, P., 1972, Fatty acids, Part 34. Behaviour of all the methyl octadecenoates and octadecynoates in argentation chromatography and gas liquid chromatography, *Chem. Phys. Lipids* **8:**117.

Baumann, W. J., and Ulshoefer, H. W., 1968, Characteristic absorption bands and frequency shifts in the infrared spectra of naturally-occurring long-chain ethers, esters and ether esters of glycerol and various diols, *Chem. Phys. Lipids* **2:**114.

Bergelson, L. D., Dyatlovitskaya, E. V., and Voronkova, V. V., 1964, Complete structural analysis of fatty acid mixtures by thin-layer chromatography, *J. Chromatogr.* **15:**191.

Bohlman, F., and Hanel, P., 1969, Ueber die Polyine aus *Cicuta virosa L.*, *Chem. Ber.* **102:**3293.

Bohlman, F., Schuman, D., Bethke, H., and Zdero, C., 1967, Ueber die Massenspektren von acetylcarbonsauere Ester. *Chem. Ber.* **100:**3706.

Bottcher, C. J. F., Woodford, F. P., Boelsma-Van Haute, E., and Van Gent, C. M., 1959, Methods for the analysis of lipid extracts from human arteries and other tissues, *Recl. Trav. Chim. Pays-Bas.* **78:**794.

Bottino, N. R., Vandenberg, G. A., and Reiser, R., 1967, Pancreatic lipases selectivity of fatty acids, *Lipids* **2:**489.

Breckenridge, H. C., and Kuksis, A., 1968, Structure oi bovine milk·fat triglycerides. I. Short and medium chain lengths, *Lipids* **3:**291.

Brennan, P. J., Griffin, P. I. S., Loesel, O. M., and Tyrell, D., 1975, The lipids of fungi, in: *Progress in the Chemistry of Fats and Other Lipids*, Vol. 14 (R. T. Holman, ed.), Part 2, pp. 49–89, Pergamon Press, Oxford.

Brett, D., Howling, D., Morris, L. J., and James, A. T., 1971, Specificity of the fatty acid desaturases. The conversion of saturated to monoenoic acids, *Arch. Biochem. Biophys.* **143:**535.

Brockerhoff, H., 1971, Stereospecific analysis of triglycerides, *Lipids* **6:**942.

Brockerhoff, H., and Ackman, R. G., 1967, Positional distribution of isomers of monoenoic fatty acids in animal glycerolipids, *J. Lipid Res.* **8:**661.

Brockerhoff, H., and Jensen, R. G., 1974, *Lipolytic Enzymes,* Academic Press, New York.

Brown, C. A., 1967, Simple component apparatus for rapid process determination of unsaturation via hydrogenation on a micro and ultramicro scale, *Anal. Chem.* **39:**1882.

Campbell, I. N., and Naworal, J., 1969, Mass spectra discrimination between monoenoic and cyclopropanoid, and between *normal*, *iso*, and *anteiso* fatty acid methyl esters, *J. Lipid Res.* **10:**589.

Capella, P., and Zorzut, C. M., 1968, Determination of double bond position in monounsaturated fatty acid esters by mass spectrometry of their trimethylsilyloxy derivatives, *Anal. Chem.* **40:**1458.

Cason, J., Sumrell, G., Allen, C. F., Gillies, G. A., and Elberg, S., 1953, Certain characteristics of the fatty acids from the lipides of the tubercle bacillus, *J. Biol. Chem.* **205:**435.

Carroll, K. K., 1976, Column chromatography of neutral glycerides and fatty acids, in: *Lipid Chromatographic Analysis* (G. V. Marinetti, ed.), pp. 173–214, Dekker, New York.

Chapman, D., 1965, *The Structure of Lipids,* Methuen, London.

Christie, W. W., 1968, Chromatography of the isomeric methylene-interrupted methyl *cis,cis*-octadecadienoates. 2. Gas–liquid chromatography, *J. Chromatogr.* **37:**27.

Christie, W. W., 1970, Cyclopropane and cyclopropene fatty acids, in: *Topics in Lipid Chemistry*, Vol. 1 (F. D. Gunstone, ed.), pp. 1–49, Logos Press, London.

Christie, W. W., 1972, The preparation of alkyl esters from fatty acids and lipids, in: *Topics in Lipid Chemistry*, Vol. 3 (F. D. Gunstone, ed.), pp. 171–197, Wiley, New York.

Christie, W. W., and Holman, R. T., 1966, Mass spectrometry of lipids. I. Cyclopropane fatty acid esters, *Lipids* **1**:176.

Christie, W. W., and Holman, R. T., 1967, Synthesis and characterization of the complete series of methylene-interrupted *cis,cis*-octadecadienoic acid, *Chem. Phys. Lipids* **1**:407.

Christie, W. W., Gunstone, F. D., Ismail, I. A., and Wade, L., 1968, The synthesis and chromatographic and spectroscopic properties of the cyclopropane esters derived from all the methyl octadecanoates (Δ^2–Δ^{17}), *Chem. Phys. Lipids* **2**:196.

Christie, W. W., Rebello, D., and Holman, R. T., 1969, Mass spectrometry of derivatives of cyclopentenyl fatty acids, *Lipids* **4**:229.

Cochrane, G. C., 1975, A review of the analysis of free fatty acids (C_2–C_6), *J. Chromatogr. Sci.* **13**:440.

Conacher, H. B. S., and Gunstone, F. D., 1969, Fatty acids. Part 21. The rearrangement of methyl 12,13-epoxyoleate by boron trifluoride with formation of cyclopropane esters, *Chem. Phys. Lipids* **3**:203.

Cooper, M. J., and Anders, M. W., 1974, Determination of long chain fatty acids as 2-naphthacyl esters by high pressure liquid chromatography and mass spectrometry, *Anal. Chem.* **46**:1849.

Dinh-Nguyen, P. N., 1968, Contribution à l'étude de la spectrometrie de masse: Utilisation des esters méthyliques de monoacides à longue chaine normale marques au deuterium et au carbone-13, *Ark. Kemi* **28**:289.

Dinh-Nguyen, P. N., Ryhage, R., Ställberg-Stenhagen, S., and Stenhagen, E., 1961, Mass spectrometry studies. VIII. A study of the fragmentation of normal long chain methyl esters and hydrocarbons under electron impact with the aid of deuterium substituted compounds, *Ark. Kemi* **18**:393.

Drodz, J., 1975, Chemical derivatization in gas chromatography, *J. Chromatogr.* **113**:303.

Dudley, P. A., and Anderson, R. E., 1975, Separation of polyunsaturated fatty acids by argentation thin layer chromatography, *Lipids* **10**:113.

Dutton, H. J., 1968, Hydrogenation of fats, in: *Progress in the Chemistry of Fats and Other Lipids*, Vol. 9 (R. T. Holman, ed.), pp. 351–376, Pergamon Press, Oxford.

Eglinton, G., Hunneman, D. H., and McCormick, A., 1968, Gas chromatographic–mass spectrometric studies of long chain hydroxy acids. III. The mass spectra of the methyl esters, trimethylsilyl ethers of aliphatic hydroxy acids: A facile method of double bond location, *Org. Mass Spectrom.* **1**:593.

Emken, E. A., 1971, Determination of *cis* and *trans* in monoene and diene fatty esters by gas chromatography, *Lipids* **6**:686.

Emken, E. A., 1972, *Cis* and *trans* analysis of fatty esters by gas chromatography: Octdecenoate and octadecadienoate isomers, *Lipids* **7**:459.

Emken, E. A., and Dutton, H. J., 1974, Sequential gas chromatographic procedure for microanalysis of monoenoic double bond position in hydrogenated oils, *Lipids* **9**:272.

Emken, E. A., Scholfield, C. R., Davison, V. L., and Frankel, E. N., 1967, Separation of conjugated methyl octadecadienoate and trienoate geometric isomers by silver resin columns and preparative gas–liquid chromatography, *J. Am. Oil Chem. Soc.* **44**:373.

Fischer, W., 1973, The suitability of lipases from *Rhizopus arrhizus* for analysis of fatty acid distribution in dihexosyl diglycerides, phospholipids and plant sulfolipids, *Hoppe-Seyler's Z. Physiol. Chem.* **354**:1115.

Fischmeister, I., 1975, Infrared absorption spectroscopy of normal and substituted long-chain fatty acids and esters in the solid state, in: *Progress in the Chemistry of Fats and Other Lipids*, Vol. 14 (R. T. Holman, ed.), Part 3, pp. 93–162, Pergamon Press, Oxford.

Foglia, T. A., Heller, P., and Dooley, C. J., 1976, Analysis of α-branched fatty acids by gas chromatography and mass spectrometry, *J. Am. Oil Chem. Soc.* **53**:45.

Freeman, N. K., 1968, Applications of infrared absorption spectroscopy in the analysis of lipids, *J. Am. Oil Chem. Soc.* **45**:798.

Frost, D. J., and Barzilay, J., 1971, Proton magnetic resonance identification of nonconjugated *cis*-unsaturated fatty acids and esters, *Anal. Chem.* **43**:1316.

Frost, D. J., and Gunstone, F. D., 1975, The PMR analysis of non-conjugated alkenoic and alkynoic acids and esters, *Chem. Phys. Lipids* **15**:53.

Gardner, H. W., and Weisleder, D., 1970, Lypoxygenase from *Zea mays*: 9-D-Hydroxytrans-10-*cis*-12-octadecadienoic acid from linoleic acid, *Lipids* **5**:678.

Glass, R. L., 1971, Alcoholysis, saponification and the preparation of fatty acid methyl esters, *Lipids* **6**:919.

Glass, R. L., Krick, T. P., Sand, D. M., Rahn, C. H., and Schlenk, H., 1975, Furanoid fatty acids from fish lipids, *Lipids* **10**:695.

Gloster, J., and Fletcher, R. F., 1966, Quantitative analysis of serum lipids with thin-layer chromatography, *Clin. Chim. Acta* **13**:235.

Golovnya, R. V., Uralets, V. P., and Kuzmenko, T. E., 1976, Characterization of fatty acid methyl esters by gas chromatography on siloxane liquid phases, *J. Chromatogr.* **121**:118.

Goodridge, A. G., 1972, Regulation of fatty acid synthesis in the liver of prenatal and early postnatal chicks: Hepatic concentrations of individual free fatty acids and other metabolites, *J. Biol. Chem.* **248**:1930.

Graff, G., Szczepanik, P., Klein, P. D., Chipault, J. R., and Holman, R. T., 1970, Identification and characterization of fully deuterated fatty acids isolated from *Scenedesmus obliquus* cultured in deuterium oxide, *Lipids* **5**:786.

Groff, T. M., Rakoff, H., and Holman, R. T., 1968, Mass spectrometry of lipids. Isomeric methyl nonynoates and the corresponding nonenoates and dideuteronenoates, *Ark. Kemi* **29**:179.

Gunstone, F. D., and Inglis, R. P., 1971, NMR spectra of fatty acids and related compounds, in: *Topics in Lipid Chemistry*, Vol. 2 (F. D. Gunstone, ed.), pp. 287–307, Logos Press (Wiley-Interscience), New York.

Gunstone, F. D., and Jacobsberg, F. R., 1972, The synthesis, silver ion chromatographic and NMR spectroscopic properties of the nine 9,12-diunsaturated n-C_{18} acids, *Chem. Phys. Lipids* **9**:112.

Gunstone, F. D., Ismail, I. A., and Lie Ken Jie, M., 1967, Fatty acids. Part 16. Thin-layer and gas–liquid chromatographic properties of the *cis* and *trans* methyl octadecenoates and of some acetylenic esters, *Chem. Phys. Lipids* **1**:337.

Hamberg, M., and Samuelsson, B., 1967, On the specificity of the oxygenation of unsaturated fatty acids catalyzed by soybean lipoxidase, *J. Biol. Chem.* **242**:5329.

Hammarstrom, S., 1974, Microdetermination of stereoisomers of 2-hydroxy and 3-hydroxy fatty acids, in: *Methods of Enzymology*, Vol 35 (J. M. Lowenstein, ed.), *Lipids* Part B, pp. 326–334, Academic Press, New York.

Hansen, R. P., 1967, 11-Cyclohexylundecanoic acid: Its occurrence in bovine rumen bacteria, *Chem. Ind.* (*London*) **1967**:1640.

Hilditch, T. P., and Williams, P. N., 1964, *The Chemical Constitution of Natural Fats*, 4th ed., Chapman and Hall, London.

Hintze, U., Roper, H., and Gercken, G., 1973, Gas chromatography–mass spectrometry of C_1–C_{20} fatty acid benzyl esters, *J. Chromatogr.* **87**:482.

Hitchcock, C., and Nichols, B. W., 1971, *Plant Lipid Biochemistry*, Academic Press, New York.

Hoffmann, G., and Meijboom, P. W., 1969, Identification of 11,15-octadecadienoic acid from beef and mutton tallow, *J. Am. Oil Chem. Soc.* **46**:620.

Hofstetter, H. H., Sen, N., and Holman, R. T., 1965, Characterization of unsaturated fatty acids by gas–liquid chromatography, *J. Am. Oil Chem. Soc.* **42**:537.

Holman, R. T., and Rahm, J. J., 1966, Analysis and characterization of polyunsaturated fatty acids, in: *Progress in the Chemistry of Fats and Other Lipids*, Vol. 9 (R. T. Holman, ed.), pp. 15–90, Pergamon Press, Oxford.

Holman, R. T., Egwim, P. O., and Christie, W. W., 1969, Substrate specificity of soybean lipoxidase, *J. Biol. Chem.* **224**:1149.

Hooper, N. K., and Law, J. H., 1968, Mass spectrometry of derivatives of cyclopropane fatty acids, *J. Lipid Res.* **9**:270.

Hopkins, C. Y., 1965, Nuclear magnetic resonance in fatty acids and glycerides, in: *Progress in the Chemistry of Fats and Other Lipids*, Vol. 18 (R. T. Holman, ed.), pp. 215–252, Pergamon Press, Oxford.

Hopkins, C. Y., 1968, High resolution NMR spectroscopy and some examples of its use, *J. Am. Oil Chem. Soc.* **45**:778.

Hopkins, C. Y., 1972, Fatty acids with conjugated unsaturation, in: *Topics in Lipid Chemistry*, Vol. 3 (F. D. Gunstone, ed.), pp. 37–87, Wiley, New York.

Hopkins, C. Y., and Chisholm, M. J., 1968, A survey of the conjugated fatty acids of seed oils, *J. Am. Oil Chem. Soc.* **45**:176.

Hornstein, I., Alford, J. A., Elliott, L. E., and Crowe, P. F., 1960, Determination of free fatty acids in fat, *Anal. Chem.* **32**:540.

Howling, D., Morris, L. J., and James, A. T., 1968, The influence of chain length on the dehydrogenation of saturated fatty acids, *Biochim. Biophys. Acta* **152**:224.

Hrivnak, J., Sojak, L., Beska, E., and Janak, J., 1972, Capillary gas chromatography of free saturated C_2–C_6 fatty acids, *J. Chromatogr.* **68**:55.

Jacob, J., and Grimmer, G., 1968, Structure and amount of positional isomers of monounsaturated fatty acids in human depot fats, *J. Lipid Res.* **9**:730.

Jacob, J., and Poltz, J., 1975, Composition of uropygial gland secretions of birds of prey, *Lipids* **10**:1.

Jaeger, H., Kloer, H. U., and Ditschuneit, H., 1976, Automated glass capillary gas–liquid chromatography of fatty acid methyl esters with reference to *cis* and *trans* isomers, *J. Lipid Res.* **17**:185.

James, A. T., 1960, Qualitative and quantitative determination of the fatty acids by gas–liquid chromatography, in: *Methods of Biochemical Analysis*, Vol. 8 (D. Glick, ed.), pp. 1–59, Interscience, New York.

Jamieson, G. R., 1970, Structure determination of fatty esters by gas–liquid chromatography, in: *Topics in Lipid Chemistry*, Vol. 1 (F. D. Gunstone, ed.), pp. 107–159, Logos Press, London.

Jamieson, G. R., and Reid, E. H., 1976, Gas–liquid chromatography characteristics of some long-chain acetylenic methyl esters, *J. Chromatogr.* **128**:193.

Jangaard, P. M., 1965, A rapid method for concentrating unsaturated fatty acid methyl esters in marine lipids as an aid in their identification by GLC, *J. Am. Oil Chem. Soc.* **42**:845.

Jensen, R. G., Gordon, D. T., Heinermann, W. H., and Holman, R. T., 1972, Specificity of *Geotrichum candidum* lipases with respect to double bond position in triglycérides containing *cis*-octadecenoic acids, *Lipids* **7**:738.

Jensen, R. G., Gordon, D. T., and Scholfield, C. R., 1973, Hydrolysis of linoleate geometric isomers by *Geotrichum candidum* lipase, *Lipids* **8**:323.

Johnson, A. R., Fogerty, A. C., Pearson, J. A., Shenstone, F. S., and Bersten, A. M., 1969, Fatty acid desaturase systems of hen liver and their inhibition by cyclopropene fatty acids, *Lipids* **4**:265.

Johnson, C. B., and Wong, E., 1975, Esterification and etherification by silver oxide–organic halide reaction gas chromatography, *J. Chromatogr.* **109**:403.

Karlsson, K. A., 1974, Resolution and chromatographic configuration analysis of 2-hydroxy fatty acids, *Chem. Phys. Lipids* **12**:65.

Kates, M., 1972, Techniques of lipidology: Isolation, analysis and identification of lipids, in: *Laboratory Techniques in Biochemistry and Molecular Biology*, Vol. 3 (T. S. Work and E. Work, eds.), pp. 269–600, North-Holland/American Elsevier, New York.

Kates, M., Hancock, A. J., Ackman, R. G., and Hooper, S. N., 1972, Preparation and characterization of the DDD(RRR)-stereoisomer of 4,8,12,16-tetramethylheptadecanoic acid and 5,9,13,17-tetramethyloctadecanoic acid, *Chem. Phys. Lipids* **8**:32.

Kleiman, R., and Spencer, G. F., 1973, Gas chromatography–mass spectrometry of methyl esters of unsaturated oxygenated fatty acids, *J. Am. Oil Chem. Soc.* **50**:31.

Kleiman, R., Spencer, G. F., and Earle, F. R., 1969, Boron trifluoride as catalyst to prepare methyl esters from oils containing unusual acyl groups, *Lipids* **4**:118.

Kleiman, R., Earle, F. R., Tallent, W. H., and Wolff, I. A., 1970, Retarded hydrolysis by pancreatic lipase of seed oils with *trans*-3 unsaturation, *Lipids* **5**:513.

Kleiman, R., Spencer, G. F., Tjarks, L. W., and Earle, F. R., 1971, Oxygenated *trans*-3-olefinic acids in a *Stenachaenium* seed oil, *Lipids* **6**:617.

Koritala, S., and Rohwedder, W. K., 1972, Formation of an artifact during methylation of conjugated fatty acids, *Lipids* **7**:274.

Kuksis, A., 1971, Progress in the analysis of lipids. IX. Gas chromatography, Part 1, *Fette Seifen Anstrichm.* **71**:130.

Kuksis, A., 1977, Gas–liquid chromatography of free fatty acids, in: *Separation and Purification Methods*, Vol. 6 (E. S. Perry, C. J. Van Oss, and E. Grushka, eds.), Marcel Dekker, New York (in press).

Kuksis, A., and Ludwig, J., 1966, Fractionation of triglyceride mixtures by preparative gas chromatography, *Lipids* **1**:202.

Kuksis, A., Breckenridge, W. C., Marai, L., and Stachnyk, O., 1968, Quantitative gas chromatography in the structural characterization of glyceryl phosphatides, *J. Am. Oil Chem. Soc.* **45**:537.

Kuksis, A., Stachnyk, O., and Holub, B. J., 1969, Improved quantitation of plasma lipids by direct gas–liquid chromatography, *J. Lipid Res.* **10**:660.

Kuksis, A., Myher, J. J., Marai, L., Yeung, S. K. F., Steiman, I., and Mookerjea, S., 1975, Distribution of newly formed fatty acids among glycerolipids of isolated perfused rat liver, *Can. J. Biochem.* **53**:509.

Kuksis, A., Myher, J. J., Marai, L., and Geher, K., 1976, Estimation of plasma free fatty acids as the trimethylsilyl (TMS) esters, *Anal. Biochem.* **70**:302.

Laine, R. A., 1974, Identification of 2-hydroxy fatty acids in complex mixtures of fatty acid methyl esters by mass chromatography, *Biomed. Mass Spectrometr.* **1**:10.

Lam, C. H., and Lie Ken Jie, M. S. F., 1975, Fatty acids. IV. Synthesis of all the dimethylene-interrupted methyl octadecadiynoates and a study of their gas–liquid chromatographic properties, *J. Chromatogr.* **115**:559.

Lam, C. H., and Lie Ken Jie, M. S. F., 1976*a*, Fatty acids. VII. The gas–liquid chromatographic properties of all dimethylene interrupted methyl *cis,cis*-octadecadienoates, *J. Chromatogr.* **117**:365.

Lam, C. H., and Lie Ken Jie, M. S. F., 1976*b*, Fatty acids. VIII. Gas-liquid chromatographic properties of all dimethylene interrupted methyl *trans,trans*-octadecadienoates, *J. Chromatogr.* **121**:303.

Langenbeck, U., Mohring, H. U., and Dieckmann, K. P., 1975, Gas chromatography of α-keto acids as their *O*-trimethyl-silylquioxalinol derivatives, *J. Chromatogr.* **115**:65.

Lankin, V. Z., Anikeeva, S. P., Ananenko, A. A., and Veltischev, Y. Y., 1974, Kolichestvennoe opredelenie neeterifitsirovannykh shirnykh kislot syrorotki krovi metodom gazozhidkostnof khromatografii, *Vopr. Med. Khim.* **20**:435.

Lemarchal, P., and Munsch, N., 1965, Etude sur le métabolisme de l'acide par des homogenats de foie de rat, *C.R. Acad. Sci. Ser. D* **260**:714.

Lemieux, R. U., and von Rudloff, E., 1955, Periodate-permanganate oxidation: Oxidation of olefins, *Can. J. Chem.* **33**:1701.

Lie Ken Jie, M. S. F., 1975*a*, Fatty acids. II. The synthesis and gas–liquid chromatographic behaviour of five trimethylene-interrupted C_{18}-diunsaturated fatty acids, *J. Chromatogr.* **109**:81.

Lie Ken Jie, M. S. F., 1975*b*, Fatty acids. III. Further study of the gas–liquid chromatographic properties of all of the methyl undecynoates and methyl *cis*-undecenoates, *J. Chromatogr.* **111**:189.

Lie Ken Jie, M. S. F., and Lam, C. H., 1974, Fatty acids. I. Synthesis of all the methyl undecynoates and the methyl *cis*-undecenoates and a study of their gas–liquid chromatographic properties, *J. Chromatogr.* **97**:165.

Lin, S.-N., and Horning, E. C., 1975*a*, Convenient determination of α-tocopherol and free fatty acids in human plasma by glass open tubular capillary column gas chromatography, *J. Chromatogr.* **112**:465.

Lin, S.-N., and Horning, E. C., 1975*b*, Analysis of long chain acids of human plasma phosphatidylcholines (lecithin) and cholesteryl esters by glass open tubular capillary column gas chromatography for stroke patients and for normal subjects, *J. Chromatogr.* **112**:483.

Lough, A. K., 1973, The chemistry and biochemistry of phytanic, pristanic and related acids, in: *Progress in the Chemistry of Fats and Other Lipids,* Vol. 14 (R. T. Holman, ed.), pp. 1–48, Pergamon Press, Oxford.

Luddy, F. E., Barford, R. A., Herb, S. F., Magidman, P., and Riemenschmeider, R. W., 1964, Pancreatic lipase hydrolysis of triglycerides by a semimicro technique, *J. Am. Oil Chem. Soc.* **41**:693.

Luddy, F. E., Barford, R. A., Herb, S. F., and Magidman, P., 1968, A rapid quantitative procedure for the preparation of methyl esters of butteroil and other fats, *J. A. Oil Chem. Soc.* **45**:549.

Lundberg, W. O., and Jarvi, P., 1970, Peroxidation of polyunsaturated fatty compounds, in: *Progress in the Chemistry of Fats and Other Lipids,* Vol. 9 (R. T. Holman, ed.), pp. 377–406, Pergamon Press, Oxford.

MacLean, I., Eglinton, G., Douraghi-Zadeh, K., Ackman, R. C., and Hooper, S. N., 1968, Correlation of stereoisomerism in present day and geologically ancient isoprenoid fatty acids, *Nature (London)* **218:**1019.

Mani, V. V. S., and Lakshminarayana, G., 1969, The thin-layer chromatographic behaviour of cyclopentanyl and cyclopentenyl fatty acid methyl esters. *J. Chromatogr.* **39:**182.

Markley, K. S., 1960, Isomerism, in: *Fatty Acids: Their Chemistry, Properties, Production and Uses* (K. S. Markley, ed.), Part 1, pp. 251–283, Interscience, New York.

Markley, K. S., 1967, Identification of fatty acids, in: *Fatty Acids: Their Chemistry, Properties, Production and Uses* (K. S. Markley, ed.), 2nd ed., Part 4, Interscience, New York.

Markley, K. S., 1968, Isomerism, in: *Fatty Acids: Their Chemistry, Properties, Production and Uses* (K. S. Markley, ed.), 2nd ed., Part 5, pp. 3282–3290, Interscience, New York.

McCloskey, J. A., 1969, Mass spectrometry of lipids and steroids in: *Methods in Enzymology*, Vol. 14 (S. P. Colowick and N. O. Kaplan, eds.), *Lipids* (J. M. Lowenstein, ed.), pp. 382–450, Academic Press, New York.

McCloskey, J. A., 1970, Mass spectrometry of fatty acid derivatives, in: *Topics in Lipid Chemistry*, Vol. 1 (F. D. Gunstone, ed.), pp. 369–440, Logos Press, London.

McCloskey, J. A., and Law, J. H., 1967, Ring location in cyclopropane fatty acid esters by a mass spectrometric method, *Lipids* **2:**225.

McCloskey, J. A., and McClelland, M. J., 1965, Mass spectra of O-isopropylidene derivatives of unsaturated fatty esters, *J. Am. Chem. Soc.* **87:**5090.

McLafferty, F. W., 1959, Mass spectrometric analyses: Molecular rearrangements, *Anal. Chem.* **1959:**82.

Mikolajczak, K. L., and Smith, C. R., Jr., 1967, Optically active trihydroxy acids of *Chemaepeuce* seed oils, *Lipids* **2:**261.

Mikolajczak, K. L., Rogers, M. F., Smith, C. R., Jr., and Wolff, I. A., 1967, An octadecatrienoic acid from *Lamium purpureum* L. seed: Isolation of 5,6-allenic and *trans*-16-olefinic unsaturation, *Biochem. J.* **105:**1245.

Minnikin, D. E., 1972, Ring location in cyclopropane fatty acid esters by boron trifluoride-catalyzed methoxylation followed by mass spectroscopy, *Lipids* **7:**398.

Miwa, T. K., Mikolajczak, K. L., Earle, F. R., and Wolff, I. A., 1960, Gas chromatographic characterization of fatty acids. Identification constants for mono- and dicarboxylic methyl esters, *Anal. Chem.* **32:**1739.

Morris, L. J., 1966, Separations of lipids by silver ion chromatography, *J. Lipid Res.* **7:**717.

Morris, L. J., 1970, Mechanism and stereochemistry of fatty acid metabolism, *Biochem. J.* **118:**681.

Morris, L. J., and Crouchman, M. L., 1969, The stereochemistry of enzymic hydration and chemical cleavage of D-(+)-*cis*-12,13-epoxyoleic acid (vernolic acid), *Lipids* **4:**50.

Morris, L. J., and Crouchman, M. L., 1972, The absolute optical configurations of the isomeric 9,10-epoxystearic,9,10-dihydroxystearic and 9,10,12-trihydroxystearic acids, *Lipids* **7:**372.

Morris, L. J., and Nichols, B. W., 1970, Argentation thin-layer chromatography of lipids, in: *Progress in Thin-Layer Chromatography and Related Methods*, Vol. 1 (A. Niederwieser and G. Pataki, eds.), pp. 75–93, Ann Arbor-Humphrey Science Publishers, Ann Arbor, Mich.

Morris, L. J., and Wharry, D. M., 1965, Chromatographic behaviour of isomeric long-chain aliphatic compounds. I. Thin-layer chromatography of some oxygenated fatty acid derivatives, *J. Chromatogr.* **20:**27.

Morris, L. J., and Wharry, D. M., 1966, Naturally occurring epoxy acids. IV. The absolute optical configuration of vernolic acid, *Lipids* **1:**41.

Morris, L. J., Wharry, D. M., and Hammond, E. W., 1968, Chromatographic behaviour of isomeric long-chain aliphatic compounds. III. Thin-layer chromatography of positional isomers of substituted fatty acids and alcohols, *J. Chromatogr.* **33:**471.

Morrison, W. R., and Smith, L. M., 1964, Preparation of fatty acid methyl esters and dimethylacetals from lipids with boron trifluoride-methanol, *J. Lipid Res.* **5:**600.

Murata, T., 1975, Chemical ionization–mass spectrometry. I. Application to analysis of fatty acids, *Anal. Chem.* **47:**573.

Murawski, U., Egge, H., Gyorgi, P., and Zilliken, F., 1971, Identification of non-methylene-interrupted *cis,cis*-octadecadienoic acids in human milk, *FEBS Lett.* **18:**290.

Myher, J. J., Marai, L., and Kuksis, A., 1974, Identification of fatty acids by GC-MS using polar siloxane liquid phases, *Anal. Biochem.* **62:**188.

Nelson, G. J., 1972, Quantitative analysis of blood lipids, in: *Blood Lipids and Lipoproteins: Quantitation, Composition, and Metabolism* (G. J. Nelson, ed.), pp. 25–73, Interscience, New York.

Nicolaides, N., Apon, J. M. B., and Wong, D. H., 1976, Further studies of the saturated methyl branched fatty acids of vernix caseosa lipid, *Lipids* **11:**781.

Niehaus, W. G., Jr., and Ryhage, R., 1967, Determination of double bond positions in polyunsaturated fatty acids using combination gas chromatography–mass spectrometry, *Tetrahedron Lett.* 5021.

Niehaus, W. G., and Schroepfer, G. J., 1965, The reversible hydration of oleic acid to 10-hydroxystearic acid, *Biochem. Biophys. Res. Commun.* **21:**271.

Niehaus, W. G., Kisic, A., Torkelson, A., Bednarczyk, D. J., and Schroepfer, G. J., 1970, Stereospecific hydration of *cis*- and *trans*-9,10-epoxyoctadecenoic acids, *J. Biol. Chem.* **245:**3190.

Nutter, L. J., and Privett, O. S., 1966, Phospholipase A properties of several snake venom preparations, *Lipids* **1:**258.

O'Brien, J. S., and Rouser, G., 1964, Analyses of hydroxy fatty acids by gas–liquid chromatography, *Anal. Biochem.* **7:**288.

O'Connor, R. T., 1968, Spectral properties, in: *Fatty Acids: Their Chemistry, Properties, Production, and Uses* (K. S. Markley, ed.), 2nd ed., Part 5, pp. 3315–3417, Interscience, New York.

Odham, G., 1967, Studies on the fatty acids in the feather waxes of some water-birds, *Fette Seifen Anstrichm.* **69:**164.

Odham, G., and Stenhagen-Ställberg, E., 1972, Fatty acids, in: *Biochemical Applications of Mass Spectrometry* (G. R. Waller, ed.), pp. 211–228, Interscience, New York.

Odham, G., and Waern, K., 1968, Synthesis of methyl 2D,4D,6D- and 2L,4D,6D-trimethyldodec-11-enoate, *Ark. Kemi* **29:**563.

Organiscian, D. T., 1974, The collection and elution of radioactive fatty acid methyl ester during quantitative gas–liquid chromatography, *Prep. Biochem.* **4:**89.

Ory, R. L., Kiser, J., and Pradel, P. A., 1969, Studies on positional specificity of the castor bean acid lipase, *Lipids* **4:**261.

Oshima, M., and Ariga, T., 1975, ω-Cyclohexyl fatty acids in acidophilic thermophilic bacteria: Studies on their presence, structure, and biosynthesis using precursors labelled with stable isotopes and radioisotopes, *J. Biol. Chem.* **250:**6963.

Ottenstein, D. M., and Supina, W. R., 1974, Improved columns for the separation of C_{14}–C_{20} fatty acids in the free form, *J. Chromatogr.* **91:**119.

Ottenstein, D. M., Bartley, D. A., and Supina, W. R., 1976, Gas chromatographic separation of *cis-trans* isomers: Methyl oleate/methyl elaidate, *J. Chromatogr.* **119:**401.

Pawlowski, N. E., 1972, Mass spectra of methyl sterculate and malvalate and 1,2-dialkylcyclopropenes, *Chem. Phys. Lipids* **13:**164.

Pawlowski, N. E., Nixon, J. E., and Sinnhuber, R. O., 1972, Assay of cyclopropenoid lipids by nuclear magnetic resonance, *J. Am. Oil Chem. Soc.* **49:**387.

Pearl, M. B., Kleiman, R., and Earle, F. R., 1973, Acetylenic acids of *Alvaradoa amorphoides* seed oil, *Lipids* **8:**627.

Pelick, N., Henly, R. S., Sweeny, R. F., and Miller, M., 1963, Special methods of purifying fatty acids, *J. Am. Oil Chem. Soc.* **40:**419.

Perkins, E. G., and Iwaoka, W. T., 1973, Purification of cyclic fatty acid esters: A GC-MS study, *J. Am. Oil Chem. Soc.* **50:**44.

Perry, T. L., Houser, S., Diamond, S., Bullis, B., Mok, C., and Melangon, S. B., 1970, Volatile fatty acids in normal human physiological fluids, *Clin. Chim. Acta* **29:**369.

Phillipou, G., Bigham, D. A., and Seamark, R. F., 1975, Subnanogram detection of *t*-butyldimethylsilyl fatty acid esters by mass fragmentography, *Lipids* **10:**714.

Plattner, R. D., Spencer, G. F., and Kleiman, R., 1976, Double bond location in polyenoic fatty esters through partial oxymercuration, *Lipids* **11:**222.

Polgar, N., 1971, Natural alkyl-branched long-chain acids, in: *Topics in Lipid Chemistry* (F. D. Gunstone, ed.), pp. 207–246, Interscience, New York.

Powell, R. G., Smith, C. R., Jr., and Wolff, I. A., 1967, Geometric configuration and etherification reactions of some naturally occurring 9-hydroxy-10,12- and 13-hydroxy-9,11-octadecadienoic acids, *J. Org. Chem.* **32:**1442.

Privett, O. S., 1966, Determination of the structure of unsaturated fatty acids via degradative methods, in: *Progress in the Chemistry of Fats and Other Lipids*, Vol. 9 (R. T. Holman, ed.), pp. 91–117, Pergamon Press, London.

Privett, O. S., 1968, Preparation of polyunsaturated fatty acids from natural sources, in: *Progress in the Chemistry of Fats and Other Lipids*, Vol. 9 (R. T. Holman, ed.), pp. 407–452, Pergamon Press, New York.

Privett, O. S., and Nickell, E. C., 1963, Preparation of highly purified fatty acids via liquid–liquid partition chromatography, *J. Am. Oil Chem. Soc.* **40:**189.

Raju, P. K., and Reiser, R., 1966, Gas–liquid chromatographic analysis of cyclopropene fatty acids, *Lipids* **1:**10.

Rapport, M. M., and Alonzo, N., 1955, Photometric determination of fatty acid ester groups in phospholipids, *J. Biol. Chem.* **217:**193.

Rodrigues De Miranda, J. F., and Eikelboom, T. D., 1975, Thin-layer chromatographic separation of free fatty acids: Analysis and purification of radioactively labelled fatty acids, *J. Chromatogr.* **114:**274.

Roehm, J. N., and Privett, O. S., 1969, Improved method for determination of the position of double bonds in polyenoic fatty acid fractions, *J. Lipid Res.* **10:**245.

Rohwedder, W. K., Mabrouk, A. F., and Selke, E., 1965, Mass spectrometric studies of unsaturated methyl esters, *J. Phys. Chem.* **69:**1711.

Ryhage, R., and Stenhagen, E., 1960, Mass spectrometry in lipid research, *J. Lipid Res.* **1:**361.

Ryhage, R., and Stenhagen, E., 1963, Mass spectrometry of long-chain esters, in: *Mass Spectrometry of Organic Ions* (F. W. McLafferty, ed.), p. 399, Academic Press, New York.

Ryhage, R., Ställberg-Stenhagen, S., and Stenhagen, E., 1961, Mass spectrometry studies. VII. Methyl esters of alpha, beta-unsaturated long chain acids. On the occurrence of C_{27}-phthienoic acid, *Ark. Kemi* **18:**179.

Schauenstein, E., 1967, Autoxidation of polyunsaturated esters in water: Chemical structure and biological activity of the products, *J. Lipid Res.* **8:**417.

Schlenk, H., 1970, Odd numbered polyunsaturated fatty acids, in: *Progress in the Chemistry of Fats and Other Lipids*, Vol. 9 (R. T. Holman, ed.), Part 5, pp. 589–605, Pergamon Press, Oxford.

Schlenk, H., and Gellerman, J. L., 1960, Esterification of fatty acids with diazomethane on a small scale, *Anal. Chem.* **32:**1412.

Schogt, J. C. M., and Haverkamp-Begeman, P., 1965, Isolation of 11-cyclohexylundecanoic acid from butter, *J. Lipid Res.* **6:**466.

Scholfield, C. R., 1975*a*, High performance liquid chromatography of fatty methyl esters: Analytical separations, *J. Am. Oil Chem. Soc.* **52:** 36.

Scholfield, C. R., 1975*b*, High performance liquid chromatography of fatty methyl esters: Preparative separations, *Anal. Chem.* **47:**1417.

Scholfield, C. R., and Dutton, H. J., 1970, Gas chromatographic equivalent chain lengths of isomeric methyl octadecenoates and octadecynoates, *J. Am. Oil Chem. Soc.* **47:**1.

Scholfield, C. R., and Dutton, H. J., 1971, Equivalent chain lengths of methyl octadecadienoates and octadecatrienoates, *J. Am. Oil Chem. Soc.* **48:**228.

Scholfield, C. R., Butterfield, R. O., Peters, H., Glass, C. A., and Dutton, H. J., 1967, Counter-current distribution of alkali-isomerized methyl linolenate with an argentation system, *J. Am. Oil Chem. Soc.* **44:**50.

Schroepfer, G. J., Niehaus, W. G., and McCloskey, J. A., 1970, Enzymatic conversion of linoleic acid to 10D-hydroxy-12-*cis*-octadecenoic acid, *J. Biol. Chem.* **245:**3798.

Sen Gupta, A. K., and Peters, H., 1969, Synthesis and properties of some new C_{15}-ω-aliphatic and cyclic acids, *Chem. Phys. Lipids* **3:**371.

Shenstone, F. S., 1971, Ultraviolet and visible spectroscopy of lipids, in: *Biochemistry and Methodology of Lipids* (A. R. Johnson and J. G. Davenport, eds.), pp. 219–242, Interscience, New York.

Sheppard, A. J., and Iverson, J. L., 1975, Esterification of fatty acids for gas–liquid chromatographic analysis, *J. Chromatogr. Sci.* **13:**448.

Smith, C. R., Jr., 1970*a*, Occurrence of unusual fatty acids in plants, in: *Progress in the Chemistry of Fats and Other Lipids*, Vol. 11 (R. T. Holman, ed.), Part 1, pp. 137–177, Pergamon Press, Oxford.

Smith, C. R., Jr., 1970*b*, Optically active long-chain compounds and their absolute configuration, in: *Topics in Lipid Chemistry*, Vol. 1 (F. D. Gunstone, ed.), pp. 277–368, Logos Press, London.

Smith, T. M., and White, H. B., 1966, Two-column gas–liquid chromatography of fatty acid methyl esters, *J. Lipid Res.* **7**:327.

Spener, F., and Mangold, H. K., 1974, New cyclopentenyl fatty acids in Flacourtiaceae: Straight-chain fatty acids and cyclic fatty acids in lipids during maturation of the seeds, *Biochemistry* **13**:2241.

Sprecher, W. W., Maier, R., Barber, M., and Holman, R. T., 1965, Structure of an optically active allene containing tetraesters triglyceride isolated from the seed oil of *Sapium sebiferum*, *Biochemistry* **4**:1856.

Stoffel, W., and Michaelis, G., 1976, Chemical syntheses of novel fluorescent-labelled fatty acids, phosphatidylcholines and cholesterol esters, *Hoppe-Seyler's Z. Physiol. Chem.* **357**:7.

Strocchi, A., and Piretti, M., 1968, Separation and identification of geometrical isomers of 9,12-octadecadienoic and 9,12,15-octadecatrienoic acids, *J. Chromatogr.* **36**:181.

Sun, K. K., and Holman, R. T., 1968, Mass spectrometry of lipid molecules, *J. Am. Oil Chem. Soc.* **45**:810.

Svennerholm, L., 1968, Distribution and fatty acid composition of phosphoglycerides in normal human brain, *J. Lipid Res.* **9**:570.

Tallent, W. H., and Kleiman, R., 1968, Bis(trimethylsilyl)acetamide in the silylation of lipolysis products of gas–liquid chromatography, *J. Lipid Res.* **9**:146.

Tanaka, K., and Yu, G. M., 1973, A method for the separate determination of isovalerate and α-methylbutyrate by use of GLC-mass spectrometry, *Clin. Chim. Acta* **43**:151.

Tatsumi, K., 1974, Stereochemical aspects of synthetic and naturally occurring 2-hydroxy fatty acids: Their absolute configurations and assays of optical purity, *Arch. Biochem. Biophys.* **165**:656.

Thenot, J.-P., Horning, E. C., Stafford, M., and Horning, M. G., 1972, Fatty acid esterification with *N,N*-dimethylformamide dialkyl acetals for GLC analysis, *Anal. Lett.* **5**:217.

Therriault, D. G., 1963, Fractionation of lipids by countercurrent distribution, *J. Am. Oil Chem. Soc.* **40**:395.

Tulloch, A. P., 1964, Gas liquid chromatography of the hydroxy-acetoxy- and oxo-stearic acid methyl esters, *J. Am. Oil Chem. Soc.* **41**:833.

Tulloch, A. P., and Mazurek, M., 1976, ^{13}C nuclear magnetic resonance spectroscopy of saturated, unsaturated and oxygenated fatty acid methyl esters, *Lipids* **11**:228.

Uchida, K., 1974, Occurrence of saturated and monounsaturated fatty acids with unusually-long chains (C_{20}–C_{30}) in *Lactobacillus heterohiochii*, an alcoholphilic bacterium, *Biochim, Biophys. Acta* **348**:86.

Van Deenen, L. L. M., and De Haas, G. H., 1963, The substrate specificity of phospholipase A, *Biochim. Biophys. Acta* **70**:538

Vulliet, P., Markey, S. P., and Tornabene, T. G., 1974, Identification of methoxyester artifacts produced by methanolic-HCl solvolysis of the cyclopropane fatty acids of the genus *Yersinia*, *Biochim, Biophys. Acta* **348**:299.

Wadke, M., Brunengraber, H., Lowenstein, J. M., Dolhun, J. J., and Arsenault, G. P., 1973, Fatty acid synthesis by the liver perfused with deuterated and tritiated water, *Biochemistry* **12**: 2619.

Warthen, J. D., 1975, Separation of *cis* and *trans* isomers by reverse phase high pressure liquid chromatography, *J. Am. Oil Chem. Soc.* **52**:151.

Weihrauch, J. L., Brewington, C. R., and Schwartz, D. P., 1974, Trace constituents in milk fat: Isolation and identification of oxofatty acids, *Lipids* **9**:883.

Wells, M. A., and Hanahan, D. J., 1969, Phospholipase A from *Crotalus adamanteus* venom, in: *Methods in Enzymology*, Vol. 14 (J. M. Lowenstein, ed.), pp. 178–184, Academic Press, New York.

Wolff, I. A., and Miwa, T. K., 1965, Effect of unusual acids on selected seed oil analyses, *J. Am. Oil Chem. Soc.* **42**:208.

Wood, R., 1967, GLC and TLC analyses of isopropylidene derivatives of isomeric polyhydroxy acids derived from positional and geometrical isomers of unsaturated fatty acids, *Lipids* **2**:199.

Wood, R., and Reiser, R., 1965, Cyclopropane fatty acid metabolism: Physical and chemical identification of propane ring metabolic products in the adipose tissue, *J. Am. Oil Chem. Soc.* **42**:315.

Wood, R., Bever, E. L., and Snyder, F., 1966, The GLC and TLC resolution of diastereoisomeric polyhydroxystearates and assignment of configurations, *Lipids* **1**:399.

Wright, A., Dankert, M., Fennessey, P., and Robins, P. W., 1967, Characterization of a polyisoprenoid compound functional in O-antigen biosynthesis, *Proc. Natl. Acad. Sci. USA* **57**:1798.

Zeman, I., and Pokorny, J., 1963, Gas chromatographic analysis of cyclopentenyl fatty acids, *J. Chromatogr.* **10**:15.

Zimmerman, D. C., and Vires, B. A., 1970, Specificity of flaxseed lipoxidases, *Lipids* **5**:392.

Chapter 2

Synthesis and Analysis of Stable Isotope- and Radioisotope-Labeled Fatty Acids

Edward A. Emken

2.1. Introduction

This chapter will touch on points of general consideration for preparing labeled fatty acids and on selected methods used to label fatty acids. Representative syntheses of some labeled fatty acids will be described, and typical analytical methods for determining isotope purities and label position will be covered.

General reviews on deuterium isotope labeling and analysis may be found in Thomas's book *Deuterium Labeling in Organic Chemistry* (1971), and in Fetizon and Gramain's review "Recent Methods of Deuteration" (1969). A general review on ^{13}C-labeled compounds is found in "Stable Isotope Tracers in the Life Sciences and Medicine" by Matwiyoff and Ott (1973). The preparation of many stable isotope-labeled saturated fatty acids has been summarized by Dinh-Nguyen (1964, 1968).

Synthesis, analysis, and other topics related to ^{14}C-labeled compounds have been reviewed by Murray and Williams (Part I, 1958), Calvin (1949), Raaen *et al.* (1968), Nevenzel *et al.* (1957), Catch (1961), Tolbert and Siri (1960), and Ronzio (1954). Reviews on the synthesis of tritium-labeled compounds are Evans's *Tritium and Its Compounds* (1966) and Murray and Williams's *Organic Synthesis with Isotopes* (Part II, 1958). Basic laboratory techniques are found in *Techniques of Radiobiochemistry* (Aronoff, 1956), "Techniques of Lipidology: Isolation, Analysis, and Identification of Lipids" (Kates, 1972), and *Radiotracer Methodology in Biological Science* (Wang and Willis, 1965).

The preparation of radioisotope-labeled fatty acids has been summarized by Stoffel and Bierwirth (1963), Stoffel (1964), Osbond (1966), Snyder and Piantadosi

Edward A. Emken • Northern Regional Research Center Agricultural Research Service, U.S. Department of Agriculture, Peoria, Illinois 61604.

(1966), Mangold (1968), Marcel and Holman (1968), Kunau (1973), and Mounts (1973).

2.2. Synthesis of Stable Isotope-Labeled Fatty Acids

Deuterated fats have been used as tracers and analytical aids in a variety of experiments. These experimental uses included lipid metabolism studies (both *in vivo* and *in vitro*), mechanistic studies, indentification of ionization fragments in mass spectrometry, identification of lipid structure by nuclear magnetic resonance (NMR) and infrared (IR) spectrometry, and research involving interaction, configuration, and stereoisomerism of lipids.

The approach taken to prepare deuterated fatty acids is dependent on several interrelated factors which must be weighed against one another. These factors are (1) amount of fatty acid needed, (2) required isotopic purity of the fatty acid, (3) position of the deuterium label, (4) cost of the deuterium needed in the reaction, (5) efficiency and specificity of the deuterium incorporation method, (6) yield of the synthetic sequence after deuterium is incorporated, and (7) choice of alternative synthetic reactions. The importance of these factors varies with the purpose of the experiment and the information desired. Certainly the synthesis of 100 g of deuterated fats for use in a metabolic experiment would require different considerations than synthesis of 1 mg of deuterated fats for mass spectrometry studies.

The most economical source of deuterium is deuterium oxide ($2.80/mole D_2), followed by deuterium gas ($23.00/mole D_2) and acetic acid-d_4 ($28.00/mole D_2). Other compounds such as methyl-d_3 iodide ($310.00/mole D_2) are much more expensive. Expensive deuterating agents such as deuterohydrazine ($500/mole D_2), sodium borodeuteride ($275/mole D_2), and lithium deuteride ($90/mole D_2) are useful for large-scale preparations only when they provide a selective means of introducing deuterium which cannot be readily achieved by other methods. When small quantities of a deuterated fatty acid are required, the convenience of some of the more expensive deuterated reagents can be well worth their cost.

2.2.1. Experimental Techniques

Glassware contains a considerable amount of adsorbed water. If high isotopic purity is essential, all adsorbed water must be driven from the glass surface. A recommended procedure is to bake the glassware at 250°C for 48 hr, cool while flushing with dry nitrogen, rinse with 99% D_2O, and then flame out under a high vacuum for 2–3 hr (Eisch and Kaska, 1966).

Solvents likewise must be free of all water. The simplest procedure is to select reaction solvents which can be dried by azeotropic distillation. Solvents should not be stored over drying agents because traces of water adsorbed on the drying agents can back-exchange into the solvent. Solvents which have an exchangeable proton must obviously be avoided unless they are first completely exchanged with deuterium oxide; this exchange can be a tedious and expensive task.

The difficulty of keeping water out of a reaction is so great that, even when all precautions are taken, the isotopic purity of the product is often not more than 90–95%. Fortunately, the need for 99% isotopic purity does not occur often, and very reliable work can be done with much less isotopic purities.

If large volumes of deuterium gas are needed, the time required to set up adequate equipment for the electrolysis of deuterium oxide may be justified. The availability of commercial hydrogen generators such as Elhygen (Milton Roy Company, St. Petersburg, Florida) and Aerograph model A-650 (Varian) has somewhat simplified this procedure.

2.2.2. Saturated Fatty Acids

Saturated aliphatic carboxylic acids are the easiest fatty acids to deuterate, and many synthetic preparations involve this class of fats.

Dinh-Nguyen (1964, 1968) has summarized the preparation of many deuterated fatty acids using techniques involving hydrogen–deuterium exchange and anodic condensation. The condensation of aliphatic acids via the Kolbe reaction is illustrated in the preparation of methyl octadecanoate-18,18,18-d_3.

2.2.2.1. *Kolbe Reaction*

$$CD_3CO_2D + CH_3O_2C\text{–}(CH_2)_{16}CO_2H \xrightarrow[-2e]{2Na} CD_3\text{–}(CH_2)_{16}CO_2CH_3$$

Metallic sodium (230 mg) is reacted with 30 ml of methanol and then added dropwise to 345 mg of perdeuteroacetic acid in 5 ml of methanol. This solution is added to 1.64 g of dimethyl octadecane-1,18-dioate in 40 ml methanol. The total mixture is transferred into the electrolyzer with 20 ml of methanol. A 1.5-A current is passed through the solution for 20 min at 45°C. The solution is cooled, filtered, and concentrated to 5 ml. This solution is placed on a column of 35 g of neutral aluminum oxide and eluted with petroleum ether. A yield of 305 mg of methyl octadecanoate is obtained.

The electrolytic cell consists of a rotating anode of platinum sheet, which is continually scraped clean with a broom of 97:3 platinum–indium wires, and a sodium-retaining cathode of mercury. Experimental details of the Kolbe reaction are discussed by Swann (1948).

Dinh-Nguyen (1964, 1968) has reported the preparation of a series of *gem*-d_2 methyl octadecanoates (see Table I) by the electrolysis of RCD_2CO_2H with $R'CH_2CO_2H$. This synthesis appears satisfactory for saturated acids, but it has been reported to give a complex mixture of products when used to prepare aliphatic acetylenic acids (Klok *et al.*, 1974).

Other methods for preparation of deuterated methyl stearate have included the reduction of oleic acid or linoleic acid with tetradeuterohydrazine, which adds two deuteriums to each double bond (Scholfield *et al.*, 1961; Rohwedder *et al.*, 1967; Morris *et al.*, 1967, 1968; Dinh-Nguyen, 1968). Aylward and Narayana Rao (1956, 1957) have reported the reaction conditions which affect hydrazine reduction. The preparation of *erythro*- and *threo*-octadecanoic acid-9,10-d_2 and methyl octadecanoate-9,10,12,13-d_4 from methyl linoleate by hydrazine reduction will be

Table I. Deuterium-Labeled Saturated Fatty Acids

Compound	Reference
$CH_3(CH_2)_9CD_2CO_2H$ dodecanoic acid-2,2-d_2	Dinh-Nguyen (1964)
$CD_3(CH_2)_{12}CO_2H$ tetradecanoic acid-14,14,14-d_3	Dinh-Nguyen (1964)
$CD_3(CD_2)_{14}CO_2H$ perdeuterohexadecanoic acid	Wendt and McCloskey (1970)
$CH_3(CH_2)_{14}CD_2CO_2H$ heptadecanoic acid-2,2-d_2	Dinh-Nguyen (1964)
$CH_3(CH_2)_9CD_2(CH_2)_5CO_2H$ heptadecanoic acid-7,7-d_2	Dinh-Nguyen (1964)
$CH_3(CH_2)_{15}CD_2CO_2H$ octadecanoic acid-2,2-d_2	Dinh-Nguyen (1964)
$CH_3(CH_2)_{14}CD_2CH_2CO_2H$ octadecanoic acid-3,3-d_2	Dinh-Nguyen (1968)
$CH_3(CH_2)_8CD_2(CH_2)_7CO_2H$ octadecanoic acid-9,9-d_2	Dinh-Nguyen (1968)
$CH_3(CH_2)_7CD_2(CH_2)_8CO_2H$ octadecanoic acid-10,10-d_2	Dinh-Nguyen (1968)
$CH_3(CH_2)_6CD_2(CH_2)_9CO_2H$ octadecanoic acid-11,11-d_2	Dinh-Nguyen (1968)
$CH_3(CH_2)_5CD_2(CH_2)_{10}CO_2H$ octadecanoic acid-12,12-d_2	Dinh-Nguyen (1968)
$CH_3(CH_2)_4CD_2(CH_2)_{11}CO_2H$ octadecanoic acid-13,13-d_2	Dinh-Nguyen (1968)
$CH_3(CH_2)_3CD_2(CH_2)_{12}CO_2H$ octadecanoic acid-14,14-d_2	Dinh-Nguyen (1968)
$CH_3(CH_2)_2CD_2(CH_2)_{13}CO_2H$ octadecanoic acid-15,15-d_2	Dinh-Nguyen (1968)
$CH_3CH_2CD_2(CH_2)_{14}CO_2H$ octadecanoic acid-16,16-d_2	Dinh-Nguyen (1968)
$CH_3CD_2(CH_2)_{15}CO_2H$ octadecanoic acid-17,17-d_2	Dinh-Nguyen (1968)
$CH_3(CH_2)_{14}(CD_2)_2CO_2H$ octadecanoic acid-2,2,3,3-d_4	Dinh-Nguyen (1964)
$CH_3(CH_2)_8CHD(CH_2)_7CO_2H$ octadecanoic acid-9(10)-d_1	Rohwedder *et al.* (1967)
$CH_3(CH_2)_{10}CHDCHD(CH_2)_4CO_2H$ octadecanoic acid-6,7-d_2	Rohwedder *et al.* (1967)
$CH_3(CH_2)_7CHDCHD(CH_2)_7CO_2H$ octadecanoic acid-9,10-d_2	Rohwedder *et al.* (1967)
$CH_3(CH_2)_{14}CHDCHDCO_2H$ octadecanoic acid-2,3-d_2	Dinh-Nguyen (1968)
$CH_3(CH_2)_{10}CHDCHD(CH_2)_4CO_2H$ octadecanoic acid-6,7-d_2	Dinh-Nguyen (1968)
$CH_3(CH_2)_8CHDCHD(CH_2)_6CO_2H$ octadecanoic acid-7,8-d_2	Dinh-Nguyen (1968)
$CH_3(CH_2)_7CHDCHD(CH_2)_7CO_2H$ octadecanoic acid-9,10-d_2	Dinh-Nguyen (1968)

Table I—Continued

$CH_3(CH_2)_6CHDCHD(CH_2)_6CO_2H$ octadecanoic acid-8,9-d_2	Dinh-Nguyen (1968)
$CH_3(CH_2)_5CHDCHD(CH_2)_9CO_2H$ octadecanoic acid-11,12-d_2	Dinh-Nguyen (1968)
$CH_3(CH_2)_3CHDCHD(CH_2)_{11}CO_2H$ octadecanoic acid-13,14-d_2	Dinh-Nguyen (1968)
$CH_2DCHD(CH_2)_{15}CO_2H$ octadecanoic acid-17,18-d_2	Dinh-Nguyen (1968)
$CH_3(CH_2)_7CHDCD_2(CH_2)_7CO_2H$ octadecanoic acid-9,(9),10,(10)-d_3	Rohwedder *et al.* (1967)
$CH_3(CH_2)_4CHDCHDCH_2CHDCHD(CH_2)_7CO_2H$ octadecanoic acid-9,10,12,13-d_4	Rohwedder *et al.* (1967)
$CH_3(CH_2)_5CHDCHDCHDCHD(CH_2)_7CO_2H$ octadecanoic acid-9,10,11,12-d_4	Rohwedder *et al.* (1967)
$CH_3(CH_2)_7CD_2CD_2(CH_2)_7CO_2H$ octadecanoic acid-9,9,10,10-d_4	Rohwedder *et al.* (1967)
$CH_3CH_2(CHDCHDCH_2)_3(CH_2)_6CO_2H$ octadecanoic acid-9,10,12,13,15,16-d_6	Rohwedder *et al.* (1967)
$CD_3(CD_2)_{16}CO_2H$ perdeuterooctadecanoic acid	Rohwedder *et al.* (1967)
$CD_3(CH_2)_{17}CO_2H$ nonadecanoic acid-19,19,19-d_3	Dinh-Nguyen (1968)
$CH_3(CH_2)_{15}CD_2(CH_2)_2CO_2H$ eicosanoic acid-4,4-d_2	Dinh-Nguyen (1964)
$CH_3(CH_2)_{14}CD_2(CH_2)_3CO_2H$ eicosanoic acid-5,5-d_2	Dinh-Nguyen (1964)
$CH_3(CH_2)_9CD_2(CH_2)_8CO_2H$ eicosanoic acid-10,10-d_2	Dinh-Nguyen (1964)
$CH_3(CH_2)_{14}CD_2(CH_2)_4CO_2H$ heneicosanoic acid-6,6-d_2	Dinh-Nguyen (1964)

used to illustrate the synthesis. Ethanol, which is used in nondeuterated hydrazine reductions, cannot be used as a solvent with deuterohydrazine because of its exchangeable hydrogen. Dry acetonitrile has been reported (Koch, 1969*a*, *b*) as a suitable solvent, but we observed that the solubility of $N_2D_4 \cdot D_2O$ in CH_3CN was limited. Dry dioxane appears to be a better choice of solvent.

2.2.2.2. Deuterohydrazine Reduction of Methyl Esters

Methyl linoleate (150 mg) is dissolved in 3 ml of dry dioxane and 0.5 ml of $N_2D_4 \cdot D_2O$ is added. The mixture is stirred rapidly at ~50°C and a slow stream of dioxane-saturated O_2 is bubbled into the mixture. The mixture is stirred for 3 hr, and then additional $N_2D_4 \cdot D_2O$ is added with continual stirring until GC analysis shows 80–85% reduction. The mixture is acidified with 0.1 N HCl, diluted with 5 ml H_2O, and extracted two times with 2-ml portions of petroleum ether. Recovery of about 50 mg of methyl stearate-9,10,12,13-d_4 can be expected.

Methyl octadecanoate-9,10,12,13,15,16-d_6 (Rohwedder *et al.*, 1967) has also been prepared by the reduction of linolenic acid with hydrazine hydrate-d_6. However, an easier method is to use tris(triphenylphosphine)-rhodium(I) chloride to deuterate methyl linolenate (Birch and Walker, 1966).

2.2.2.3. Deuterohydrazine Reduction of Fatty Acids

Dinh-Nguyen (1968) first saponified 30 mg of an unsaturated methyl ester by heating a solution of 3 mg of lithium hydroxide in 3 ml of methanol for 15 min. The solution is saturated with carbon dioxide and evaporated to dryness. The white residue is washed with acetone and then with petroleum ether, and dried. The residue is mixed with 0.4 ml deuterated hydrazine hydrate and heated at 60–80°C for 20–48 hr. Lack of foaming indicates the end of the reaction. The residue is acidified with dilute HCl and extracted with ether.

2.2.2.4. Reduction with Azodicarboxylate

Potassium azodicarboxylate (KADC) has been reported (Dinh-Nguyen, 1968; Van Tamelen *et al.*, 1961; Koch, 1969*a*, *b*) as an alternative to hydrazine reduction. A typical procedure consists of adding 220 mg of methyl linoleate to 60 ml of dry pyridine. KADC (4.0 g) is then added along with 1 ml of acetic acid-1-d_1. The mixture is stirred for 6 hr while an additional 5 ml of acetic acid-1-d_1 is added. Acidification with dilute HCl followed by extraction with petroleum ether resulted in the recovery of 202 mg of product containing 48% stearate-d_4, 44% octadecenoate-d_2, and 8% linoleate-d_0.

Methyl stearate-9(10)-d_1 has been synthesized by reaction of methyl oleate with disiamylborane followed by the addition of acetic acid-1-d_1 (Rohwedder *et al.*, 1967).

Heterogeneous catalysts such as nickel, platinum, palladium, cobalt, and rhodium cannot be used to prepare deuterated fatty acids from unsaturated fats because of extensive H–D exchange which normally occurs (Dutton *et al.*, 1968).

Methyl-d_3 fatty esters can easily be prepared from fatty acids by esterification of corresponding fatty acids with CD_3OH using acid catalysts such as BF_3 or HCl.

Table I lists a complete series of labeled saturated fatty acids containing deuterium at various positions.

2.2.3. Monounsaturated Fatty Acids

Methyl oleate-9(10)-d_1 is obtained from the reaction of methyl stearolate with disiamylborane followed by reaction with acetic acid-1-d_1. Methyl oleate-9,10-d_2 is formed in a similar manner from deuterodisiamylborane and methyl stearolate followed by reaction with acetic acid-1-d_1. The experimental details are similar to those used for the preparation of stearate-9(10)-d_1 (Rohwedder *et al.*, 1967).

Morris *et al.* (1967, 1968) have partially reduced octadecadienoic acid with deuterohydrazine to prepare *erythro*- and *threo*-9-octadecenoic acid-12,13-d_2 and *erythro*- and *threo*-9-octadecenoic acid-15,16-d_2. Unwanted octadecenoic acid isomers were removed by low-temperature argentation TLC.

A convenient method of preparing methyl *cis*-octadecenoates-d_2 (Emken *et al.*, 1976) has been to reduce the corresponding octadecynoate over Lindlar's catalyst (Lindlar and Dubuis, 1973). Deuterated all-*cis* polyunsaturated esters can also be prepared by deuterating acetylenic fatty esters using Lindlar's catalyst.

2.2.3.1. Reduction of Alkynoates with Lindlar's Catalyst

Methyl stearolate is easily prepared in large quantities by bromination–debromination of methyl oleate (Butterfield and Dutton, 1968). The methyl stearolate is then reduced over Lindlar's catalyst using deuterium gas (Emken *et al.*, 1976).

Methyl oleate-9,10-d_2 prepared this way contains 2–4% *trans* isomer. According to Steenhoek *et al.* (1971), the *trans* isomer is formed by isomerization of the *cis* bond rather than by direct formation from the acetylenic bond. Steenhoek *et al.* (1971) have reported the optimum conditions for *cis* hydrogenation and have discussed experimental problems with the preparation and use of Lindlar's catalyst. They indicate that a key step in the preparation of active Lindlar's catalyst is to use freshly precipitated $CaCO_3$.

Unfortunately, methyl oleate-9,10-d_2 prepared by deuteration of methyl stearolate contains approximately 15% of an oleate-d_1 species. The hydrogen is thought to come from H–D exchange with the quinoline used to poison the catalyst (Calf *et al.*, 1968).

2.2.3.2. Octadecenoate Synthesis by the Wittig Reaction

Tucker *et al.* (1971) have prepared oleic-11,11-d_2 acid by using the Wittig reaction to couple nonanal-2,2-d_2 with methyl 8-formyloctanoate.

The Wittig reaction has been used by DeJarlais and Emken (1976) to synthesize methyl 9-octadecenoate-8,8,11,11-d_4 from 1-bromononane-2,2-d_2 and methyl 8-formyloctanoate-2,2-d_2 and to synthesize methyl 9-octadecenoate-13,13,14,14-d_4 from 1-chlorononane-4,4,5,5-d_4 and methyl 8-formyloctanoate. They also prepared methyl 9-octadecenoate-8,8,13,13,14,14-d_6 by coupling methyl 8-formyloctanoate-2,2-d_2 with 1-chlorononane-4,4,5,5-d_4 with the Wittig reaction.

An example of the experimental details for a Wittig reaction is given for 9-octadecenoate. In a 50-ml three-necked flask equipped with dropping funnel, mechanical stirrer, and thermometer are placed 4 g of 1-nonyl-4,4,5,5-d_4-triphenylphosphonium iodide, 0.5 g of molecular sieve (3A), and 20 ml of dry THF. The contents are cooled to 1°C, and 7.8 ml of 1 N *t*-butyl lithium in pentane is added. The deep-red solution is stirred for 5 min, and then a 10% molar excess of methyl 8-formyloctanoate is added over 20 min. The orange solution is warmed to 35°C and stirred for 3 hr. The mixture is added to 200 ml of 0.2 N HCl and extracted three times with 25-ml portions of petroleum ether. Evaporation of the washed and dried petroleum ether extracts gives a 27% yield of *cis*- and *trans*-9-octadecenoate isomers containing 27% *trans*-9-octadecenoate-13,13,14,14-d_4 isomer. The *cis* and *trans* isomers were separated by silver-resin chromatography (Emken *et al.*, 1964).

2.2.3.3. *Deuteration with Tris(triphenylphosphine)-rhodium(I) Chloride*

The use of tris(triphenylphosphine)-rhodium(I) chloride (Wilkinson's catalyst) as a catalyst for deuteration of methyl oleate and methyl linoleate has been described by Birch and Walker (1966). This application was the result of work by Wilkinson's group (Osborn *et al.*, 1966; Young *et al.*, 1965) in which the catalytic properties were described and the lack of hydrogen–deuterium exchange was noted. This catalyst has had extensive use in the author's laboratory and elsewhere (Morandi and Jensen, 1969).

An important feature of Wilkinson's catalyst is its sensitivity to oxygen. Oxygen destroys its catalytic activity for double-bond reduction, and consequently all solvents must be carefully degassed.

The following procedure gives details for the deuteration of 1-chloro-4-nonyne. The 1-chlorononane-4,4,5,5-d_4 was utilized in the Wittig reaction (described earlier) to prepare methyl 9-octadecenoate-8,8,13,13,14,14-d_6 and methyl 9-octadecenoate-13,13,14,14-d_4 (DeJarlais and Emken, 1976).

Dry benzene (1 liter) in a 2-liter flask is degassed by evacuating the flask to 100 mm while stirring magnetically for 1 min. Oxygen-free nitrogen is then bled into the flask and stirred for 1 min. This cycle is repeated four times. The flask is connected to a manometric system, evacuated to 100 mm, repressurized to 760 mm with deuterium, and stirred for 10–15 min. This cycle is repeated, and then 10 g of tris(triphenylphosphine)-rhodium(I) chloride is added without stirring. The flask is flushed with deuterium gas and stirred under 1 atm until deuterium uptake stops. The solution at this time should be a pale amber or iced-tea color. (A red color indicates absorption of oxygen by the catalyst.) Next, 75 ml of 1 chloro-4-nonyne is injected into the reaction flask. Deuterium uptake is immediate and rapid.

After 4–5 hr, deuterium uptake ceases and the reaction mixture is diluted with 1 liter of petroleum ether. The solution is applied to a silica gel column (200 g, 5.5 × 30 cm) packed by slurrying with isooctane. The column is eluted with 1 liter of benzene–petroleum ether (1:1). Evaporation of the colorless eluate gave 68 g of product. Mass spectrometry indicated that 93% of the product contained four deuterium atoms per molecule.

2.2.4. *Polyunsaturated Fatty Acids*

Tucker *et al.* (1970) have prepared methyl 9,12-octadecadienoate-11,11-d_2 by coupling 1-bromo-2-octyne-1,1-d_2 with 9-decynoic acid. The deuterium in this synthesis is introduced by reacting the Grignard of an alkyl chloride with perdeuterated paraformaldyde. The synthesis has the advantage of placing the deuterium at the 11 position, which has less biological impact in many systems than placing the deuteriums on the double bond.

Many of the existing synthetic routes used for unlabeled polyunsaturated fat synthesis (Osbond *et al.*, 1961; Christie and Holman, 1967; Steenhoek *et al.*, 1971; Gensler and Bruno, 1963; Kunau 1971*a,b*; Howton and Stein, 1969) can be modified to allow deuterium incorporation.

Table II. Deuterium-Labeled Unsaturated Fatty Acids

Compound	Reference
$CD_3(CD_2)_7CD{=}CD(CD_2)_5CO_2H$ perdeutero-*cis*-7-hexadecenoic acid	Graff *et al.* (1970)
$CD_3(CD_2)_5CD{=}CD(CD_2)_7CO_2H$ perdeutero-*cis*-9-hexadecenoic acid	Graff *et al.* (1970); Wendt and McCloskey (1970)
$CD_3(CD_2)_8CD{=}CDCD_2CD{=}CDCD_2CO_2H$ perdeutero-*cis*-3,*cis*-6-hexadecadienoic acid	Graff *et al.* (1970)
$CD_3(CD_2)_4CD{=}CDCD_2CD{=}CD(CD_2)_5CO_2H$ perdeutero-*cis*-7,*cis*-10-hexadecadienoic acid	Graff *et al.* (1970)
$CD_3CD_2(CD{=}CDCD_2)_3(CD_2)_4CO_2H$ perdeutero-*cis*-7,*cis*-10,*cis*-13-hexadecatrienoic acid	Graff *et al.* (1970)
$CH_3(CH_2)_7CD{=}CD(CH_2)_7CO_2H$ *cis*-9-octadecenoic acid-9,10-d_2	Emken *et al.* (1976)
$CH_3(CH_2)_7CD{=}CD(CH_2)_7CO_2H$ *trans*-9-octadecenoic acid-9,10-d_2	Emken *et al.* (1976)
$CH_3(CH_2)_6CD_2CH{=}CH(CH_2)_7CO_2H$ *cis*-9-octadecenoic acid-11,11-d_2	Tucker *et al.* (1971)
$CH_3(CH_2)_6CD_2CH{=}CHCD_2(CH_2)_6CO_2H$ *cis*-9-octadecenoic acid-8,8,11,11-d_4	DeJarlais and Emken (1976)
$CH_3(CH_2)_6CD_2CH{=}CHCD_2(CH_2)_6CO_2H$ *trans*-9-octadecenoic acid-8,8,11,11-d_4	DeJarlais and Emken (1976)
$CH_3(CH_2)_3CD_2CD_2(CH_2)_2CH{=}CH(CH_2)_7CO_2H$ *cis*-9-octadecenoic acid-13,13,14,14-d_4	DeJarlais and Emken (1976)
$CH_3(CH_2)_3CD_2CD_2(CH_2)_2CH{=}CHCD_2(CH_2)_6CO_2H$ *cis*-9-octadecenoic acid-8,8,13,13,14,14-d_6	DeJarlais and Emken (1976)
$CD_3(CD_2)_7CD{=}CD(CD_2)_7CO_2H$ perdeutero-*cis*-9-octadecenoic acid	Graff *et al.* (1970)
$CD_3(CD_2)_5CD{=}CD(CD_2)_9CO_2H$ perdeutero-*cis*-11-octadecenoic acid	Graff *et al.* (1970)
$CH_3(CH_2)_4CH{=}CHCD_2CH{=}CH(CH_2)_7CO_2H$ *cis*-9,*cis*-12-octadecadienoic acid-11,11-d_2	Tucker *et al.* (1970)
$CD_3(CD_2)_4(CD{=}CDCD_2)_2(CD_2)_6CO_2H$ perdeutero-*cis*-9,*cis*-12-octadecadienoic acid	Graff *et al.* (1970), Wendt and McCloskey (1970)
$CD_3CD_2(CD{=}CDCD_2)_3(CD_2)_7CO_2H$ perdeutero-*cis*-9,*cis*-12,*cis*-15-octadecatrienoic acid	Graff *et al.* (1970)

Deuterium-labeled unsaturated fatty acids synthesized by various routes are listed in Table II.

Table III summarizes general reactions which are useful for deuterium incorporation into fatty acids or precursors used for fatty acid synthesis.

2.2.5. Perdeutero Fatty Acids

Mixtures of perdeutero fatty acids can be extracted from algae grown in D_2O. *Scenedesmus obliques* and *Chlorella vulgaris* are the algae frequently used. Their lipid content can be varied by light conditions, and up to 70% of the dry weight has been reported as lipid (Mangold, 1968). The fatty acid composition varies depending on the light conditions used to grow the algae. Fatty acids available from these algae

Table III. General Synthetic Reactions for Deuterium Labeling

1. Hydrazine (Rohwedder *et al.*, 1967; Dinh-Nguyen, 1968; Koch, 1969*a,b*)

$$R{-}CH{=}CH{-}R' + N_2D_4 \longrightarrow R{-}CHD{-}CHD{-}R' + N_2$$

2. Potassium azodicarboxylate (Koch, 1969*a,b*; Dinh-Nguyen, 1968)

$$KO_2CN{=}NCO_2K + DOAC \longrightarrow CO_2 + N_2D_2 + K^+$$

$$R{-}CH{=}CH{-}R' + N_2D_2 \longrightarrow R{-}CHD{-}CHD{-}R' + N_2$$

3. Electrolysis (Dinh-Nguyen, 1968)

$$RCD_2CO_2H + HO_2C(CH_2)_nCO_2CH_3 \xrightarrow[\text{coupling}]{\text{anodic}\ -2e^-} RCD_2(CH_2)_nCO_2CH_3$$

4. Platinum–deuterium exchange (Thomas, 1971; Dinh-Nguyen *et al.*, 1972; Hsiao *et al.*, 1974)

$$C_nH_xCO_2H \xrightarrow[\text{Pt, NaO}_2]{D_2,\ \text{NaOD,}} C_nD_xCO_2D$$

$$C_nH_xCO_2H \xrightarrow[D_2,\ 195°C]{\text{Pd on carbon}} C_nD_xCO_2D$$

5. Tris(triphenylphosphine)-rhodium(I) chloride (Emken *et al.*, 1976; Birch and Walker, 1966)

$$R{-}C{\equiv}C{-}R' \xrightarrow[\text{Rh(Ph}_3\text{P)}_3\text{Cl}]{D_2,\ \text{RT, 1 atm}} R{-}CD_2{-}CD_2{-}R'$$

6. Pyridine–D_2O exchange (Tucker *et al.*, 1971)

$$RCH_2CHO \xrightarrow[\text{pyridine}]{D_2O} RCD_2CHO$$

7. Lindlar's catalyst (Emken *et al.*, 1976)

$$R{-}C{\equiv}C{-}R' \xrightarrow[\text{quinoline, } D_2\quad \text{RT, 1 atm}]{\text{Lindlar's catalyst}} \underset{cis}{R{-}CD{=}CD{-}R'}$$

8. Copper chromite (Koritala and Selke, 1971)

$$R{-}C{\equiv}C{-}R' \xrightarrow[\text{Cu-Cr catalyst}]{D_2,\ 190°C,\ \text{1 atm}} \underset{cis}{R{-}CD{=}CD{-}R'}$$

9. Chromium carbonyl (Frankel *et al.*, 1969)

$$R{-}CH{=}CH{-}CH{=}CH{-}R' \xrightarrow[\text{Cr(CO)}_3]{D_2} R{-}CHD{-}\underset{cis}{CH{=}CH}{-}CHD{-}R'$$

10. Disiamylborane (Rohwedder *et al.*, 1967; Thomas, 1971)

$$R{-}C{\equiv}C{-}R' \xrightarrow[\text{disiamylborane}]{} \xrightarrow{CH_3CO_2D} \underset{cis}{R{-}CH{=}CD{-}R'} + \underset{cis}{R{-}CD{=}CH{-}R'}$$

11. Hydrogenolysis of alkyl halides

$RX \xrightarrow[\text{DOAC}]{\text{Zn}} RD$ (Craig and Fowler, 1961; Kovachic and Leitch, 1961)

$RX \xrightarrow{R_3SnD} RD$ (Kuhlein *et al.*, 1968)

$RCH{=}CHCH(Br)R \xrightarrow[\text{2. } D_2O]{\text{1. Mg}} RCH{=}CHCHD{-}R$ (Friedrich and Thieme, 1973)

Table III—Continued

12. *Para*-perdeuteroformaldehyde addition (Tucker *et al.*, 1970)

$$\mathrm{RX} \xrightarrow{\mathrm{Mg^{\circ}}} \mathrm{RMgX} \xrightarrow{(\mathrm{CD_2O})_n} \mathrm{RCD_2OH}$$

13. Zinc–chromyl chloride reduction (Castro and Stephens, 1964)

$$\mathrm{R{-}C{\equiv}C{-}R'} \xrightarrow[\mathrm{DCl,\ D_2O}]{\mathrm{CrCl_2,\ Zn}} \underset{trans}{\mathrm{R{-}CD{=}CD{-}R'}}$$

14. Zinc–cuperic sulfate reduction (Nicholas and Carroll, 1968)

$$\mathrm{DC{\equiv}CD} \xrightarrow[\mathrm{DCl}]{\mathrm{CuSO_4,\ Zn}} \underset{cis}{\mathrm{DCH{=}HCD}}$$

15. Desulfurization of thioketals with nickel (Thomas, 1971; Tokes *et al.*, 1968)

$$\mathrm{R{-}\underset{\underset{O}{\|}}{C}{-}R'} \xrightarrow[\text{Ethyl dithiol}]{} \xrightarrow[\mathrm{Ni}]{\mathrm{D_2}} \mathrm{R{-}CD_2R'}$$

16. Lithium aluminum hydride reduction (Magoon and Slaugh, 1967; Slaugh, 1966)

$$\mathrm{R{-}C{\equiv}C{-}R' + LiAlH_4} \xrightarrow[\mathrm{THF,\ 190^{\circ}C}]{\mathrm{800\ psi\ D_2}} \underset{trans}{\mathrm{R{-}CH{=}CD{-}R'}} + \underset{trans}{\mathrm{R{-}CD{=}CH{-}R'}}$$

$$\mathrm{R{-}C{\equiv}CR' + LiAlH_4} \xrightarrow[\mathrm{Reflux}]{\mathrm{THF/Diglyme}} \xrightarrow[\mathrm{D_2O}]{} \underset{trans}{\mathrm{R{-}CH{=}CD{-}R'}} + \underset{trans}{\mathrm{R{-}CD{=}CH{-}R'}}$$

17. Lithium diethylaluminum hydride (Eisch and Kaska, 1966)

$$\mathrm{R{-}C{\equiv}C{-}R' + Li(Et)_2AlH} \longrightarrow \xrightarrow{\mathrm{D_2O}} \underset{cis}{\mathrm{R{-}CH{=}CD{-}R'}} + \underset{cis}{\mathrm{R{-}CD{=}CH{-}R'}}$$

include 16:0, 16:1, 16:2, 16:3, 18:0, 18:1, 18:2, and 18:3 (see Table II). Perdeutero fatty acids are available commercially from Merck, Sharp, and Dohme of Canada, Ltd. Crespi *et al.* (1960) have described the methodology for growing perdeutero algae, and .Graff *et al.* (1970) have characterized the perdeutero fatty acids by gas chromatography, infrared spectrometry, NMR spectrometry, and mass spectrometry.

The basic limitations to using perdeutero fatty acids are their cost ($235/g stearic-$d_{35}$ acid) and the isotope effects which occur in biological systems, especially when the studies involve enzymatic reactions. For the occasional user of perdeuterated fats, their preparation by biosynthesis is best left to the specialist in this area.

Perdeutero saturated fatty acids have also been synthetically prepared by H–D exchange using D_2 and palladium on charcoal at 195°C (Hsiao *et al.*, 1974) or by using D_2, platinum, sodium peroxide, and sodium deuterioxide at 240°C (Dinh-Nguyen *et al.*, 1972). These exchange methods are necessarily limited to preparation of saturated fatty acids and are an outgrowth of earlier hydrocarbon exchange methods. The methods are simple enough for the average laboratory to use for preparing perdeutero saturated fatty acids.

Table II lists a variety of deuterated unsaturated fatty acids prepared by chemical and biochemical synthesis.

2.2.6. ^{13}C-Labeled Fatty Acids

^{13}C-labeled fatty acids can be synthesized using essentially the same synthetic routes which have been established for ^{14}C-labeled fatty acids. (^{14}C syntheses are reviewed in Section 2.3.1 of this chapter.) A ^{13}C isotope source is simply substituted for the ^{14}C-containing reagent, which is usually $^{14}CO_2$, $K^{14}CN$, $^{14}CH_3I$, or other simple molecules. The development of ^{13}C as tracer has been limited because of the high cost of reagents containing high isotope enrichment. At the present time, 90% oleic-1-^{13}C acid costs about $700/g. Recent developments in ^{13}C technology, particularly at the Los Alamos Scientific Laboratory, have substantially lowered the cost of ^{13}C-enriched material, and, in addition, 95–99% enriched material is now available. These developments are predicted to spur the use of ^{13}C-labeled compounds in biological and biochemical tracer experiments.

The increasing availability of ^{13}C NMR instrumentation has renewed interest in ^{13}C as a means for determining structure, conformation, and environmental factors

Table IV. ^{13}C-Labeled Fatty Acids

Compound	Reference
$CH_3(CH_2)_{15}{}^{13}CO_2H$ heptadecanoic acid-1-^{13}C	Dinh-Nguyen (1968)
$CH_3(CH_2)_{15}{}^{13}CH_2CO_2H$ octadecanoic acid-2-^{13}C	Dinh-Nguyen (1968)
$CH_3(CH_2)_{15}{}^{13}CH_2CH_2CO_2H$ nonadecanoic acid-3-^{13}C	Dinh-Nguyen (1968)
$^{13}CH_3({}^{13}CH_2)_{14}{}^{13}CO_2H$ hexadecanoic acid-U-^{13}C	Matwiyoff and Ott (1973)
$^{13}CH_3({}^{13}CH_2)_5{}^{13}CH{=}{}^{13}CH({}^{13}CH_2)_7{}^{13}CO_2H$ *cis*-9-hexadecenoic acid-U-^{13}C	Matwiyoff and Ott (1973)
$^{13}CH_3({}^{13}CH_2)_2({}^{13}CH{=}{}^{13}CH{}^{13}CH_2)_2({}^{13}CH_2)_6{}^{13}CO_2H$ *cis*-9,*cis*-12-hexadecadienoic acid-U-^{13}C	Matwiyoff and Ott (1973)
$^{13}CH_3({}^{13}CH_2)_{16}{}^{13}CO_2H$ octadecanoic acid-U-^{13}C	Matwiyoff and Ott (1973)
$^{13}CH_3({}^{13}CH_2)_7{}^{13}CH{=}{}^{13}CH({}^{13}CH_2)_7{}^{13}CO_2H$ *cis*-9-octadecenoic acid-U-^{13}C	Matwiyoff and Ott (1973)
$^{13}CH_3({}^{13}CH_2)_4({}^{13}CH{=}{}^{13}CH{}^{13}CH_2)_2({}^{13}CH_2)_6{}^{13}CO_2H$ *cis*-9,*cis*-12-octadecadienoic acid-U-^{13}C	Matwiyoff and Ott (1973)
$^{13}CH_3{}^{13}CH_2({}^{13}CH{=}{}^{13}CH{}^{13}CH_2)_3({}^{13}CH_2)_6{}^{13}CO_2H$ *cis*-9,*cis*-12,*cis*-15-octadecatrienoic acid-U-^{13}C	Matwiyoff and Ott (1973)
$CH_3({}^{13}CH_2CH_2)_6{}^{13}CO_2H$ tetradecanoic acid-^{13}C	Cronan and Batchelor (1973)
$CH_3{}^{13}CH_2(CH_2{}^{13}CH_2)_2CH{=}{}^{13}CHCH_2({}^{13}CH_2CH_2)_2{}^{13}CO_2H$ 7-tetradecenoic acid-^{13}C	Cronan and Batchelor (1973)
$CH_3({}^{13}CH_2CH_2)_7{}^{13}CO_2H$ hexadecanoic acid-^{13}C	Cronan and Batchelor (1973)
$CH_3{}^{13}CH_2(CH_2{}^{13}CH_2)_2CH{=}{}^{13}CHCH_2({}^{13}CH_2CH_2)_3{}^{13}CO_2H$ 9-hexadecenoic acid-^{13}C	Cronan and Batchelor (1973)
$CH_3{}^{13}CH_2(CH_2{}^{13}CH_2)_2CH{=}{}^{13}CHCH_2({}^{13}CH_2CH_2)_4{}^{13}CO_2H$ 11-octadecenoic acid-^{13}C	Cronan and Batchelor (1973)

Table V. ^{14}C-Labeled Fatty Acids

Compound	Reference
$CH_3(CH_2)_{10}{}^{14}CO_2H$ dodecanoic acid-1-^{14}C	Snyder and Piantadosi (1966)
$CH_3(CH_2)_{12}{}^{14}CO_2H$ tetradecanoic acid-1-^{14}C	Snyder and Piantadosi (1966)
$CH_3(CH_2)_9{}^{14}CH_2(CH_2)_4CO_2H$ hexadecanoic acid-6-^{14}C	Snyder and Piantadosi (1966)
$CH_3(CH_2)_4{}^{14}CH_2(CH_2)_9CO_2H$ hexadecanoic acid-11-^{14}C	Snyder and Piantadosi (1966)
$CH_3(CH_2)_{16}{}^{14}CO_2H$ octadecanoic acid-^{14}C	Snyder and Piantadosi (1966)
$CH_3(CH_2)_{11}{}^{14}CH_2(CH_2)_4CO_2H$ octadecanoic acid-6-^{14}C	Snyder and Piantadosi (1966)
$CH_3(CH_2)_2(CH{=}CHCH_2)_2(CH_2)_6{}^{14}CO_2H$ 9,12-hexadecadienoic acid-1-^{14}C	Marcel and Holman (1968)
$CH_3(CH_2)_3(CH{=}CHCH_2)_2(CH_2)_5{}^{14}CO_2H$ 8,11-hexadecadienoic acid-1-^{14}C	Marcel and Holman (1968)
$CH_3(CH_2)_3(CH{=}CHCH_2)_2(CH_2)_6{}^{14}CO_2H$ 9,12-heptadecadienoic acid-1-^{14}C	Marcel and Holman (1968)
$CH_3(CH_2)_3(CH{=}CHCH_2)_3CH_2{}^{14}CO_2H$ 4,7,10-pentadecatrienoic acid-1-^{14}C	Stoffel (1965)
$CH_3(CH_2)_4(CH{=}CHCH_2)_2(CH_2)_4{}^{14}CO_2H$ 7,10-hexadecadienoic acid-1-^{14}C	Stoffel (1965)
$CH_3(CH_2)_7CH{=}CH(CH_2)_7{}^{14}CO_2H$ 9-octadecenoic acid-1-^{14}C	Stoffel (1965)
$CH_3(CH_2)_5CH{=}CH(CH_2)_9{}^{14}CO_2H$ *trans*-11-octadecenoic acid-1-^{14}C	Christie *et al.* (1973)
$CH_3(CH_2)_7CH{=}{}^{14}CH(CH_2)_7CO_2H$ *cis*-9-octadecenoic acid-9-^{14}C	Barley *et al.* (1973)
$^{14}CH_3(CH_2)_7CH{=}CH(CH_2)_7CO_2H$ *cis*-9-octadecenoic acid-18-^{14}C	Pichat *et al.* (1969)
$CH_3(CH_2)_3(CH{=}CHCH_2)_2(CH_2)_7{}^{14}CO_2H$ 10,13-octadecadienoic acid-1-^{14}C	Marcel and Holman (1968)
$CH_3(CH_2)_5(CH{=}CHCH_2)_2(CH_2)_5{}^{14}CO_2H$ 8,11-octadecadienoic acid-1-^{14}C	Marcel and Holman (1968), Budny and Sprecher (1971)
$CH_3(CH_2)_6(CH{=}CHCH_2)_2(CH_2)_4{}^{14}CO_2H$ 7,10-octadecadienoic acid-1-^{14}C	Marcel and Holman (1968)
$CH_3(CH_2)_7(CH{=}CHCH_2)_2(CH_2)_3{}^{14}CO_2H$ 6,9-octadecadienoic acid-1-^{14}C	Marcel and Holman (1968)
$CH_3(CH_2)_4(CH{=}CHCH_2)_2(CH_2)_6{}^{14}CO_2H$ 9,12-octadecadienoic acid-1-^{14}C	Stoffel (1964), Osbond (1966)
$CH_3(CH_2)_4CH{=}CHCH_2CH{=}{}^{14}CH(CH_2)_7CO_2H$ 9,12-octadecadienoic acid-9-^{14}C	Barley *et al.* (1973)
$CH_3(CH_2)_4C{\equiv}CCH_2CH{=}{}^{14}CH(CH_2)_7CO_2H$ 9-octadecene-12-ynoic acid-9-^{14}C	Barley *et al.* (1973)
$CH_3CH_2(CH{=}CHCH_2)_3(CH_2)_6{}^{14}CO_2H$ 9,12,15-octadecatrienoic acid-1-^{14}C	Stoffel (1964), Osbond (1966)
$CH_3(CH_2)_4(CH{=}CHCH_2)_3(CH_2)_3{}^{14}CO_2H$ 6,9,12-octadecatrienoic acid-1-^{14}C	Stoffel (1964), Osbond (1966), Marcel and Holman (1968)

(*Continued*)

Table V—Continued

$CH_3(CH_2)_2(CH{=}CHCH_2)_3(CH_2)_5{}^{14}CO_2H$ 8,11,14-octadecatrienoic acid-1-^{14}C	Stoffel (1964)
$CH_3CH_2(CH{=}CHCH_2)_4(CH_2)_2{}^{14}CH_2CO_2H$ 6,9,12,15-octadecatetraenoic acid-2-^{14}C	Osbond (1966)
$CH_3CH_2(CH{=}CHCH_2)_4(CH_2)_3{}^{14}CO_2H$ 6,9,12,15-octadecatetraenoic acid-1-^{14}C	Stoffel (1965)
$CH_3(CH_2)_3(CH{=}CHCH_2)_2(CH_2)_8{}^{14}CO_2H$ 11,14-nonadecadienoic acid-1-^{14}C	Marcel and Holman (1968)
$CH_3(CH_2)_4(CH{=}CHCH_2)_2(CH_2)_7{}^{14}CO_2H$ 10,13-nonadecadienoic acid-1-^{14}C	Marcel and Holman (1968)
$CH_3(CH_2)_5(CH{=}CHCH_2)_2(CH_2)_6{}^{14}CO_2H$ 9,12-nonadecadienoic acid-1-^{14}C	Marcel and Holman (1968)
$CH_3(CH_2)_4(CH{=}CHCH_2)_2(CH_2)_8{}^{14}CO_2H$ 11,14-eicosadienoic acid-1-^{14}C	Stoffel (1965), Budny and Sprecher (1971)
$CH_3(CH_2)_4(CH{=}CHCH_2)_2(CH_2)_7{}^{14}CH_2CO_2H$ 11,14-eicosadienoic acid-2-^{14}C	Osbond (1966)
$CH_3(CH_2)_4(CH{=}CHCH_2)_2(CH_2)_6{}^{14}CH_2CH_2CO_2H$ 11,14-eicosadienoic acid-3-^{14}C	Osbond (1966)
$CH_3(CH_2)_5(CH{=}CHCH_2)_2(CH_2)_7{}^{14}CO_2H$ 10,13-eicosadienoic acid-1-^{14}C	Budny and Sprecher (1971)
$CH_3(CH_2)_5(CH{=}CHCH_2)_2(CH_2)_5{}^{14}CH_2CH_2CO_2H$ 10,13-eicosadienoic acid-3-^{14}C	Budny and Sprecher (1971)
$CH_3(CH_2)_4(CH{=}CHCH_2)_3(CH_2)_5{}^{14}CO_2H$ 8,11,14-eicosatrienoic acid-1-^{14}C	Stoffel (1964), Marcel and Holman (1968), Budny and Sprecher (1971)
$CH_3(CH_2)_5(CH{=}CHCH_2)_3(CH_2)_4{}^{14}CO_2H$ 7,10,13-eicosatrienoic acid-1-^{14}C	Budny and Sprecher (1971)
$CH_3(CH_2)_4(CH{=}CHCH_2)_3(CH_2)_3{}^{14}CH_2{}^{14}CH_2CO_2H$ 8,11,14-eicosatrienoic acid-2,3-^{14}C	Osbond (1966)
$CH_3(CH_2)_4(CH{=}CHCH_2)_3(CH_2)_4{}^{14}CH_2CO_2H$ 8,11,14-eicosatrienoic acid-2-^{14}C	Osbond (1966)
$CH_3CH_2(CH{=}CHCH_2)_3(CH_2)_7{}^{14}CH_2CO_2H$ 11,14,17-eicosatrienoic acid-2-^{14}C	Osbond (1966)
$CH_3(CH_2)_4(CH{=}CHCH_2)_4(CH_2)_2{}^{14}CO_2H$ 5,8,11,14-eicosatetraenoic acid-1-^{14}C	Stoffel (1964), Osbond (1966), Sprecher (1971)
$CH_3CH_2(CH{=}CHCH_2)_4(CH_2)_5{}^{14}CO_2H$ 8,11,14,17-eicosatetraenoic acid-1-^{14}C	Stoffel (1965)
$CH_3CH_2(CH{=}CHCH_2)_4(CH_2)_4{}^{14}CH_2CO_2H$ 8,11,14,17-eicosatetraenoic acid-1-^{14}C	Osbond (1966)
$CH_3(CH_2)_2(CH{=}CHCH_2)_4(CH_2)_2{}^{14}CH_2CH_2CO_2H$ 7,10,13,16-eicosatetraenoic acid-1-^{14}C	Sprecher (1971)
$CH_3CH_2(CH{=}CHCH_2)_5(CH_2)_3{}^{14}CH_2CO_2H$ 7,10,13,16,19-docosapentaenoic acid-2-^{14}C	Osbond (1966)

affecting molecular configuration and associations such as occur in lipid membrane structures.

The advantage of the ^{13}C isotope is that it is less likely to be involved in isotope effects than tritium or deuterium. There is little chance that the ^{13}C label will be lost due to exchange, which can be a problem with tritium and deuterium isotopes.

Biosynthesis is the preferred method for the preparation of uniformly labeled complex molecules. Uniformly labeled ^{13}C fatty acids can be obtained by extracting the lipids from algae or photosynthetic bacteria which have been grown in a $^{13}CO_2$ atmosphere (Matwiyoff and Ott, 1973). This type of preparation is similar to that described for the biosynthesis of perdeuterated fatty acids but suffers from the same problems of low lipid content in algae unless light conditions are controlled (Mangold, 1968). As with perdeuterated fatty acids, the biosynthetic preparation of ^{13}C-labeled fatty acids is best left to the specialist in the field.

Escherichia coli can be grown on acetate-1-^{13}C, but only about 10% of the ^{13}C is incorporated into lipid material (Roberts *et al.*, 1963). However, Cronan and Batchelor (1973) have reported a mutant of *E. coli* which incorporated 70–95% of the added acetate-1-^{13}C into long-chain fatty acids.

The fatty acids produced by incorporation of acetate-1-^{13}C using *E. coli* are not uniformly labeled but instead have every other carbon labeled with ^{13}C, which is an advantage for ^{13}C NMR analysis.

A list of some ^{13}C-labeled fatty acids which have been reported is given in Table IV. The ^{14}C-labeled fatty acids listed in Table V can obviously be synthesized with a ^{13}C label, but more care must be taken to ensure high isotopic purity.

2.3. Synthesis of Radioisotope-Labeled Fatty Acids

A much larger variety of ^{14}C- and ^{3}H-labeled fatty acids have been synthesized than of deuterated fatty acids. Three basic reasons for this difference are (1) ease of synthesis and analysis of compounds, (2) availability and sensitivity of analytical equipment and methods, and (3) cost of analytical equipment.

^{14}C- and ^{3}H-labeled fatty acids are easier to prepare than deuterated fatty acids since the preparations do not need to have high isotopic purity. For example, it makes little difference if only part of the fatty acid molecules are tritiated with one, two, or three tritium atoms per molecule, but a deuterated fatty acid preparation containing a gaussian scatter of deuterium atoms is difficult to analyze quantitatively.

Most ^{14}C-labeled fatty acids can be synthesized from variations of three basic reactions: (1) carboxylation with $^{14}CO_2$, (2) addition of CN^- to an alkyl bromide, and (3) coupling of labeled intermediates by various methods such as the Wittig reaction.

The cost of automated commercial scintillation counters is much less than that of a mass spectrometer–computer system, and methods for sample preparation and counting are simpler and require less time for radioisotope-labeled fatty acids. In addition, less radioisotope-labeled fatty acid is required for biological experiments because of the great sensitivity of the radioisotope tracer methods.

These factors have all contributed to a greater use of radioisotope-labeled fats, and, consequently, more radioisotope-labeled fatty acid preparations have been reported.

2.3.1. ^{14}C-Labeled Fatty Acids

The commercial availability of many common 1-carboxyl-labeled saturated and unsaturated fatty acids makes a review of their specific synthesis unnecessary. A few commercial suppliers of 1-carboxyl-^{14}C-labeled fatty acids and other radiochemicals including tritiated fatty acids are New England Nuclear Corporation, Boston, Massachusetts 02118; ICN Chemical and Radioisotope Division, Irvine, California 92664; Amersham/Searle Corp., Arlington Heights, Illinois 60005; Applied Science Laboratories, Inc., State College, Pennsylvania 16801; and Dhom Products, Ltd., North Hollywood, California 91601.

Unusual fatty acids, such as those having an odd-numbered carbon chain or unsaturated positional isomer, cannot normally be purchased and must be synthesized.

The synthesis of carboxyl-labeled saturated fatty acids from the reaction of the appropriate alkyl halide with $K^{14}CN$ or $Na^{14}CN$ followed by hydrolysis with aqueous ethanolic sodium hydroxide is well known. Similarly, the carboxylation of the Grignard of alkyl halides with $^{14}CO_2$ is a classical synthesis of carboxyl-labeled fatty acids. The appropriate halide can be obtained from the unlabeled fatty acid by the Hunsdiecker decarboxylation or similar reaction. Osbond (1966) has reviewed the synthesis of ^{14}C-labeled fatty acids by this synthetic sequence. If the unusual saturated fatty acid is not available, then classical organic synthetic methods are available for synthesizing the alkyl halide.

The preparation of ^{14}C-labeled unsaturated fatty acids is more of a problem, but, again, if the unlabeled fatty acid is available it can be decarboxylated by electrolysis to the terminal alkyl alcohol (Gunstone *et al.*, 1974). The 1-alkyl bromide can then be prepared by reacting with $Ph_3P{-}Br_2$ or some similar method. The unsaturated ^{14}C-labeled fatty acid is prepared by reacting the Grignard of 1-alkyl bromide with $K^{14}CN$ or $^{14}CO_2$.

A similar approach used by Howton *et al.* (1952) is to first brominate the unsaturated fatty acid, then decarboxylate via the Hunsdiecker degradation reaction. The unsaturated 1-bromoalkene is regenerated with zinc, and it is then carboxylated with $^{14}CO_2$ by a Grignard reaction. The problem with this sequence is that up to 25% of the product can contain isomerized double bonds formed during debromination.

The synthesis of several *cis*-1-^{14}C-labeled polyunsaturated fatty acids has been reported by Stoffel (1964, 1965). Carboxyl-labeled *cis* polyunsaturated fatty acids which have been synthesized are summarized in Table V.

The general synthetic sequence used is shown below:

$$\mathrm{R{-}(C{\equiv}C{-}CH_2)_m{-}Br + BrMg{-}C{\equiv}C{-}(CH_2)_n{-}Cl} \xrightarrow[\text{THF or } CH_2Cl_2]{\text{CuCN}}$$

$$\mathrm{R{-}(C{\equiv}C{-}CH_2)_{m+1}{-}(CH_2)_{n-1}{-}Cl} \xrightarrow[\text{Lindlar's catalyst}]{H_2}$$

$$\mathrm{R{-}(CH{=}CH{-}CH_2)_{m+1}{-}(CH_2)_{n-1}{-}Cl} \xrightarrow[(CH_3)_2SO]{Na^{14}CN}$$

$$\mathrm{R{-}(CH{=}CH{-}CH_2)_{m+1}{-}(CH_2)_{n-1}{-}{}^{14}CN} \xrightarrow[CH_3OH]{H^+}$$

$$\mathrm{R{-}(CH{=}CH{-}CH_2)_{m+1}{-}(CH_2)_{n-1}{-}{}^{14}CO_2CH_3}$$

The acetylenic starting materials are chosen to give the desired chain length and double-bond placement.

Marcel and Holman (1968) have prepared ^{14}C-carboxyl-labeled polyunsaturated fatty acids by first using the acetylenic route of Osbond *et al.* (1961) to form the desired polyunsaturated fatty acid having one less carbon. The unlabeled fatty acid is then converted to its mesylate, which is reacted with $K^{14}CN$. Methanolysis of the alkenyl cyanide results in the desired ^{14}C-labeled *cis* polyunsaturated fatty acids. No *cis-trans* isomerization, no double-bond migration, less difficulty in the addition of KCN, and better conversion of the nitrile to the carboxylic acid are cited as advantages of this procedure.

The following preparation of 6,9,12-octadecatrienoic acid-1-^{14}C (Marcel and Holman, 1968) is typical of many other syntheses which utilize $K^{14}CN$ in the preparation of 1-^{14}C-alkyldienoic and -alkyltrienoic acids.

Heptadeca-5,8,11-triynoic acid is first prepared by the coupling of the di-Grignard of ω-hexynoic acid with undeca-2,5-diynyl bromide using cuprous cyanide. The heptadecatriynoic acid is then esterified and hydrogenated over Lindlar's catalyst to methyl heptadeca-5,8,11-trienoate. The methyl ester is reduced with $LiAlH_4$. The resulting heptadeca-5,8,11-trienol (10.3 g) is dissolved in 0.75 ml of pyridine, and methanesulfonyl chloride (7.5 g) is added. The mixture is stirred for 5 hr. After extraction, 13.2 g of heptadeca-5,8,11-trienyl mesylate is isolated. Next, a mixture of KCN and $K^{14}CN$ (10 mmol, 2 mCi) is dissolved in 15 ml of water and added to 1.68 g of the heptadeca-5,8,11-trienyl mesylate in 60 ml ethanol. The solution is refluxed for 16 hr and the ethanol is evaporated. After extraction of the residue with diethyl ether, recovery of 1.2 g of heptadeca-5,8,11-trienyl nitrile-1-^{14}C is obtained. The nitrile (100 mg) is dissolved at 0°C in 20 ml HCl–methanol (25 : 75) and stirred for 2 hr. Water (0.2 ml) is added, and the mixture is allowed to stand at room temperature for 16 hr. The methyl esters are extracted with hexane, washed, dried, and chromatographed on activated magnesium silicate. Recovered methyl esters are saponified with 4% KOH in 10% aqueous methanol by refluxing 40 min under nitrogen. The 6,9,12-octadecatrienoic-1-^{14}C acid (specific activity 0.17 mCi/mM) is extracted with diethyl ether and dried. If necessary, further purification can be carried out with silica gel chromatography.

Sprecher (1971) and Budny and Sprecher (1971) have used the general procedures of Osbond *et al.* (1961) and Stoffel (1964) for the synthesis of eight all-*cis* polyunsaturated ^{14}C-labeled fatty acids (see Table V).

The Wittig reaction has been used to couple phosphoranes with ^{14}C-labeled methyl azelaldehydate to give methyl oleate-9-^{14}C, methyl linoleate-9-^{14}C, and methyl crepenynate-9-^{14}C. Barley *et al.* (1973) used the following sequence for synthesizing methyl crepenynate-9-^{14}C:

$$CH_3(CH_2)_4C{\equiv}C{-}CH_2CH_2Br \xrightarrow{(PH_3)P} CH_3(CH_2)_4C{\equiv}C(CH_2)_2P^+(Ph)_3Br^-$$

$$\xrightarrow{n\text{-butyl lithium}} CH_3(CH_2)_4C{\equiv}CCH_2CH{=}P(Ph)_3 \xrightarrow[OH^{14}C(CH_2)_7CO_2Me]{}$$

$$CH_3(CH_2)_4C{\equiv}CCH_2CH{=}{}^{14}CH(CH_2)_7CO_2Me$$

Methyl linoleate-9-^{14}C is prepared from methyl crepenynate-9-^{14}C by reducing the triple bond with hydrogen and Lindlar's catalyst.

Methyl oleate-18-^{14}C has been prepared by Pichat *et al.* (1969) according to the following reaction sequence, which also uses the Wittig reaction:

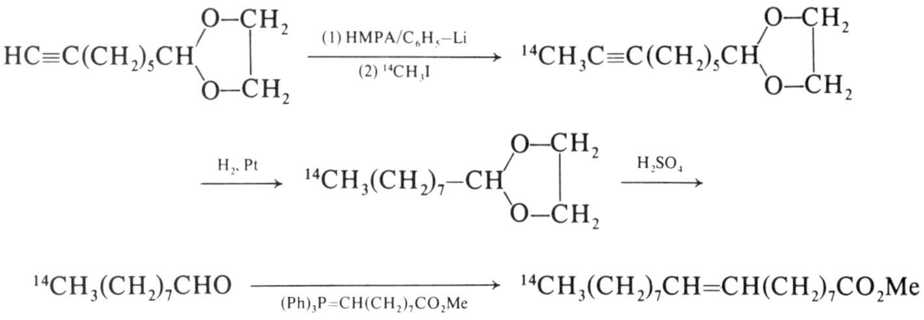

$$HC{\equiv}C(CH_2)_5CH\langle\begin{matrix}O-CH_2\\ |\\ O-CH_2\end{matrix} \xrightarrow[(2)\ ^{14}CH_3I]{(1)\ HMPA/C_6H_5-Li} {}^{14}CH_3C{\equiv}C(CH_2)_5CH\langle\begin{matrix}O-CH_2\\ |\\ O-CH_2\end{matrix}$$

$$\xrightarrow{H_2,\ Pt} {}^{14}CH_3(CH_2)_7-CH\langle\begin{matrix}O-CH_2\\ |\\ O-CH_2\end{matrix} \xrightarrow{H_2SO_4}$$

$$^{14}CH_3(CH_2)_7CHO \xrightarrow[(Ph)_3P=CH(CH_2)_7CO_2Me]{} {}^{14}CH_3(CH_2)_7CH{=}CH(CH_2)_7CO_2Me$$

HMPA = hexamethylphosphoramide

An example of the experimental details for a Wittig reaction has previously been given for the preparation of methyl oleate-d$_4$. Since the stereospecificity of the Wittig reaction is not absolute, all products prepared by this reaction contain *trans* isomers and must be purified by argentation chromatography (Emken *et al.*, 1964).

Some of the ^{14}C-labeled fatty acids which have been synthesized are listed in Table V. This list is not complete, but it does cover a variety of labeled fatty acids prepared by representative methods.

2.3.2. ^{3}H-Labeled Fatty Acids

The Wilzbach gas exposure technique as used by Nystrom *et al.* (1959) involves exposure of fatty acids to tritium gas and is a historic development in the labeling of saturated fatty acids. It was also used for the synthesis of tritium-labeled oleic acid by Dutton *et al.* (1962). This method results in both the exchange of tritium for hydrogen on the double bonds and the addition of tritium to double or triple bonds.

2.3.2.1. *Stearate-^{3}H by Wilzbach Technique*

Nystrom *et al.* (1959) have labeled methyl palmitate and methyl stearate by the Wilzbach exposure technique. Methyl stearate (1 g) is sealed in a 1 × 10 cm reaction tube with about 1 Ci of tritium gas. The tube is warmed to melt the methyl stearate and then rotated to deposit a thin film of the fatty ester on the wall of the tube. Exposure of the methyl stearate to the tritium source for 13 days results in the incorporation of 7.1 mCi of tritium.

The product is purified by saponification and extraction of the unsaponifiables. Labile tritium is removed by distilling a total of 1.5 liters of anhydrous ethanol from the soaps in 50-ml portions. The soaps are then acidified and extracted with diethyl ether.

Methyl palmitate, methyl laurate, and methyl myristate have also been labeled by this procedure. The methyl esters were free of radiation damage and decomposition products.

Diimide reduction of unsaturated fatty acids to form tritiated methyl stearate has been reported by Koch (1969*a*, *b*). In this method, tritiated water is used to obtain tritiated hydrazine or tritiated potassium azodicarboxylate. These reagents are then used to reduce oleic acid to tritiated stearic acid.

2.3.2.2. Preparation of Stearic-9(10)-^{3}H Acid by Diimide Reduction

Methyl oleate (100 mg) is dissolved in 6 ml of dry pyridine and heated to 80–100°C under nitrogen. Potassium azodicarboxylate (0.67 mmol) is mixed with 80 ml of acetic anhydride and 12 μl tritiated water (specific activity 2.63 mCi/mmol) in 3 ml of dry pyridine. One milliliter of the diimide solution is added dropwise every 2.5 hr to the methyl oleate solution. The solvent is evaporated from the reaction mixture, and the ester is saponified with 4 ml of 10% KOH in 10% aqueous methanol. Water is added, and the mixture is extracted with hexane and then acidified with 1 N HCl. The fatty acid is extracted with diethyl ether, washed, and dried. Recovery of a total of 85 mg of stearic-9(10)-^{3}H acid has been obtained, with a specific activity of 0.547 mCi/mmol. If necessary, the stearic-9(10)-^{3}H acid can be further purified by usual methods to remove unhydrogenated oleic acid.

Morris (1967) has reported the synthesis of 9-octadecenoic acid-*erythro*-12,13-^{3}H by using tritiated hydrazine to partially reduce linoleic acid and then separating the 9 and 12 positional monoene isomers by TLC. The 9-octadecenoic acid-12L-^{3}H and 9-octadecenoic acid-12D-^{3}H have been prepared by an extension of Schroepfer and Block's (1965) procedure.

Tritium adds to the acetylenic bond in methyl stearolate to form *cis*- and *trans*-octadecenoate-9,10-^{3}H plus small amounts of tritiated methyl stearate (Dutton *et al.*, 1962).

Mounts *et al.* (1971) have reported a convenient method of preparing methyl *cis*-9-octadecenoate-9,10-^{3}H by the reduction of methyl stearolate in the presence of a copper-chromite catalyst, tritiated water, and hydrogen gas. This procedure avoids the use of tritium gas and the accompanying manometric equipment. This technique has been utilized to label a series of positional octadecenoic acids by reduction of the corresponding octadecynoic acids (Mounts, 1976).

2.3.2.3. Reduction of Alkynoates with Copper-Chromite

Copper-chromite catalyst (100 mg) is stirred with 16 μl of tritiated water and 1 atm hydrogen gas for 1 hr at 195°C. Methyl stearolate (1.0 g) is injected into the reaction flask and stirred for 2 hr at 1 atm hydrogen and 195°C. Koritala and Selke (1971) have demonstrated that no isomerization of methyl oleate or double-bond migration occurs at 150°C; however, 25% *trans* isomerization occurred in 3 hr at 200°C. The reaction mixture was filtered and 0.95 g of product was recovered.

Reductive ozonolysis has demonstrated no double-bond migration, and IR analysis indicated less than 1% *trans* isomer. Radiochemical analysis showed that the product had a specific activity of 5.27 mCi/mmol with 95% of the tritium located at the 9,10 position.

Methyl oleate-10-^{3}H and methyl linoleate-10-^{3}H have been prepared by using the Wittig reaction to couple the appropriate tritiated nonyl-1-^{3}H-phosphorane with methyl azelaldehydate (Barley *et al.*, 1973). The reaction sequence is similar to the one given earlier for the synthesis of methyl crepenynate-9-^{14}C. The major difference is the use of unlabeled methyl azelaldehydate and the use of the appropriate nonyl-^{3}H-phosphorane.

A synthesis for *cis* polyunsaturated tritium-labeled fatty acid has been reported by Stoffel (1965). An example of the overall synthetic sequence for methyl 6,9,12-linolenate-^{3}H follows:

$$HC{\equiv}C(CH_2)_3OH \xrightarrow[PtO_2]{^3H_2} \xrightarrow[PBr_3]{} {^3H_2}CHC{^3H_2}(CH_2)_3Br \xrightarrow[2.\ hydrolysis]{1.\ \text{(tetrahydropyranyl)}O{-}CH_2C{\equiv}CLi}$$

$${^3H_2}CHC{^3H_2}(CH_2)_3C{\equiv}CCH_2OH \xrightarrow[PBr_3]{} {^3H_2}CHC{^3H_2}(CH_2)_3C{\equiv}CCH_2Br$$

$$\xrightarrow[2.\ HC{\equiv}C{-}CH_2OH]{1.\ Mg^{\circ}} \xrightarrow{CuCN} {^3H_2}CHC{^3H_2}(CH_2)_3C{\equiv}CCH_2C{\equiv}CCH_2OH \xrightarrow{PBr_3}$$

$$\xrightarrow[2.\ CuCN]{1.\ BrMgC{\equiv}C(CH_2)_3CO_2MgBr} {^3H_2}CHC{^3H_2}(CH_2)_3(C{\equiv}C{-}CH_2)_3(CH_2)_3CO_2H$$

$$\xrightarrow[2.\ 5\%\ HCl\text{-}MeOH]{1.\ \text{Lindlar's catalyst},\ H_2} \underset{\text{methyl 6,9,12-linolenate-}^3\text{H}}{{^3H_2}CHC{^3H_2}(CH_2)_3(CH{=}CH{-}CH_2)_3(CH_2)_3CO_2CH_3}$$

Tritiated fatty acids which have been synthesized by this route are 9,12-octadecadienoic, 6,9,12-octadecatrienoic, 11,14-eicosadienoic, and 8,11,14-eicosatrienoic acid.

2.3.2.4. *Reduction of Polyacetylenic Acids with Lindlar's Catalyst*

An alternative route to tritiated *cis* fatty acids involves the synthesis of the acetylenic or polyacetylenic fatty acids followed by half-hydrogenation using Lindlar's catalyst and tritium gas (Osbond *et al.*, 1961; Sgoutas and Kummerow, 1964). This method places the tritium on the double bond, where it may exchange or may result in unwanted isotope effects. Sgoutas *et al.* (1965) have established the stability of tritium on the double bond during metabolism in rats.

The following synthesis of linoleate-9,10,12,13-^{3}H has been described by Sgoutas and Kummerow (1964) using Lindlar's catalyst to introduce the tritium. Octadeca-9,12-diynoic acid (0.5 g) and 0.1 g of Lindlar's catalyst are placed in a flask fitted with a break-seal joint. The flask is flushed with an inert gas, cooled with liquid nitrogen, and evacuated to 0.001 mm Hg. Hexane (25 ml) is added along with 0.4 ml of a 5% quinoline–hexane solution. Tritium gas (1 ml, 1 °C) and hydrogen (300 ml) are added by means of a Toepler pump, and the flask is sealed. The mixture is stirred for 6 hr at room temperature, and the excess tritium and hydrogen are pumped off. The solution is filtered, and the hexane is evaporated.

Anhydrous ethanol (600 ml total) is mixed with the product in 50-ml portions and distilled off to remove labile tritium. The residue is crystallized from hexane at −60 °C, methylated with 5% HCl-methanol, and distilled at 0.1 mm. The purified methyl linoleate-^{3}H was found to have a specific activity of 35.7 mCi/mmol.

2.3.2.5. Reduction of Polyacetylenic Acids with Disiamylborane

The reduction of polyacetylenic fatty acids with disiamylborane and tritiated acetic acid has been used to prepare several tritiated polyunsaturated fatty acids labeled at the carbon–carbon double bond.

The partial reduction of methyl 10,13-octadecadiynoate to the corresponding octadecadienoic acid has been described by Sgoutas *et al.*, (1969). The procedure consists of dissolving 75 mg of a $NaBH_4$-NaB^3H_4 mixture (16.4 mCi/mmol) in 4 ml of diglyme freshly distilled from $LiAlH_4$. 2-Methyl-2-butene (350 mg) is added, the flask is cooled to 0°C, and 0.3 ml of freshly distilled BF_3-etherate is added dropwise while the reaction mixture is stirred. Moisture must be excluded, and the reaction must be conducted under an atmosphere of inert gas.

After the reaction mixture has been stirred at 0–5 °C for 2 hr, the methyl 10,13-octadecadiynoate (290 mg) is slowly added. The solution is stirred for 30 min at 0°C and then allowed to warm to room temperature while being stirred for 2 hr. The mixture is again cooled to 0°C, and ethylene glycol is added to decompose excess hydride. Glacial acetic acid (0.5 ml) is added, and the mixture is stirred overnight at room temperature. Water is added, and the product is extracted with hexane. Purification is by silica gel chromatography using 10% Et_2O in hexane for elution. A total of 231 mg of methyl ester has been recovered having a specific activity of 9.3 mCi/mmol. Further purification by preparative GLC has resulted in a specific activity of 8.85 mCi/mmol. Analysis by argentation TLC showed only methyl *cis,cis*-octadecadienoate to be present.

Table VI. ^3H-Labeled Fatty Acids

$CH_3(CH_2)_{10}CO_2H$ dodecanoic acid-^3H	Nystrom *et al.* (1959)
$CH_3(H_2)_6CH^3H(CH_2)_3CO_2H$ dodecanoic acid-5L-^3H	Schroepfer and Block (1965)
$CH_3(CH_2)_{12}CO_2H$ tetradecanoic acid-^3H	Nystrom *et al.* (1959)
$CH_3(CH_2)_{14}CO_2H$ hexadecanoic acid-^3H	Nystrom *et al.* (1959)
$CH_3(CH_2)_7CH^3HCH^3H(CH_2)_7CO_2H$ octadecanoic acid-9,10-^3H	Koch (1969*b*)
$CH_3(CH_2)_{16}CO_2H$ octadecanoic acid-^3H	Nystrom *et al.* (1959)
$CH_3(CH_2)_6CH^3H(CH_2)_9CO_2H$ octadecanoic acid-11L-^3H	Egmond *et al.* (1973)
$CH_3(CH_2)_8C^3HH(CH_2)_7CO_2H$ octadecanoic acid-9D-^3H	Morris *et al.* (1968)
$CH_3(CH_2)_8CH^3H(CH_2)_7CO_2H$ octadecanoic acid-9L-^3H	Morris *et al.* (1968)
$CH_3(CH_2)_5C^3HH(CH_2)_{10}CO_2H$ octadecanoic acid-12D-^3H	Morris *et al.* (1968)

(*Continued*)

Table VI—Continued

$CH_3(CH_2)_5CH^3H(CH_2)_{10}CO_2H$ octadecanoic acid-12L-^3H	Morris *et al.* (1968)
$CH_3(CH_2)_5CH^3HCH_2CH{=}CH(CH_2)_7CO_2H$ 9-octadecenoic acid-12L-^3H	Morris (1967)
$CH_3(CH_2)_5C^3HHCH_2CH{=}CH(CH_2)_7CO_2H$ 9-octadecenoic acid-12D-^3H	Morris (1967)
$CH_3(CH_2)_4CH^3HCH^3HCH_2CH{=}CH(CH_2)_7CO_2H$ 9-octadecenoic acid-*erythro*-12,13-^3H	Morris (1967)
$CH_3(CH_2)_7CH{=}CH(CH_2)_7CO_2H$ 9-octadecenoic acid-^3H	Dutton *et al.* (1962)
$CH_3(CH_2)_7C^3H{=}C^3H(CH_2)_7CO_2H$ 9-octadecenoic acid-9,10-^3H	Mounts *et al.* (1971)
$CH_3(CH_2)_7C^3H{=}CH(CH_2)_7CO_2H$ 9-octadecenoic acid-10-^3H	Barley *et al.* (1973)
$CH_3(CH_2)_8C^3H{=}C^3H(CH_2)_6CO_2H$ 8-octadecenoic acid-8,9-^3H	Mounts (1976)
$CH_3(CH_2)_6C^3H{=}C^3H(CH_2)_8CO_2H$ 10-octadecenoic acid-10,11-^3H	Mounts (1976)
$CH_3(CH_2)_5C^3H{=}C^3H(CH_2)_9CO_2H$ 11-octadecenoic acid-11,12-^3H	Mounts (1976)
$CH_3(CH_2)_4C^3H{=}C^3H(CH_2)_{10}CO_2H$ 12-octadecenoic acid-12,13-^3H	Mounts (1976)
$CH_3(CH_2)_4CH{=}CHCH_2C^3H{=}CH(CH_2)_7CO_2H$ 9,12-octadecadienoic acid-10-^3H	Barley *et al.* (1973)
$CH_3(CH_2)_4CH{=}CHCH_2C^3H{=}C^3H(CH_2)_7CO_2H$ 9,12-octadecadienoic acid-9,10-^3H	Sgoutas *et al.* (1969)
$CH_3(CH_2)_4CH{=}CHCH^3HCH{=}CH(CH_2)_7CO_2H$ 9,12-octadecadienoic acid-11L$_s$-^3H	Egmond *et al.* (1973)
$HC^3H_2C^3H_2(CH_2)_3(CH{=}CHCH_2)_2(CH_2)_6CO_2H$ 9,12-octadecadienoic acid-17,18-^3H	Osbond (1966), Stoffel (1965)
$CH_3(CH_2)_4C^3H{=}C^3HCH_2C^3H{=}C^3H(CH_2)_7CO_2H$ 9,12-octadecadienoic acid-9,10,12,13-^3H	Osbond (1966), Sgoutas and Kummerow (1964), Sgoutas *et al.* (1969)
$CH_3(CH_2)_3C^3H{=}C^3HCH_2C^3H{=}C^3H(CH_2)_8CO_2H$ 10,13-octadecadienoic acid-10,11,13,14-^3H	Sgoutas *et al.* (1969)
$CH_3(CH_2)_4(C^3H{=}C^3HCH_2)_3(CH_2)_3CO_2H$ 6,9,12-octadecatrienoic acid-6,7,9,10,12,13-^3H	Sgoutas *et al.* (1969)
$HC^3H_2C^3H_2(CH_2)_3(CH{=}CHCH_2)_3(CH_2)_3CO_2H$ 6,9,12-octadecatrienoic acid-17,18-^3H	Osbond (1966), Stoffel (1965)
$CH_3(CH_2)_5CH{=}CHCH_2C^3H{=}C^3H(CH_2)_7CO_2H$ 9,12-nonadecadienoic acid-9,10-^3H	Sgoutas *et al.* (1969)
$CH_3(CH_2)_4(C^3H{=}C^3HCH_2)_4CH_2CO_2H$ 4,7,10,13-nonadecatetraenoic acid-4,5,7,8,10,11,13,14-^3H	Osbond (1966)
$HC^3H_2C^3H_2(CH_2)_3(CH{=}CHCH_2)_2(CH_2)_8CO_2H$ 11,14-eicosadienoic acid-19,20-^3H	Osbond (1966), Stoffel (1965)
$HC^3H_2C^3H_2(CH_2)_3(CH{=}CHCH_2)_3(CH_2)_5CO_2H$ 8,11,14-eicosatrienoic acid-19,20-^3H	Osbond (1966), Stoffel (1965)
$CH_3(CH_2)_4(C^3H{=}C^3HCH_2)_4(CH_2)_2CO_2H$ 5,8,11,14-eicosatetraenoic acid-5,6,8,9,11,12,14,15-^3H	Osbond (1966)

Methyl esters of unsaturated fatty acids labeled with tritium in the methyl group have been prepared by Mounts and Dutton (1967) for use as tracers in chromatographic analysis and other methods. The procedure uses boron trifluoride and methanol-2-^{3}H to esterify the unsaturated fatty acid. This method is simpler and less hazardous than other procedures which use diazomethane-^{14}C or tritiated diazomethane and gives products with high specific activity. In addition, the radioactive methanol can be recovered for reuse.

Tritium-labeled fatty acids available from chemical or biochemical synthesis are listed in Table VI. The list covers tritium-labeled fatty acids synthesized by a wide variety of methods.

2.3.3. Biosynthesis of ^{14}C- and ^{3}H-Labeled Lipids

Randomly labeled fatty acids have been produced by growing *Chlorella pyrenoidosa* for 5 weeks in cultures shaken under an atmosphere of N_2 and 5–10% $^{14}CO_2$ (Mangold and Schlenk, 1957). The extracted lipids contained labeled 16:0, 16:1, 16:2, 16:3, 18:0, 18:1, 18:2, 18:3, and 20:0.

Ochromonas danica, which contains over 30% 18:3, 20:3, 20:4, and 22:5, has been grown on acetate-^{14}C to yield the corresponding ^{14}C-labeled fatty acids (Gellerman and Schlenk, 1965). This biosynthetic method appears to be an expedient procedure for the preparation of these ^{14}C long-chain polyunsaturated fatty acids.

Christie *et al.* (1973) have prepared *trans*-11-octadecenoic-1-^{14}C acid from linoleic-1-^{14}C by incubation with sheep rumen microorganisms. Overall yields of 60% based on the linoleic-1-^{14}C have been reported. The specific activity of the preparation was reduced by 50% because of unlabeled endogenous fats. The preparation did result in 15% of the C_{10} homologue and approximately 5% of the C_9 and C_{12} homologues.

Incubation of 1-^{14}C-labeled fats with liver slices has been reported by Imaizumi *et al.* (1972) as a method for preparing ^{14}C-labeled phosphatidylcholine (PC). Similarly, potato slices have been used to incorporate 1-^{14}C-labeled fatty acids into PC (Sugano and Yamamoto, 1974). PC isolated from the potato slices contained less endogenous PC and had a specific activity 55–350 times higher than PC isolated from liver slices.

Mounts and Dutton (1964) have described an efficient procedure for producing high-specific-activity, randomly labeled fatty acids by short-term exposure of a nearly mature plant to high levels of $^{14}CO_2$ which would be lethal if used over the lifetime of the plant.

2.3.3.1. Biosynthesis of ^{14}C-Labeled Lipids

For biosynthesis of ^{14}C-labeled lipids, plants whose seeds contain a high percentage of the desired fatty acids are chosen (Mounts and Dutton, 1964). For example, flax, soybean, safflower, and perilla have been successfully used. The plant is grown to the seed-setting stage and then transferred to the chamber shown in Fig. 1. Pressure in the chamber is reduced to 740 mm Hg, and 0.5 mCi $^{14}CO_2$ (0.34 ml gas)

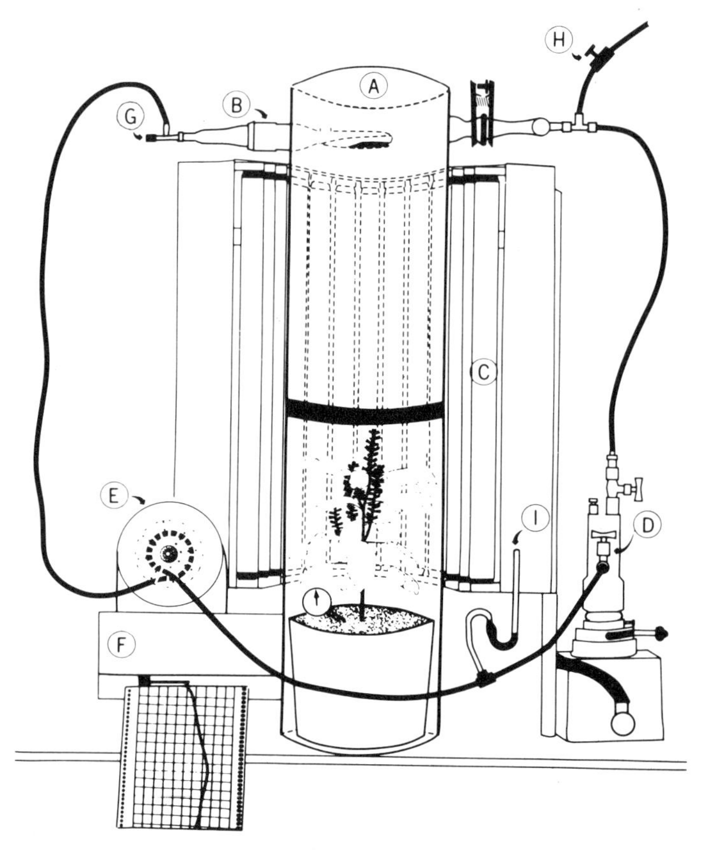

Fig. 1. Growth chamber for $^{14}CO_2$ incorporation into seed lipids. A, Sealed exposure vessel; B, $^{14}CO_2$ source; C, fluorescent tubes; D, continuous-flow ion chamber; E, recycling pump; F, recorder; G, rubber septum; H, vacuum line; I, manometer.

is introduced into the chamber. The chamber is irradiated with 26 fluorescent lights (20 W) and cooled by spraying water on the sides. Illumination for 4 hr has been found to result in over 90% absorption of $^{14}CO_2$. During a 12-hr dark period, 30% of the $^{14}CO_2$ was respired. A second 4-hr illumination period causes reabsorption of most of the $^{14}CO_2$. The chamber is then flushed with air and placed in a battery jar to retain any dead leaves, and the plant is allowed to mature under natural sunlight for 12 days.

Seeds are collected, ground in a micro-Waring Blender with 20 ml of diethyl ether, and extracted two times with 20-ml portions of diethyl ether. The ether extracts are dried and filtered, and the ether is evaporated.

Hirsch (1960) found that transesterification of the residue gave 314 mg of methyl esters, from which 134 mg of linolenate-^{14}C (specific activity 0.64 μCi/mg) was separated by rubber-column chromatography.

^{14}C uniformly labeled fatty esters available from biosynthetic methods are listed in Table VII.

Table VII. ^{14}C-Uniformly Labeled Fatty Acids

Hexadecanoic acid-U-^{14}C	Mounts and Dutton (1964), Mangold and Schlenk (1957)
9-Hexadecenoic acid-U-^{14}C	Mangold and Schlenk (1957)
9,12-Hexadecadienoic acid-U-^{14}C	Mangold and Schlenk (1957)
9,12,15-Hexadecatrienoic acid-U-^{14}C	Mangold and Schlenk (1957)
Octadecanoic acid-U-^{14}C	Mounts and Dutton (1964), Mangold and Schlenk (1957)
9-Octadecenoic acid-U-^{14}C	Mounts and Dutton (1964), Mangold and Schlenk (1957)
9,12-Octadecadienoic acid-U-^{14}C	Mounts and Dutton (1964), Mangold and Schlenk (1957)
9,12,15-Octadecatrienoic acid-U-^{14}C	Mounts and Dutton (1964), Mangold and Schlenk (1957), Gellerman and Schlenk (1965)
8,11,14-Eicosatrienoic acid-U-^{14}C	Gellerman and Schlenk (1965)
4,7,10,13,16-Eicosapentaenoic acid-U-^{14}C	Gellerman and Schlenk (1965)
5,8,11,14,17-Eicosapentaenoic acid-U-^{14}C	Gellerman and Schlenk (1965)
7,10,13,16-Docosatetraenoic acid-U-^{14}C	Gellerman and Schlenk (1965)

2.3.3.2. *Biosynthesis of ^{3}H-Labeled Lipids*

The biosynthesis of linoleic acid-11L-^{3}H from stearic acid-11L-^{3}H using algae has been reported by Egmond *et al.* (1973). The procedure involves synthesis of dodecanoic acid-5L-^{3}H by the methods of Schroepfer and Block (1965) and Hamberg and Samuelson (1967). The dodecanoic acid is then anodic coupled with the half ester of suberic acid to form methyl stearate-11L-^{3}H. Saponification followed by incubation of the tritiated stearic acid with *Chlorella vulgaris* produces a mixture of fatty esters after extraction and methanolysis. Isolation on silver nitrate TLC plates gives methyl linoleate-11L-^{3}H containing about 40% of the initial radioactivity and having a specific activity of 0.2 Ci/mol. The stereospecific D and L enantiomers of labeled octadecanoic acid-9-^{3}H and octadecanoic acid-12-^{3}H have been synthesized (Morris *et al.*, 1967, 1968) by using procedures similar to that of Schroepfer and Block (1965).

2.3.3.3. *Biosynthesis of Labeled Phosphatidylcholine*

The following synthesis of ^{32}P-, ^{14}C-, and ^{3}H-labeled phosphatidylcholine (PC) has been reported by Van den Bosch and Van Deenen (1966). ^{32}P-labeled PC is hydrolyzed by phospholipase A_2 from *Crotalus adamanteus* and the resulting 1-acylglycerol-3-(^{32}P)phosphorylcholine (^{32}P-lysolecithin) is purified by TLC. The ^{32}P-lysolecithin is ultrasonically dispersed with 4 μCi linoleic acid-1-^{14}C in 0.5 ml Krebs–Ringer solution containing 10 μmol ATP and 0.2 μmol coenzyme A. This mixture is incubated with fresh liver homogenate in bicarbonate (100 mg) for 2 hr. The lipids are extracted with diethyl ether and purified by TLC.

Glycerol-3-phosphorylcholine is reacted with cadmium chloride and acylated with oleoyl chloride. The product is extracted, purified, and hydrolyzed with phospholipase A_2. The lysolecithin is hydrogenated over PtO_2 to give 1-stearylglycerol-3-phosphorylcholine. This product is then reacted with cadmium chloride and acylated with stearoyl-1-^{14}C chloride. After purification, the 1-stearyl-2-(1-^{14}C)stearoyl-glycerol-3-phosphorylcholine has a specific activity of 380 μCi/mmole, with 96% of the radioactivity in the 2 position.

Stearoyl-^{3}H-2-stearylglycerol-3-phosphorylcholine is prepared in a similar manner except that the 1-oleoylglycerol-3-phosphorylcholine is tritiated over Adam's catalyst. The specific activity of the purified product has been found to be 417 mCi/mmole, with 92% of the radioactivity in the 1 position.

2.4. *Analysis of Radioisotope-Labeled Fats*

The methods used to analyze radioisotope-labeled lipids have been summarized by Christie (1973) and are briefly reviewed in this section.

2.4.1. *Liquid Scintillation Counting*

Basically, scintillation counting involves degrading high-frequency energy from radioactive isotope emissions by a scintillator which converts the energy into a light flash. The light flash is amplified and detected by a photomultiplier tube. The method is applicable to a large range of sample sizes, and it is reproducible, efficient, and sensitive. It can resolve energy levels from different isotopes and is applicable for α, β, and γ emitters.

Fatty acids are generally labeled with ^{3}H or ^{14}C. These isotopes are both β emitters and have a β energy of 0.018 MeV for ^{3}H and 0.156 MeV for ^{14}C. The relatively low β energy of ^{3}H makes it difficult to analyze quantitatively for this isotope by methods other than scintillation counting.

Scintillation counting is not without its problems. Quenching of the light flashes can occur due to external, thermal, internal, and chemical interactions between the solvent molecules. Color quenching occurs when a color compound absorbs a portion of the photon energy from each light flash. Chemical and color quenching are the two most likely problems to be encountered in lipid work. If these effects occur, they must be recognized and the data corrected. Procedures for correcting data are normally given in the manufacturer's instrument manual (Zimmerman, 1967).

For nonaqueous soluble lipid extracts where quenching is not a problem, a typical scintillation solvent is toluene to which has been added 0.5% 2,5-diphenyl-oxazole (PPO) and 0.02% 2-*p*-phenylene-bis(5-phenyloxazole) (POPOP) or 0.02% 2-*p*-phenylene-bis(4-methyl-5-phenyloxazole) (dimethyl POPOP). This scintillation solvent system is suitable for counting most samples containing both ^{3}H- and ^{14}C-labeled lipids.

The multitude of other factors which must be considered when using scintillation counting makes any further discussion of this topic pointless from a practical standpoint. Instead, the reader is referred to the many articles and books available, of which a few are referenced (Horrocks and Chin-Tzu, 1971; Raaen *et al.*, 1968; Blanchard *et al.*, 1968; Arnold, 1963; Birks, 1964; Bonner, 1954; Bell and Hayes, 1958).

2.4.2. Radio-Gas Chromatography

Chromatographic techniques which have proven especially useful in lipid work are combined gas chromatography/scintillation counting and radio-thin-layer chromatography (Dutton, 1965). In the gas chromatography/scintillation counting method, the basic technique is to sequentially collect many fractions of effluent from the GC into the scintillation solvent. A diagram of the equipment used to collect the effluent is shown in Fig. 2 (Thomas and Dutton, 1970). The fractions are counted

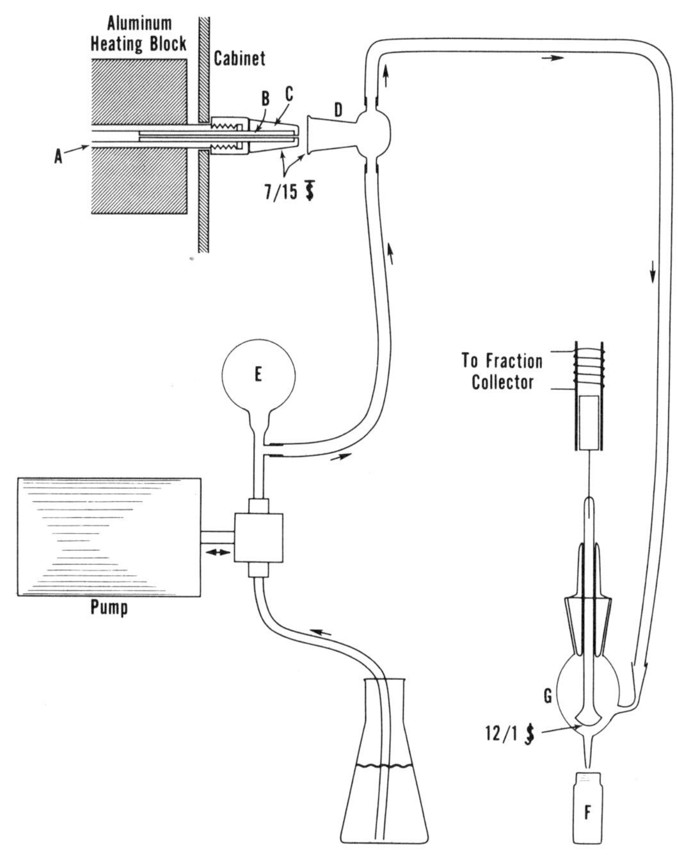

Fig. 2. Apparatus for collection of radioactive effluent from gas chromatograph. A, Stainless-steel collection tube attached to detector outlet; B, aluminum tube sealed to A with a teflon washer; C, 7/15 $ male joint machined from teflon ($ indicates standard tapered glass joints); D, glass tee connected to rubber tubing; E, damping reservoir to smooth solvent flow; F, collection vial; G, solenoid-operated glass valve to stop solvent flow between fractions.

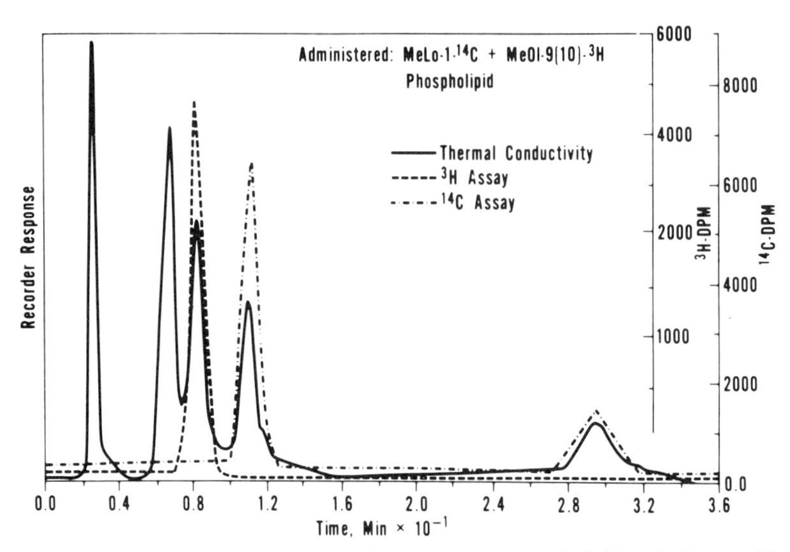

Fig. 3. Radio-gas chromatograph of ^{3}H- and ^{14}C-labeled egg phospholipid methyl esters. The curve was reconstructed from counting sequentially collected effluent fractions from GC by using the apparatus in Fig. 2.

and a radio-gas chromatogram is constructed from the data. This radio-gas chromatogram is then compared to the usual gas chromatogram in order to identify those compounds containing radioactivity. Figure 3 shows the radio-gas chromatogram of phospholipid methyl esters containing Me Lo-1-^{14}C and Me Ol-9(10)-^{3}H.

An alternative system for monitoring effluent from a gas chromatograph uses a flowthrough ionization chamber. An ionization chamber suitable for gas chromatography of fatty methyl esters has been constructed (Dutton, 1961; Nelson *et al.*, 1963). The chamber uses a teflon insulator which eliminates stress currents

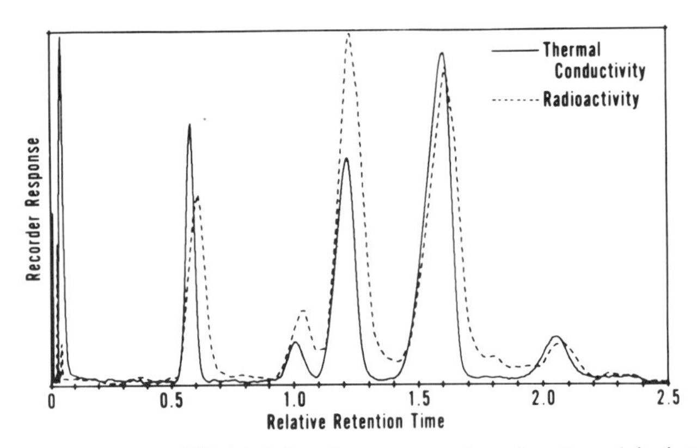

Fig. 4. Radio-gas chromatogram of ^{3}H-labeled soybean esters using a flowthrough ionization chamber. ^{3}H-labeled soybean esters were prepared by growing soybeans hydroponically in a sealed plant growth chamber with tritiated water added to the nutrient solution.

that occur in ceramic insulators at high temperatures. Figure 4 shows a radio chromatogram of ^{3}H-labeled soybean oil esters. Of the two chromatography monitoring techniques, the scintillation counting method has several times greater sensitivity.

2.4.3. Thin-Layer Chromatography

Thin-layer chromatography (TLC) can be coupled with autoradiography or strip scanners using windowless Geiger-Muller tubes (Snyder, 1968) to check the purities of ^{14}C- and ^{3}H-labeled compounds. TLC separations of ^{14}C-labeled fatty esters can be analyzed by placing the developed thin-layer chromatograms in contact with a photographic film such as Eastman Kodak Company's Royal Blue X-ray film (Richardson *et al.,* 1963) or Eastman's No-Screen medical X-ray safety film (Mangold *et al.*, 1962). This technique can usually detect about 7×10^{-3} μCi/cm^{2} ^{14}C by using 1-day exposure times.

Tritium-labeled compounds on thin-layer chromatograms can be detected by autoradiography using photoplates prepared with Kodak-type NTB nucleartrack emulsions (Sheppard and Tsien, 1963). This film is also suitable for ^{14}C detection. Appendix III in Raaen *et al.* (1968) lists other suppliers of autoradiography film.

An alternate procedure for detecting radioactivity on TLC plates is to scrape thin, uniform horizontal bands of silica gel from the plate into scintillation vials. The radioactivity is measured and the R_f's are plotted to correlate radioactivity with known standards and to substantiate the purity of the sample. A problem with this technique is the artificially high measurements which can occur due to the binders, fluorescent indicators, plasticizers, and basic compounds, which may be in the silica gel layer or in the developing and extracting solvents.

2.4.4. Determination of Radiochemical Purity

All radioisotope-labeled fatty acids should be checked for radiochemical purity whether they are purchased or synthesized. Fatty acids can be conveniently esterified by diazomethane to facilitate their analysis by GC. The radioanalytical methods previously described, such as TLC and GC, should be routinely used. Snyder and Piantadosi (1966) have reviewed these procedures for checking radioisotopic purity and have discussed other techniques and methods. In addition, chemical degradation of the fatty acid can be a valuable aid in determining the position of the radioisotope label since synthetic techniques can introduce isotope labels into unwanted positions. This problem is especially prevalent in tritium-labeled fatty acids.

Basic techniques for determining radioisotope purity are to analyze the preparation by TLC and radio-gas chromatography or GC-scintillation counting. The radioactivity in the sample should coincide with the R_f and retention time determined for unlabeled standards. In addition, radioisotope-labeled fatty acids can be diluted with a nonlabeled fatty acid and the specific activity determined. If the specific activity remains constant after appropriate purification techniques and derivatizations, good radiochemical purity is likely. A further test is to add a fatty acid which has a different radioisotope label. The $^{3}H/^{14}C$ ratio of the mixture can be determined, and, if the ratio is constant after purification and analysis by TLC and GC, high radiochemical purity is indicated. In determining purities, one must continually guard against the incorporation of quenching material, fluorescent compounds, or compounds which can induce chemiluminescence. Continual checks after every step against standards and blanks are necessary until a procedure is well proven.

2.4.5. Chemical Degradation Methods

^{14}C-carboxyl-labeled fatty acids can be degraded using the Hunsdiecker reaction, which involves preparing the silver salt of the fatty acid (Calvin, 1949). Susan and Nystrom (1971*a*, *b*) and Numrich and Nystrom (1971) have discussed applications and modifications of this method to short-chain aliphatic acids.

The Schmidt reaction has also been reported for short-chain aliphatic acids (Phares, 1951; Mosbach *et al.*, 1951) and can be used to achieve stepwise degradation of saturated acids.

The Darzen–Mentzer method has been investigated as a means for stepwise degradation of palmitic and other aliphatic carboxylic acids (Dauben *et al.*, 1953). The method has been applied to unsaturated fatty acids by first hydrogenating to the saturated acid (Gibble *et al.*, 1956).

The Lemieux and von Rudloff periodate–permanganate oxidative cleavage procedure has been adapted for the cleavage of oleic and elaidic acids at the point of unsaturation (Jones and Stolp, 1958). The method involves first cleaving the tritiated double bonds and then measuring radioactivity in the mono- and dibasic acid fragments (Dutton *et al.*, 1962). Reductive ozonolysis (Stein and Nicolaides, 1962) using ozone and triphenylphosphine followed by GC analysis of the aldehyde and aldehyde–ester cleavage products is presently the method of choice for determining double-bond position.

Privett (1966) and Conacher (1976) have summarized the techniques involved in the location of double bonds in unsaturated fatty acids. Ozonolysis followed by GC analysis of the fragmentation products is straightforward with monoenoic acids, but partial reduction with hydrazine to the monoenes is first required with polyunsaturates. To apply ozonolysis to the determination of ^{14}C position in unsaturated fatty acids, the monoenoic acid must first be extensively isomerized and then the radioactivity of the fragments formed by ozonolysis is determined in order to identify the position of the ^{14}C label. If this approach is applied to polyunsaturates, it is again necessary to partially reduce the polyunsaturate to a monoene either before or after extensive double-bond migration. Perchloric acid has been reported as a good catalyst for promoting the necessary bond migration. This procedure is satisfactory for locating ^{14}C labels, but it cannot be used with ^{3}H-labeled fatty acids because of the loss and redistribution of the tritium label during the isomerization process.

A simple procedure which involves bromination of the double bond in unsaturated fatty acids followed by debromination is useful for establishing the exclusive presence of tritium on the double bond (Mounts *et al.*, 1971).

Tritium-labeled terminal methyl groups in fatty acids have been determined by the Kuhn–Roth procedure (Ginger, 1944). This procedure involves oxidation of saturated fatty acids with sulfuric acid and potassium dichromate. The experimental details are as follows.

The fatty acid (20 mg) is dissolved in 2 ml of concentrated H_2SO_4 and warmed slightly. The solution is cooled to 0–2 °C, and 5 N chromic acid (5 ml) is added. It is refluxed for 1.5 hr and cooled, and 0.5 ml of 20% hydrazine hydrate is added to reduce excess chromic acid. The acetate formed is distilled off and titrated. A silicic acid column can be used to further purify the acetate.

Rognstad *et al.* (1968) have reported that the Kuhn–Roth procedure results in loss of tritium from the methyl end of fatty acids and that some acetate is produced from the center portion of the fatty acid molecule.

2.4.6. Storage of Radioisotope-Labeled Fatty Acids

Self-radiolysis of ^{14}C- and ^{3}H-labeled fatty acids makes their prolonged storage difficult. Regardless of the method and precautions taken, some decomposition can be certain. The primary decomposition products resemble free-radical-propagated reactions and probably are the result of free-radical generation due to radiation absorption. Common products which form during self-radiolysis are (1) fragmentation products due to homolytic scission of the carbon chain (these products increase if the sample is in a solid state); (2) dimerization, cyclization, and polymerization products; (3) free acids from methyl esters; (4) hydrocarbons by decarboxylation; (5) saturated fatty acids by hydrogen addition to unsaturated fatty acids; (6) mixtures of *cis* and *trans* isomers by isomerization of geometrically pure samples.

Methods for minimizing these unwanted reactions are (1) store in solvent such as cyclohexane or pentane (benzene enhances *cis-trans* isomerization); (2) dilute as much as practical with an inert carrier; (3) keep as cold as possible without solidifying (fragmentation is enhanced in the solid state); (4) store under an inert gas or *in vacuo*.

The problems associated with self-radiolysis are discussed and expanded by Colosimo and Guarino (1972), Evans and Stanford (1963), and Raaen *et al.* (1968).

2.5. Analysis of Stable Isotope-Labeled Fatty Acids

The emphasis in this section will be to illustrate a few typical analytical techniques with examples of the type of data that can be obtained for fatty acids labeled with deuterium or ^{13}C.

2.5.1. Mass Spectrometry

The usual method for producing ions for mass spectrometry analysis is by electron impact. This method is the oldest, most widely used, and best developed. Very good quantitative analyses of the deuterium and ^{13}C distribution in labeled fatty esters have been reported by this method. In the case of d_1- and d_2-labeled fats, corrections must be applied for the 1.1% natural abundance of ^{13}C. For example, the $[M + 1]^+$ peak (the peak at 1 mass unit higher than the molecular ion) for methyl stearate has an intensity equal to approximately 21% of the molecular ion, and the $[M + 2]^+$ peak has an intensity equal to approximately 2.5% of the molecular ion. Dinh-Nguyen (1968) has published the spectra of all the *gem*-dideuterated methyl stearates plus some spectra of other deuterated and ^{13}C-labeled fatty esters. However, for unsaturated fatty esters, the molecular ion peak is less than 1% of the total ions produced (i.e., only 1% of the signal is in the molecular ion region). This weak molecular ion signal can limit the accuracy of the system when limited sample is available.

The chemical ionization (CI) method of producing ion yields is a much stronger molecular ion signal because of reduced fragmentation of the molecule. The stronger signal results in a greater signal-to-background noise ratio and consequently higher sensitivity and accuracy when limited sample size is available. Application of CI mass spectrometry to fatty acids has been reported by Murata *et al.* (1975), but quantitation of ^{2}H- or ^{13}C-labeled fatty acids has not been well studied. The difficulty with quantitative CI mass spectrometry arises from the potential formation of non-reproducible amounts of $[M-1]^+$, $[M]^+$, and $[M+1]^+$ ions.

Field ionization also produces less fragmentation and a stronger parent peak, but the state of the art has not developed to the point where this method is easily applied to gas chromatography/mass spectrometry (GC/MS) work or to the quantitative analysis of deuterated fatty esters.

The coupling of a gas chromatograph to the inlet of a mass spectrometer and the use of on-line computer technique have greatly simplified analysis by mass spectroscopy. Details of these techniques are covered in various review articles (McFadden, 1973; Perkins, 1975). GC/MS involves measuring the mass spectrum of each peak many times as it emerges from the gas chromatograph. This method requires rapid cycling of the MS magnet or rapid stepping of the acceleration voltage while holding the magnetic field constant. Wendt and McCloskey (1970) published GC/MS data for the perdeuterated fatty acids isolated from *Scenedesmus obliques*, and McCloskey (1975) has reviewed the mass spectra and GC characteristics of perdeuterated fatty acids. The shorter GC retention time for perdeuterated fatty esters is sufficient to permit their separation from undeuterated fatty esters using normal packed columns (McCloskey, 1970).

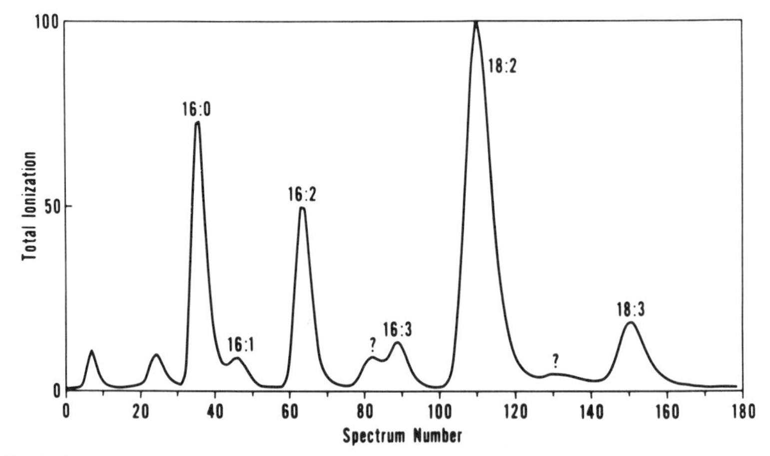

Fig. 5. GC/MS analysis of ^{13}C-labeled *Chlorella pyrenoidosa* methyl esters isolated from algae grown in the presence of 90% $^{13}CO_2$ under high-intensity light conditions. Isotopic purity is estimated at 88.8%. Lyophilized cells were a gift from the Los Alamos Scientific Laboratory.

A typical GC/MS curve of a ^{13}C-labeled fatty ester mixture isolated from *Chlorella pyrenoidosa* grown in the presence of 90.5% $^{13}CO_2$ is shown in Fig. 5.

Individual mass spectra from this curve are shown in Figs. 6–8 for 16:0-^{13}C, 18:2-^{13}C, and 18:3-^{13}C. The algae fatty esters contained 88.8% ^{13}C. Electron impact was used, and the characteristic small molecular ion signal is apparent for the unsaturated fatty esters (Ryhage and Stenhagen, 1963).

A second GC/MS method commonly referred to as multiple-ion mass spectroscopy (MIMS) concentrates on measuring only selected mass units instead of the entire mass spectra for each GC peak. This technique is accomplished by rapid-stepping the accelerator voltage while holding the magnetic field at one setting (Rohwedder, 1975). In this case, the ion current at each mass unit is measured for a

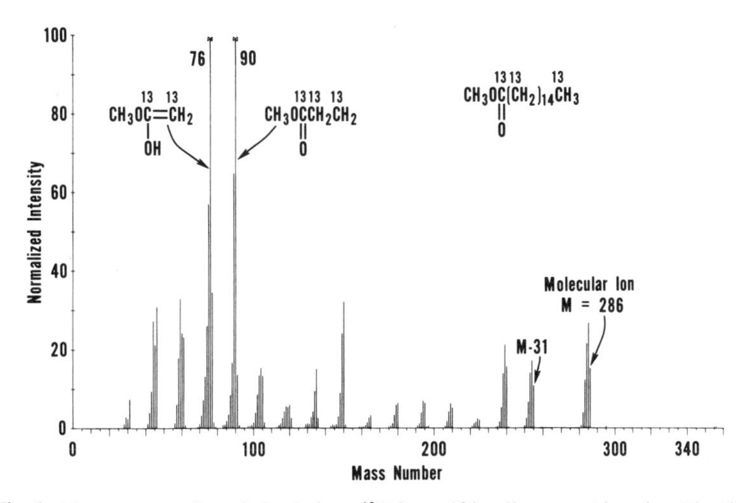

Fig. 6. Mass spectra of methyl palmitate-^{13}C from *Chlorella pyrenoidosa* (see Fig. 5).

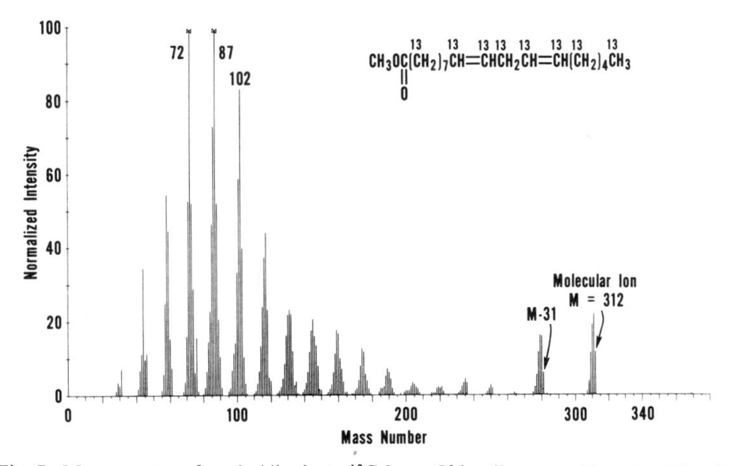

Fig. 7. Mass spectra of methyl linoleate-^{13}C from *Chlorella pyrenoidosa* (see Fig. 5).

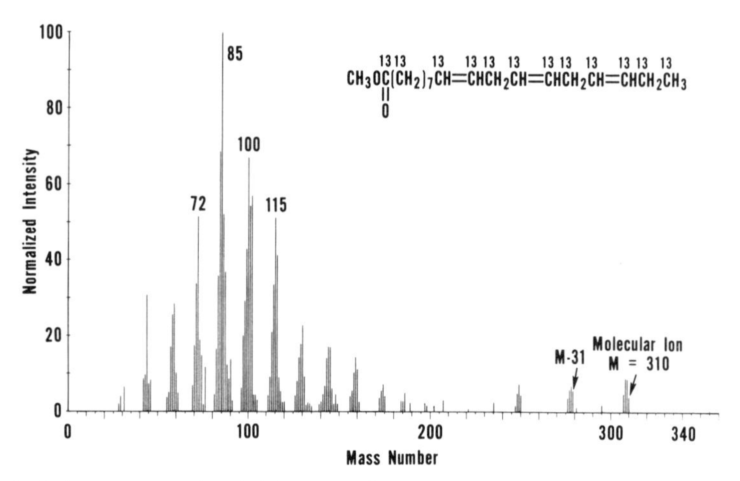

Fig. 8. Mass spectra of methyl linolenate-^{13}C from *Chlorella pyrenoidosa* (see Fig. 5).

predetermined time and then the accelerator voltage is stepped to the next mass unit where the operation is repeated.

Figure 9 shows a MIMS curve plotted from data on a mixture of oleate-d_2, elaidate-d_4, and oleate-d_6. The GC retention times for oleate-d_2 and oleate-d_6 follow the reported pattern of shorter retention times with increasing deuterium content. MIMS has been used to quantitate the amounts of two or three deuterated fatty esters in the same GC peak, which allows dual- or triple-isotope experiments to be used. This dual-isotope technique has been particularly useful for obtaining metabolic data on the rate of incorporation of deuterated fats into human blood lipids (Emken *et al.*, 1976).

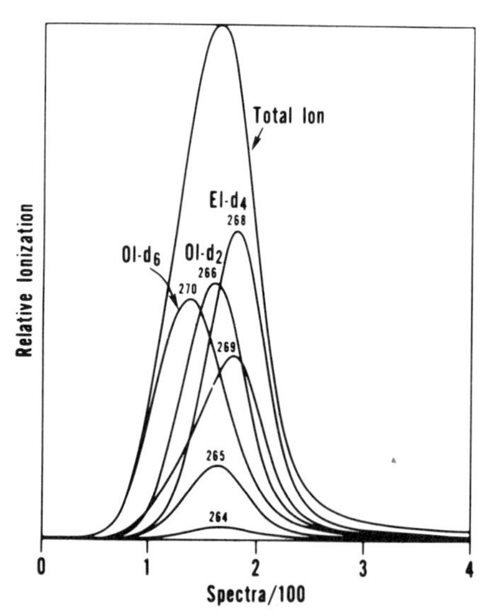

Fig. 9. Multiple-ion mass spectroscopy data of an oleate-d_2, elaidate-d_4, oleate-d_6 mixture. Numbers above peaks indicate mass of the fragmentation ions which were monitored.

2.5.2. Infrared Spectrometry

Infrared (IR) absorption spectra of deuterated fats have apparently been limited to methyl octadecanoates and methyl octadecenoates. The lack of published data on deuterated polyunsaturated fatty acids is probably due to the limited availability of these deuterated unsaturated fatty acids.

Rohwedder *et al.* (1967) have published the spectra of methyl stearate-6,7-d_2, methyl stearate-9,9,10,10-d_4, and methyl stearate-9,10,12,13-d_4. Figure 10 shows the IR spectra of these deuterated stearates as compared to unlabeled methyl stearate. The main feature of these spectra is the shift of the CH_2-stretching band from 2925

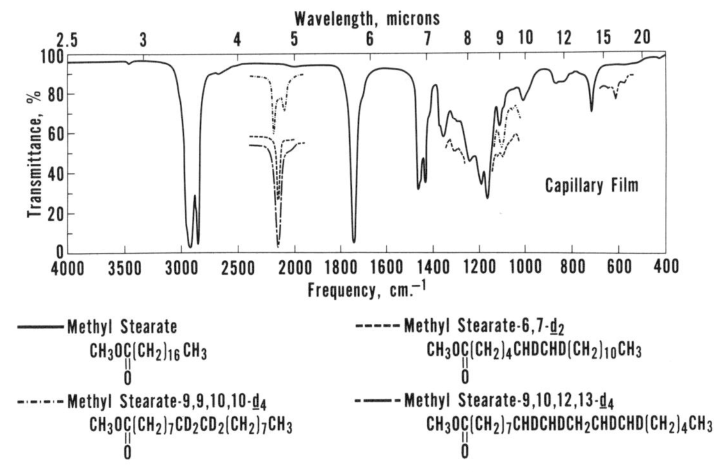

Fig. 10. Infrared spectra of deuterated methyl stearates compared to methyl stearate-d_0.

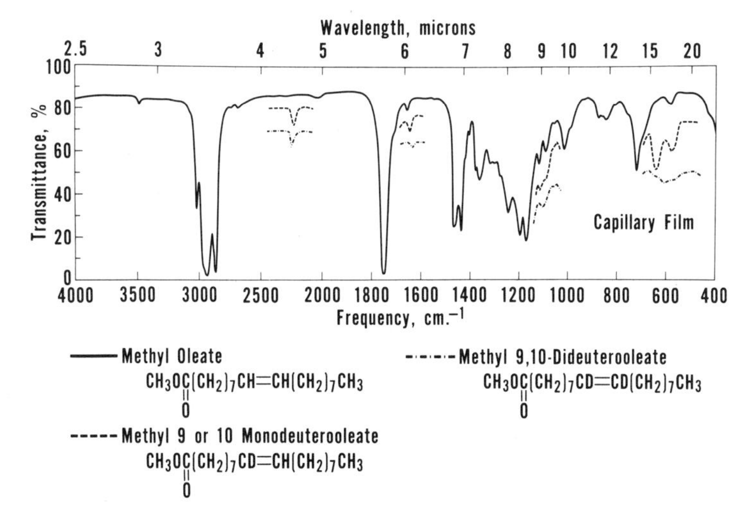

Fig. 11. Infrared spectra of deuterated methyl oleate compared to methyl oleate-d_0.

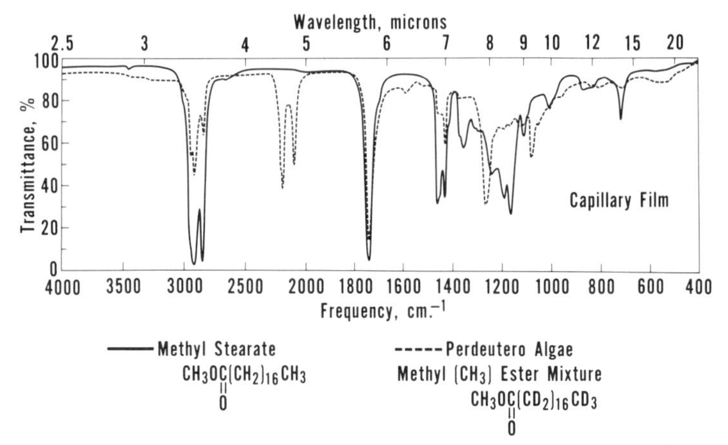

Fig. 12. Infrared spectra of perdeutero methyl esters from *Scenedesmus obliques*.

cm^{-1} (asym) and 2850 cm^{-1} (sym) to 2191 cm^{-1} (asym) and 2102 cm^{-1} (sym) for the CH_2 group.

IR spectra of methyl oleate-9(10)-d_1 and methyl oleate-9,10-d_2 are shown in Fig. 11. The characteristic absorption bands found for the deuterostearates are present in these spectra also.

The spectra of perdeutero methyl esters from *Scenedesmus obliques* are shown in Fig. 12. The peaks in the 2800–3000 cm^{-1} region are from the CH_3OH used to esterify the perdeutero fatty acids.

The IR spectra of positionally isomeric methyl *gem*-dideuterooctadecanoates has been reported by Dinh-Nguyen and Fischmeister (1970*a*,*b*). All their dideuterostearate spectra were similar to those in Fig. 11 except for the 2,2-dideutero- and 18,18,18-trideuterostearates. In the methyl stearate-2,2-d_2 spectra, the absorption bands are at higher frequencies (2198 and 2118 cm^{-1}), and the methyl stearate-18,18,18-d_3 spectra have three bands, with the strongest band split into two bands at 2210 and 2201 cm^{-1}.

The solid-state IR spectra of perdeuterostearate, perdeuterobehenate, stearate-2,2,3,3,4,4,5,5-d_8, and stearate-2,2,3,3-d_4 have also been reported by Dinh-Nguyen and Fischmeister (1970*a*, *b*).

2.5.3. Nuclear Magnetic Resonance

Proton nuclear magnetic resonance (NMR) spectroscopy applied to deuterated fats has been used to identify the position of the deuteriums on the fatty acid chain. The method is particularly useful if the deuteriums are α or β to the carboxylic group or if they are on the double bond. The drawback to this technique is that evidence for deuterium substitution is negative since deuterium does not produce an absorption. When deuterium is substituted for protons on methylene groups which are not resolved, the reduction in the proton signal is usually not definitive. NMR has been used to estimate the deuterium content of fatty acids, but the method does not usually have high accuracy.

Deuterium NMR of fatty acid suffers from the same limitations as proton NMR. Signals from many of the methylenes in a fatty acid molecule have similar chemical shifts, and consequently the exact position of the deuterium on the fatty acid cannot be identified. Deuterium NMR has been used in various membrane studies in an attempt to study the structure of both natural and synthetic membranes (Saito *et al.*, 1973; Oldfield *et al.*, 1971, 1972; Salsbury *et al.*, 1972; Stockton *et al.*, 1976; Seelig and Seelig, 1974).

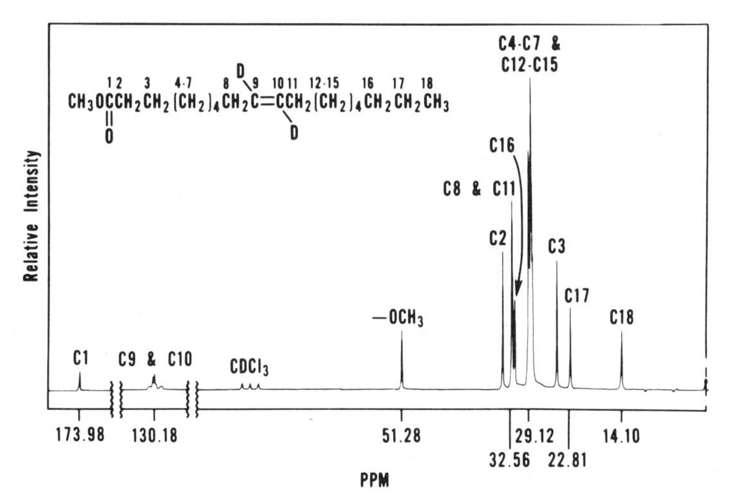

Fig. 13. ^{13}C NMR spectra of methyl elaidate-9,10-d_2. Spectra were measured with a Bruker WH90 Fourier transform NMR spectrometer, 8K data memory, 1000 scans, chemical shifts given relative to TMS.

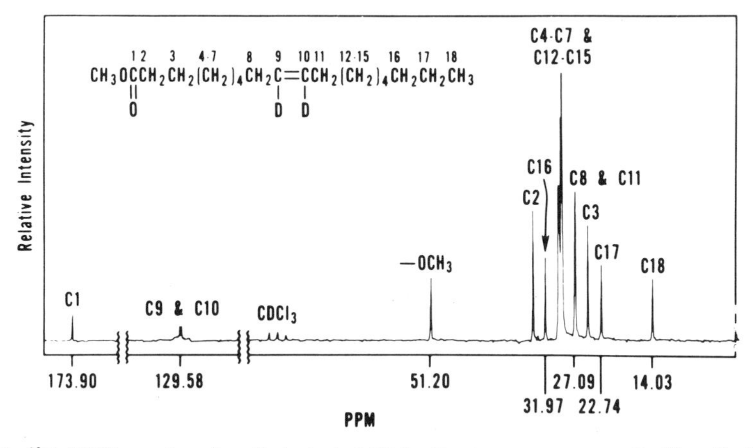

Fig. 14. ^{13}C NMR spectra of methyl oleate-9,10-d_2. Spectra were measured with a Bruker WH90 Fourier transform NMR spectrometer, 8K data memory, 1000 scans, chemical shifts given relative to TMS.

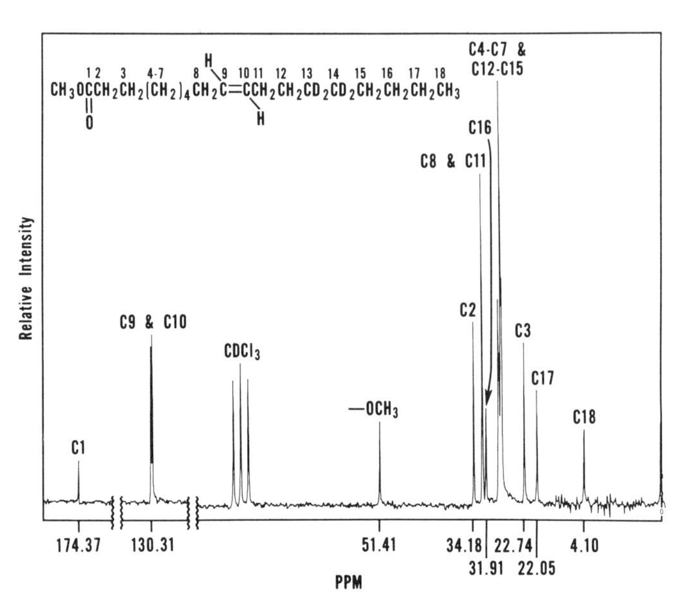

Fig. 15. ^{13}C NMR spectra of methyl elaidate-13,13,14,14-d_4. Spectra were measured with a Bruker WH90 Fourier transform NMR spectrometer, 8K data memory, 64,000 scans, chemical shifts given relative to TMS.

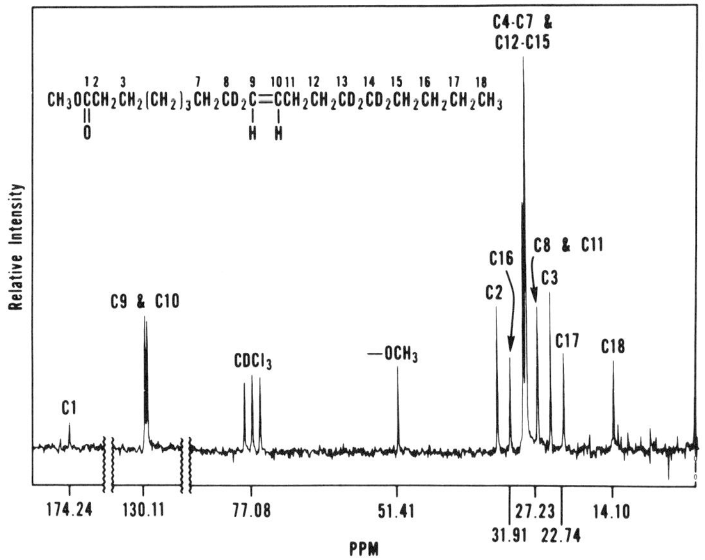

Fig. 16. ^{13}C NMR spectra of methyl oleate-8,8,13,13,14,14-d_6. Spectra were measured with a Bruker WH90 Fourier transform NMR spectrometer, 8K data memory, 10,000 scans, chemical shifts given relative to TMS.

^{13}C NMR coupled with Fourier transformation is a fast-emerging technique which is capable of locating double-bond position in the fatty acid molecule as well as identifying many of the protons on methylene carbons.

Figures 13–16 show typical ^{13}C NMR spectra for methyl elaidate-d_2, methyl oleate-d_2, methyl elaidate-d_4, and methyl oleate-d_6. A feature of ^{13}C NMR is the large chemical shifts relative to proton NMR. Chemical shifts have been assigned to the ^{13}C NMR spectra of methyl stearate, methyl oleate, and methyl petroselinate by Tulloch and Mazurek (1976). The chemical shifts are quite sensitive to structural changes and environmental situations. For this reason, recent investigations have used ^{13}C NMR in an attempt to elucidate the physical structuring of membranes (Stoffel *et al.*, 1972). Bus and Frost (1974) have assigned ^{13}C NMR chemical shifts to 16:0, 18:0, 18:1, 18:2, 18:3, and 20:4 spectra by enriching specific positions with ^{13}C. They also have made ^{13}C NMR assignments to spectra from several PC and sphingomyelin samples containing various fatty acids. The use of ^{13}C NMR and proton NMR for locating double bonds has been reviewed by Bus and Frost (1976).

The scope of the various NMR techniques and their applications is much too large to cover fully, and the reader is referred to those papers referenced in this section and to books by Axenrod and Webb (1974), Clerc *et al.* (1973), and Jackman and Cotton (1975).

2.6. Summary

Syntheses, methods, and experimental details for preparation of ^{2}H-, ^{13}C-, ^{3}H-, and ^{14}C-labeled fatty acids have been reviewed. Many of the reactions discussed in this chapter have been evaluated in the author's laboratory. An attempt has been made to describe in greater detail those reactions and syntheses which are particularly useful for unique transformations or which appear to have been successfully utilized in numerous other laboratories. For example, reactions which selectively label or allow labeling without isotope exchange have been discussed in detail. Details on selective biosynthetic methods have been included because of their potential preparative value or stereospecificity.

The analytical methodology for isotope determination in fatty acids is too extensive to review thoroughly. Consequently, this chapter has been restricted to a short description of methods and techniques normally used for analyzing isotope-labeled fatty acids.

It is hoped that the coverage of synthetic and analytical methods in this chapter will aid less experienced researchers in selecting methods for preparing and using isotope-labeled fatty acids.

2.7. References

Arnold, J. R., 1963, Liquid scintillation counting of tritium, in: *Advances in Tracer Methodology*, Vol. 1 (S. Rothchild, ed.), pp. 69–76, Plenum, New York.

Aronoff, S., 1956, *Techniques of Radiobiochemistry*, The Iowa State College Press, Ames, Ia.

Axenrod, T., and Webb, G. A., 1974, *Nuclear Magnetic Resonance Spectroscopy of Nuclei Other Than Protons*, Wiley, New York.

Aylward, F., and Narayana Rao, C. V., 1956, Use of hydrazine as a reducing agent for unsaturated compounds. I. The hydrogenation of oleic acid, *J. Appl. Chem.* **6**:248.

Aylward, F., and Narayana Rao, C. V., 1957, Use of hydrazine as a reducing agent for unsaturated compounds. III. Hydrogenation of linoleic acid, *J. Appl. Chem.* **7**:134.

Barley, G. C., Jones, Sir Ewart R. H., Thaller, V., and Vere Hodge, R. A., 1973, Natural acetylenes.

Part XXXIX. Synthesis of methyl (1,9-^{14}C)-, (9-^{14}C)-, and (10-^{3}H)-crepenynate, methyl (9-^{14}C)- and (10-^{3}H)-linoleate, and methyl (9-^{14}C)- and (10-^{3}H)-oleate, *J. Chem. Soc. Perkin Trans.* **1(2):**151.

Bell, C. G., Jr., and Hayes, F. N., 1958, *Liquid Scintillation Counting*, Pergamon Press, New York.

Birch, A. J., and Walker, K. A. M., 1966, Aspects of catalytic hydrogenation with a soluble catalyst, *J. Chem. Soc. C* **21:**1894.

Birks, J. B., 1964, *The Theory and Practice of Scintillation Counting*, Pergamon Press, New York.

Blanchard, F. A., Wagner, M. R., and Takahashi, 1968, Liquid scintillation counting: Automated mathematical fitting and use of channels ratio methods by computer program, in: *Advances in Tracer Methodology* (S. Rothchild, ed.), pp. 133–144, Plenum, New York.

Bonner, J. F., 1954, Determination of radioactivity by scintillation counting, in: *Technique of Organic Chemistry*, Vol. 1 (A. Weissberger, ed.), Part III, pp. 2491–2515, Interscience, New York.

Budny, J., and Sprecher, H., 1971, A study of some of the factors involved in regulating the conversion of octadeca-8,11-dienoate to eicosa-4,7,10,13-tetraenoate in the rat, *Biochim. Biophys. Acta* **239:**190.

Bus, J., and Frost, D. J., 1974, ^{13}CMR analysis of methyl octadecenoates, *Recl. Trav. Chim. Pays-Bas* **93:**213.

Bus, J., and Frost, D. J., 1976, Determination of the positions of double bonds in unsaturated fatty acids by ^{13}C and proton NMR spectrometry, in: *Lipids*, Vol. 2: *Technology* (R. Paoletti, G. Jacini, and R. Porcellati, eds.), Raven Press, New York.

Butterfield, R. O., and Dutton, H. J., 1968, High-yield preparation of methyl stearolate, *J. Am. Oil Chem. Soc.* **45:**635.

Calf, G. E., Garnett, J. L., and Pickles, V. A., 1968, Catalytic deuterium exchange reactions with organic XI: Pyridine, the quinolines, azines, and aniline on unsupported groups VIII transition metals, *Aust. J. Chem.* **21:**961.

Calvin, M., 1949, *Isotopic Carbon*, Wiley, New York.

Castro, C. E., and Stephens, R. D., 1964, The reduction of multiple bonds by low-valent transition metal ions: The homogeneous reduction of acetylenes by chromous sulfate, *J. Am. Chem. Soc.* **86:**4358.

Catch, J., 1961, *Carbon-14 Compounds*, Butterworth, Washington, D.C.

Christie, W. W., 1973, *Lipid Analysis*, pp. 282–297, Pergamon Press, New York.

Christie, W. W., and Holman, R. T., 1967, Synthesis and characterization of the complete series of methylene-interrupted *cis,cis*-octadecadienoic acids, *Chem. Phys. Lipids* **1:**407.

Christie, W. W., Hunter, M. L., and Harfoot, C. G., 1973, The biosynthetic preparation of 1-^{14}C-*trans*-11-octadecenoic acid, *J. Labelled Compd.* **9:**483.

Clerc, J. T., Pretsch, E., and Sternhill, S., 1973, *Carbon-13 Nuclear Resonance Spectroscopy*, Akad. Verlag, Frankfurt, Germany.

Colosimo, M., and Guarino, A., 1972, Self-radiolysis of tritiated compounds. IV. Unsaturated fatty acid esters, *J. Labelled Compd.* **8:**257.

Conacher, H. B. S., 1976, Chromatographic determination of *cis-trans* monoethylenic unsaturation in fats and oils—A review, *J. Chromatogr. Sci.* **14:**405.

Craig, D., and Fowler, R. B., 1961, Deuterio 1,3-butadienes derived by reductive dechlorination, *J. Org. Chem.* **26:**713.

Crespi, H. L., Conrad, S. M., Uphaus, R. A., and Katz, J. J., 1960, Cultivation of microorganisms in heavy water, *Ann. N.Y. Acad. Sci.* **84:**648.

Cronan, J. E., Jr., and Batchelor, J. G., 1973, An efficient biosynthetic method to prepare fatty acyl chains highly enriched with ^{13}C, *Chem. Phys. Lipids* **11:**196.

Dauben, W. G., Hoerger, E., and Petersen, J. W., 1953, Distribution of acetic acid carbon in high fatty acids synthesized from acetic acid by the intact mouse, *J. Am. Chem. Soc.* **75:**2347.

DeJarlais, W. J., and Emken, E. A., 1976, Syntheses of tetra- and hexadeuterated octadecenoates, *Lipids* **11:**594.

Dinh-Nguyen, N., 1964, Organic syntheses with heavy isotopes. Part I. Some deuterium-substituted normal long-chain saturated hydrocarbons, acids and esters, *Ark. Kemi* **22:**151.

Dinh-Nguyen, N., 1968, Contribution à l'étude de la spectrometrie de masse: Utilisation des esters méthyliques de monoacides à longue chaine normale marques au deuterium et au carbone-13, *Ark. Kemi* **28:**289.

Dinh-Nguyen, N., and Fischmeister, I., 1970*a*, The solid state infrared absorption spectra of the positionally isomeric methyl *gem*-dideutero-octadecanoates, *Ark. Kemi* **32:**181.

Dinh-Nguyen, N., and Fischmeister, I., 1970*b*, The solid infrared absorption spectra of deuterated long chain fatty acids and esters, *Ark. Kemi* **32:**205.
Dinh-Nguyen, N., Raal, A., and Stenhagen, E., 1972, Perdeuteriated organic compounds. I. Normal long-chain saturated deuteriocarbons, monocarboxylic acids and methyl esters, *Chem. Scr.* **2:**171.
Dutton, H. J., 1961, Monitoring eluates from chromatography and countercurrent distribution for radioactivity, *J. Am. Oil Chem. Soc.* **38:**631.
Dutton, H. J., 1965, Some techniques of radioactive gas chromatography for lipid research, in: *Advances in Tracer Methodology*, Vol. 2 (S. Rothchild, ed.), pp. 123–134, Plenum, New York.
Dutton, H. J., Jones, E. P., Davison, V. L., and Nystrom, R. F., 1962, Labeling fatty acids by exposure to tritium gas. III. Methyl stearolate and methyl linolenate, *J. Org. Chem.* **27:**2648.
Dutton, H. J., Scholfield, C. R., Selke, E., and Rohwedder, W. K., 1968, Double bond migration, geometric isomerization, and deuterium distribution during heterogeneous catalytic deuteration of methyl oleate, *J. Catal.* **10:**316.
Egmond, M. R., Vliegenthart, J. F. G., and Boldingh, J., 1973, Synthesis of 11(n-8)L_s tritium-labelled linoleic acid, *Biochim. Biophys. Acta* **316:**1.
Eisch, J. J., and Kaska, W. C., 1966, Stereochemistry and orientation in the reactions of 1-phenylpropyne with diisobutyl aluminum hydride, *J. Am. Chem. Soc.* **88:**2213.
Emken, E. A., Scholfield, C. R., and Dutton, H. J., 1964, Chromatographic separation of *cis* and *trans* fatty esters by argentation with a macroreticular exchange resin, *J. Am. Oil Chem. Soc.* **41:**388.
Emken, E. A., Rohwedder, W. K., Dougherty, R., Mackin, J., Iacono, J. M., and Dutton, H. J., 1976, Dual-labeled technique for human lipid metabolism studies using deuterated fatty acid isomers, *Lipids* **11:**135.
Evans, E. A., 1966, *Tritium and Its Compounds*, Van Nostrand, Princeton, N.J.
Evans, E. A., and Stanford, F. G., 1963, Decomposition of tritium-labelled organic compounds, *Nature (London)* **197:**551.
Fetizon, M., and Gramain, J. C., 1969, Recent methods of deuteration, *Bull. Soc. Chim. Fr.* **1969:**651.
Frankel, E. N., Selke, E., and Glass, C. A., 1969, Homogeneous hydrogenation of diolefins catalyzed by tricarbonyl chromium complexes. II. Deuteration, *J. Org. Chem.* **34:**2628.
Friedrich, K., and Thieme, H. K., 1973, Activated ethylenes. Synthesis of electronegatively substituted 2-chloro alkenes, *Synthesis* **2:**111.
Gellerman, J. L., and Schlenk, H., 1965, Preparation of fatty acids labeled with C^{14} from *Ochromonas danica*, *J. Protozool.* **12:**178.
Gensler, W. J., and Bruno, J. J., 1963, Synthesis of unsaturated fatty acids: Positional isomers of linoleic acids, *J. Org. Chem.* **28:**1254.
Gibble, W. P., Kurtz, E. B., Jr., and Kelley, A. E., 1956, A semi-micro procedure for the separation and degradation of long-chain fatty acids, *J. Am. Oil Chem. Soc.* **33:**66.
Ginger, L. G., 1944, The chemistry of the lipids of tubercle bacilli. LXXI. The determination of terminal methyl groups in branched chain fatty acids, *J. Biol. Chem.* **156:**453.
Graff, G., Szczepanik, P., Klein, P. D., Chipault, J. R., and Holman, R. T., 1970, Identification and characterization of fully deuterated fatty acids isolated from *Scenedesmus obliquus* cultured in deuterium oxide, *Lipids* **5:**786.
Gunstone, F., Scrimgeour, C., and Vedanayagem, S., 1974, New procedure for converting unsaturated fatty acids to their nor-alcohols required as intermediates for the preparation of carboxy-labelled acids, *J. Chem. Soc. Chem. Commun.* **1974:**916.
Hamberg, M., and Samuelson, B., 1967, On the mechanism of the biosynthesis of prostaglandins E_1 and $F_{1\alpha}$, *J. Biol. Chem.* **242:**5336.
Hirsch, J., 1960, Chromatography of lipids on non-polar stationary phases with automatic recording, *Colloq. Int. Centre Natl. Rech. Sci. (Paris)*, pp. 17–21.
Horrocks, D. L., and Chin-Tzu, P., 1971, *Organic Scintillators and Liquid Scintillation Counting*, Academic Press, New York.
Howton, D. R., and Stein, R. A., 1969, Ahmad-Strong synthesis of 8-, 9-, and 10-pentadecynoic acids, *J. Lipid Res.* **10:**631.
Howton, D. R., Davis, R. H., and Neuenzel, J. C., 1952, Decarboxylation and reconstitution of linoleic acid, *J. Am. Chem. Soc.* **74:**1109.
Hsiao, C. Y. Y., Ottaway, C. A., and Wetlaufer, D. B., 1974, Preparation of fully deuterated fatty acids by simple method, *Lipids* **9:**913.
Imaizumi, K., Cho, S., Sugano, M., and Wada, M., 1972, The effect of overnight fasting on the synthesis of glycerolipids by liver slice, *Agr. Biol. Chem.* **36:**1783.

Jackman, L. M., and Cotton, F. A., 1975, *Dynamic Nuclear Magnetic Resonance Spectroscopy*, Academic Press, New York.

Jones, E. P., and Stolp, J. A., 1958, Periodate-permanganate oxidations for determining location and amount of unsaturation in monounsaturated fatty acids, *J. Am. Oil Chem. Soc.* **35**:71.

Kates, M., 1972, Techniques of lipidology: Isolation, analysis and identification of lipids, Part II, in: *Laboratory Techniques in Biochemistry and Molecular Biology* (T. S. Work and E. Work, eds.), pp. 470–501, American Elsevier, New York.

Klok, R., Egmond, G. J. N., and Pabon, H. J. J., 1974, Synthesis of 19-*cis*-docosenoic, 17-*cis*-eicosenoic and 15-*cis*-octadecenoic acid, *Recl. Trav. Chim. Pays-Bas* **93**:222.

Koch, G. K., 1969*a*, Specific ^{3}H-labelling by diimide reduction of unsaturated bonds. I. Mechanism of reduction with hydrazine in aprotic solvents, *J. Labelled Compd.* **5**:99.

Koch, G. K., 1969*b*, Specific ^{3}H-labelling by diimide reduction of unsaturated bonds. II. Methods and applications of labelling, *J. Labelled Compd.* **5**:110.

Koritala, S., and Selke, E., 1971, Selective hydrogenation with copper catalysts. IV. Reaction of stearolate, oleate and conjugated esters with deuterium, *J. Am. Oil Chem. Soc.* **48**:222.

Kovachic, D., and Leitch, L. C., 1961, Organic deuterium compounds. XIII. Synthesis of deuteriated olefins, *Can. J. Chem.* **39**:363.

Kuhlein, K., Neumann, W. P., and Mohring, H., 1968, A versatile method for preparation of C-deuterated compounds, *Angew. Chem. Int. Ed. Engl.* **7**:455.

Kunau, W.-H., 1971*a*, Chemical synthesis of highly unsaturated fatty acids. I. Preparation of $(n - 4)$, $(n - 1)$-alkadiynoic acids, *Chem. Phys. Lipids* **7**:101.

Kunau, W.-H., 1971*b*, Chemical synthesis of highly unsaturated fatty acids. II. Preparation of substituted propargyl halides, *Chem. Phys. Lipids* **7**:108.

Kunau, W.-H., 1973, Synthesis of unsaturated fatty acids, *Chem. Phys. Lipids* **11**:254.

Lindlar, H., and Dubuis, R., 1973, Palladium catalyst for partial reduction of acetylenes, in: *Organic Syntheses*, Collective Vol. 5, pp. 880–885, Wiley, New York.

Magoon, E. F., and Slaugh, L. H., 1967, Reduction of acetylenes and conjugated diolefins by lithium aluminum hydride, *Tetrahedron* **23**:4509

Mangold, H. K., 1968, Preparation of labelled lipids, *J. Labelled Compd.* **4**:3.

Mangold, H. K., and Schlenk, H., 1957, Preparation and isolation of fatty acids randomly labeled with C^{14}, *J. Biol. Chem.* **229**:731.

Mangold, H. K., Kammereck, R., and Malins, D. C., 1962, *Microchemical Techniques* (N. D. Cheronis, ed.), pp. 697–714, Interscience, New York.

Marcel, Y. L., and Holman, R. T., 1968, Synthesis of ^{14}C-labelled polyunsaturated fatty acids, *Chem. Phys. Lipids* **2**:173.

Matwiyoff, N. A., and Ott, D. G., 1973, Stable isotope tracers in the life sciences and medicine, *Science* **181**:1125.

McCloskey, J. A., 1975, Gas chromatography–mass spectrometry of esters of perdeuterated fatty acids, in: *Methods in Enzymology*, Vol. 35 (J. M. Lowenstein, ed.), Part B, p. 341, Academic Press, New York.

McFadden, W., 1973, *Techniques of Combined Gas Chromatography/Mass Spectrometry: Applications in Organic Analysis*, Wiley, New York.

Morandi, J. R., and Jensen, H. B., 1969, Homogeneous catalytic deuteration of olefinic double bonds, *J. Org. Chem.* **34**:1889.

Morris, L. J., 1967, The mechanism of ricinoleic acid biosynthesis in *Ricinus communis* seeds, *Biochem. Biophys. Res. Commun.* **29**:311.

Morris, L. J., Harris, R. V., Kelly, W., and James, A. T., 1967, The stereospecificity of desaturations of long-chain fatty acids in *Chlorella vulgaris, Biochem. Biophys. Res. Commun.* **28**:904.

Morris, L. J., Harris, R. V., Kelly, W., and James, A. T., 1968, The stereochemistry of desaturations of long-chain fatty acids in *Chlorella vulgaris, Biochem. J.* **109**:673.

Mosbach, E. H., Phares, E. F., and Carson, S. F., 1951, Degradation of isotopically labeled citric, α-ketoglutaric and glutamic acids, *Arch. Biochem. Biophys.* **33**:179.

Mounts, T. L., 1973, Tritium labeling of lipids, *Lipids* **8**:190.

Mounts, T. L., 1976, Double bond position affects metabolism of *cis*-cotadecenoates, *Lipids* **11**:676.

Mounts, T. L., and Dutton, H. J., 1964, Efficient production of biosynthetically labeled fatty acids, *J. Am. Oil Chem. Soc.* **41**:537.

Mounts, T. L., and Dutton, H. J., 1967, Methyl esters of unsaturated fatty acids labeled with tritium in the methyl group, *J. Labelled Compd.* **3:**343.

Mounts, T. L., Emken, E. A., Rohwedder, W. K., and Dutton, H. J., 1971, Metabolism of labeled isomeric octadecenoates by the laying hen, *Lipids* **6:**912.

Murata, T., Takahashi, S., and Takeda, T., 1975, Chemical ionization–mass spectrometry, I. Application to analysis of fatty acids, *Anal. Chem.* **47:**573.

Murray, A., III, and Williams, D. L., 1958, *Organic Synthesis with Isotopes*, Parts I and II, Interscience, New York.

Nelson, D. C., Ressler, P. C., Jr., and Hawes, R. C., 1963, Performance of an instrument for simultaneous gas chromatographic and radioactivity analysis, *Anal. Chem.* **35:**1575.

Nevenzel, J. C., Riley, R. F., Howton, D. R., and Steinberg, G., 1957, *Bibliography of Syntheses with Carbon Isotopes*, United States Atomic Energy Commission Report, UCLA-395.

Nicholas, P. P., and Carroll, R. T., 1968, The chlorination of olefins with cuperic chloride: A comparative study of *trans*-ethylene-d_2 and *cis*- and *trans*-2-butene, *J. Org. Chem.* **33:**2345.

Numrich, R., and Nystrom, R. F., 1971, Small-scale degradation of carbon-14 labelled carboxylic acids. III. Degradation of acetic acid-^{14}C, *J. Labelled Compd.* **7:**283.

Nystrom, R. F., Mason, L. H., Jones, E. P., and Dutton, H. J., 1959, Labelling fatty acids by exposure to tritium gas. I. Saturated methyl esters, *J. Am. Oil Chem. Soc.* **36:**212.

Oldfield, E., Chapman, D., and Derbyshire, W., 1971, Deuteron resonance: A novel approach to the study of hydrocarbon chain mobility in membrane systems, *FEBS Lett.* **16:**102.

Oldfield, E., Chapman, D., and Derbyshire, W., 1972, Lipid mobility in acholeplasma membranes using deuteron magnetic resonance, *Chem. Phys. Lipids* **9:**69.

Osbond, J. M., 1966, The synthesis of naturally-occurring and labelled 1,4-polyunsaturated fatty acids, in: *Progress in Chemistry of Fats and Other Lipids*, Vol. 9 (R. T. Holman, ed.), Part 1, pp. 121–157, Pergamon Press, New York.

Osbond, J. M., Philpott, P. G., and Wickens, J. C., 1961, Essential fatty acids. Part I. Synthesis of linoleic, γ-linolenic, arachidonic, and docosa-4,7,10,13,16-pentaenoic acid, *J. Chem. Soc.* **1961:**2779.

Osborn, J. A., Jardine, F. H., Young, J. F., and Wilkinson, G., 1966, The preparation and properties of tris(triphenylphosphine) halogenorhodium(I) and some reactions thereof including catalytic homogeneous hydrogenation of olefins and acetylenes and their derivatives, *J. Chem. Soc. A* **1966(12):**1711.

Perkins, E. G., 1975, Gas chromatography–mass spectrometry of lipids, in: *Analysis of Lipids and Lipoproteins* (E. G. Perkins, ed.), pp. 183–203, The American Oil Chemists' Society, Champaign, Ill.

Phares, E. F., 1951, Degradation of labeled propionic and acetic acids, *Arch. Biochem. Biophys.* **33:**173.

Pichat, L., Guermont, J. P., and Levron, J. C., 1969, Methylations de composés acétyléniques métallés en milieu hexamethylphosphotriamide et emploi de la reaction de Wittig pour la synthèse d'acides gras marqués au ^{14}C. I. Synthèse de l'acide oléique (^{14}C-18), *Bull. Soc. Chim. Fr.* **1969:**1198.

Privett, O. S., 1966, Determination of the structure of unsaturated fatty acids via degradative methods, in: *Progress in the Chemistry of Fats and Other Lipids*, Vol. 9 (R. T. Holman, ed.), Part 1, p. 91, Pergamon Press, New York.

Raaen, V. F., Ropp, G. A., and Raaen, H. P., 1968, *Carbon-14*, McGraw-Hill, New York.

Richardson, G. S., Weliky, I., Batchelder, W., Griffin, M., and Engel, L. L., 1963, Radioautography of ^{14}C- and ^{3}H-labeled steroids on thin-layer chromatograms, *J. Chromatogr.* **12:**115.

Roberts, R. B., Abelson, P. H., Cowie, P. B., Bolton, E. T., and Britten, R. J., 1963, *Studies on Biosynthesis in Escherichia coli*, Carnegie Institution, Washington, D.C.

Rognstad, R., Woronsberg, J., and Katz, J., 1968, Acetyl group transfer in lipogenesis. I. Studies involving the degradation of fatty acids by the Kuhn–Roth and related methods, *Arch. Biochem. Biophys.* **127:**429.

Rohwedder, W. K., 1975, Mass spectrometry of lipids, in: *Analysis of Lipids and Lipoproteins* (E. G. Perkins, ed.), pp. 170–182, The American Oil Chemists' Society, Champaign, Ill.

Rohwedder, W. K., Scholfield, C. R., Rakoff, Henry, Nowakowska, Janina, and Dutton, H. J., 1967, Infrared analysis of methyl stearates containing deuterium, *Anal. Chem.* **39:**820.

Ronzio, A. R., 1954, Microsyntheses with tracer elements, in: *Technique of Organic Chemistry*, Vol. 6 (A. Weissberger, ed.), pp. 367–409, Interscience, New York.

Ryhage, R., and Stenhagen, E., 1963, Mass spectrometry of long-chain esters, in: *Mass Spectrometry of Organic Ions* (F. W. McLafferty, ed.), pp. 399–452, Academic Press, New York.

Saito, H., Schrier-Muccillo, S., and Smith, I. C. P., 1973, High resolution deuterium magnetic resonance—An approach to the study of molecular organization in biological membranes and model systems, *FEBS Lett.* **33**:281.

Salsbury, N. J., Dorke, A., and Chapman, D., 1972, Deutron magnetic resonance studies of water associated with phospholipids, *Chem. Phys. Lipids* **8**:142.

Scholfield, C. R., Jones, E. P., Nowakowska, J., Selke, E., and Dutton, H. J., 1961, Hydrogenation of linolenate. II. Hydrazine reduction, *J. Am. Oil. Chem. Soc.* **39**:208.

Schroepfer, G. J., and Block, K., 1965, The stereospecific conversion of stearic acid to oleic acid, *J. Biol. Chem.* **240**:54.

Seelig, A., and Seelig, J., 1974, The dynamic structure of fatty acyl chains in a phospholipid bilayer measured by deuterium magnetic resonance, *Biochemistry* **13**:4839.

Sgoutas, D. S., and Kummerow, F. A., 1964, Chemical synthesis of tritium-labeled linoleic acid, *Biochemistry* **3**:406.

Sgoutas, D. S., Kim, M. J., and Kummerow, F. A., 1965, Radiohomogeneity of H^3- and C^{14}-labeled linoleic acid *in vivo*, *J. Lipid Res.* **6**:383.

Sgoutas, D. S., Sanders, H., and Yang, E. M., 1969, Tritioboration and synthesis of tritium-labeled polyunsaturated fatty acids, *J. Lipid Res.* **10**:642.

Sheppard, H., and Tsien, W. H., 1963, Autoradiography of tritium-containing thin-layer chromatograms, *Anal. Chem.* **35**:1992.

Slaugh, L. H., 1966, Lithium aluminum hydride, a homogeneous hydrogenation catalyst, *Tetrahedron* **22**:1741.

Snyder, F., 1968, Thin-layer chromatography radioassay, in: *Advances in Tracer Methodology: A Review* (S. Rothchild, ed.), pp. 81–104, Plenum, New York.

Snyder, F., and Piantadosi, C., 1966, Labeling and radiopurity of lipids, in: *Advances in Lipid Research*, Vol. 4 (R. Paoletti and D. Kritchevsky, eds.), pp. 257–283, Academic Press, New York.

Sprecher, H., 1971, The synthesis of 1-^{14}C-arachidonate and 3-^{14}C-docosa-7,10,13,16-tetraenoate, *Lipids* **6**:889.

Steenhoek, A., Van Wijngaarden, B. H., and Pabon, H. J. J., 1971, Optimization, mechanism, and kinetics of the hydrogenation of skipped polyynoic acids to all *cis* skipped polyenoic acids, *Recl. Trav. Chim. Pays-Bas.* **90**:961.

Stein, R. A., and Nicolaides, N., 1962, Structure determination of methyl esters of unsaturated fatty acids by gas–liquid chromatography of the aldehydes formed by triphenylphosphine reduction of the ozonides, *J. Lipid Res.* **3**:476.

Stockton, G. W., Polnaszek, C. F., Tulloch, A. P., Hasan, F., and Smith, I. C. P., 1976, Molecular motion and order in single-bilayer vesicles and multilamellar dispersions of EGG lecithin and lecithin cholesterol mixtures: A deuterium nuclear magnetic resonance study of specifically labeled lipids, *Biochemistry* **15**:954.

Stoffel, W., 1964, Synthese von (1-^{14}C)-markierten all-*cis*-Polyenfettsauren, *Justus Liebigs Ann. Chem.* **673**:26.

Stoffel, W., 1965, Chemical synthesis of ^{3}H- and 1-^{14}C-labeled polyunsaturated fatty acids, *J. Am. Oil Chem. Soc.* **42**:583.

Stoffel, W., and Bierwirth, E., 1963, Synthesis of 1-^{14}C-labelled polyunsaturated fatty acids, *Angew. Chem. Int. Ed. Engl.* **2**:94.

Stoffel, W., Zierenberg, O., and Tunggal, B. D., 1972, ^{13}C-Nuclear magnetic resonance spectroscopic studies on saturated, mono-, di-, and polyunsaturated fatty acids, phospho- and sphingolipids, *Hoppe-Seyler's Z. Physiol. Chem.* **352**:1962.

Sugano, M., and Yamamoto, M., 1974, A simple biosynthetic method for preparation of high specific radioactivity phosphatidylcholine, *Agr. Biol. Chem.* **38**:1255.

Susan, A. B., and Nystrom, R. F., 1971*a*, Small-scale degradation of carbon-14 labelled carboxylic acids. I. Modification of the Hunsdieker reaction, *J. Labelled Compd.* **7**:269.

Susan, A. B., and Nystrom, R. F., 1971*b*, Small-scale degradation of carbon-14 labelled carboxylic acids. II. Carbon-14 labelled propionic and *n*-butyric acids, *J. Labelled Compd.* **7**:275.

Swann, S., Jr., 1948, Electrolytic reactions, in: *Technique of Organic Chemistry*, Vol. II (A. Weissberger, ed.), Interscience, New York.

Thomas, A. F., 1971, *Deuterium Labeling in Organic Chemistry*, Meredith Corporation, New York.

Thomas, P. J., and Dutton, H. J., 1970, Preparation and counting of lipophilic samples, in: *The Current Status of Liquid Scintillation Counting* (E. D. Bransome, Jr., ed.), pp. 164–169, Grune and Stratton, New York.

Tokes, L., Jones, G., and Djerassi, C., 1968, Mass spectrometry in structural and stereochemical problems. CLXI. Elucidation of the course of the characteristic ring D fragmentation of steroids, *J. Am. Chem. Soc.* **90:**5465.

Tolbert, M. B., and Siri, W. E., 1960, Determination of radioactivity, in: *Technique of Organic Chemistry*, Vol. 1 (A. Weissberger, ed.), pp. 3335–3448, Part IV, Interscience, New York.

Tucker, W. P., Trove, S. B., and Kepler, C. R., 1970, The synthesis of 11,11-dideuterolinoleic acid, *J. Labelled Compd.* **7:**11.

Tucker, W. P., Trove, S. B., and Kepler, C. R., 1971, The synthesis of 11,11-dideuterooleic acid, *J. Labelled Compd.* **7:**137.

Tulloch, A. P., and Mazurek, M., 1976, ^{13}C Nuclear magnetic resonance spectroscopy of saturated, unsaturated, and oxygenated fatty acid methyl esters, *Lipids* **11:**228.

Van den Bosch, H., and Van Deenen, L. L. M., 1966, Synthesis of ^{32}P-, ^{14}C-, and ^{3}H-labeled lecithins and their use in studies on lipid metabolism, in: *Advances in Tracer Methodology*, Vol. 3 (S. Rothchild, ed.), pp. 61–69, Plenum, New York.

Van Tamelen, E. E., Dewey, R. S., and Timmons, R. J., 1961, The reduction of olefins by means of azodicarboxylic acid *in situ*, *J. Am. Chem. Soc.* **83:**3725.

Wang, C. H., and Willis, D. L., 1965, *Radiotracer Methodology in Biological Science*, Prentice-Hall, Englewood Cliffs, N.J.

Wendt, G., and McCloskey, J. A., 1970, Mass spectrometry of perdeuterated molecules of biological origin fatty acid esters from *Scenedesmus obliques*, *Biochemistry* **9:**4854.

Young, J. F., Osborn, J. A., Jardine, F. H., and Wilkinson, G., 1965, Hydride intermediates in homogeneous hydrogenation reactions of olefins and acetylenes using rhodium catalysts, *J. Chem. Soc. Chem. Commun.* **1965(1):**131.

Zimmerman, M. E., 1967, *Preparation of Samples for Liquid Scintillation Counting*, Nuclear-Chicago Corporation, Des Plaines, Ill.

Chapter 3

Separation and Determination of the Structure of Acylglycerols and Their Ether Analogues

John J. Myher

3.1. Introduction

This chapter describes the techniques that are used for the separation and structural characterization of neutral glycerolipids having one to three molecules of fatty acid (esters) or alcohol (ethers) combined with the hydroxyl groups of glycerol. The emphasis is placed on practical considerations, and theoretical discussions are either brief or absent. Although references quoted in this chapter often do not indicate the original source of a procedure or idea, such information can be found within the comprehensive reviews that are cited in each section.

3.2. Nomenclature

The rules of stereospecific numbering were put forward by the IUPAC–IUB Commission on Biochemical Nomenclature (Anonymous, 1967). In order to designate the stereochemistry of glycerol derivatives, the carbon atoms of glycerol are numbered stereospecifically. If a glycerol molecule is drawn in the Fischer projection with the secondary hydroxyl group to the left of the central carbon, the carbons are stereospecifically numbered 1, 2, and 3 from top to bottom. The prefix "*sn*" (for *s*tereospecifically *n*umbered), used immediately preceding the term "glycerol" and separated from it by a hyphen, differentiates such numbering from conventional numbering conveying no steric information. On the other hand, the prefix "*rac*-" precedes the full name if the product is an equal mixture of both antipodes, and the prefix "X-" is used if the configuration of the compound is either unknown or unspecified.

Any glycerolipid will possess mirror-image asymmetry if the substituents at the

John J. Myher • Banting and Best Department of Medical Research, University of Toronto, Toronto, Ontario, M5G 1L6, Canada.

sn-1 and *sn*-3 position are different. Mirror-image molecules or enantiomers possess equal but opposite optical rotations; however, if both substituents are long-chain acyl groups, the optical rotation is extremely small (Schlenk, 1965).

3.3. Selected Methods for Modifying or Degrading Glycerolipids

Frequently, chemical or enzymatic methods of degradation or modification can be used to produce a product that is much more amenable to structural analysis than the original molecule. The analysis of the molecular species of phospholipid, for example, is made much easier once the glycerol–phosphate bond has been cleaved because the resultant diacylglycerols can be analyzed by thin-layer chromatography (TLC), gas–liquid chromatography (GLC), and gas chromatography/mass spectrometry (GC/MS). In the following section, some of the common degradative methods are described.

3.3.1. Methanolysis

Determination of the fatty acid methyl esters (FAMEs) resulting from acid- or base-catalyzed methanolysis of glyceryl esters and of the dimethyl acetals (DMAs) resulting from the acid-catalyzed methanolysis of the alk-1-enylglycerol moiety plays a very important role in the analysis of any lipid. This subject is discussed in detail in Chapter 1.

3.3.2. Saponification

Saponification has been used in the past to generate glycerol or glyceryl ethers from neutral or polar lipids (Thompson and Kapoulas, 1969), but this procedure has largely been replaced by reductive hydrogenolysis with $LiAlH_4$ or Vitride* reagent (see later). Saponification may be carried out according to the procedure outlined in Chapter 1 of this volume.

3.3.3. Phospholipase C Hydrolysis

Representative 1,2(2,3)-diacylglycerols may be released from phospholipids by phospholipase C (Kuksis, 1972*a*). The enzyme from *Clostridium welchii* is effective for the hydrolysis of lecithin or, if lysolecithin or sphingomyelin is present, for the hydrolysis of phosphatidylethanolamine. In the presence of lysolecithin or sphingomyelin, the enzyme from *Bacillus cereus* can also be used for the hydrolysis of phosphatidylserine or inositol. Where applicable, enzyme hydrolysis is the preferred method for the preparation of diacylglycerols (or ether analogues) from phospholipids. The reaction conditions are mild, and at worst only 1% isomerization of 1,2- to 1,3-diacylglycerols takes place during the hydrolysis (Renkonen, 1971). Although enzymatic hydrolysis of the ether analogues is somewhat slower

* Trademark of the National Patent Development Company. Supplied by Eastman Kodak Co., Rochester, New York.

(Renkonen, 1971), the method is suitable for alk-1-enylacylglycerol phospholipids. Some of the procedures used are as follows:

1. Phosphatidylcholine (4.5–5.0 mg) in 2 ml of diethyl ether is added to 2 ml of 1 M tris buffer (pH 7.3) containing 1.5 μl of a 45% $CaCl_2$ solution. Phospholipase C from *Clostridium welchii* (1 mg) is added to a 1-dram vial, and the reaction is allowed to proceed at room temperature for 3 hr while being mixed. The reaction mixture is then extracted three times with diethyl ether (Wood and Snyder, 1969).
2. Phosphatidylethanolamine (4.5 mg) and sphingomyelin (3.0 mg) are added to 0.5 ml of 0.2 M phosphate buffer (pH 7.0) containing 1.0×10^{-3} M 2-mercaptoethanol and 4.0×10^{-4} M $ZnCl_2$. Crude *Bacillus cereus* phospholipase C (0.5 ml) is added, and the incubation is allowed to proceed 1–2 hr at 37°C under vigorous agitation. The reaction mixture is then extracted three times with diethyl ether (Wood and Snyder, 1969).
3. Phosphatidylethanolamine (5.1 mg) in 2 ml of diethyl ether is added to 0.5 ml of 0.1 M tris buffer (pH 7.4) with 10 mM $CaCl_2$ containing 2 units of *Bacillus cereus* phospholipase C. The mixture is incubated at 29°C for 90 min and then extracted with diethyl ether (Yeung and Kuksis, 1974).

3.3.4. Acetolysis

Acetolysis may also be used to cleave the glycerophosphate bonds of diacylglycerophospholipids (Kuksis, 1972*a*; Renkonen, 1971), but the conditions are too severe to be applicable to alk-1-enyl-acylglycerophospholipids. The reaction leads mainly to 1,2-diacyl-3-acetyl-*sn*-glycerol, but an appreciable amount of the 1,3 isomer also results from isomerization of the 1,2 isomers. The species, however, are generally representative, and the acetates of the 1,2 isomers can be purified by TLC.

Procedure: The reaction is carried out (Renkonen, 1965) by dissolving 0.5–50 mg of phospholipid in 2 ml of acetic anhydride–acetic acid (2 : 3, v/v) and heating the mixture in a sealed tube for approximately 4 hr at 145°C. The reaction mixture is evaporated to dryness, and the 1,2-diacylglycerol acetates are purified by TLC (see section 3.7.1.1).

3.3.5. Thermal Cleavage

Diacylglycerols are released from phospholipids by heating the latter at 250°C in diphenyl ether (Horning *et al.*, 1969). A low yield of 1,2-diacylglycerols plus some 1,3 isomers was obtained when the reaction was allowed to take place for 1 min at 250°C, but if the reaction was allowed to proceed for 5 min the yield of diacylglycerol was much better. However, the 1,3 isomer was then the major product.

3.3.6. Reductive Hydrogenolysis

Reductive hydrogenolysis with $LiAlH_4$ or Vitride reagent [$NaAlH_2$-$(OCH_2OCH_3)_2$, 70% in benzene] is a very important method for releasing 1-alkyl-

or 1-alk-1-enylglycerol from neutral or polar ether lipids (Schmid *et al.*, 1975) or for releasing glycerol from glyceryl esters. A better recovery of alk-1-enylglycerol was obtained when Vitride reagent was used rather than $LiAlH_4$ (Snyder *et al.*, 1971; Snyder, 1976). Esterified fatty acids are reduced to the corresponding fatty alcohols. Reduction with Vitride reagent may be carried according to the procedure of Snyder *et al.* (1971) as modified by Blank *et al.* (1975).

Procedure: The sample (up to 10 mg) dissolved in 2.5 ml diethyl ether containing 2% (v/v) benzene is mixed with 0.5 ml of Vitride reagent, and the solution is heated at 37°C for 30 min in a closed tube. After the solution has been allowed to cool, 10 ml of 20% ethanol in water is cautiously added (dropwise at first) and extracted three times with 6 ml diethyl ether. Ten milliliters of hexane is added to the pooled ether extract and dried over sodium sulfate. The extract is decanted or filtered and evaporated to dryness. Since acids bring about the decomposition of the 1-alk-1-enylglycerols, acids are avoided in the procedure. The products are analyzed by methods outlined later.

$LiAlH_4$ reduction, although less convenient, leads to the same products as reduction with Vitride. The following procedure according to Thompson and Lee (1965) can be scaled down for smaller amounts of lipid (Wood and Snyder, 1968).

Procedure: Ten milliliters of a diethyl ether suspension of $LiAlH_4$ (30 mg/ml) is added to 5 ml of dry diethyl ether. This solution is cooled by immersion of the flask in a dry ice–acetone bath, and 50–100 mg of lipid dissolved in 10 ml of dry diethyl ether is cautiously added, with stirring. The solution is gradually brought to room temperature and then refluxed for 30 min. Finally, the excess reagent is destroyed at dry ice–acetone temperature by the cautious addition of water, and the products are extracted with diethyl ether.

3.3.7. Grignard Hydrolysis

Partial hydrolysis with Grignard reagents such as ethyl magnesium bromide can be used to generate representative 1,2(2,3)- and 1,3-diacylglycerols from triacylglycerols (Yurkowski and Brockerhoff, 1966; Brockerhoff, 1971). Ethyl magnesium bromide is the preferred reagent for this reaction because the tertiary alcohols are more readily separated from the diacylglycerols by TLC than those generated when methyl magnesium bromide is used. The resultant diacylglycerols may be analyzed by GLC or by the method of stereospecific analysis. The subject of Grignard hydrolysis and stereospecific analysis is discussed in detail in Chapter 4.

3.3.8. Oxidative Cleavage of Double Bonds

Reductive ozonolysis which yields simple fragments that can be analyzed by GLC has been used to locate the position of double bonds occurring in natural alk-1-enyl- and alkylglycerols (Ramachandran *et al.*, 1968; Schmid *et al.*, 1969; Snyder and Blank, 1969). The reaction of the isopropylidene derivatives with ozone and the reduction with Lindlar's catalyst (lead-poisoned palladium) can be carried out under the conditions described by Privett *et al.* (1963) for FAMEs. This subject is also treated in more detail in Chapter 1.

Alternatively, the location of double bonds may be determined by the analysis of the products resulting from oxidation with a combined permanganate–periodate reagent (Hanahan *et al.*, 1963). In this way, it has been determined that the olefinic bond of monooctadecenylglycerol from dogfish liver or bovine erythrocytes is at the 9 : 10 position of the alkyl chain.

3.3.9. Hydrogenation

Hydrogenation of unsaturated triacylglycerols or their ether analogues has been used to improve the resolution during GLC (see later) and to prevent the decomposition of polyunsaturated species during high-temperature GLC (Litchfield, 1972). The procedure used for the catalytic hydrogenation of triacylglycerols with hydrogen and platinum oxide (Adam's catalyst) is the same as that used for FAMEs (see Chapter 1).

3.4. Derivatization Procedures

3.4.1. Acetates

Neutral glycerol lipids having one or more free hydroxyl groups (monoacylglycerols, monoalkylglycerols, diacylglycerols, etc.) may be converted to the corresponding acetate (Ac) derivatives using acetic anhydride–pyridine (10 : 1, v/v). Since this ratio is not very critical, the reagent volumes may be measured by drops (e.g., from a pasteur pipette).

Procedure: To 0–10 mg neutral lipid are added 250 μl (or 20 drops) of acetic anhydride and 25 μl (or 2 drops) of dry pyridine. This is mixed thoroughly, and the solution is heated for 1 hr at 80°C. After the mixture has been chilled in ice, 350 μl methanol (or 28 drops) is added and the solution is allowed to stand for 5 min at room temperature. The acetates may be obtained by evaporating this mixture directly or by extracting with 5 ml ether plus 0.5 ml H_2O. In the latter procedure, the ether solution is dried with Na_2SO_4 and taken to dryness. Acetates are stable derivatives and are suitable for TLC, GLC, and GC/MS.

3.4.2. Isopropylidenes

The acid-catalyzed condensation of a ketone with 1,2- or 1,3-diols results in the formation of a cyclic ketal. Isopropylidene (IP) derivatives are formed from monoacyl- and monoalkylglycerols by condensation with acetone. The content of the recent literature indicates that isopropylidenes are the derivatives of choice for the analysis of monoalkylglycerols. It should be noted, however, that monoacylglycerols are likely to isomerize in an acidic environment. The following procedure is essentially that described by Hanahan *et al.* (1963).

Procedure: To 1–200 mg of the sample in 5 ml dry acetone is added 25 μl of

12 N perchloric acid, and the mixture is refluxed for 2 hr (or heated in a sealed tube at 60°C). After the reaction mixture has cooled, 10 ml ether and 7 ml H_2O are added. The ether layer is taken off, and the aqueous phase is extracted twice more with 7-ml aliquots of ether. The combined ether extract is then washed with water, evaporated to dryness, and redissolved in an appropriate solvent. Alternatively, according to Wood (1967), 10 mg of sample is allowed to react for 15 min at room temperature with 1 ml acetone and 1 μl 60% perchloric acid. The perchloric acid is then neutralized with excess ammonium hydroxide and the sample is extracted with diethyl ether. Isopropylidene derivatives of monoacyl- and monoalkylglycerols have been found suitable for TLC, GLC, and GC/MS.

3.4.3. Trimethylsilyl Ethers

Various procedures have been used to make the trimethylsilyl ether (TMS) derivatives of mono- and diacylalkylglycerols. In the case of the acylglycerols, however, care should be taken not to cause isomerization. Watts and Dils (1969) reported that isomerization of 1- and 2-monoacylglycerols takes place upon silylation with hexamethyldisilazane (HMDS) and trimethylchlorosilane (TMCS, 2:1, v/v), but other reports (Wood *et al.*, 1965; Myher and Kuksis, 1974) indicated that there is no isomerization when pyridine–HMDS–TMCS is used. The extra peaks observed by Watts and Dils (1969) and attributed to isomerization can also be explained by the presence of partially silylated products. However, it has been reported that the use of TRI SIL/BSA (Pierce Chemicals), a formulation containing pyridine, *N,O*-bis(trimethylsilyl)acetamide (BSA), and TMCS, causes isomerization of 1- and 2-monoacylglycerols (Myher and Kuksis, 1974) and 1,2(2,3)- and 1,3-diacylglycerols (Myher and Kuksis, 1975), while the use of pyridine–HMDS–TMCS (12:5:2, v/v) does not cause isomerization.

Procedure: To 0–10 mg neutral lipid are added 350 μl (or 22 drops) pyridine, 150 μl (or 12 drops) HMDS, and 60 μl (or 5 drops) TMCS, and the solution is thoroughly mixed. The reaction is usually complete within 15 min at room temperature. An aliquot (1.2 μl) of this mixture may be used directly for GLC or GLC/MS. Alternatively, the reaction mixture may be taken to dryness and the TMS ethers extracted with approximately 5 ml of petroleum spirit. The petroleum spirit extract is washed with 1 ml of water, dried over Na_2SO_4, and taken to dryness under nitrogen. TMS ethers are suitable derivatives for GLC and GC/MS, but since they slowly hydrolyze even in nonacidic solvent systems such as chloroform–methanol (O'Brien and Klopfenstein, 1971) TMS ethers are not very suitable for TLC.

3.4.4. Tertiarybutyldimethylsilyl Ethers

Tertiarybutyldimethylsilyl (*t*-BDMS) derivatives have recently been used for fatty acids (Phillipou *et al.*, 1975), steroids and prostaglandins (Kelly and Taylor, 1976), and diacylglycerols (Myher *et al.*, 1977*a*). These derivatives are approximately 10^4 times more stable than the corresponding TMS ethers (Sommer, 1965) and do not hydrolyze provided that strongly acidic conditions are avoided. Corey and Venkateswarlu (1972) were able to increase the rate of formation of the *t*-BDMS ethers by using imidazole as a catalyst and dimethylformamide as solvent.* The following procedure is based on that of Corey and Venkateswarlu (1972).

Procedure: A mixture made of 0–10 mg neutral lipid and 150 μl *t*-butyldimethylchlorosilane/imidazole reagent (Applied Science) is heated at 80°C for 20 min. After cooling, the reaction mixture is thoroughly mixed with 5 ml light petroleum spirit and washed three times with 0.5 ml water. The petroleum extract is then dried over Na_2SO_4 and taken to dryness under a stream of nitrogen. No isomerization of mono- or diacylglycerols has been observed (as determined by GC/MS) to take place during derivatization. The *t*-BDMS ethers of acylglycerols are suitable for TLC, GLC, and GC/MS.

3.4.5. Other Derivatives

For the preparation of derivatives such as methoxy ethers, trifluoroacetates, or any other derivatives referred to later, the reader may consult the original references that are cited.

3.5. Determination of the Structural Components of Glycerolipids

This section includes the identification and quantitation of the structural components that can be derived from the more complex neutral glycerolipids. This includes the glycerol backbone, fatty acids (see Chapter 1), fatty alcohols, and alkyl- or alk-1-enylglycerols.

3.5.1. Glycerol

Glycerol may be determined by GLC as the triacetyl (Horrocks and Cornwell, 1962) or trimethylsilyl (Roberts, 1967) derivative following the release of glycerol by saponification, basic alcoholysis, reductive hydrogenation, or, in the case of phospholipids, acetolysis followed by saponification (Roberts, 1967). The following procedure was introduced by Holla *et al.* (1964).

Procedure: The acylglycerol (30–100 mg) and 30 mg internal standard are added to a round-bottom flask and dissolved in 20 ml of anhydrous diethyl ether. A solution of 200 mg $LiAlH_4$ in 30 ml ether is cautiously added to the glyceride solution until the bubbling stops. After the addition of excess $LiAlH_4$ (50%), the solution is refluxed for $1\frac{1}{2}$ hr. The excess $LiAlH_4$ is then decomposed by the cautious addition of acetic anhydride until all bubbling stops. After 25 ml acetic anhydride and 30 ml xylene have been added, the ether is removed by evaporation and the solution is refluxed for a further 6 hr. The solution is then filtered, evaporated to dryness on a flask evaporator, and dissolved in ether for GLC. Methyl eicosanoate or methyl erucate has been used as internal standard. The triacetyl derivative of glycerol has the same retention time as octadecanyl acetate when chromatographed on an ethylene glycol succinate (EGS) column at 205°C.

* Applied Science supplies a *t*-BDMS reagent kit containing ampules of 1 mmol *t*-butyldimethylchlorosilane and 2–5 mmol imidazole dissolved in 1 ml of dimethylformamide.

The following procedure can be used to generate glycerol triacetate from small samples (<1 mg) of neutral acylglycerols (J. J. Myher and A. Kuksis, unpublished data).

Procedure: To the acylglycerol sample (0.5–5 mg) is added 150 μl of a solution of sodium methoxide (0.5 N) in anhydrous methanol. It is mixed thoroughly and allowed to stand for 5 min at room temperature. Then 1 drop (10 μl) of acetyl chloride is added in order to acidify the solution, and the solvent is evaporated under a stream of nitrogen. The FAMEs are then recovered by extracting the residue twice with 2 ml of hexane (or pentane), and the resulting residue is acetylated by adding 250 μl acetic anhydride plus 50 μl pyridine and heating for $1\frac{1}{2}$ hr at 80°C. To the chilled solution (0°C) is added 150 μl methanol. This is left for 10 min at room temperature. Four milliliters of hexane is added, and the solution is washed twice with 0.2 ml distilled water and dried over sodium sulfate. The triacetin is then concentrated by cautious evaporation of solvent (no heating!) under a stream of nitrogen.

A useful internal standard for the above procedure is 1,2-diheptadecanoyl-3-methoxypropane. Methyl heptadecanoate generated during the methanolysis can be used to quantitate the FAMEs, and 1,2-diacetoxy 3-methoxypropane generated during the acetylation step can be used to quantitate the triacetin. GLC of the triacetin and FAME fractions may be carried out on a 10% EGSS-X column at 190°C. The following relative retention times apply: 1,2-diacetoxy-3-methoxy propane, 0.655; methyl palmitate, 1.000; and triacetin, 1.83. Triacetin may be further identified by mass spectrometry; the *m/e* values of the major ion fragments in the mass spectrum of triacetin along with the relative intensities (in parentheses) are as follows: *m/e* 158 (1), 145 (22), 116 (10), 115 (7), 103 (22), 43 (100). This procedure results in a positive identification of glycerol and can also be used to identify other diols or triols that in esterified form may be mistaken for glyceryl esters. With mass spectrometric analysis of the triacetin, it is also possible to measure the incorporation of stable isotopes (^{2}H,^{13}C) into the glycerol backbone of glyceryl esters (Kuksis *et al.*, 1975*b*; Kuksis and Myher, 1976).

Glycerol may also be selectively determined by enzymatic methods following its release from acylglycerols by saponification (Witter and Whitner, 1972). Two enzymatic combinations are popular: The glycerol residue resulting from saponification is incubated 10–30 min at 30°C with ATP, NAD, $MgCl_2$, and a buffered solution of glycerol kinase. After the absorption at 340 nm has been read, the solution is incubated a further 30 min with glycerol dehydrogenase. The change of absorption at 340 nm represents the reduction of glycerol phosphate (Litchfield, 1972).

An alternate procedure requires a measurement at 340 nm after the addition of ATP, phosphoenolpyruvate, NADH, $MgCl_2$, pyruvate kinase, and lactate dehydrogenase. Glycerol kinase is added, and after 10 min a reading is made at 340 nm. The difference is proportional to the quantity of glycerol (Litchfield, 1972).

Released glycerol can also be oxidized with periodate to produce formaldehyde that can be determined by the chromotropic acid method (Blankenhorn *et al.*, 1961). The procedure, although very sensitive, is not specific for glycerol but gives a reaction for all 1,2-diols.

3.5.2. Fatty Alcohols

The fatty acid moieties of glyceryl esters are best determined as FAMEs (see Chapter 1), but they may also be determined via the alcohol derivatives generated by reductive hydrogenation with $LiAlH_4$ or Vitride reagent. They are easily chromatographed quantitatively on polar columns (EGS, EGSS-X) as acetates (Horrocks and Cornwell, 1962), trifluoroacetates (TFAs), TMS ethers (Wood, 1968), or *t*-BDMS ethers (J. J. Myher and A. Kuksis, unpublished data). The Ac and *t*-BDMS derivatives are very stable derivatives that may be separated by TLC on silica gel and silica gel impregnated with silver nitrate. As in the case of FAMEs (see Chapter 1), argentation TLC is required to remove the ambiguities that remain if GLC alone is used for the identification of the alcohols. The *t*-BDMS ethers are particularly suitable for GC/MS; the abundant $[M - 57]^+$ ion, due to loss of a butyl radical, makes molecular weight determination possible even for polyenoic species such as 22 : 6. In this respect, the *t*-BDMS ethers of fatty alcohols are better than FAMEs.

3.5.3. Monoalkyl- and Monoalk-1-enylglycerols

Subsequent to isolation by TLC and derivatization, the monoalkyl- and monoalk-1-enylglycerols can be identified and quantitated by GLC (see later). However, if only the total amount of these lipids is required, it is possible to use a spectrophotometric procedure such as that developed by Blank *et al.* (1975). The phosphoryl base moiety and the acyl groups of a phospholipid can be removed either by phospholipase C hydrolysis (*Bacillus cereus*) followed by saponification or by reduction with Vitride reagent. After separation of the alk-1-enyl- and alkylglycerols by TLC (see later), they are converted to the corresponding glycolic aldehydes by oxidation with periodate. The aldehydes are then reacted with fuchsin reagent (Gray, 1969), and the resulting colored complexes are measured spectrophotometrically. The alkylglycolic aldehydes resulting from periodate oxidation can also be quantitated by reaction with *p*-nitrophenylhydrazine followed by spectrophotometric measurement at 380 nm (Gelman and Gilbertson, 1969; Ferrell *et al.*, 1969; Schwartz *et al.*, 1969). The alk-1-enylglycerols may also be converted to dimethylacetals (DMAs) by acidic methanolysis (see Chapter 1) and quantitated by GLC.

3.6. Extraction and Purification of Lipids

Tissues may be extracted according to the procedure of Folch *et al.* (1957) or that of Bligh and Dyer (1959) using 20–50 vol of $CHCl_3$–CH_3OH–H_2O (2 : 1 : 0.8, v/v/v). Nonlipid components may be removed by aqueous partition following the addition of 1 vol $CHCl_3$ and 1 vol H_2O, but the losses of certain polar lipids that occur may be avoided if the solvent of the original extract is evaporated and the residue passed through a column of Sephadex G-25 (Rouser *et al.*, 1967; Nelson, 1975). Detailed discussions of the problems associated with tissue extraction may be found elsewhere (Nelson, 1975; Radin, 1969; Burton, 1974). After the removal of nonlipids by Sephadex chromatography, the total lipids can be further fractionated by selective elution from a column of DEAE-cellulose (acetate form) (Rouser *et al.*,

1967; Burton, 1974). The neutral lipids (excluding free fatty acids) are eluted first with 8–10 vol of chloroform, and further elution with a series of solvent systems results in a number of polar lipid fractions. The DEAE-cellulose fractions can be resolved further by TLC.

3.7. *Thin-Layer Chromatography*

TLC, by virtue of its sensitivity, selectivity, and rapidity, is the method of choice in many laboratories for the separation of lipid classes. Not only is TLC useful for lipid fractionation, but it is also a good method for monitoring samples obtained from preparative separations or the products of reaction mixtures. For details concerning theory, apparatus, and general technique, the reader is referred to the TLC handbook edited by Stahl (1969).

3.7.1. *Plate Preparation*

3.7.1.1. *Silica Gel*

An aqueous slurry of silica gel, made by mixing 40 g silica gel H and 105 ml distilled water, is spread on five 20 × 20 cm plates using a spreader (e.g., Desaga) adjusted to give an adsorbent thickness of 0.5 mm. After the plates have been allowed to air dry at room temperature, they are activated for 1–2 hr at 120°C. They are then taken from the oven and used as soon as they have cooled to room temperature. If a suitable desiccator is not available, only unactivated plates should be stored. Approximately 10 mg of lipid can usually be separated on one plate, but up to 100 mg may be spotted per plate if the mixture includes only compounds having very different R_f values.

3.7.1.2. *Silica Gel Impregnated with Boric Acid*

Preparative plates are made as before by spreading a slurry made by mixing 50 g silica gel G and a solution of 2.6 g boric acid in 105 ml distilled water. Air-dried plates are activated at 120°C for 1 hr before use.

3.7.1.3. *Silica Gel Impregnated with Silver Nitrate*

The procedures used for argentation TLC differ little from those of the more conventional procedures. Generally, the adsorbent consists of 5–20% silver nitrate and 95–80% silica gel G (or the equivalent). For example, 10% $AgNO_3$-impregnated plates are made by spreading a slurry made with 50 g silica gel G and a solution consisting of 5.6 g silver nitrate and 105 ml distilled water. This slurry is sufficient to coat five 20 × 20 cm plates at a thickness of 0.5 mm. The plates must be stored in darkness and are activated for 1 hr at 115°C before use. Morris (1966) cautions that laboratory fumes may also cause the deterioration of impregnated layers.

3.7.2. Detection of Neutral Lipids

Neutral lipids are easily stained nondestructively with a solution of 2′,7′-dichlorofluorescein–CH_3OH–H_2O (0.05 : 75 : 25, w/v/v) so that the bands are visible under UV irradiation. However, if the TLC plate is then exposed to NH_3 vapor, the lipids can be seen in visible light as pink bands. Other stains such as rhodamine B can be used, but iodine should be avoided in preparative work with lipids containing unsaturated fatty acids. Although as little as 1 μg lipid can be detected by charring with sulfuric acid, sulfuric acid–potassium dichromate, or perchloric acid, this procedure is destructive and therefore not useful for preparative work.

3.7.3. Recovery of Lipids from Silica Gel

A number of procedures are available for elution of lipids from silica gel. Triacylglycerols are readily eluted with diethyl ether, but for more polar materials such as monoacylglycerols, diethyl ether–methanol (80:20, v/v) is better. All neutral lipids, however, may be completely eluted with $CHCl_3$–CH_3OH (2:1, v/v) as follows.

Procedure: The gel (a band from one plate) is extracted with 4 ml $CHCl_3$-MeOH (2 : 1, v/v), and a short centrifuge spin is used to speed the settling of the gel. The extractant is decanted, and the procedure is repeated with two further aliquots of $CHCl_3$. The combined extract is washed with 4 ml dilute aqueous ammonia (0.5 M) in order to remove the dye (2′, 7′-dichlorofluorescein) and silica gel, and then with 2 ml of distilled water. The $CHCl_3$ extract is passed through a small column of sodium sulfate and evaporated to dryness. This procedure is also suitable for extracting neutral lipids from gel impregnated with silver nitrate.

3.7.4. Separations on Silica Gel

This section deals with the TLC separation of neutral lipids. Phospholipids will be the subject of a later volume in this series. At present, however, the reviews of Renkonen and Luukkonen (1976) and Skipski and Barclay (1969) may be consulted for details concerning the TLC separation of phospholipids. The analysis of neutral lipids by TLC has also been described in numerous reviews (Blank and Snyder, 1975; Snyder, 1973; Mangold, 1969; Skipski and Barclay, 1969; Mahadevan, 1976).

The R_f values listed in Table I for a number of solvent systems illustrate the fact that any of the solvent systems described for neutral lipids is suitable for the separation of mono-, di-, and triacylglycerols. Since the presence of ether analogues, nonglycerol lipids such as cholesterol, or free fatty acids may interfere with the isolation of a pure lipid class, the choice of a particular solvent system will depend on the composition of the sample. Preliminary tests on the total lipid extract such as the analysis of the total derived FAMEs (see Chapter 1) and a search for glyceryl ethers (see Section 3.5.3) will provide important information regarding the complexity of the sample. Components that are incompletely resolved in one solvent system may be eluted and chromatographed in a different solvent system. For example, triacylglycerols and their ether analogues may be more completely resolved when rerun in hexane–diethyl ether (95:5, v/v). Some R_f values in this system are as follows: cholesterol ester, 0.83; fatty acid methyl ester, 0.53; trialkylglycerol, 0.44;

dialkylacylglycerol, 0.34; alk-1-enyldiacylglycerol, 0.16; alkyldiacylglycerol, 0.23; and triacylglycerol, 0.16 (Snyder, 1973). Effective resolution of 1,2-alk-1-enylacylglycerol acetates (R_f 0.63), 1,2-alkylacylglycerol acetates (R_f 0.57), and 1,2-diacylglycerol acetates (R_f 0.49) has been obtained by TLC on silica gel developed first in hexane–ether (1:1, v/v) and then in toluene (Renkonen, 1971). Cholesterol can be removed from diacylglycerols either directly in solvent system 5 (see Table I) or by separation in solvent system 4 following acetylation of the mixture. If desired, the separation of 1,2 and 1,3 isomers of diacylglycerol and the ether analogues can be prevented by the use of hexane–diethyl ether–aqueous ammonia (40:60:1, v/v/v) (Snyder, 1973). Although the separation of 1- and 2-monoacylglycerols has been achieved on plain silica gel using a 12-component solvent system (Kuntz, 1973), the acidity could induce isomerization. The use of borate-impregnated plates is recommended for the resolution of the isomeric monoacylglycerols.

The chain length of the fatty acids making up an acylglycerol can have a considerable effect on the R_f value. Acylglycerols containing short-chain fatty acids

Table I. R_f Values for Various Neutral Lipid TLC Solvent Systems

	Solvent system[b]					
Lipid class[a]	1	2	3	4	5	6
Hydrocarbons	0.9–1.0	0.98	0.98	—	—	0.92
Sterol esters	0.90	0.94	0.95	0.87	—	0.85
Wax esters	0.90	0.88	0.93	—	—	
Fatty acid methyl esters	0.65	0.77	0.73	—	—	0.74
Dimethyl acetals	—	0.73	0.66			
Aldehydes	0.55	—	0.64	—	0.81	0.70
Alkenyldiacylglycerols	0.65	0.82		—	—	—
Alkyldiacylglycerols	0.55	0.78	0.82	—	—	—
Triacylglycerols	0.35	0.60	0.66	0.67	0.86	0.70
Cholesterol acetate	—	—	—	0.68	—	—
Fatty acids	0.18	0.39	0.30	0.54	0.53	0.64
1,3-Diacylglycerol acetate	—	—	—	0.47	—	—
1,2-Diacylglycerol acetate	—	—	—	0.38	—	—
Alcohols	0.15	0.30	0.17	—	0.49	0.49
Sterols	0.10	0.19	0.11	0.21	0.44	0.37
1-Monoacylglycerol diacetates	—	—	—	0.21	—	—
1,3-Alkylacylglycerol	—	—	—	—	0.75	—
1,3-Diacylglycerol	0.08	0.21	—	0.20	0.70	0.47
1,2-Alkylacylglycerol	—	—	—	—	0.66	—
1,2-Diacylglycerol	0.08	0.15	—	0.15	0.61	0.42
Monoalkenylglycerol					0.26	—
Monoalkylglycerol	0.00	0.03	—		0.22	0.13
Monoacylglycerol	0.00	0.02	—	0.02	0.18	0.11

[a] In most cases, R_f values refer to lipids having hydrocarbon, fatty acid, alkyl ether, alcohol, or aldehyde chain lengths of 16 carbons, for example, hexadecane or tripalmitin.

[b] Solvent 1: petroleum ether (60–70°C)–diethyl ether–acetic acid (90:10:1, v/v/v). Solvent 2: petroleum ether (60–70°C)–diethyl ether–acetic acid (80:20:1, v/v/v). Solvent 3: hexane–diethyl ether–acetic acid (80:20:1, v/v/v). Solvent 4: heptane–isopropyl ether–acetic acid (60:40:4, v/v/v). Solvent 5: chloroform–methanol–acetic acid (98:2:1, v/v/v). Solvent 6: first development, isopropylether–acetic acid (96:4, v/v), second development, petroleum ether–diethylether–acetic acid (90:10:1, v/v/v). The data for solvents 1 and 2 were taken from Kates (1972), for solvent 3 from Blank and Snyder (1975), for solvent 4 from Kuksis and co-workers, for solvent 5 from Snyder (1973), and for solvent 6 from Skipski and Barclay (1969).

Table II. R_f and ECL Values for the Acetate Derivatives of Some Short-Chain Diacylglycerols

Species	R_f[a]	ECL[b]
1-Octadecanoyl-2-acetoyl-*rac*-glycerol	0.239	18.000
1-Hexadecanoyl-2-butanoyl-*rac*-glycerol	0.266	16.975
1-Tetradecanoyl-2-hexanoyl-*rac*-glycerol	0.276	16.667
1-Dodecanoyl-2-octanoyl-*rac*-glycerol	0.288	16.475
1-Decanoyl-2-decanoyl-*rac*-glycerol	0.291	16.466

[a] R_f values of diacylglycerols for TLC on silica gel in heptane–isopropyl ether–acetic acid (60 : 40 : 4, v/v/v).

[b] ECL values for acetate derivatives of diacylglycerols relative to the set of saturated monoacylglycerol diacetates. GLC separations made on a 183-cm 3% EGSS-X column at 240°C (J. J. Myher and A. Kuksis, unpublished results).

are more polar than their long-chain counterparts, and have smaller R_f values. The triacylglycerols of bovine milk, for example, have been separated on silica gel into fractions made up of long-, medium-, and short-chain triacylglycerols (Kuksis and Breckenridge, 1968). However, in some solvent systems the presence of free fatty acids can interfere with the separation of certain short-chain triacylglycerols. This problem is illustrated by the fact that the R_f value of monoacetyldiacylglycerols in solvent 4 (see Table I) is actually smaller than the R_f value of the long-chain free fatty acids. Parodi (1975) isolated the monoacetyldiacylglycerols from milk fat using hexane–ethyl acetate (88 : 12, v/v), and the free fatty acids remained below the 1,2-diacylglycerols. Table II contains the R_f values of a series of synthetic 1,2(2,3)-diacylglycerols that have the same total carbon number, and it is apparent that, like the triacylglycerols, the R_f values decrease as the chain length of the shortest fatty acid decreases. The effect of additional polar functional groups is illustrated by the fact that triacylglycerols containing hydroxy fatty acids (up to three hydroxyls per fatty acid) have much lower R_f values (Mikolajczak and Smith, 1967).

A number of substituted alkylglyceryl ethers have also been resolved by TLC. Muramatsu and Schmid (1972) used TLC on silica gel H in hexane–diethyl ether (50 : 50, v/v) to separate 1-*O*-(2-ketoalkyl)-2,3-isopropylidene glycerol (R_f 0.73), 1-*O*-acetoxyalkyl)-2,3-isopropylidene glycerol (R_f 0.86), 1-*O*-(2-acetoxyalkyl)-2,3-diacetoxyglycerol (R_f 0.61), 1-*O*-(2-hydroxyalkyl)-2,3-isopropylidene glycerol (R_f 0.42), and 1-*O*-(2-hydroxyalkyl)glycerol ($R_f < 0.1$). Hallgren and Ställberg (1967) also resolved the 1-*O*-(2-methoxyalkyl)glycerols (R_f 0.41) from the 1-*O*-alkylglycerols (R_f 0.56) by TLC on silica gel G developed in trimethylpentane–ethyl acetate–methanol (50 : 40 : 7, v/v/v). Kasama *et al.* (1973) found that 1-*O*-(*O*-acylalkyl)-2,3-diacylglycerols were not resolved from triacylglycerols by TLC when double development in hexane–diethyl ether–acetic acid (90 : 10 : 1, v/v/v) was used but ran lower (R_f 0.15) than triacylglycerols (R_f 0.27) when double development in benzene was used.

3.7.5. Separations on Silica Gel Impregnated with Borate or Arsenite

Excellent separation of the isomers of monoacyl- or monoalkylglycerols may be achieved on silica gel impregnated with boric acid (Thomas *et al.*, 1965) or arsenite

Table III. R_f Values on Silica Gel Impregnated with Boric Acid

Lipid class	R_f value[a]
Triacylglycerols	0.85
Tertiary alcohols[b]	0.67
1,3-Alkylacylglycerol	0.63
1,3-Diacylglycerol	0.57
1,2-Alkylacylglycerol	0.50
Fatty acids	0.36
2-Monoalkylglycerol	0.13
2-Monoacylglycerol	0.11
1-Monoalkylglycerol	0.07
1-Monoacylglycerol	0.05

[a] R_f values for TLC on silica gel 6 containing 5% boric acid (w/w) in chloroform–acetone (96:4, v/v) according to Thomas *et al.* (1965) and J. J. Myher and A. Kuksis (unpublished data).
[b] Long-chain tertiary alcohols resulting from the reaction of ethyl magnesium bromide with fatty acid esters.

(Wood and Snyder, 1967*a*). The isomerization of the monoacylglycerols is prevented in these systems due to the formation of chelate complexes. Table III contains the R_f values for a number of compounds separated on boric acid (5%, w/w) impregnated silica gel G in $CHCl_3$–acetone (96:4, v/v). In this system, 2-monoacylglycerols run higher than the 1 isomer and the alkylglycerols are similarly separated, with the ethers running somewhat higher than the corresponding esters. The R_f values of the monoacylglycerols may be increased by using chloroform with higher concentrations of acetone or by the addition of methanol (Thomas *et al.*, 1965). The elution order of 1- and 2-alkylglycerol is reversed on arsenite-impregnated silica gel, and when $CHCl_3$–ethanol (90:10, v/v) is used as a solvent the ratio of R_f values for the two isomers is 1.3 on borate/silica gel and 0.56 on arsenite/silica gel (Snyder, 1973).

The borate system also leads to an excellent separation of 1,3- and 1,2-diacylglycerols as well as a separation of 1,3- and 1,2-acylalkylglycerols. Alternatively, Pollack *et al.* (1971) prevented isomerization by using a four-directional development procedure incorporating trimethyl borate for the resolution of complex mixtures of lipids containing neutral lipids, glycolipids, and phospholipids.

3.7.6. Separations on Silica Gel Impregnated with Silver Nitrate

Argentation chromatography, especially argentation TLC, is one of the most important techniques used by the glyceride chemist. Separations effected by argentation chromatography are due to the formation of a reversible complex between silver ions and compounds containing ethylenic or acetylenic bonds. Following the original suggestion of Nichols (1952), the argentation principle has been used in conjunction with countercurrent distribution, liquid chromatography (LC), GLC, and TLC. Of these procedures, argentation TLC is the most useful. Morris (1966) used argentation TLC for the separation of saturated and unsaturated esters, including the separation of oleate and elaidate, and Barrett *et al.* (1962) first described the separations of glyceride mixtures on silver nitrate-impregnated thin layers.

3.7.6.1. *Triacylglycerols*

Triacylglycerols previously separated by conventional TLC can be further resolved by argentation TLC into bands having the same total number of double bonds, but the choice of solvent system will depend somewhat on the complexity of the sample. Triacylglycerols, for example, may be resolved on 10–20% silver nitrate/silica gel using hexane–diethyl ether (70:30, v/v) or chloroform–methanol (100:0 to 94:6) (Roehm and Privett, 1970). In general, up to ten bands can be resolved on a single plate, and sometimes further resolution can be achieved if the material from a particular band is rerun in the same solvent system. Polyunsaturated triacylglycerols that are crowded near the origin when $CHCl_3$–MeOH (99.2:0.8, v/v) is used may be separated by rerunning in a more polar system such as $CHCl_3$–MeOH (95:5, v/v). The separation factors are determined mainly by the formation constants of the silver ion–olefin complexes, and the elution order from top to bottom of triacylglycerols* containing saturated, monoenoic, linoleic, and linolenic acids is as follows: 000, 001, 011, 002, 111, 012, 112, 022, 003, 122, 013, 222, 113, 023, 123, 223, 033, 133, 233, 333 (Gunstone and Padley, 1965). This sequence can be predicted if each fatty acid is assigned a complexing power as follows: saturated = 0, monoene = 1, diene = $2 + a$, and triene = $4 + 4a$, with $a < 1$, and the total complexing power for a triacylglycerol (P) is calculated by summing the values for the individual fatty acids. For example, we could predict that 011 with $P = 2$ would run above 002 with $P = 2 + a$. In general, it appears that double bonds that are closer to the methyl-terminal end of the fatty acid have a greater complexing power than double bonds that are farther from the end. In addition, there is a stronger interaction between silver ions and *cis* double bonds than between silver ions and *trans* double bonds.

Although only triacylglycerols based on the common fatty acids have been considered above, the same principles can be applied to triacylglycerols containing more unusual fatty acids. However, it should be noted that there is a reaction of the cyclopropene ring of cyclopropene fatty acids with $AgNO_3$ that leads to a mixture of products. Ether analogues, where one of the acyl groups has been replaced by an ether linkage, could also be separated by argentation TLC.

3.7.6.2. *Diacylglycerols and Ether Analogues*

Argentation TLC can also be used to resolve diacylglycerols that are isolated in free form, derived from phospholipids (phospholipase C, acetolysis), or derived from triacylglycerols by partial hydrolysis (pancreatic lipase, Grignard) (Kuksis, 1972*a*). The technique for the separation of diacylglycerols by argentation TLC, first described by Barrett *et al.* (1962), was subsequently used to study the effect of diet on the composition of rat liver lecithin (Van Golde and Van Deenan, 1966; Van

* Defined by a three-number code, each number signifies the degree of unsaturation of each of the three fatty acids contained in the triacylglycerol molecule.

Golde *et al.*, 1968). The diacylglycerols from rat liver lecithin were eluted with $CHCl_3$–ethanol (93 : 7, v/v) in the order (from top to bottom of the plate) 00, 01, 11, 02, 12, 03, 03, and 04,* where the upper 03 species contains 20 : 3 ω9 and the lower 03 species contains 20 : 3 ω7. Thus the fractionation of diacylglycerols on $AgNO_3$/silica gel depends on the total number of double bonds, the distribution of double bonds in the 1 and 2 positions of glycerol, and the double bond position within the component fatty acids.

Since diacylglycerols may isomerize during TLC, it is better to derivatize them prior to chromatography. Acetates (Renkonen, 1966; Kuksis and Marai, 1967) or *t*-BDMS ethers (J. J. Myher and A. Kuksis, unpublished data) are suitable for this purpose. The acetates are well resolved on 20% $AgNO_3$/silica gel G and developed in chloroform–methanol (99.0 : 1.0, v/v) (Kuksis *et al.*, 1968*a*). The elution order of the former list of species is the same as that observed for the underivatized diacylglycerols. In addition, the R_f values decrease in the series 04, 14, 05, 06. Species containing up to 12 double bonds have been detected in the lecithin of Baltic herring (Renkonen, 1968).

Alk-1-enylacylglycerol acetates or alkylacylglycerol acetates can also be separated according to their degree of unsaturation by argentation TLC in benzene–chloroform (9 : 1, v/v) (Renkonen, 1971) or in chloroform–methanol (98 : 2, v/v) (Yeung and Kuksis, 1974). It is interesting to note that the "saturated"† alk-1-enylacylglycerol acetates run lower on argentation plates than the saturated alkylacylglycerol acetates, whereas on plain silica gel the reverse order of elution is observed. In comparison, diacylglycerol acetates having the same degree of unsaturation elute midway on $AgNO_3$/silica gel between these two types of ether lipids.

3.7.6.3. Monoacylglycerols and Ether Analogues

Because it is possible to resolve monoacyl- and monoalkylglycerol derivatives on the basis of carbon number and unsaturation by GLC, argentation TLC is not generally required. However, based on the experience obtained with FAMEs and glyceryl esters, there should be no trouble resolving any stable derivative such as the acetates or isopropylidenes. For example, Hallgren and Ställberg (1967) separated the saturated and monounsaturated 2-methoxyalkylglycerols by argentation TLC using trimethylpentane–ethyl acetate–methanol (50 : 40 : 7, v/v/v), and Wood and Snyder (1966) separated monoalkylglycerols with 0, 1, and 2 double bonds on 20% $AgNO_3$/silica gel G in chloroform–ethanol (90 : 10, v/v).

3.7.7. Reversed-Phase TLC

Separations within a lipid class according to carbon number and the number of double bonds can be achieved by reversed-phase partition TLC (Skipski and Barclay, 1969). For reversed-phase TLC, the silica gel is coated with a hydrophobic

* Similar to the triacylglycerols, the diacylglycerols having different degrees of unsaturation are described by a two-number code indicating the number of double bonds in each of the two fatty acids.

† The double bond of the vinyl linkage is not included in the total degree of unsaturation when describing molecular species of alk-1-enylacylglycerols.

substance such as paraffin oil, silicone oil, or undecane, and development is carried out with a hydrophilic solvent system. In order to avoid the overlap of certain "critical pairs" such as tripalmitoyl-/trioleoylglycerol and trimyristoyl-/trilinoleoylglycerol (Mangold, 1969), the sample should first be separated according to unsaturation by means of argentation TLC. Each of the fractions could then be separated by reversed-phase TLC according to carbon number, with the higher-molecular-weight molecules lower on the plate.

Impregnated adsorbent layers can be prepared by allowing plates coated with silica gel G or Kieselguhr G to stand in a 5% solution of paraffin oil or silicone oil in petroleum spirit until the solution has ascended through the layer (Mangold, 1969). After the solvent has evaporated, the plates are developed in the same direction as that used for the impregnation. Triacylglycerols and derivatized mono- or diacylglycerols may be fractionated on the basis of chain length using solvents of acetic acid or acetonitrile containing 1–10% water and 80–100% saturated with the impregnation agent. Monoalkylglycerols of uniform degree of unsaturation may also be separated on siliconized silica gel G in acetone–water (65 : 35, v/v) (Hanahan *et al.*, 1963).

Unsaturated lipids may be visualized with iodine vapor. If silicone oil has been used, charring with dichromate–sulfuric acid is possible, but, in the case of preparative work, detection is a greater problem. An application of alcoholic 2′,7′-dichlorofluorescein followed by a short exposure to water vapor can be used to detect bands on Kieselguhr G (but not silica gel G) impregnated with paraffin. GLC has largely replaced this technique for the quantitative analysis of neutral lipids according to carbon number.

3.7.8. Quantitation

One of the most reliable methods of quantitation of neutral lipids is by GLC (see Section 3.9.5). Since any single TLC system does not guarantee the complete separation of all neutral lipid classes, the additional separation by GLC may be necessary to insure the resolution of all classes. Figure 1 illustrates how a variety of lipid classes may be measured quantitatively even without TLC separation. Fatty acid esters can also be quantitated as FAMEs by GLC (see Chapter 1) after the addition of methyl heptadecanoate to the sample and methanolysis.

Alternatively, neutral lipids, charred with sulfuric acid or sulfuric acid–potassium dichromate at 185°C, may be quantitated by photodensitometry (Privett *et al.*, 1965; Nutter and Privett, 1968; Blank and Snyder, 1975). In order to obtain meaningful results, it is necessary to use uniform adsorbent layers and to ensure that the charring reagent is sprayed evenly. Appropriate standards must also be run on the same plate.

3.8. Liquid Chromatography

Liquid chromatography (LC) in various forms has long been used for the fractionation of crude lipid extracts. Chromatography on Sephadex G-25 has replaced the aqueous wash as the method of choice for the removal of nonlipid

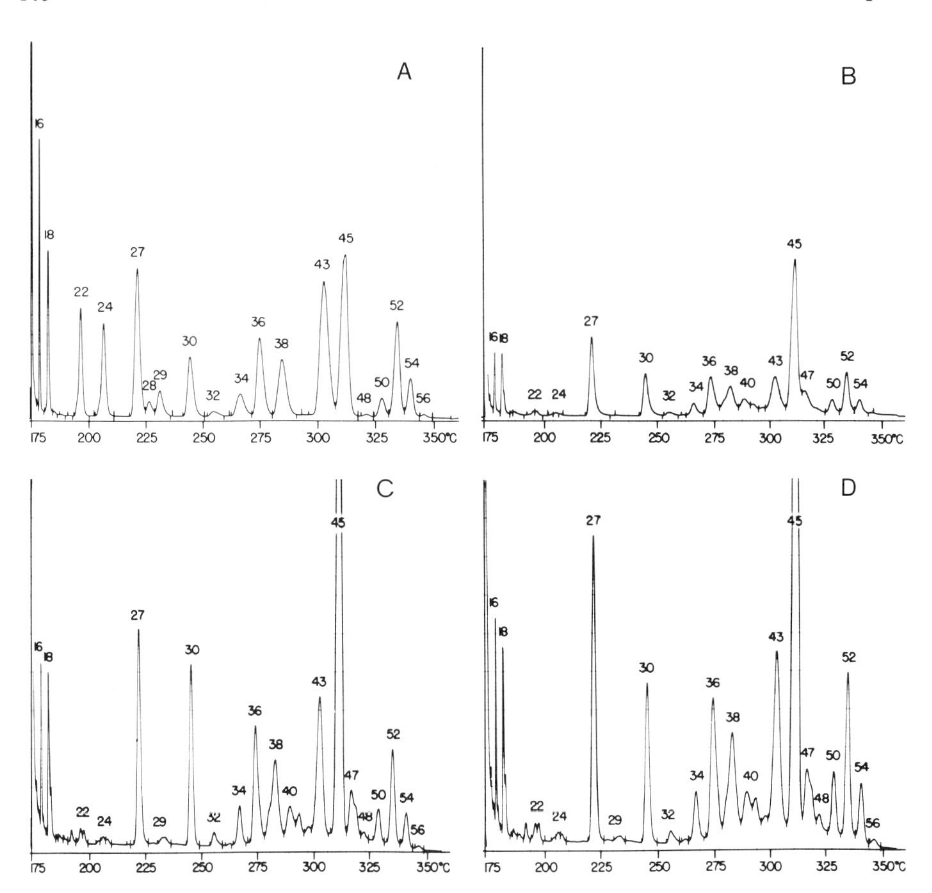

Fig. 1. GLC elution patterns of a synthetic mixture of neutral lipids and free fatty acids (A) and the lipids of human blood plasmas after phospholipase C digestion (B, C, D). Peaks 16 and 18, TMS esters of C_{16} and C_{18} fatty acids; peaks 22 and 24, TMS ethers of monoacylglycerols of C_{16} and C_{18} fatty acids; peaks 27, 28, and 29, TMS ethers of cholesterol, campesterol, and β-sitosterol; peak 30, tridecanoylglycerol internal standard; peaks 34, 36, and 38, TMS ethers of diacylglycerols corresponding in retention time to triacylglycerols with a total of 34, 36, and 38 acyl carbons; peaks 43, 45, and 47, cholesteryl esters of C_{16}, C_{18}, and C_{20} fatty acids, respectively; peaks 48–54, triacylglycerols of 48–54 total acyl carbons per molecule. Instrument: Hewlett-Packard model 5700 with unheated on-column injector, dual columns, differential electrometer, electronic peak area integrator, and an automatic liquid sample injector. Columns: stainless steel tubes, 50 cm × 2 mm inside diameter, packed with 3% OV-1 on Gas Chrom Q (100–120 mesh). Carrier gas, nitrogen at 80 ml/min. Detector, 350°C. Column temperature programmed at 4°C/min. Sample, 1 μl of a 1% solution of total lipid in silylation mixture. Attenuation, 1/100 full sensitivity. Vertical lines intersecting the baseline are the event marks of the peak area integrator.

components in total lipid extracts (Nelson, 1975) and chromatography on silicic acid or Florisil may be used to further separate the sample into several neutral and polar lipid fractions (Rouser, 1973; Carroll, 1976). These crude fractions are subsequently separated by TLC.

Although high-resolution liquid chromatography (HRLC) may approach the efficiency of open tubular gas chromatography (Kirkland, 1972), the application to lipids has been rather limited. This is partly due to the inadequacy of UV and refractive index type detectors, as well as complications arising out of the need for two or three different solvent gradients. To date, a flame ionization detector in conjunction with a moving-wire sample transport system appears to be the best universal detector (Privett and Erdahl, 1975; Aitzetmüller, 1975).

Recently, however, advances have been made in coupling a liquid chromatographic system with a mass spectrometric detector. Scott *et al.* (1974) used a wire transport system to convey samples to a quadrupole mass spectrometer, whereas McLafferty and co-workers have introduced 1% of the LC effluent directly into a chemical ionization mass spectrometer (Arpino *et al.*, 1974*a*, *b*; McLafferty *et al.*, 1975). Horning *et al.* (1974) and Carroll *et al.* (1975) have described a system that allows the entire effluent stream from the chromatograph into an atmospheric pressure ionization mass spectrometer where ions are formed in a corona discharge ion source. Such combinations could become powerful analytical tools on a par with GC/MS systems.

The potential of HRLC for neutral lipid separations can be illustrated by some of the interesting applications that have been reported. Nickel and Privett (1967), for example, presented an improved column procedure for the separation of triacylglycerols according to the chain length of the fatty acids using a column of silanized Celite, and Joustra *et al.* (1967) separated tristearoyl-, tricaproyl-, and triacetoylglycerol on a column of Sephadex LH-20 (a hydroxypropyl ether of Sephadex G-25) using elution with chloroform. However, Sephadex LH-20 is too polar and cannot be used for reversed-phase separations of neutral lipids (Sjövall *et al.*, 1968). Ellingboe *et al.* (1968, 1970) have described reversed-phase separations on columns of hydroxy (C_{15}–C_{18}) alkoxypropyl Sephadex, and Nyström and Sjövall (1975) have reviewed the preparation and use of these Sephadex derivatives.

An improved method for the separation of triacylglycerols has been described by Lindqvist *et al.* (1974). Using a linear gradient of two solvents (I, isopropanol–chloroform–heptane–water, 115:15:2:35, v/v/v/v; II, heptane–acetone–water, 4:15:1, v/v/v) with a column temperature of 40°C, they were able to obtain an excellent separation of saturated C_9–C_{56} triacylglycerols. A typical run takes about 20 hr and uses 3 liters of eluent. Since unsaturated triglycerides are eluted the equivalent of 1.42 CH_2 units earlier for each double bond in the molecule, each of the peaks observed for the separation of butterfat contained triacylglycerols of several carbon numbers and degrees of unsaturation.

Free 1,2-diacyl-*sn*-glycerols could not be analyzed on modified Sephadex columns without extensive isomerization, but Curstedt and Sjövall (1974) solved the problem by using the TMS ethers, which also have good characteristic mass spectra. 1,2-Diacyl-3-TMS-*sn*-glycerols were separated on hydroxyalkoxy Sephadex prepared from Sephadex LH-20 and C_{11}–C_{14} alkyl epoxide (Curstedt and Sjövall, 1974). The sample, dissolved in acetone, was injected 5–10 mm below the gel surface and eluted using a solvent system of acetone–water–heptane (87:13:10, v/v/v) containing 1% pyridine to prevent acid-catalyzed hydrolysis (Sjövall *et al.*, 1968); tests indicated that 94.8 ± 0.3% of the TMS ethers remained unchanged after passing through the column. Curstedt and Sjövall (1974) separated the 1,2-diacylglycerols derived by phospholipase C hydrolysis of rat bile phosphatidylcholines and found that diacylglycerols having the same number of double bonds per molecule were eluted in order of increasing chain length. Each additional double bond shifted the retention volume the equivalent of two methylene units. Thus four peaks were observed: two small peaks corresponding to 38 : 6* and to 32 : 2 + 34 : 3 + 36 : 4, and two major peaks corresponding to 32 : 1 + 34 : 2 + 36 : 3 and to 32 : 0 + 34 : 1 + 36 : 2. The contents of each peak were then separated according to carbon number by GLC.

For further details on the theory, technique, and application of liquid column chromatography, a number of reviews may be consulted (Privett and Erdahl, 1975; Nyström and Sjövall, 1975; Rouser, 1973; Aitzetmüller, 1975; De Stefano and Kirkland, 1975*a*, *b*).

3.9. Gas Chromatography

GLC is a very powerful tool capable of providing resolutions based on molecular weight, degree of unsaturation, and to some extent geometric configuration. Many ambiguities that remain after simple TLC are resolved by GLC. This section contains a general description of the nature of GLC resolutions, as well as a more detailed account of the separations possible within single lipid classes. For further information, the reader may consult a number of comprehensive reviews (Kuksis, 1967, 1975, 1976). In addition, the literature of gas chromatography is reviewed every 2 years (e.g., Cram and Juvet, 1976).

3.9.1. Column Packings

The preparation of a good GLC column has been made considerably easier by the commercial availability of high-quality column packings from GLC supply companies. At a cost only slightly greater than that charged for stock items, a number of these companies will custom make any combination of liquid phase (at any percent loading) and solid support carried by the company. Among the many high-quality solid supports commercially available are the following: Gas Chrom Q, Supelcoport, and Chromosorb W HMDS. Currently, mesh sizes 80–100 and 100–120 are probably the most suitable.

If one looks through the catalogue of a GLC supplier, he is faced with a large number of liquid phases. Fortunately, there are not that many fundamentally different liquid phases that are important for the resolution and quantitation of the neutral glycerol lipids. For the purpose of lipid analysis, the two most important properties of a liquid phase are its polarity and its upper temperature limit. Liquid-phase polarity can be defined in terms of a set of McReynolds constants (McReynolds, 1970), which are now often being quoted by suppliers; the polarity is greater for the liquid phase having the higher value for these constants. Since the GLC of higher-molecular-weight compounds requires higher column temperatures, an upper temperature limit is effectively an upper molecular weight limit. Consequently, only the low-polarity liquid phases such as the dimethylsiloxanes

* The molecular species are indicated by a two-number code, total acyl carbon number : total number of double bonds.

(JXR, OV-1, SE-30, SP-2100) which have temperature limits of 350–375°C are suitable for the GLC of high-molecular-weight triacylglycerols. A polar polyester liquid phase such as ethyleneglycosuccinate–methylsilicone (EGSS-X) has an upper limit of 250°C and cannot be used for the GLC of high-molecular-weight triacylglycerols. Cyanopropylphenyl siloxanes such as Silar 5CP (50% cyanopropyl, 50% phenyl) can be used up to 375°C, have moderate polarity, and are useful for the separation of diacylglycerols (see later).

The requirements for a good liquid phase for GC/MS are even more restrictive than those required for GLC using conventional detectors. The reason for this is that excessive column bleed produces a background mass spectrum that can mask that of the eluted compound, but if the ions do not occur at the same *m/e* values as those of the compound under study they can be subtracted from the total mass spectrum. Thus for some applications a high bleed may be tolerated if the GLC separation of a more selective liquid phase is necessary. Another disadvantage of high column bleed is the requirement for more frequent bakeouts and ion source cleanings. Although the methyl and phenyl siloxanes such as OV-1, SE-30, and OV-17 are the most commonly used liquid phases, the more stable polar phases such as the cyanopropylphenyl siloxanes (e.g., Silar 5CP) have also been used.

3.9.2. General Separation Characteristics

Any of the liquid phases that have been used for the separation of neutral glycerolipids will separate a homologous series according to molecular weight, but separations based on the degree of unsaturation, the nature of the functional groups within a molecule, and those based on the isomeric form of the molecule depend on the nature of the liquid phase. In general, the ability of liquid phase to effect a separation of acylglycerols or FAMEs according to unsaturation appears to be linked to the polarity as measured by the McReynolds constants. On relatively nonpolar phases (e.g., SE-30, OV-1) there is actually some reversal so that the more unsaturated molecules are eluted earlier than the saturated compounds. As the polarity of the liquid phase increases, the reversal is counterbalanced by polar interaction with the double bonds (e.g., OV-17), and, in the case of the medium-polarity (e.g., Silar 5CP) and high-polarity (e.g., EGSS-X, Silar 10C, OV-275) liquid phases, the molecules are eluted in increasing order of unsaturation, with some retention modulation due to the differences of the various unsaturated isomers (ω value, *cis/trans*, conjugation). If a polar liquid phase is used, the introduction of more polar functional groups into a molecule will cause it to be eluted later. Diacylglycerol acetates, for example, when chromatographed on Silar 5CP, have retention times longer by a factor of 2.3 than those of the corresponding trimethylsilyl ethers (Myher and Kuksis, 1975).

The separation factor of two successive homologues decreases as the column temperature is increased (James and Martin, 1956; Landowne and Lipski, 1961) and is lower on a polar than on a nonpolar liquid phase (James and Martin, 1952; Haken, 1974). Although the homologue separation factor of short-chain acid esters is somewhat dependent on the esterified alcohol (Haken, 1974), there does not appear to be any significant dependence on the nature of the derivative for glyceryl esters of long-chain fatty acids (Myher and Kuksis, 1974).

3.9.3. Equivalent Chain Length

Equivalent chain length (ECL) values have been defined in terms of the saturated fatty acid methyl esters (FAMEs), but it is also possible to define a set of equivalent chain lengths for each lipid class in terms of the saturated homologues of that class. This becomes necessary because it is impractical to use FAMEs as retention standards for compounds of high molecular weight such as diacyl- or triacylglycerols and it is impractical to coin another term whenever a different set of reference compounds is used. Therefore, just as the saturated straight-chain FAMEs have ECL values that are equal in value to the number of carbons in the acyl chain, the equivalent chain length for saturated acylglycerols is defined as being equal to the sum of the acyl carbons. By convention, neither the carbons of glycerol nor the carbons in any derivatization agent are included. Relative to the set of saturated 1,2-diacylglycerol TMS ethers, for example, 1-palmitoyl-2-stearoyl-3-TMS glycerol would have an ECL value of 34.00. No confusion will take place if the reference set of compounds is always properly defined. ECL values are normally read off a graph of log (retention time) vs. carbon number of the saturated reference compounds. Mathematically, the ECL value (ECL_x) of an unknown compound is given by

$$ECL_x = ECL_M + \frac{\log R_x - \log R_M}{\log R_N - \log R_M}(n - m)$$

where m and n are the carbon numbers of two saturated reference compounds, M and N, respectively, and the R values are the retention times of the subscripted compounds.

3.9.4. Separation of Neutral Lipids

After separation by TLC, the glycerolipids can be analyzed further by GLC. Before particular lipid classes are discussed, the reader may gain some perspective with regard to the relative elution order by inspection of Fig. 1. The neutral lipids are eluted in the order fatty acid TMS esters, monoacylglycerol TMS ethers, sterol TMS ethers, tridecanoylglycerol (internal standard), diacylglycerol TMS ethers plus ceramide TMS ethers, sterol esters, and triacylglycerols; each lipid class is also separated according to carbon number. Therefore, we see that although fatty acid derivatives may be chromatographed at temperatures below 200°C, long-chain triacylglycerols require temperatures greater than 300°C (Kuksis, 1973).

3.9.4.1. Monoacylglycerols and Ether Analogues

After derivatization, monoacylglycerols and their ether analogues are readily resolved by GLC (Huebner, 1959). Good separations of monoacyl- and monoalkylglycerols have been obtained with the TMS ethers (Wood *et al.*, 1965; Myher and Kuksis, 1974), acetates (Huebner, 1959; Kuksis *et al.*, 1973; Myher and Kuksis, 1974), trifluoroacetates (Wood and Snyder, 1966) and isopropylidenes (Hanahan *et al.*, 1963; Snyder, 1976). Further details are also available in a number of reviews (Kuksis, 1976; Snyder, 1976; Viswanathan, 1974; Schmid *et al.*, 1975).

Since there is only one fatty acid group per molecule, the GLC separations of monoacylglycerol derivatives parallel the GLC separations of FAMEs. Monoacyl-, monoalkyl-, or monoalk-1-enylglycerol derivatives are readily resolved on the basis of carbon number even on short (50 cm) nonpolar columns (Fig. 1), and, when run as the TMS ethers, 1-alk-1-enylglycerols are eluted ahead of the alkylglycerol of corresponding carbon number (separation factor = 1.12) (Myher and Kuksis, 1974). However, in order to achieve more effective resolutions of natural mixtures of monoacylglycerols, polar columns must be used. Excellent separations of monoacylglycerol acetates have been obtained on EGSS-X (Kuksis *et al.*, 1973) or on Silar 5CP (Myher and Kuksis, 1974), and the TMS ethers have been resolved on Silar 5CP (Myher and Kuksis, 1974). Monoalkylglycerols have been chromatographed on EGSS-X as TMS ethers (Wood and Snyder, 1966) or as isopropylidenes (Snyder, 1976), and on Silar 5CP as TMS ethers or acetates (Myher and Kuksis, 1974).

Silar 5CP has a higher thermal stability than the other polar liquid phases such as EGSS-X and is therefore preferable for operation at elevated temperatures (Myher and Kuksis, 1974), especially for GC/MS where low bleed is even more desirable. Figure 2 illustrates the separation of the TMS ethers of the monoacylglycerols derived from cod liver oil by Grignard hydrolysis on Silar 5CP at 220°C. When the 1(3) or 2 isomers are chromatographed individually, good separation is obtained on the basis of unsaturation as well as molecular weight. The similarity in the elution pattern to that obtained for FAMEs (at 185°C) on Silar 5CP can be seen by comparing the ECL (relative to the respective saturated homologues) values in Table IV obtained for the monoacylglycerols and for FAMEs. It can also be seen that a similar separation is obtained for the acetate derivatives even though a higher temperature (248°C) is required. When chromatographed as the TMS ethers on Silar 5CP, the 1(3) isomers have retention times that are 1.13 times those of the corresponding 2 isomers. Figure 2C shows the elution profile for a mixture of 1(3) and 2 isomers. The overlapping pairs such as 1-18:1/2-18:2* and 1-18:0/2-18:1 can be resolved by GC/MS (see later), but prior separation of the isomers by TLC on silica gel impregnated with borate is the easiest procedure. The acetates of the isomeric monoacylglycerols are not resolved on Silar 5CP, whereas the 1(3) and 2 isomers of monoalkylglycerols are resolved as acetates (separation factor = 1.07) but not as TMS ethers. These results can be explained in terms of the greater polarity of ester bonds compared to that of ether bonds, as well as the influence of positional placement of these bonds. Interactions with the liquid phase due to chain length and unsaturation apparently do not depend significantly on the position of glycerol substitution, and a separation of isomers can be expected only when the polarity of the derivative bond (TMS ether or acetic acid ester) is sufficiently different from that of the native bond (fatty acid ester or alkyl ether). Since bonds at the primary position of glycerol have a greater interaction with the liquid phase than those at the secondary position, the isomer having the less polar bond in the secondary position will be eluted later. This explanation is not only consistent with the data obtained on Silar 5CP but also should apply to all polar liquid phases.

The TMS ethers of monoalkyl- and monoalk-1-enylglycerol are also readily separated on the basis of chain length and unsaturation (see ECL data in Table V).

* The monoacylglycerol species are identified by the following combination, isomer-number of acyl carbons : number of double bonds.

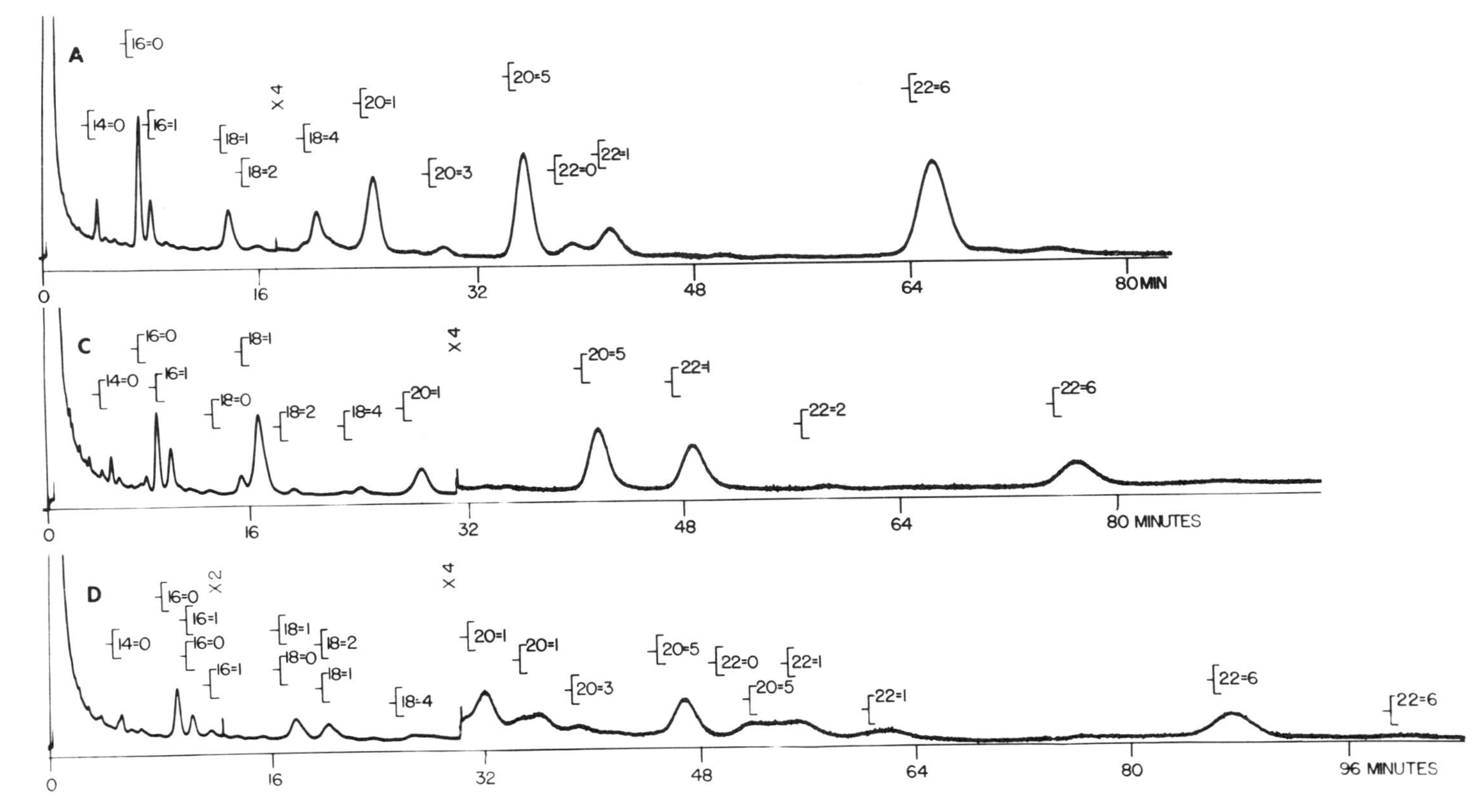

Fig. 2. GLC elution patterns of TMS ethers of monoacylglycerols on Silar 5CP. A, 2-Monoacylglycerols; B, 1(3)-monoacylglycerols; C, mixture of A and B. Peaks identified according to carbon number of acyl radicals. Instrument, F and M model 402 biomedical gas chromatograph equipped with 180 cm × 5 mm outside-diameter glass columns containing 3% Silar 5CP on Gas Chrom Q (100–120 mesh). Injector and oven heaters, 220°C; detector, 250°C. Carrier gas, helium, 40 ml/min. Sample, 1 μl of a 1% solution of the TMS ethers of monoacylglycerols in the reaction mixture.

Table IV. GLC Data for Derivatives of Monoacylglycerols and Fatty Acids[a]

Number of carbons : number of double bonds	Relative retention times[b]			Equivalent chain lengths[c]		
		Monoacylglycerols			Monoacylglycerols	
	FAMEs	TMS ethers	Acetate	FAMEs	TMS ethers	Acetates
	(185°C)	(220°C)	(248°C)	(185°C)	(220°C)	(248°C)
14:0	0.230	0.318	0.393	14.00	14.00	14.00
16:0	0.482	0.558	0.627	16.00	16.00	16.00
16:1	0.543	0.630	0.699	16.31	16.41	
18:0	1.000	1.000	1.000	18.00	18.00	18.00
18:1	1.115	1.100	1.101	18.27	18.30	18.38
18:2	1.319	1.276	1.258	18.73	18.83	18.96
18:3	1.655	1.540	1.470	19.35	19.48	19.63
18:4	1.834	1.628	1.565	19.63	19.68	19.88
20:0	2.075	1.792	1.595	20.00	20.00	20.00
20:1	2.323	1.951	1.749	20.29	20.28	20.36
20:5	4.050	2.885	2.445	21.81	21.61	21.29
22:0	4.304	3.211	2.544	22.00	22.00	22.00
22:1	4.746	3.384	2.737	22.23	22.17	22.29
22:6	9.242	5.420	4.223	24.06	23.80	24.14

[a] Data taken from Myher and Kuksis (1974). Separations were made on a 180-cm 3% Silar 5CP column.
[b] FAMEs, fatty methyl esters; TMS, trimethylsilyl.
[c] Equivalent chain lengths here are relative to the saturated homologues of the respective compound type at the indicated temperatures.

Although the alkyl- and alk-1-enylglycerols are not separated from each other on Silar 5CP, both are eluted approximately 3 ECL units ahead of the corresponding 1-acylglycerol, and, when chromatographed at the same temperature, the acetates of a particular alkyl- or acylglycerol are eluted approximately 9 ECL units later than the

Table V. ECL Values for Derivatives of Monoacyl- and Monoalkylglycerols on Silar 5CP[a]

Carbon number	TMS ethers		Acetates			
	220°C[b]		220°C[b]		235°C[c]	
	1 alkyl	1 acyl	1 alkyl	1 acyl	1 alkyl	1 acyl
14	14.00	17.00	20.89	23.81	14.00	16.48
16	16.00	19.01	22.90	25.83	16.00	18.51
18	18.00	21.02	24.92	27.85	18.00	20.54
20	20.00	23.02	26.94	29.88		

[a] Data taken from Myher and Kuksis (1974) for GLC on 180-cm 3% Silar 5CP columns. ΔECL (1-acyl-1-alkyl)-trimethylsilyl ethers = 3.01 (220°C); ΔECL (1-acyl-1-alkyl)acetates = 2.93 (220°C); ΔECL (1-acyl-1-alkyl)-acetates = 2.51 (235°C); ECL of C_{28} hydrocarbon (octacosane) = 18.44 (220°C) relative to trimethylsilyl ethers of 1-alkylglycerols; ECL of C_{34} hydrocarbon (tetratriacontane) = 17.08 (235°C) relative to trimethylsilyl ethers of 1-alkylglycerols.
[b] Relative to the trimethylsilyl ethers of 1-alkylglycerols at 220°C.
[c] Relative to the trimethylsilyl ethers of 1-alkylglycerols at 235°C.

TMS ether. If necessary, the identity of any particular GLC peak can be confirmed by GC/MS (Myher *et al.*, 1974).

Glyceryl ethers having various substituents on the alkyl chain have also been separated by gas chromatography. Standard 1-(9/10-hydroxy)octadecylglycerol and the 1-(11/12-hydroxy)alkylglycerols obtained after $LiAlH_4$ or Vitride reduction of the major ether lipids of the pink portion of rabbit harderian gland were separated on a 183-cm 10% EGSS-X column at 200°C as the TMS, TMS/isopropylidene, and acetate/isopropylidene derivatives (Kasama *et al.*, 1973; Rock and Snyder, 1975). The relative retention times (R_T) for the various derivatives of 1-(9/10-hydroxy)-octadecylglycerol are as follows: TMS, $R_T = 1.0$; TMS/isopropylidene, $R_T = 2.9$; and acetate/isopropylidene, $R_T = 9.9$. A number of derivatives of 1-*O*-(2-hydroxy-alkyl)- and 1-*O*-(2-ketoalkyl)glycerol have also been synthesized and characterized by GLC, IR, and MS (Muramatsu and Schmid, 1972). Relative to the set of saturated FAMEs, 1-*O*-hexadecyl-2,3-isopropylideneglycerol, 1-*O*-(2-ketohexa-decyl)-2,3-isopropylideneglycerol, 1-*O*-(2-hydroxyhexadecyl)-2,3-isopropylidene-glycerol, 1-*O*-(2-acetoxyhexadecyl)-2,3-isopropylideneglycerol, and 1-*O*-(2-acetoxy-alkyl)-2,3-diacetoylglycerol have ECL values of 20.55, 21.95, 22.25, 22.70, and 24.60, respectively, when chromatographed on a 6% SE-30 column and 20.50, 24.90, 25.45, and 24.60 (value for the -2,3-diacetoyl derivative not determined), respectively, when chromatographed on a 10% SP-1000 column.

Hallgren and Ställberg (1967) have chromatographed the isopropylidene derivatives of the alkyl- and 2-methoxyalkylglycerols obtained from Greenland shark liver oil on a 1% Apiezon L plus 0.1% polyethylene glycol column at 218°C and on a 1% EGS column at 194°C. In general, the unsaturated species were found to be eluted earlier than the corresponding saturated species on Apiezon L, and later on EGS. 1-*O*-(2-Methoxyhexadecyl)isopropylideneglycerol is eluted later than 1-*O*-(hexadecyl)isopropylideneglycerol by 0.97 ECL unit on Apiezon L and 2.72 ECL units on EGS.

Long-chain cyclic acetals of glycerol may be found naturally or as artifacts produced from 1-alk-1-enylglycerol lipids during isolation (Su *et al.*, 1974). Baumann *et al.* (1972) have identified the four isomers formed by the acid-catalyzed cyclization of 1-*O*-*cis*-alk-1-enyl-*sn*-glycerol as *cis*-2-alkyl-5-hydroxy-1,3-dioxane (I), *trans*-2-alkyl-5-hydroxy-1,3-dioxane (II), *cis*-2-alkyl-4-hydroxymethyl-1,3-dioxolane (III), and *trans*-2-alkyl-4-hydroxymethyl-1,3-dioxolane (IV). The acetates of these isomers can be separated into three fractions by TLC on silica gel H in hexane–diethyl ether. The R_f values for the acetates of I, II, III, and IV are 0.35, 0.68, 0.46, and 0.46, respectively. The acetates of all four isomers were resolved by GLC when a 190-cm 18% HI-EFF-2BP (ethylene glycol succinate) column was used. The relative retention times for the acetate derivatives of I, II, III, and IV are 1.42, 1.0, 1.09, and 1.21, respectively.

It is also possible to separate the 1-*S*-alkylglycerol ethers from the corresponding 1-*O*-alkylglycerol ethers when they are chromatographed as the isopropylidene derivatives (Wood *et al.*, 1969*a*). Relative to the isopropylidene derivatives of the saturated 1-*O*-alkylglycerols, the ECL values of the isopropylidene derivatives of 1-*S*-hexadecylglycerol are approximately 18.0, 19.6, and 19.2 on SE-30, EGSS-X, and EGS, respectively. Therefore, GLC on EGS maximizes the separation of the 1-*O*-alkyl- and 1-*S*-alkylglycerol ethers. These compounds may also be separated by

TLC on silica gel G developed in hexane–diethyl ether (9 : 1, v/v); the 1-*S*-alkyl ethers run just above the 1-*O*-alkyl ethers.

3.9.4.2. *Diacylglycerols and Ether Analogues*

Prior to GLC, the diacylglycerol, alk-1-enylacylglycerol, alkylacylglycerol, and dialkylglycerol subclasses should be separated by TLC. Each of these can exist in a number of isomeric forms having substituents placed in position 1,2, 2,3, or 1,3 of *sn*-glycerol, and, although enantiomeric 1,2 and 2,3 isomers have not been resolved by any physical chemical technique, they may be separately analyzed by the method of stereospecific analysis (see Chapter 4). The combined 1,2(2,3) isomers are readily separated from the 1,3 isomers by TLC. There are also reverse isomers such as 1-palmitoyl-2-stearoyl-*sn*-glycerol and 1-stearoyl-2-palmitoyl-*sn*-glycerol that are generally not resolved by TLC or GLC.

Diacylglycerol derivatives (TMS ethers, acetates) may be chromatographed on short or long columns using either nonpolar or polar liquid phases (Kuksis, 1976). Free diacylglycerols, however, are subject to dehydration, isomerization, or trans-esterification. If hot injectors (300°C) and short columns are used, even phospholipids may be chromatographed; the pyrolysis products, consisting mainly of diacylpropene diols (Kuksis, 1967; Perkins and Johnston, 1969), give rise to well-defined peaks indicative of the molecular weight composition of the original phospholipids. However, prior cleavage of the phosphoryl grouping by phospholipase C (Renkonen, 1965), acetolysis (Renkonen, 1965), or silolysis (Horning *et al.*, 1969) is preferable.

Nonpolar columns have been used to separate diacylglycerol derivatives according to carbon number (Fig. 1) and, although polyunsaturated species are eluted ahead of the corresponding saturated compound, the resolution is not good enough to separate a complex mixture containing saturated and unsaturated compounds. Consequently, molecular species are determined by means of argentation TLC and GLC (Kuksis *et al.*, 1968*b*). The 1,2(2,3) and 1,3 isomers of diacylglycerols have also been separated on nonpolar GLC columns. Huebner (1959) described a separation of the isomeric diacylglycerol acetates equivalent to approximately 0.6 carbon on a 61-cm column packed with 23% silicone vacuum grease on Celite 545, but effective separation could not be obtained on a 70-cm 1% OV-1 column (Wood *et al.*, 1969*b*). Similarly, Horning *et al.* (1969) and O'Brien and Klopfenstein (1971) observed a separation of isomeric diacylglycerol TMS ethers equivalent to approximately 0.35 carbon.

The various types of ether analogues are readily separated with respect to carbon number on SE-30 or OV-1 GLC columns, and, as we can see from the ECL values in Table VI, it is also possible to achieve some resolution between the various types. The TMS ethers of the dialkylglycerols overlap with diacylglycerols having two fewer carbon numbers, but the alkylacylglycerols are well resolved from the diacylglycerols. In some respects, the acetate derivatives are superior, since combinations of diacyl-, alkylacyl-, and dialkylglycerols or diacyl-, alkylacyl-, and alk-1-enylglycerols may be resolved. Since polyunsaturated species, however, are eluted as much as 0.7 ECL unit earlier than the saturated components (Renkonen, 1967*a*), the separation of natural mixtures is not as effective. The resolution of a natural mixture could be improved by hydrogenation, but then alk-1-enyl- and alkyl-

Table VI. ECL Values for Saturated[a] C_{34} Disubstituted Glycerols on Methylsiloxane

Compound (derivative)	ECL[b]
Dialkylglycerol (TMS)[c]	34.0
Alkylacylglycerol (TMS)[d]	34.9
Diacylglycerol (TMS)[c]	36.6
Dialkylglycerol (Ac)[c]	34.7
Alk-1-enylacylglycerol (Ac)[e]	34.7
Alkylacylglycerol (Ac)[e]	35.3
Diacylglycerol (Ac)[c, e]	36.0

[a] Excepting the double bond of the alkyl-1-enyl moiety.
[b] ECL values are relative to the TMS ethers of the dialkylglycerols and in all cases are for the species having a carbon number of 34. The values which were compiled from the references indicated were measured under a number of conditions on either SE-30 or OV-1. When run simultaneously with the above compounds, a C_{36} triacylglycerol has ECL = 35.2.
[c] Wood *et al.* (1969).
[d] J. J. Myher and A. Kuksis (unpublished data).
[e] Renkonen (1967*a*).

acylglycerols would become indistinguishable. Consequently, prior TLC resolution of the ether analogues is still the preferred procedure. Further separation of the molecular species of alkylacyl- and alk-1-enylglycerol acetates by argentation TLC may also be achieved (Renkonen, 1967*a*).

The first effective GLC separations of saturated and unsaturated diacylglycerol derivatives were made with ethylene glycol succinate polyester liquid phases (Kuksis, 1971). Column life was improved when polyesters of succinic acid and higher diols were used, but the use of these liquid phases resulted in the loss of resolution between saturated and monoenoic diacylglycerols that had the same carbon number (Kuksis, 1972*b*). Silar 5CP (50% cyanopropylphenyl siloxane, 50% methylsiloxane) provides improved column stability and gives good GLC resolution of all saturated and unsaturated diacylglycerol TMS ethers (Myher and Kuksis, 1975). Although the TMS ethers of the 1,2(2,3) isomers are eluted earlier than the 1,3 isomers (separation factor = 1.087), the isomers should first be separated by TLC in order to avoid overlapping of critical pairs such as the TMS ethers of 1,2(2,3)-palmitoyl-linoleoylglycerol and 1,3-palmitoyloleoylglycerol. Figure 3 shows the GLC elution patterns obtained for the TMS ethers of the 1,2(2,3)- and 1,3-diacylglycerols resulting from the partial Grignard hydrolysis of linseed oil triacylglycerols. In spite of the complexity of these mixtures, most species are resolved. Only the TMS ethers of oleoyl-linolenoyl- and dilinoleoylglycerols remain unresolved, whereas the TMS ethers of stearoyl-linoleoylglycerol and dioleoylglycerol are partially resolved. Reverse isomers such as the TMS ethers of 1-oleoyl-2-linoleoyl-*rac*-glycerol and 1-linoleoyl-2-oleoyl-*rac*-glycerol and enantiomeric pairs are also unresolved. Diacylglycerols obtained from slected phospholipids by phospholipase C digestion have also been chromatographed as the TMS ethers on Silar 5CP. Even though 1-stearoyl-2-linoleoyl-*sn*-glycerol is only partially resolved from 1-palmitoyl-2-

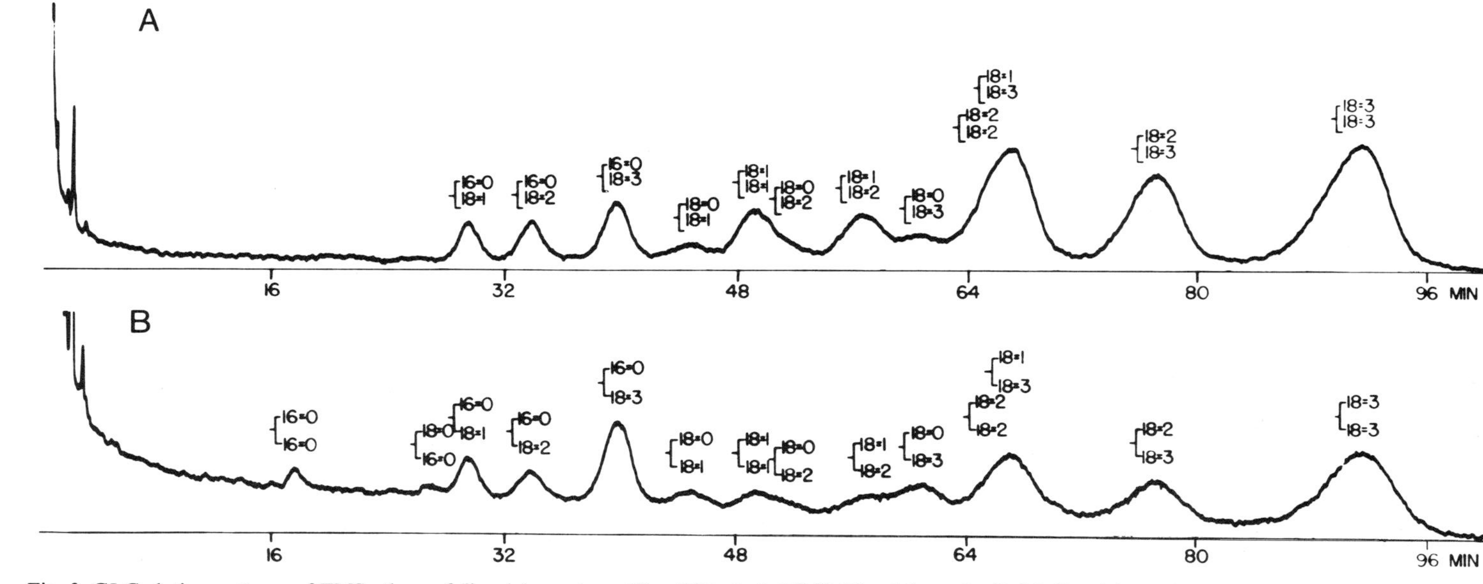

Fig. 3. GLC elution patterns of TMS ethers of diacylglycerols on Silar 5CP. A, 1, 2(2,3)-Diacylglycerols; B, 1,3-diacylglycerols. Instrument, F and M model 402 biomedical gas chromatograph with dual columns and hydrogen flame ionization detectors and a Honeywell 1-mV recorder. Column, 180 cm × 3 mm inside-diameter glass tube packed with 3% Silar 5CP on Gas Chrom Q (100–120 mesh). Injector and column, 270°C; detector, 290°C. Carrier gas, helium, 30 ml/min. Sample, 1 μl of a 1% solution in petroleum ether. Attenuation, 300 times full sensitivity.

Table VII. Relative GLC Retention Data for the TMS Ethers of Diacylglycerols[a]

Diacylglycerol[b] species	Relative retention times		Equivalent chain length	
	Experimental	Calculated[c]	Experimental	Calculated[d]
16:0 16:0	1.000	1.000	32.00	32.00
16:0 18:0	1.517	1.517	34.00	34.00
16:0 18:1	1.669	1.669	34.46	34.25
16:0 18:2	1.918	1.919	35.12	34.73
16:0 18:3	2.261	2.264	35.91	35.28
16:0 20:4	3.07		37.40	37.15
18:0 18:0	2.301	2.301	36.00	36.00
18:0 18:1	2.531	2.531	36.46	36.25
18:1 18:1	2.794	2.784	36.93	36.50
18:0 18:2	2.908	2.910	37.12	36.73
18:1 18:2	3.227	3.202	37.62	36.98
18:0 18:3	3.448	3.433	37.94	37.28
18:2 18:2	3.721	3.682	38.31	37.46
18:1 18:3	3.794	3.779	38.40	37.53
18:2 18:3	4.369	4.344	39.08	38.01
18:0 20:4	4.66		39.40	39.15
18:3 18:3	5.163	5.127	39.88	38.56
18:0 20:0		3.84		

[a] Data taken from Myher and Kuksis (1975) and apply to individual sets of 1,2(2,3) or 1,3 isomers where 1,2(2,3)-dipalmitoyl- and 1,3-dipalmitoylglycerol, respectively, are used as the retention time reference. Retention times of the TMS ethers were determined on 3% Silar 5CP at 270°C.

[b] Diacylglycerols described by the component fatty acids. Enantiomers and reverse isomers are not distinguished.

[c] Calculated by multiplying the appropriate F factors as defined in the text.

[d] Calculated by summing the ECL values of the component fatty acids chromatographed as FAMEs on 3% Silar 5CP at 180°C.

arachidonoyl-*sn*-glycerol, it is possible to estimate the quantities of each. Because of its long retention time (>2 hr) and low abundance, the peak due to 1-stearoyl-2-docosahexaenoyl-*sn*-glycerol was not observed.

The relative retention times and equivalent chain lengths obtained for the TMS ethers of a number of diacylglycerols are shown in Table VII. Since it was found that each of the two acyl chains in any selected diacylglycerol molecule contributes an independent increment to the total relative retention time, it is possible to calculate the relative retention time of any diacylglycerol by multiplying the appropriate separation factors attributed to each fatty acid. These separation factors were further divided into increments due to the degree of unsaturation and chain length. Relative to a saturated diacylglycerol, the incremental factors were found to be $F1 = 1.100$, $F2 = 1.493$, and $F3 = 1.517$ for fatty acids having one, two, and three double bonds, respectively. Furthermore, a factor F of 1.517 is needed for an increase in chain length of two carbon numbers. For example, relative to 1,2-dipalmitoylglycerol, the relative retention time (R) of 1-oleoyl-2-linolenoyl-*rac*-glycerol is determined by the following equation:

$$\begin{aligned} R &= F1 \times F2 \times F \times F \\ &= 1.100 \times 1.493 \times 1.517 \times 1.517 \\ &= 3.779 \end{aligned}$$

Similarly, the ECL values of any diacylglycerol species are calculated by summing the ECL increments that correspond to the retention factors defined above. The good agreement between calculated and experimental values shown in Table VII proves that the calculation provides a reliable means of obtaining the relative retention times of species for which synthetic standards are not available. Table VII also shows a comparison of the experimental ECL values with those obtained by summing the ECL values (on Silar 5CP, 180°C) of the appropriate FAMEs. The poor agreement in this case has been attributed to differences in the temperature dependence of the factors associated with unsaturation and chain length.

Figure 4 shows the GLC profile obtained when the acetate derivatives of the short-chain 2,3-diacylglycerols that were derived from bovine milk fat triacylglycerols by partial Grignard hydrolysis were chromatographed on a 3% EGSS-X column at 250°C (Kuksis *et al.*, 1973). In addition to separation factors based on carbon number and unsaturation, separation factors based on chain length (at constant total carbon number) were also observed. Thus, within a particular carbon number, the species containing caprylic, caproic, and butyric acids are resolved from one another and are eluted in the stated order. The ECL values (relative to 1-octadecanoyl-2,3-diacetoyl-*rac*-glycerol) obtained for a number of synthetic short-chain diacylglycerol acetates having the same carbon number (Table II) decrease as the minimum chain length increases. This effect can be attributed to the increase in the polarity of the ester bond that accompanies a decrease in the fatty acid chain length, and a numerical chain length polarity factor can be defined for each ester group as being equal to the reciprocal of the chain length of the corresponding fatty acid. For a diacylglycerol acetate containing fatty acids with chain lengths R_1, R_2,

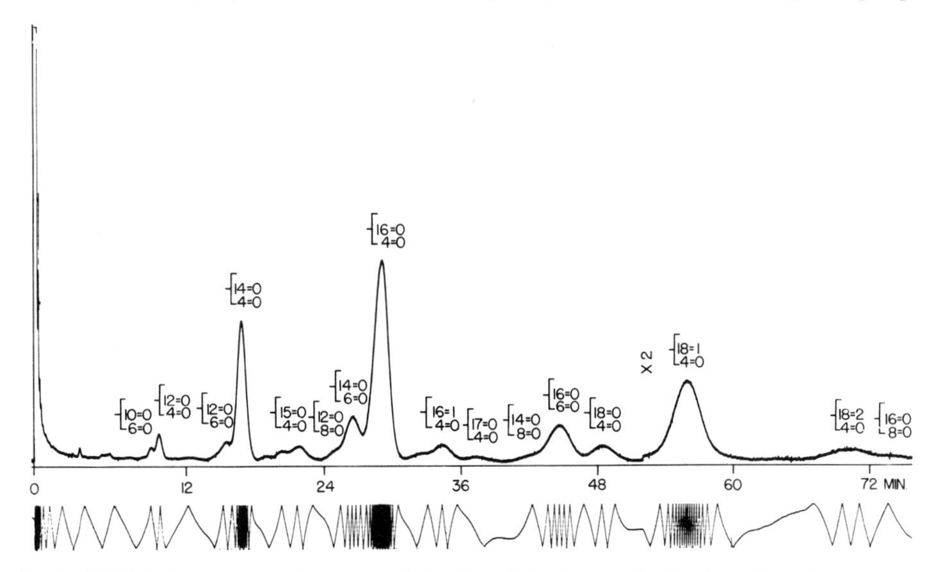

Fig. 4. GLC elution pattern of the acetate derivatives of the short-chain 2,3-diacylglycerols recovered from Grignard degradation of bovine milk fat. Instrument, F and M model 402 biomedical gas chromatograph with dual columns and hydrogen flame ionization detectors. Column, 180 cm × 3 mm inside-diameter glass tube packed with 3% EGSS-X on Gas Chrom Q (100–120 mesh). Injector and column, 235°C; detector, 250°C. Carrier gas, nitrogen, 30 ml/min. Sample, 1 μl of a 1% solution in petroleum ether. The markings below the time axis are disk integrator tracings.

and R_3 (including the acetate moiety), the total chain length polarity factor (P) for a particular diacylglycerol is then defined by the following equation:

$$P = \frac{1}{R_1} + \frac{1}{R_2} + \frac{1}{R_3}$$

Indeed, if the ECL values of a series of 1,2-diacylglycerol acetates having the same carbon number are plotted against P, an almost linear correlation is observed. Similar GLC results are obtained when Silar 5CP is used as the liquid phase.

3.9.4.3. *Triacylglycerols and Ether Analogues*

Gas–liquid chromatography is still the best method available for obtaining the carbon number profile of a triacylglycerol mixture, and a number of detailed reviews have been written on this subject (Kuksis, 1976; Litchfield, 1972).

Because of the requirement for column temperatures in excess of 300°C, only stable, nonpolar liquid phases (e.g., OV-1, SE-30, OV-17, Dexil 300) coated (1–3%) on high-quality supports are suitable for the GLC separation of long-chain triacylglycerols. This restriction to nonpolar liquid phases means that the triacylglycerols are separated by carbon number only (Fig. 1). Long-chain triacylglycerols are most efficiently recovered when a short column (30–50 cm) and temperature programming are used, but good results have also been obtained with longer columns operated in either temperature-programmed or isothermal mode. However, because of the length of time required, isothermal operation becomes impractical for mixtures containing a wide range of carbon numbers. Schomberg *et al.* (1976) have demonstrated the use of a capillary column for the separation of triacylglycerols. Using a 20-m column coated with Poly-S179 and temperature programming from 300°C to 400°C at 4°/min, they were able to effect some separation of the triacylglycerols of margarine, palm butter, and corn butter. Each carbon number was split into a number of partially resolved but still unidentified components.

Although ideal mixtures such as triolein and tristearin have been partially separated on 183-cm columns using JXR or Apiezon L (Litchfield *et al.*, 1967), complex natural mixtures are not resolved on the basis of unsaturation. Peak broadening due to the presence of various unsaturated fatty acids can be eliminated by hydrogenation of the sample. Triacylglycerols (C_{42}–C_{64}) differing by only one carbon number were well resolved when hydrogenated mullet, tuna, menhaden, and pilchard oil triacylglycerols were chromatographed on a 183-cm, 3% JXR column (Litchfield *et al.*, 1967). At lower temperatures, the separation factor for triacylglycerols differing by one carbon number is greater, and small quantities of odd-carbon triacylglycerols are readily resolved. For example, the odd-carbon species in the silver nitrate TLC fractions of a butter oil distillate containing C_{24}–C_{42} triacylglycerols are resolved even on a 50-cm JXR column (Kuksis, 1967). Hydrogenation has been used in order to prevent the occurrence of thermal decomposition of polyunsaturated species, but good separations can be obtained even for raw samples of fish oils containing polyunsaturated triacylglycerols (Kuksis, 1967).

The presence of significant amounts of branched-chain acids also tends to decrease the apparent GLC resolution of triacylglycerols, and, although branched-chain fatty acids are easily resolved by GLC (as FAMEs), their presence in triacylglycerols generally leads to peak broadening or shouldering. Litchfield *et al.* (1967), for example, found that the resolution of the hydrogenated triacylglycerols from *Sterculia foetida* containing 7.2% 8(9)-methylheptadecanoate and 46.9% 9(10)-methyloctadecanoate is very poor. On the other hand, triacylglycerols containing fatty acids with terminal cyclopentane rings are eluted sufficiently later than the straight-chain analogues to make their resolution possible when a sample of hydrogenated *Hydnocarpus wightiana* seed oil is chromatographed on a 183-cm JXR column. Considerable ambiguity, however, would arise if both cyclopentane and odd-chain fatty acids were present in the same sample.

Triacylglycerols containing acetoxy fatty acids are also readily separated by GLC (Powell *et al.*, 1969). Even though they elute later than unsubstituted triacylglycerols, the GLC profile of a mixture of both types would be difficult to unscramble and prior separation of the acetoxy-substituted triacylglycerols by TLC is preferable. Similarly, epoxy triacylglycerols or their 1,3-dioxolane derivatives have been separated according to carbon number by GLC (Fiorti *et al.*, 1969). The 1,3-dioxolane derivatives formed by condensing the epoxyglycerides with cyclopentane in the presence of boron trifluoride are thermally stable and allow a simultaneous separation of both the regular and derivatized triacylglycerols.

GLC can also be used to separate alkyldiacylglycerols according to carbon number. Wood and Snyder (1967*b*) have separated the alkyldiacylglycerols isolated from tumor tissue using a 53-cm Pyrex column packed with 3% JXR on 100–120 mesh Gas-Chrom Q. Because these compounds have one less oxygen atom than triacylglycerols having the same carbon number, they are eluted approximately 1 carbon unit earlier than the respective triacylglycerol. The resolution of species having different carbon numbers was comparable to that obtained for long-chain triacylglycerols. In general, tri(alkyl/acyl)-glycerol species having the same carbon number elute from the GLC column approximately 1 carbon unit earlier for each *O*-alkyl group present, so that a mixture of C48 glyceryl ethers and esters is eluted in the order tripalmityl-, 1,2-dipalmityl-3-palmitoyl-, 1-palmityl-2,3-palmitoyl-, and tripalmitoylglycerol (Kuksis, 1976). Blank *et al.* (1972) separated a class of 1-alkyl-diacylglycerols containing 1 mole of isovaleric acid on a 30-cm stainless steel column packed with 2.5% SE-30 on Aeropak 30. The shoulders that were observed on the three major peaks were attributed to species containing odd- and/or branched-chain *O*-alkyl groups.

It is possible to separate low-molecular-weight triacylglycerols by means of polar GLC columns. The triacylglycerols of a molecular distillate of bovine milk fat have been separated on 3% EGSS-X (Kuksis *et al.*, 1973) and 3% Silar 5CP (Kuksis, 1975). The elution pattern was quite complex, with numerous partially resolved peaks, but, similar to the GLC resolutions obtained for diacylglycerol acetates containing short-chain acids, there appeared to be resolution, within a carbon number, of species containing butyric acid and species containing caproic acid. The complexity arises because of the superimposition of this type of effect with that due to the degree of unsaturation within a molecule. The interpretation of these patterns could be made simpler if the triacylglycerols were first separated by argentation TLC.

3.9.5. Quantitative Gas Chromatography

Two different but related types of quantitation may be considered (Kuksis, 1976). It is necessary to add a known amount of an appropriate internal standard to the sample prior to injection into the GLC if the absolute quantity of an acyl- and/or alkylglycerol class is required, but an internal standard is unnecessary if only the molar or weight ratio of the various homologues and isomers within a lipid class is required. The main difficulty in the latter case is the necessity for knowing the relative response factors (detector response factors and column recovery factors inclusive) of each molecular species. Pure mixed-acid, di-, or triacylglycerols, however, either are not generally available or are prohibitively expensive, and it is necessary to depend on secondary standards. Certain natural mixtures such as the seed oil triacylglycerols that have been characterized by accepted procedures may serve as such standards, but a set of oils that has been randomized in regard to the location of any fatty acid on the glycerol backbone would be even better. Once the fatty acid composition is accurately known, the triacylglycerol composition could be accurately determined by calculation using the 1,2,3 random procedure. The composition of the diacylglycerols released by Grignard hydrolysis could also be calculated. When a meaningful standard mixture is not available, it must be assumed that the peak area for a particular compound is proportional to the weight percent of that compound within the mixture. This assumption is probably valid as long as the operating conditions are optimized so that the response of any one compound is proportional to injected mass over a sufficiently wide concentration range. The weight %/area % response ratios for a number of triacylglycerols have been found to be almost constant (Kuksis, 1976).

Neutral acylglycerols are readily quantitated via GLC by adding a known amount of an appropriate internal standard and comparing the peak area of the unknowns with that of the internal standard. Although tridecanoylglycerol is a convenient internal standard for the analysis of neutral lipid mixtures (Fig. 1), other choices are possible as long as there is no overlap with any of the unknown peaks.

The mass, $W(\mathrm{X})$, of component X is calculated using the following equation:

$$W(\mathrm{X}) = [A(\mathrm{X})/A(\mathrm{IS})]\, W(\mathrm{IS})\, F(W\mathrm{X})$$

where $A(\mathrm{X})$ and $A(\mathrm{IS})$ are the areas obtained for component X and internal standards, respectively, $W(\mathrm{IS})$ is the mass of internal standard added, and $F(W\mathrm{X})$ is the weight response factor for component X. The weight response factors are obtained from calibration runs using known amounts of component X and internal standard using the relationship

$$F(W\mathrm{X}) = [W(\mathrm{X})/W(\mathrm{IS})][A(\mathrm{IS})/A(\mathrm{X})]$$

The data in Table VIII show that the response factors obtained for a number of neutral lipids are constant within the mass range indicated.

Short columns are preferable for quantitative analysis because of the higher relative recoveries of triacylglycerols obtained on short columns (~50 cm) than on long columns (~180 cm). Although it may be argued that high relative recoveries are not necessary so long as they are reproducible, it appears that accuracy is not

Table VIII. Response Factors for Various Neutral Lipids[a]

Lipid class[b]	Tested range[c] (μg/peak)	$F(W) \pm$ SD[d]	
Fatty acid TMS ester	0.16–1.27	0.75	0.013
Monoacylglycerol TMS ether	0.63	0.71	0.007
Cholesterol TMS ether	0.127–50.8	0.71	0.007
Diacylglycerol TMS ether	0.15–3.0	0.91	0.01
Cholesterol esters	1.3–12.7	0.85	0.01
Triacylglycerol (lard)	4.75–38.0	0.98	0.012
Tridecanoylglycerol	1.0–25.0	1.00	
Ceramide TMS ether	(Assumed to have the same response factor as diacylglycerol TMS ether)		

[a] Data taken from Kuksis *et al.* (1975*a*).
[b] Species include the C_{16} and C_{18} fatty acids and their glycerol and cholesterol esters.
[c] Masses based on the underivatized form.
[d] $F(W)$ as defined in text.

possible under conditions of poor recovery. The data in Table IX and those of Bezard and Bugaut (1972) illustrate the fact that the shape of a response curve of peak area correction factor, $F(W)$, vs. amount injected, m, appears to be hyperbolic. For larger m, the $F(W)$ is almost constant, with a value close to 1.0, but as m decreases, $F(W)$ increases, gradually at first ("elbow" region), then very sharply. Reproducibility in the latter region is poor because a small change in m results in a large change in $F(W)$. The large uncertainty in measured triacylglycerol mass can be avoided if all quantitative analyses of triacylglycerols are made under conditions where $F(W) < 1.5$. Some uncertainty also arises from the fact that different response curves are obtained for each individual triacylglycerol species (Bezard and Bugaut, 1972; Kuksis, 1976) and from the observation that the individual peak response

Table IX. Response Factors, $F(W)$,[a] *for Triacylglycerols*[b,c]

Total mass injected (μg)	Average response factor [$F(W)$]
0.05	10
0.095	3.92
0.190	2.14
0.475	1.80
0.950	1.48
1.900	1.25
4.75	1.05
9.50	1.01
19.0	0.98
38.0	0.94

[a] $F(W)$ as defined in the text.
[b] Data taken from Kuksis *et al.* (1975*a*).
[c] Stripped lard: C_{50}, 13.7%; C_{52}, 60.7%; C_{54}, 22.6%; others, 3.0%.

curves obtained when natural mixtures are injected are not identical to those obtained when the pure components are injected. There appears to be an interaction that results in an improved recovery of the C_{54} peak in lard over that expected from the individual peak response curve. Response curves should therefore be obtained for triacylglycerol mixtures that have a fatty acid composition similar to that of the unknown mixture.

3.10. Mass Spectrometry

Mass spectrometry is being used ever more frequently to solve problems involving the detection and identification of biologically relevant compounds. In conjunction with GLC, the method can provide a vast amount of structural information even if the unknown compound has not been isolated in pure form. Mass spectrometry is also one of the important techniques used for the measurement of stable isotope labeling.

When molecules in a mass spectrometer ion source are subjected to some form of excitation process such as 70-eV electron impact, a number of processes take place. The primary event is the production of a singly charged molecule ion $[M]^+$. Those ions that possess sufficient excess energy decompose in a series of parallel and consecutive reactions, forming a spectrum of fragment ions that is characteristic of the original molecular structure. A fragment ion may be formed by the simple cleavage of a bond or by a process involving considerable rearrangement. For singly charged ions, the mass-to-charge ratio, *m/e*, is equal to the mass of a fragment.

The *m/e* values of ions are calculated using the atomic masses corresponding to the isotope having the highest natural abundance, e.g., ^{1}H, ^{12}C, ^{15}N, ^{16}O, and ^{28}Si. As well as ionic species (p) containing only the most abundant isotopes, there will in general be corresponding species containing one or more heavy isotopes. These latter ions having *m/e* values 1,2,3, . . . , *n* higher than those of the parent ion are referred to as p + 1, p + 2, p + 3, . . . , p + *n* ions. Thus molecular ions formed by the removal of a single electron from the sample molecule are usually labeled M (or $[M]^+$) in the literature, and isotope peaks are labeled $[M + 1]^+$, $[M + 2]^+$, $[M + 3]^+$, . . . , $[M + n]^+$. If the ions are formed in a high-pressure region such as in chemical ionization sources, ion–molecule reactions may also lead to $[M + 1]^+$ peaks. Hydrogen or hydride transfer processes leading to $[M + 1]^+$ and $[M - 1]^+$ ions, respectively, are important features of chemical ionization or field desorption sources.

For a general introduction to mass spectrometry, books by these authors may be consulted: Biemann (1962), McLafferty (1973), Hamming and Foster (1972), and Budzikiewicz *et al.* (1967). Detailed accounts of gas chromatography/mass spectrometry (GC/MS) can be found in the comprehensive reviews of McFadden (1973) and Perkins (1975), and an overall view of the mass spectrometry of lipids can be found in the reviews of Ryhage and Stenhagen (1960), Sun and Holman (1968), Odham and Stenhagen (1972), and McCloskey (1969). Furthermore, the literature of mass spectrometry is reviewed every 2 years in the journal *Analytical Chemistry* (e.g., Burlingame *et al.*, 1976).

3.10.1. *Monoacylglycerols*

The mass spectra of monoacylglycerols were first reported by Johnson and Holman (1966). Using a direct-sample insertion probe, they measured the mass spectra for a series of monoacylglycerols, monoacylglycerol diacetates, and monoacylglycerol TMS ethers. More recently, GC/MS has been used for a study of monoacylglycerol TMS ethers (Myher *et al.*, 1974). Furthermore, ^{2}H-labeled compounds have been used (Curstedt, 1974) to clarify the nature of fragments formed during electron impact ionization of monoacylglycerol TMS ethers. The use of the acetyl or trimethylsilyl derivatives prevents subsequent isomerization and allows the investigator to make full use of the resolving power of GLC.

3.10.1.1. Underivatized Compounds

Monoacylglycerols in their free hydroxyl form must be admitted via a heated direct inlet system, and they are liable to extensive isomerization in this form. Johnson and Holman (1966) found that the differences in the spectra of 1- and 2-monoacylglycerols were not sufficiently great to be of practical value.

Molecular ions are either absent or of low intensity in the spectra of the saturated monoacylglycerols. The relative ion abundance (RI) of $[M]^+$ increased with unsaturation to a maximum of 2.4 for 1–18 : 3 at high electron energy (80 eV) and 100 at low electron energy (6–13 eV). At high energies, the base peak of all saturated compounds was that of the $C_3H_7^+$ ion at *m/e* 43. A number of ions that include the fatty acid backbone are prominent, including $[RCO]^+$, $[RCOOH]^+$, $[RCOOH + 1]^+$, and $[RCOOCH_3]^+$. At low energies, the RIs of ions containing oxygen were lower and those of the hydrocarbon-type ions higher. The $[RCO]^+$ ion is the base peak for short-chain species such as monodecanoylglycerol, but the RI decreases with increasing chain length. The $[RCOOH]^+$ ion behaves in inverse fashion, becoming the base peak for long-chain saturated monoacylglycerols. In the case of the unsaturated monoacylglycerols, the base peaks in the high-energy spectra are hydrocarbon ions: *m/e* 55 $[C_4H_7]^+$ for oleate; *m/e* 67 $[C_5H_7]^+$ for linoleate; and *m/e* 79 $[C_6H_7]^+$ for linolenate. The $[RCO - 1]^+$ ion is also more abundant in the spectra of the unsaturated compounds than for the saturated homologues.

3.10.1.2. Acetates

The spectra of monoacylglycerol diacetates were reported by Johnson and Holman (1966). Although this study made use of a direct inlet system, acetates are also good derivatives for GLC. As is the case with the underivatized monoacylglycerols, the difference in the spectra for the 1(3) and 2 isomers is not large enough to be of analytical value and no molecular ions are observed at either high or low electron energies. The *m/e* 43 ion is the base peak in all high-energy spectra, and because its RI greatly decreased at low energies, it has been assigned a hydrocarbon structure. Ions resulting from loss of the long-chain acyloxy group are especially intense in the spectra of the 2-monoacylglycerols, but ions formed by the loss of an acyloxymethylene fragment are more intense in the spectra of the 1-monoacylglycerols. Whereas the RIs of these ions increase with FA chain length, the $[CH_3COOCH_2]^+$ ion at *m/e* 73 is most intense for the shorter-chain compounds.

The low-energy spectra are similar except that the ions $[M - CH_3COO]^+$ and $[M - \text{long-chain acyloxy}]^+$ are more abundant. In fact, the $[M - \text{long-chain acyloxy}]^+$ ion is the base peak for all the unsaturated 1-monoacylglycerols as well as for 2-monolinolein. The $[\text{acyl} + 1]^+$ and $[\text{acyl} - 1]^+$ ions are the base peaks for the diacetates of 2-monolinolenein and 2-monoolein, respectively.

3.10.1.3. Trimethylsilyl Ethers

The mass spectra for the TMS ethers of monoacylglycerols were also first reported by Johnson and Holman (1966). Curstedt (1974) used deuterium-labeled monoacylglycerol TMS ethers in order to determine the structural origin of hydrogen in the various ion fragments, and Myher *et al.* (1974) used the TMS ethers in a GC/MS study that included a comparison with the respective monoalkyl- and monoalk-1-enylglycerols.

Table X summarizes the RI data for the more significant ion fragments that occur in the GC/MS spectra (70 eV) of the TMS ethers of monopalmitoyl-, monostearoyl-, monooleoyl-, and monolinoleoylglycerol, and Fig. 5 illustrates two complete fragmentation patterns. In contrast to that observed for the diacetate derivatives, a small molecular ion is seen in the mass spectra for the TMS ethers of the saturated monoacylglycerols. The molecular ion is larger in the case of the unsaturated species, and in all cases the RI is larger for the 1 isomer than for the 2 isomer. The $[M - 15]^+$ fragment, due to loss of one CH_3 radical from the TMS group, is more abundant than the molecular ion and is a good indicator of the molecular weight. Although the $[M - 90]^+$ ion, due to loss of $(CH_3)_3SiOH$ from the molecular ion, is an abundant ion in the mass spectra of the unsaturated monoacylglycerols, it is of minor importance for the saturated compounds. The acyl ion, $[RCO]^+$, is more intense in the spectra of the 1(3) isomers, and the $[RCO + 74]^+$ ion is more intense in the 2 isomers. Deuterium labeling indicates that the additional 74 mass units are made up from the TMS group and not from the glycerol backbone. In the case of the 1(3)-monoacylglycerol TMS ethers, a single cleavage between glycerol carbons 2 and 3(1) leads to an $[M - 103]^+$ ion that is found almost exclusively in the spectra of the 1(3) isomers and only trace amounts ($<2\%$) of this ion are found in the spectra of the 2-isomers. Similarly, an ion at *m/e* 205 appearing mainly with spectra of 1(3)-monoacylglycerols appears to be due to cleavage between glycerol carbons 2 and 1(3).

There is no totally unique fragment for the TMS ethers of 2-monoacylglycerols, but the *m/e* 218 ion intensity is very high in the case of the 2 isomer and much weaker in the case of the 1 isomer. This ion is formed from the molecular ion by the loss of the acyloxy (RCOO) radical plus one hydrogen atom derived from the 1 or 3 position of glycerol. Similarly, loss of these same radicals from the $[M - 15]^+$ peak gives rise to an ion at *m/e* 203. The ions at *m/e* 147 $[(CH_3)_3SiOSi(CH_3)_2]^+$, *m/e* 191 $[HC(OSi(CH_3)_3)_2]^+$, and *m/e* 129 $[CH_2CHCHOSi(CH_3)_3]^+$ are generally found in the spectra of TMS ethers (Radford and DeJongh, 1972).

As noted in the GLC section, certain monoacylglycerol compounds may overlap if the mixture is sufficiently complex. The GLC of mixed 1(3)- and 2-

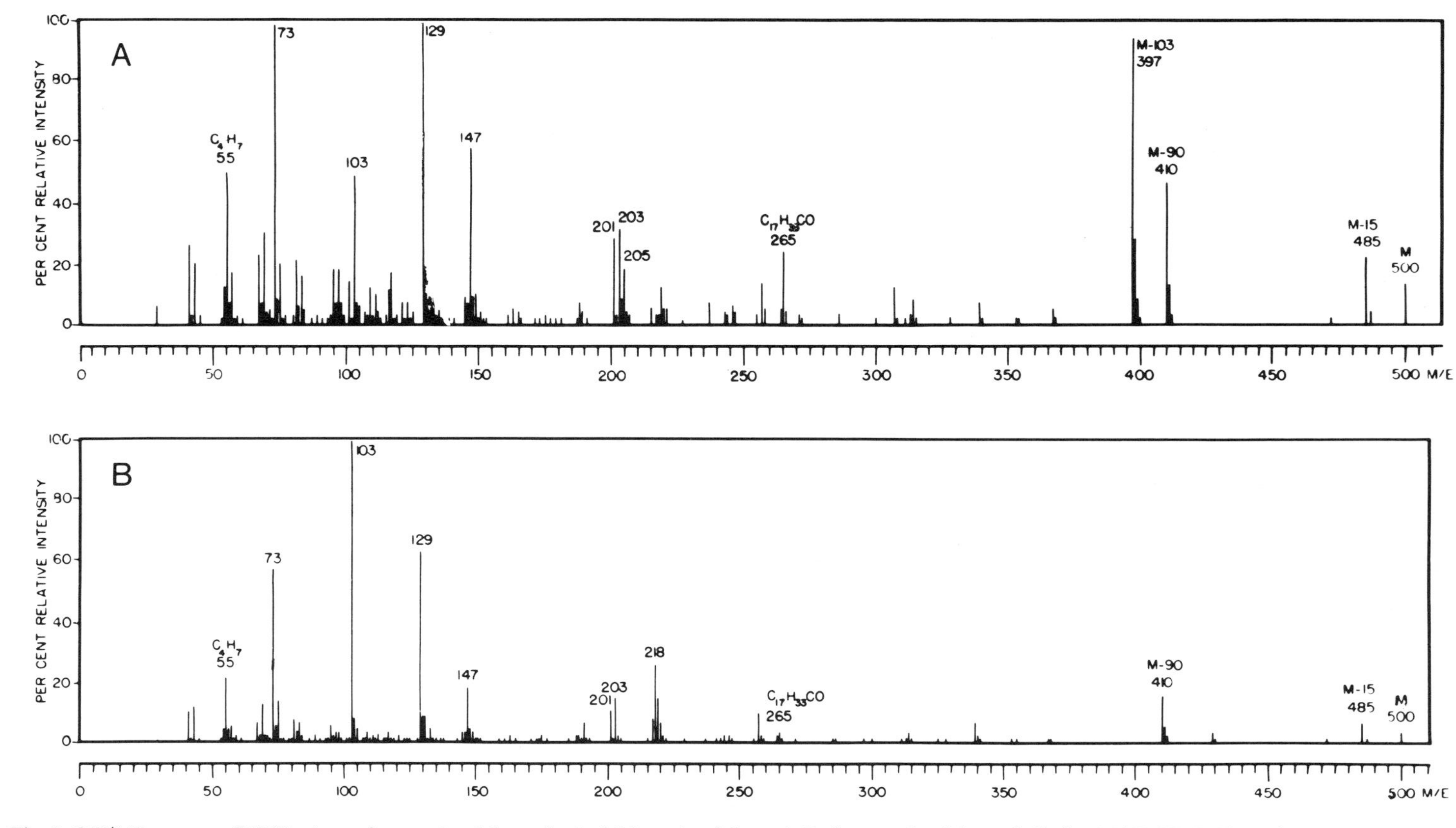

Fig. 5. GC/MS spectra of TMS ethers of monooleoylglycerols. A, 1-Monooleoylglycerol; B, 2-monooleoylglycerol. Varian MAT CH-5, 70 eV, ion source 270°C. GLC on 180 cm × 2 mm inside-diameter stainless steel column packed with 3% Silar 5CP on Gas Chrom Q (100–120 mesh). Injector, 225°C; column, 200°C; transfer line, 250°C. Carrier gas, helium, 10 ml/min. Sample, 1 μl of a 1% solution in silylation mixture.

Table X. Relative Abundance Data for the Major Ions Found in the Spectra of the TMS Ethers of Some Monoacylglycerols[a]

Ion type	Positional isomer	Relative abundance			
		16:0[b]	18:0	18:1	18:2
		(Relative peak intensity)			
$[M]^+$	1	0.8	1.3	14	12
	2		0.6	3	
$[M - 15]^+$	1	14	11	22	15
	2	11	12	5	
$[M - 90]^+$	1	2	4	47	32
	2	4	5	16	
$[M - (71 + 90)]^+$	1	2	5	7	4
	2	16	13	8	
$[M - 103]^+$	1	100	100	94	54
	2	2			
$[M - (103 + 90)]^+$	1			13	17
	2				
$[Acyl]^+$	1	20	22	24	12
	2	9	5	3	
$[Acyl - 1]^+$	1			5	27
$[M - 236]^+$	2			2	
$[Acyl - 15]^+$	1				
	2				
$[Acid + 1]^+$	1	1			
	2	1			
$[Acid]^+$	1				
	2				
$[Acid - 1]^+$	1				2
	2			1	
m/e 218	1	2	4	3	6
	2	100	100	26	
m/e 205	1	14	14	19	12
	2	2	2	1	
m/e 203	1	12	23	32	18
	2	20	17	14	
m/e 201	1	2	5	28	2
	2	2	3	10	
m/e 191	1			2	3
	2	18	14	7	
m/e 147	1	29	23	58	50
	2	28	33	18	
m/e 129	1	12	26	100	100
	2	28	71	62	
m/e 103	1	12	14	48	39
	2	28	29	100	
m/e 73	1	34	38	99	51
	2	48	35	58	
m/e 69 $[C_5H_9]^+$	1	7	10	31	5
	2	5	5	12	
m/e 67 $[C_5H_7]^+$	1	4	5	23	45
	2	4	2	7	
m/e 57 $[C_4H_9]^+$	1	27	26	17	2
	2	25	22	5	
m/e 55 $[C_4H_7]^+$	1	14	15	51	14
	2	15	13	22	

[a] Data from Myher *et al.* (1974). [b] Acyl chain identified by carbon number: unsaturation.

monoacylglycerol TMS ethers on Silar 5CP leads to overlapping of 1-18:2 with 2-18:3, 1-18:1 with 2-18:2, and 1-18:0 with 2-18:1. Complete resolution, however, can be obtained by making use of the appropriate mass spectral information and the techniques of mass chromatography or multiple ion monitoring. Both procedures provide the investigator with plots of ion current for any chosen ion vs. time. As an example of the additional resolving power of GC/MS, the following procedure effects a resolution of 1-18:2 and 2-18:3 within a complex mixture when run as the TMS ethers on Silar 5CP. A mass chromatogram for the $[M - 90]^+$ ion of 18:3 at *m/e* 406 has two peaks, the first corresponding to 2-18:3 and the second to 1-18:3. The ion provides us with a detector specific for the TMS ether of monolinolenoylglycerol, and the liquid phase provides the resolution of 1 and 2 isomers. The nature of the 1 isomer is confirmed from a chromatogram plot of the $[M - 103]^+$ ion (*m/e* 393) for 18:3. In this case, since only the spectra of 1 isomers have this ion, one peak is observed. Finally, a plot of the $[M - 103]^+$ fragment for 18:2 at *m/e* 395 contains two peaks, the first corresponding to 1-18:2 and the second to the natural stable isotope contribution at p + 2, where p is the $[M - 103]^+$ fragment in the spectrum of 1-18:3. Thus the resolution of 1-18:2 and 2-18:3 is complete, with the chromatogram plot of *m/e* 406 giving us a measure of 2-18:3 and the chromatogram plot of *m/e* 395 a measure of 1-18:2. In general, compounds that cannot be resolved by GLC alone can be resolved by GC/MS if the spectra of the overlapped compounds are sufficiently different, or compounds that have identical mass spectra but different GLC retention times can be uniquely identified.

3.10.2. Monoalkylglycerols

Glyceryl monoethers are not generally found as such in natural mixtures but are obtained from more complex neutral or polar glycerol lipids, either by saponification or by reductive hydrogenation. Mass spectrometry has played an important role in the analysis of glyceryl monoethers.

3.10.2.1. Dimethoxy Derivatives

Hallgren and Larsson (1962) reported the mass spectra of the dimethoxy derivatives of batyl (1-18:0) and selachyl (1-18:1) alcohols as well as that of 1-20:1 (the double bond was found to be $\omega 7$ by oxidative cleavage). Table XI contains abundance data obtained for the major ion fragments found in the mass spectra of these compounds. Although the molecular ion intensity is quite low for 1-18:0, it is very high in the spectra of the unsaturated compounds, 1-18:1 and 1-20:1. In contrast, the $[M - 32]^+$ ion, due to loss of CH_3OH from the molecular ion, is intense only for the saturated compound. Small ions are also found at *m/e* $[M - 46]^+$ and $[M - 64]^+$ due to the losses of CH_3OCH_3 and $2(CH_3OH)$ moieties, respectively. Ions corresponding to $[M - 77]^+$, $[M - 90]^+$, and $[M - 91]^+$ have been attributed to losses of CH_3OH plus CH_2OCH_3, $CH_2(CHOCH_3)OCH_3$ plus H, and $CH_2(CHOCH_3)OCH_3$ plus 2H, respectively. The $[M - 91]^+$ is an abundant ion only for the saturated compound. The ion fragment at *m/e* 119 $[CH_2(CH_2)_{16}CH_3]^+$ in the spectra of 1-18:0 results from cleavage of the *O*-alkyl bond, and a similar fragment with the loss of one additional hydrogen predominates in the case of both

Table XI. Abundance Data for the Major Ion Fragments Found in the Mass Spectra of the Dimethoxy Derivatives of Glycerol Monoethers[a]

	Relative abundance		
Ion type[b]	18:0[c]	18:1 ω9	20:1 ω7
$[M]^+$	2	33	49
$[M-32]^+$	18	2	4
$[M-77]^+$	1	12	17
$[M-91]^+$	25	3	4
$[M-90]^+$	5	7	10
$[M-119]^+$	6	9	5
$[M-120]^+$	trace	29	15
m/e 121	7.1	64	59
m/e 103	4.7	20	27
m/e 89	91	100	100
m/e 45	58	69	52

[a] Data from Hallgren and Larson (1962).
[b] Further description of ions is given in the text.
[c] Alkyl chain identified by carbon number: degree of unsaturation. The value indicates the location of the first carbon of the double bond when carbons are numbered from the terminal carbon of the alkyl chain.

unsaturated compounds. There are also a number of fragments that retain some of the glycerol backbone as well as one or more of the methoxy groups: *m/e* 45 $[CH_2OCH_3]^+$, *m/e* 89 $[CH_2(CHOCH_3)OCH_3]^+$, *m/e* 103 $[CH_2(CH(CH_2)OCH_3)OCH_3]^+$, and *m/e* 121 $[CH_2(CH(CH_2O)OCH_3)OCH_3 + 2H]^+$. The ions at *m/e* 103 and 121 are notably much lower for the saturated compounds than for either of the unsaturated compounds.

Methoxyalkylglycerol ethers have spectra that are quite different from those of the alkyl ethers (Hallgren and Ställberg, 1967). The base peak in the spectrum of 1-*O*-(2-methoxyhexadecyl)-2,3-di-*O*-methyl-glycerol is at *m/e* 241 as a result of a cleavage on the glycerol side of the methoxy-substituted alkyl carbon, with the charge being retained on the methoxyalkyl moiety, and a smaller fragment at *m/e* 255 has one more methylene group than the *m/e* 241 ion. The $[M-32]^+$ ion due to loss of CH_2OH can be used as a molecular weight indicator.

3.10.2.2. Isopropylidenes

Although the isopropylidene derivatives of glyceryl monoethers are frequently used, a comprehensive mass spectrometric study of these derivatives has not been made. Hallgren and Ställberg (1967) have reported the mass spectrum of the isopropylidene derivative of batyl alcohol (1-18:0), and Snyder *et al.* (1970) and Saito and Gamo (1973) have reported the mass spectrum of the isopropylidene derivative of chimyl alcohol (1-16:0). The base peak in these compounds is at *m/e* 101, which could be formed by a simple cleavage between glycerol carbons 1 and 2, with the charge being retained by the isopropylidene-containing fragment. The molecular weight can be determined from the abundant $[M-15]^+$ ion.

Table XII. Relative Abundance Data for the Major Ions Found in the Mass Spectra of the Isopropylidene Derivatives of Some Methoxy-Substituted Alkylglycerols and Batyl Alcohol[a]

	Relative abundance			
Ion type	2(CH_3O)-16:0[b]	2(CH_3O)-18:0	2(CH_3O)-18:1	18:0 (batyl)
$[M]^+$	—	<1	1.5	—
$[M - 15]^+$	24	57	14	65
$[M - 32]^+$	—	—	8	—
$[M - 131]^+$	3	6	2	—
$[M - 145]^+$	100	100	12	—
m/e 189	—	—	3	—
m/e 131	8	14	100	—
m/e 101	33	17.8	31	100

[a] Data from Hallgren and Ställberg (1967).
[b] Carbon number and unsaturation indicated as in Table XI. Methoxy substitution on carbon 2 of the alkyl chain designated by 2(CH_3O)-.

Hallgren and Ställberg (1967) have reported the mass spectra obtained for a number of glyceryl ethers that have a methoxy substituent in the alkyl chain. Table XII compares the mass spectrometric abundance data for a number of (2-methoxyalkyl)glycerols with those of batyl alcohol. The major fragments can be derived via the cleavages indicated in Fig. 6. Since the molecular ion is either small or absent in these spectra, the best indicator of molecular weight is the abundant $[M - 15]^+$ ion. The location of the methoxy group at carbon 2 of the alkyl chain can be verified from the ions resulting from a cleavage adjacent to the methoxy-bearing carbon atom, namely the $[M - 145]^+$ ion, and the ion at *m/e* 189. The latter ion has a significant abundance only in the spectra of the unsaturated compounds. In general, a saturated compound having the methoxy group situated on carbon *n* of the alkyl group would have an intense peak corresponding to $[M - 117 - 14n]^+$, and the $[M - 15]^+$ ion, being only a function of the molecular weight, does not depend on the location of the methoxy group.

Muramatsu and Schmid (1972) have reported the mass spectra of the isopropylidene derivatives of 1-*O*-2′-hydroxyhexadecylglycerol and 1-*O*-2′-ketohexadecylglycerol; the hydroxyl and keto functions were left underivatized. The spectrum of the 2-hydroxyl compound is similar to that of chimyl alcohol in that the ion at *m/e* 101 is the base peak. The $[M - 15]^+$ ion provides the molecular weight, but the $[M - 145]^+$ ion has a very low intensity and does not provide very good

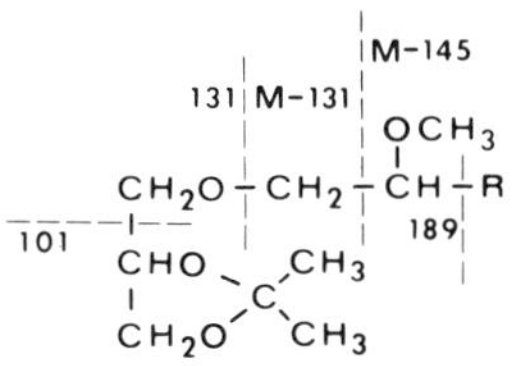

Fig. 6. Fragmentation scheme for 1-(2-methoxyalkyl)-2,3-isopropylideneglycerol.

proof for the location of the hydroxyl group. In contrast, the $[M - 145]^+$ ion for the 2-ketoalkyl compound is sufficiently abundant and can be used to locate the keto function. When the free hydroxyl group of 1-*O*-hydroxy-alkyl-2,3-isopropylidene-glycol is converted to the TMS ether, the mass spectrum is much more characteristic (Kasama *et al.*, 1973). Mass spectra have also been reported for the derivatives of synthetic 1-(9/10-hydroxy)octadecylglycerol, natural 1-(10/11-hydroxy)hexadecyl-glycerol, and natural 1-(11/12-hydroxy)octadecylglycerol. In common with other isopropylidene derivatives, the spectra of these compounds show intense *m/e* 101 ions and appreciable $[M - 15]^+$ ions. There is an intense ion* due to cleavage on the glycerol side of the substituted carbon and a smaller ion due to cleavage on the other side of the substituted carbon that provide the location of the substituent. In both cases, the charge resides on the fragment retaining the substituted carbon. In general, the placement of the substituent can be determined as follows: substract the mass of major cleavage ion from the mass of the $[M - 15]^+$ ion, then subtract 116 from the remainder and divide by 14. The alkyl carbon having the substituent is then found by adding 1 to the final result; for example, derivatized (1-11-hydroxy)octadecyl-glycerol with an $[M - 15]^+$ ion at *m/e* 457 and a major cleavage ion at *m/e* 201 has the hydroxy group placed on carbon number $(457 - 201 - 116)/14 + 1 = 11$.

The mass spectrum for 1-*O*-phytanyl-2,3-*O*-isopropylideneglycerol (Hallgren *et al.*, 1974) has *m/e* 101 as base peak and a strong $[M - 15]^+$ ion at *m/e* 397. There are also very small peaks at *m/e* 353, 378, and 252. The spectra for synthetic and natural samples were identical, but the position of the four branched methyl groups cannot be determined. However, even when the position of substitution can be derived from the mass spectrometric data, the stereochemical configuration is left undetermined.

3.10.2.3. *Acetates*

Abundance data for glyceryl ether diacetates could not be found, but some features in the high-resolution mass spectrum of chimyl diacetate have been reported (Jost, 1974). The major fragment, found at *m/e* 159 $[C_7H_{11}O_4]^+$, retains the entire glycerol backbone with its two acetyl groups. This ion is also observed in the mass spectra of monoacylglycerol diacetates. Other ions observed were as follows: *m/e* 340 $[C_{21}H_{40}O_3]^+$ due to loss of acetic acid from $[M]^+$, *m/e* 297 $[C_{19}H_{37}O_2]^+$ due to a further loss of CH_3CO from *m/e* 340, and *m/e* 280 $[C_{19}H_{36}O]^+$ due to loss of two molecules of acetic acid from $[M]^+$. Unfortunately, the use of this derivative led Jost (1974) to the conclusion that the unknown glyceryl monoether isolated from rabbit harderian gland contained an unsaturated alkyl chain. The compound has been correctly identified as a hydroxyalkylglycerol (Rock and Snyder, 1975). Upon electron impact, acetic acid is readily lost from the acetylated hydroxalkyl chain. The resulting ions found 2 *m/e* units lower than the corresponding ions found in the spectra of the saturated alkylglycerols may be mistaken for a sign of unsaturation.

3.10.2.4. *Trimethylsilyl Ethers*

Myher *et al.* (1974) have reported the GC/MS spectra obtained for the TMS ethers of a number of 1- and 2-monoalkylglycerols. Table XIII contains abundance

* This ion is probably the base peak in the spectrum of a single pure isomer.

data for the major fragments for these compounds, and Fig. 7 illustrates the complete 70-eV spectra obtained for the TMS ethers of 1- and 2-monooleylglycerol. Similar to that observed for the dimethoxy and isopropylidene derivatives, the major ion in the spectra of the TMS ethers of the 1-monoalkylglycerols at *m/e* 205 is due to

Table XIII. Relative Abundance Data for the Major Ions Found in the Mass Spectra of the TMS Ethers of Some Monoalkylglycerols[a]

Ion type	Positional isomer	Relative abundance		
		16:0[b]	18:0	18:1
$[M]^+$	1			13
	2			30
$[M-15]^+$	1	7	2	4
	2	2		3
$[M-90]^+$	1	8	2	2
	2	1		1
$[M-104]^+$	1	3	1	2
	2	10		11
$[M-(73+74)]^+$	1	13	4	1
	2	2		1
$[M-(90+90)]^+$	1	5	1	2
	2			
$[M-(103+90)]^+$	1			7
	2			25
$[M-235]^+$	1	2		
	2	2		3
m/e 218	1			
	2	100		100
m/e 205	1	100	100	100
	2	11		15
m/e 191	1			
	2	15		19
m/e 147	1	34	25	47
	2	15		28
m/e 133	1	39	27	14
	2	83		39
m/e 131	1	20	7	21
	2	12		21
m/e 130	1	37	18	12
	2	33		47
m/e 129	1	12	4	19
	2	23		48
m/e 117	1	31	22	33
	2	52		44
m/e 103	1	15	5	25
	2	28		87
m/e 73	1	34	22	59
	2	38		73
m/e 57 $[C_4H_9]^+$	1	26	18	22
	2	39		40
m/e 55 $[C_4H_7]^+$	1	12	10	39
	2	16		53

[a] Data from Myher *et al.* (1974).

[b] Carbon number and unsaturation of alkyl chain indicated as in Table XI.

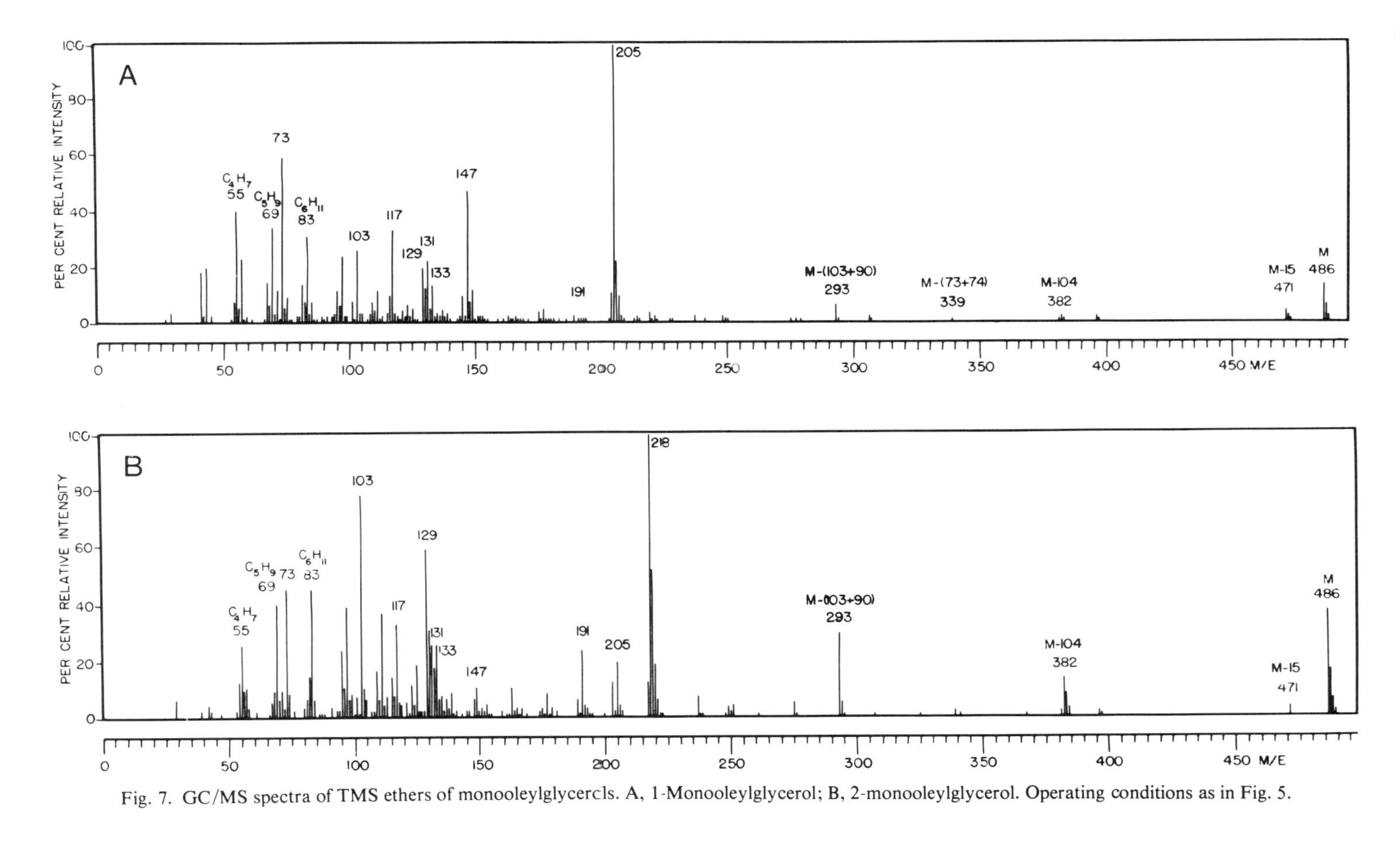

Fig. 7. GC/MS spectra of TMS ethers of monooleylglycercls. A, 1-Monooleylglycerol; B, 2-monooleylglycerol. Operating conditions as in Fig. 5.

cleavage between glycerol carbons 1 and 2. In the spectra of the 2 isomers, however, the *m/e* 205 ion is greatly reduced and the base peak occurs at *m/e* 218 due to loss of the alkoxy radical from the molecular ion. Another peak that appears to be exclusive for the 2 isomer is an ion that occurs at *m/e* 191 for all the homologues. This indicates that the alkyl group has been lost, but the precise composition of the ion is not known. The molecular ion is not observed in the spectra of the saturated compounds, but ions at $[M - 15]^+$ and $[M - 90]^+$ are indicative of the molecular weight. In the case of the 2 isomers, even these ions have a very low abundance, and the molecular weight of the saturated 2 isomers must be determined from the ion at $[M - 104]^+$ formed by loss of $CH_2OSi(CH_3)_3$ + H from the molecular ion. The molecular ion is much more intense in the spectra of the unsaturated monoalkylglycerols.

3.10.3. Monoalk-1-enylglycerols

The GC/MS for the TMS ethers of several alk-1-enylglycerols have been reported by Myher *et al.* (1974), and the mass spectra of the TMS ethers of 1-hexadec-1-enylglycerol (16:0 A) and 1-octadecadi-1,9-enylglycerol (18:1 A) are shown in Fig. 8. In contrast to the spectra of the alkylglycerols, the base peaks for these molecules are at low mass, *m/e* 73 $[Si(CH_3)_3]^+$ and *m/e* 103 $[CH_2OSi(CH_3)_3]^+$, respectively. The molecular ion is small for 16:0 A but is much more pronounced for 18:1 A. Similar to that found for the TMS ethers of the 1-alkylglycerols, a very large *m/e* 205 ion occurs in the spectra of 16:0 A. The intensity of this ion, however, is greatly reduced in the case of 18:1 A. By analogy with the spectra of the monoalkylglycerols, the *m/e* 205 ion may make it possible to distinguish between 1 and 2 isomers, but this has not been verified.

The 1-alk-1-enylglycerols also give rise to a small but characteristic fragment at *m/e* 219 that is totally absent in the spectra of the 1-alkylglycerols except as a p + 1 isotope peak from *m/e* 218 in the spectra of the 2-monoacylglycerols, and an ion representing the alkenyl chain, $[CHCH(CH_2)_nCH_3]^+$, is found 236 *m/e* units lower than the molecular ion. Both ions serve to distinguish the alk-1-enylglycerols from the alkylglycerols with which they may overlap. Acylglycerols having the same carbon number are well resolved by GLC on Silar 5CP.

Since long-chain cyclic acetals are sometimes formed as a by-product during the isolation of 1-alk-1-enylglycerols, it is important to be able to recognize these compounds should they occur. Baumann *et al.* (1972) have separated and established the structure of four isomeric long-chain cyclic acetals of glycerol, and the essential features of the 70-eV mass spectra for *cis*- and *trans*-2-pentadecyl-5-acetoxy-1,3-dioxanes and *cis*- and *trans*-2-pentadecyl-4-acetoxymethyl-1,3-dioxolanes are summarized in the following discussion. Although there is a small peak at *m/e* 356 which corresponds to the expected molecular weight, the major portion of this ion may be due to a p + 1 isotope contribution from the more abundant $[M - 1]^+$ ion that results from loss of the C-2 acetal proton. Loss of the alkyl chain, $C_{15}H_{31}$, gives rise to the base peak at *m/e* 145 for all compounds, and the ion at *m/e* 43 due to the acetyl ion is also very intense. The rearrangement ion $[M - C_{15}H_{31}CO]^+$ at *m/e* 117 is approximately five to six times more abundant for the

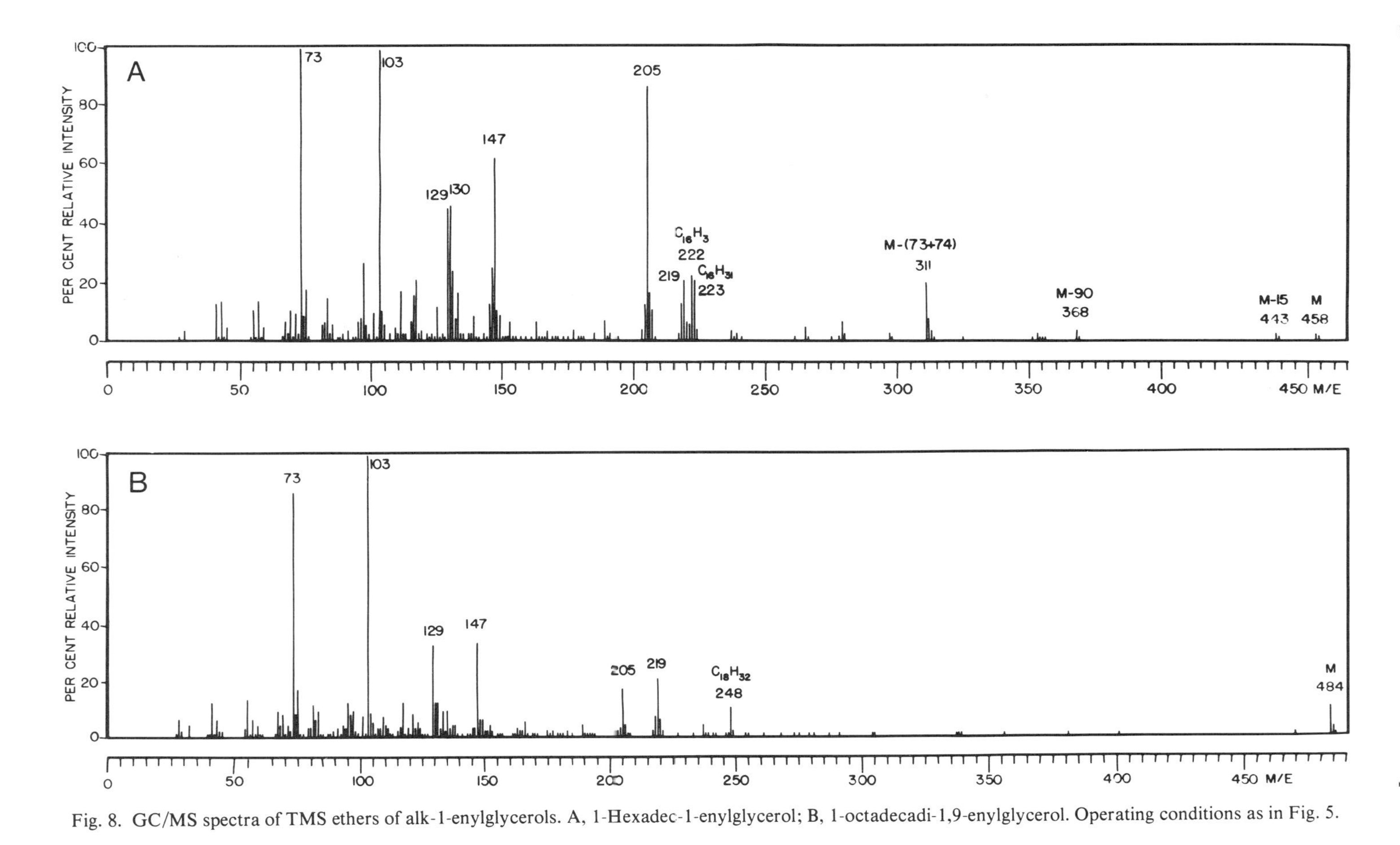

Fig. 8. GC/MS spectra of TMS ethers of alk-1-enylglycerols. A, 1-Hexadec-1-enylglycerol; B, 1-octadecadi-1,9-enylglycerol. Operating conditions as in Fig. 5.

dioxanes than for the dioxolanes. Similarly, the ion at *m/e* 86, due to the loss of CH_3COO from the *m/e* 145 ion, is also more intense for the six-membered ring compounds than for the five-membered ones.

3.10.4. Diacylglycerols

The mass spectra of diglycerides have been studied in free hydroxyl form (Morrison *et al.*, 1970), as acetates (Hasegawa and Suzuki, 1973, 1975), as TMS ethers (Barber *et al.*, 1968; Horning *et al.*, 1969, 1971; Curstedt, 1974), or as *t*-butyldimethylsilyl ethers (Myher *et al.*, 1977*a*). Except for the study of the underivatized diacylglycerols, the work involved combined GC/MS. Specifically ^2H-labeled diacylglycerols have also been used to elucidate the structures of the ions found in the spectra of the underivatized compounds (Morrison *et al.*, 1970) and in the spectra of the TMS ethers (Curstedt, 1974). In the former study, high-resolution mass measurements were also made.

3.10.4.1. Underivatized Compounds

The nature of the major fragments in the mass spectrum of 1,3-distearoylglycerol was elucidated with the aid of high-resolution mass measurements and deuterated analogues (Morrison *et al.*, 1970). Including the base peak at *m/e* 43, the ions below *m/e* 200 have hydrocarbon structures and are of little diagnostic value. The molecular ion is observed, but the intensity is less than 1% of the base peak. The major ion that still retains both acyl groups is found at *m/e* 606 (12.0%) and is formed by the loss of one molecule of H_2O that contains one hydrogen from the hydroxyl groups and one hydrogen from the acyl chain. The CH_3O group lost in the formation of the *m/e* 593 ion (1.0%) is composed of one hydrogen from the hydroxyl group and two hydrogens from the glycerol methylene group. The loss of one of the acyloxy groups leaves the abundant *m/e* 341 ion (36.0%) and is a common feature in the spectra of acylglycerols. Similar to that found for many monosubstituted glycerol compounds, there is an ion at *m/e* 327 (11.5%) due to a cleavage between glycerol carbons 1(3) and 2. There are also ions such as the acyl ion $[RCO]^+$ at *m/e* 267 (45.0%) and the acid ion $[RCOOH]^+$ at *m/e* 284 (3.0%). The spectra of 1,2(2,3)- and 1,3-diacylglycerols are indistinguishable when the compounds are introduced into the mass spectrometer in free form (Barber *et al.*, 1968).

When hot injector ports (>300°C) and short columns are used, peaks emerge after the injection of phospholipids (phosphatidylcholines, phosphatidylethanolamines) that were tentatively identified as diacylpropenediols formed by pyrolysis of the phospholipids (Kuksis *et al.*, 1967). This identification has been supported by the observations of Perkins and Johnston (1969), who found that similar mass spectra were obtained for the peaks emerging after the injection of phosphoglycerides, diacylglycerols, and 1,2-dipalmitoylpropenediol.

3.10.4.2. Acetates

Diacylglycerol acetates may be considered triacylglycerols with one short-chain fatty acid. The interpretation of their spectra depends on the earlier studies (see later)

Table XIV. Relative Abundance Data for the Major Ions Found in the Mass Spectra of Some Diacylglycerol Acetates

Ion type	Ion abundance[a]	
	1–18:0, 2–18:1[b]	2–18:2, 3–16:0[c]
$[M]^+$	—	—
$[M - 60]^+$	1	3
$[M - R_uCOO]^{+d}$	24	16
$[M - R_sCOO]^{+d}$	14	7
$[M - R_sCOOH]^+$	9	10
$[R_uCO]^+$	20	14
$[R_uCO - 1]^+$	20	45
$[R_sCO]^+$	12	4

[a] Abundances are stated as a percent of the ion intensity summation.
[b] 1-Octadecanoyl-2-oleoyl-3-acetoyl-*rac*-glycerol. Data from Hasegawa and Suzuki (1975).
[c] 1-Acetoyl-2-linoleoyl-3-palmitoyl-*sn*-glycerol. Data from Morley *et al.* (1975).
[d] R_s and R_u refer to hydrocarbon portions of the saturated and unsaturated fatty acids, respectively.

made with triacylglycerols. Hasegawa and Suzuki (1973) used GC/MS for the analysis of the diacylglycerol acetates derived from ovolecithin by means of acetolysis, and subsequently (Hasegawa and Suzuki, 1975) they compared the products obtained from lecithin by acetolysis with those obtained by phospholipase C digestion followed by acetylation. Unfortunately, the available ion abundance data are rather limited and must be obtained from the few published prints, namely the mass spectra of 1-palmitoyl-2-oleoyl-3-acetyl-*sn*-glycerol (Hasegawa and Susuki, 1973), 1-stearoyl-2-oleoyl-3-acetyl-*sn*-glycerol (Hasegawa and Suzuki, 1975), and 1-acetyl-2-linoleoyl-3-palmitoyl-*sn*-glycerol (Morley *et al.*, 1975). Table XIV summarizes the mass spectrometric abundance data for these latter two 1,2(2,3)-diacylglycerols. Since the molecular ion is either small or not detected, the total carbon number and total degree of unsaturation are usually derived from the $[M - 60]^+$ ion. The abundances of the higher-mass ions usually increase with the introduction of one to three double bonds, but their intensity will generally drop dramatically for species containing polyunsaturated fatty acids such as 20:4 or 22:6. In contrast, ions retaining one long-chain fatty acid moiety are quite abundant. These include the $[acyl]^+$ or $[acyl - 1]^+$ ions and the ions left after the loss of acyloxy or acid moieties. Hasegawa and Suzuki (1973, 1975) used the $[M - RCOOCH_2]^+$ ion as an indication of the fatty acid in the 1 position, but because of the extremely low intensity of this ion and the lack of documentation, the utility of this procedure needs further verification. In the case of the TMS and *t*-BDMS ethers, the $[M - RCOOCH_2]^+$ ion is usually much more abundant in the spectra of the 1,3-diacylglycerols.

3.10.4.3. *Trimethylsilyl Ethers*

The TMS derivatives of diacylglycerols are suitable for GC/MS and have characteristic mass spectra (Barber *et al.*, 1968). Unlike the spectra of the acetate derivatives, the spectra of the TMS ethers are different for 1,2(2,3)- and 1,3-diacylglycerols. The spectra for the TMS ethers of the reverse isomers 1-palmitoyl-2-stearoyl-glycerol and 1-stearoyl-2-palmitoyl are distinguishable when they are introduced by direct probe, but the spectra are indistinguishable when a GC/MS system is used. The GC/MS spectra obtained for the diacylglycerol TMS ethers formed by silylating the products obtained from phospholipids by hydrolysis in wet diphenyl ether at elevated temperatures are described in a number of publications (Casparrini *et al.*, 1968; Horning *et al.*, 1969, 1971). The major products of this procedure were found to be the 1,3 isomers, whereas if the phospholipids were heated in the presence of silylating reagents the 1,2 isomer was found to be the major product. Curstedt (1974) investigated the structure of the major ions through the use of a series of deuterium-labeled 1,2-dipalmitoyl-3-TMS-*rac*-glycerols. The major difference in the spectrum of the 1,3-isomer is the presence of an abundant ion at *m/e* 371 due to loss of an acyloxymethylene radical (CH_2OCOR). This type of cleavage is not significant for the 1,2(2,3) isomer even though there is still one acyl group on position 1(3) of glycerol.

Table XV summarizes abundance data taken from various sources for the TMS ethers of a number of 1,2(2,3)-diacylglycerols. Although the molecular ion is sometimes seen in trace amounts, the $[M - 15]^+$ ion due to loss of a methyl group and the $[M - 90]^+$ ion due to loss of $Si(CH_3)_3OH$ constitute the major ions that still have both acyl groups intact. These ions are important for the determination of the

Table XV. Relative Abundance Data for the Major Ions Found in the Mass Spectra of the TMS Ethers of Some Diacylglycerols

Ion type	Position 1: / Position 2:	Ion abundance[a]			
		16:0 / 16:0	18:0 / 18:1	18:1 / 18:1	16:0 / 18:2
$[M]^+$		—	<1	—	—
$[M - 15]^+$		10	6	5	5
$[M - 90]^+$		1	4	5	3
$[M - R_sCOO]^+$[b]		18	12	—	5
$[M - R_sCOOH]^+$		—	1	—	8
$[M - R_uCOO]^+$		—	32	42	30
$[M - R_uCOOH]^+$		—	—	21	—
$[R_sCO + 90]^+$		14	3	—	3
$[R_uCO + 90]^+$		—	4	—	<1
$[R_sCO + 74]^+$		39	12	—	12
$[R_uCO + 74]^+$		—	8	12	10
$[R_sCO]^+$		19	5	—	3
$[R_uCO]^+$		—	7	10	8
$[R_uCO - 1]^+$		—	5	5	14
m/e 145		58	25	not given	21
m/e 129		40	79	not given	30

[a] Abundances are stated as a percent of the ion intensity summation. Data for 1,2-dipalmitoyl-, 1-stearoyl-2-oleoyl-, 1,2-dioleoyl-, and 1-palmitoyl-2-linoleoyl-*rac*-glycerol are taken from Curstedt and Sjövall (1974), Horning *et al.* (1969), Assmann *et al.* (1973), and Curstedt and Sjövall (1974), respectively.
[b] R_s and R_u designate the hydrocarbon portions of the saturated and unsaturated fatty acids, respectively.

total carbon number and degree of unsaturation for the intact diacylglycerol. The other important ions in the >200 *m/e* region have only one fatty acid residue remaining. A major ion results from the loss of RCOO, and when an unsaturated fatty acid occurs in the diacylglycerol, the other fatty acid is also lost as RCOOH. In the case of 1-palmitoyl-2-linoleoyl-3-TMS-glycerol, the latter fragmentation mode is predominant. For mixed-acid unsaturated diacylglycerols of the type 1-(saturated)-2-(unsaturated)-glycerol, the fragment formed by the loss of RCOO or RCOOH from the 2 position predominates over that formed by loss from the 1 position. The $[RCO + 90]^+$ ion was found to correspond to $[RC(OH)OSi(CH_3)_3]^+$, whereas two types of ion contribute to the $[RCO + 74]^+$ ion. The deuterium-labeling studies indicated that in the case of 1,2-dipalmitoyl-3-TMS-*rac*-glycerol, 85–95% of the peak is due to the rearrangement ion, $[RCOOSi(CH_3)_2]^+$, and the remaining 15–5% of the peak is made up of an ion having the molecular formula $[RCOOCH_2CH(OH)CH_2]^+$. This latter ion retains the glycerol backbone intact and is a major ion in the spectra of triacylglycerols (see later). In the case of 1-palmitoyl-2-linoleoyl-3-TMS-*sn*-glycerol, 64% of the $[RCO + 74]^+$ peak that contains the palmitoyl residue and 94% of the peak that contains the linoleoyl residue are due to the $[RCOOSi(CH_3)_2]^+$ ion. Acyl ions $[RCO]^+$, corresponding to both fatty acids, are also found, and when the fatty acid is unsaturated the $[RCO - 1]^+$ ion may be even more abundant. The ions at *m/e* 145 $[CH_2(O)CHCH_2OSi(CH_3)_3]^+$ and *m/e* 129 $[CH_2CHCHOSi(CH_3)_3]^+$ retain the TMS group.

3.10.4.4. Tertiarybutyldimethylsilyl Ethers

Preliminary findings (Myher *et al.*, 1977*a*) indicate that the *t*-BDMS ethers of diacylglycerols offer many advantages for GC/MS analysis of diacylglycerols. The *t*-BDMS ethers are up to 10^4 times more stable than the corresponding TMS ethers (Corey and Venkateswarlu, 1972). The *t*-BDMS ethers of diacylglycerols can be made without causing any isomerization, and, once formed, the *t*-BDMS ethers can be manipulated in any manner that does not involve highly acidic mediums. They can, for example, be separated by argentation TLC and eluted from the gel intact without any losses due to decomposition.

Tables XVI and XVII contain the major ion abundances obtained by GC/MS for the *t*-BDMS ethers of a few synthetic diacylglycerols and the *t*-BDMS ethers of the diacylglycerols derived from rat liver lecithin. Many of the fragments are similar to those obtained for the TMS ethers, but the most important feature is the abundance peak corresponding to $[M - 57]^+$ that is also a major feature of the spectra of the *t*-BDMS derivatives of steroids, prostaglandins (Kelly and Taylor, 1976), and fatty acids (Phillipou *et al.*, 1975). The high abundance of the $[M - 57]^+$ ion allows the ready determination of carbon number of degree of unsaturation even for species containing docosahexaenoic acid (22:6). Phospholipids that are susceptible to phospholipase C hydrolysis can be converted into the corresponding diacylglycerol *t*-BDMS ethers and their composition determined by GC/MS. Since the mole percent of each carbon number can best be determined from the GLC peak areas, the mass spectrometric data are required only for the relative quantitation of molecular species within each carbon number grouping. It should be possible to carry out such species analysis on smaller amounts of sample than is now possible via argentation TLC.

Table XVI. Relative Abundance Data for the Major Ions Found in the Mass Spectra of the t-BDMS Ethers of Some Diacylglycerols[a]

		Relative abundance					
	Position 1:	16:0[b]	16:0	16:0	18:0	16:0	16:0
Ion type	Position 2:	16:0	16:0	18:0	16:0	18:1	18:2
$[M]^+$		—	—	—	—	—	—
$[M - 57]^+$		100	73	48	63	62	76
$[M - 132]^+$		14	7	5	7	6	4
$[M - R_1COO]^+$[c]		16	23	8	9	5	4
$[M - R_2COO]^+$[c]				16	15	32	15
$[M - R_1COOCH_2]^+$		24	—	—	—	—	—
$[R_1CO + 2 \times 74]^+$		27	8	6	5	5	5
$[R_2CO + 2 \times 74]^+$				1	3	3	3
$[R_1CO + 74]^+$		96	100	88	70	100	100
$[R_2CO + 74]^+$				100	100	88	72
$[R_1CO]^+$		84	20	10	6	7	7
$[R_2CO]^+$				5	12	12	16
$[R_2CO - 1]^+$		—	—	—	—	3	5
m/e 171		28	25	51	56	75	89
m/e 131		75	24	55	61	46	62

[a] Data taken from Myher *et al* (1977).
[b] 1,3-Dipalmitoylglycerol. All other diacylglycerols are the 1,2 isomers.
[c] R_1 and R_2 designate the hydrocarbon portions of the fatty acids substituted at the 1 and 2 positions, respectively.

Table XVII. Relative Abundance Data for the Major Ions Found in the Mass Spectra of the t-BDMS Ethers of Some Diacylglycerols[a]

		Relative abundance				
	Position 1:	16:0	16:0	18:1	16:0	16:0
Ion type	Position 2:	20:3 ω6	20:4	20:4	20:5	22:6
$[M]^+$		1.5	4	5	6	4
$[M - 57]^+$		55	35	47	64	43
$[M - 132]^+$		2	1	—	—	—
$[M - R_1COO]^+$[b]		—	—	2	3	2
$[M - R_2COO]^+$[b]		18	25	30	34	36
$[M - R_1COOCH]^+$		—	—	—	—	—
$[R_1CO + 2 \times 74]^+$		7	5	8	7	5
$[R_2CO + 2 \times 74]^+$		—	—	—	—	—
$[R_1CO + 74]^+$		100	87	72	88	74
$[R_2CO + 74]^+$		30	10	12	10	8
$[R_1CO]^+$		7	6	5	7	8
$[R_2CO]^+$		12	8	10	8	7
$[R_2CO - 1]^+$		6	11	12	8	6
m/e 171		77	100	100	100	100
m/e 131		52	68	66	65	35

[a] Data taken from Myher *et al.* (1977*a*).
[b] R_1 and R_2 designate the hydrocarbon portions of the fatty acids substituted at the 1 and 2 positions, respectively.

The spectra of the 1,2(2,3) and 1,3 isomers are clearly distinguishable. The $[M - RCOOCH_2]^+$ fragment found in the spectra of 1,3-dipalmitoyl-2-(*t*-BDMS)-glycerol is absent from the spectra of the 1,2 isomer. There is also a difference in the spectra of reverse isomers such as 1-palmitoyl-2-stearoyl-3-(*t*-BDMS)-*rac*-glycerol and 1-stearoyl-2-palmitoyl-3-(*t*-BDMS)-*rac*-glycerol. The abundance ratio of the ions due to losses of the acyloxy radicals from position 1 and position 2 is the best indicator of this difference. For the above two reverse isomers, the ratios of *m/e* 427 (loss of palmitoyl) to that of *m/e* 455 (loss of stearoyl) are 2.0 and 0.6, respectively.

3.10.5. Alkylacylglycerols

Satouchi and Saito (1976) have recently reported the mass spectra of the TMS ethers of several 1-alkyl-2-acylglycerols, and the abundances for the major fragments are given in Table XVIII. The molecular weight of these compounds is indicated by the $[M - 15]^+$ and $[M - 90]^+$ ions. In the case of the saturated species, the abundance of these fragments is very low, but for the monoenoic species the $[M - 90]^+$ fragment is quite abundant. There is also a low-abundance fragment, $[M - R_1O]^+$, due to loss of the alkoxy radical, as well as an ion corresponding to $[R_1O + 73]^+$. The other major ions $[M - R_2COO]^+$, $[RCO + 74]^+$, and $[RCO]^+$ are similar to those observed for the TMS ethers of diacylglycerols except that the relative abundance is lower in the case of the alkylacylglycerols. The base peak is at *m/e* 134 $[CH_2CHCH_2OSi(CH_3)_3]^+$, whereas the base peak in the spectra of the diacylglycerols is at *m/e* 129.

3.10.6. Triacylglycerols

Triacylglycerols, although larger and more complex, were studied (Ryhage and Stenhagen, 1960) much earlier than mono- or diacylglycerols. There has been a continued interest in this area, and it has been the subject of a recent comprehensive review (Hites, 1975). Although triacylglycerols are more frequently admitted to the

Table XVIII. Relative Abundance Data for the Major Ions Found in the Mass Spectra of the TMS Ethers of Some 1-Alkyl-2-acylglycerols[a]

		Relative abundance			
	Alkyl group (R_1):	16:0	18:0	16:0	16:0
Ion type	Acyl group (R_2Co):	16:0	16:0	18:0	18:1
$[M - 15]^+$		1.7	1.3	1.7	5
$[M - 90]^+$		1.1	0.6	0.9	18
$[M - R_2COO]^+$		4	3	3	4
$[M - R_1O]^+$		2	1	1	20
$[R_1O + 73]^+$		6	13	10	19
$[R_2COO + 74]^+$		2	2	2	9
$[R_2CO + 74]^+$		18	10	8	13
$[R_2CO]^+$		8	8	6	11
m/e 130		100	100	100	100

[a] Data taken from Satouchi and Saito (1976).

mass spectrometer by direct insertion probe, they may be admitted via GLC (Murata and Takahashi, 1973). Barber *et al.* (1964) reported the first comparative correlation of fragmentation pattern with triacylglycerol structure. This was followed by detailed studies using specifically deuterated triacylglycerols and triacylglycerols containing short-chain fatty acids (Lauer *et al.*, 1970; Aasen *et al.*, 1970).

The molecular ion abundance in the spectra of long-chain triacylglycerols is generally low (~0.002 × base peak) and decreases with decreasing chain length. In saturated triacylglycerols, the $[M - 18]^+$ ion that results from the loss of water from the molecular ion may be slightly more abundant, but the relative intensity decreases with the degree of unsaturation. The loss of one fatty acid residue in the form of an alkoxy radical gives rise to the intense $[M - RCOO]^+$ ion, but the $[M - RCOOH]^+$ ion, due to the loss of one fatty acid moiety, becomes more prominent than the $[M - RCOO]^+$ ion when there is at least one unsaturated fatty acid group left in the ion. Although there is some quantitative difference in the ratio of two $[M - RCOO]^+$ ions possible for each of the isomeric compounds, 1,2-distearoyl-3-oleoyl-*rac*-glycerol and 1,3-distearoyl-2-oleoyl-*rac*-glycerol, the difference is not great enough to be generally useful. The $[M - RCOOCH_2]^+$ ion that results exclusively from the loss of an acyloxymethylene radical from the primary position of glycerol can be used to locate the fatty acids esterified to the primary position. The intensity ratio $I[M - RCOO]/I[M - RCOOCH_2]$, however, depends on the chain length and varies from zero for triacetoylglycerol to 31 for trioctadecanoylglycerol. Relative to the base peak, the intensity of the $[M - RCOO]^+$ ion decreases with chain length.

There are also a number of ion types that can be used to determine the individual fatty acids that make up the triacylglycerol. The more important of these are the $[RCO]^+$ ions and the $[RCO + 74]^+$ ions. If the acyl group is unsaturated, the $[RCO - 1]^+$ ion is actually more intense than the corresponding $[RCO]^+$ ion, and in the case of trilinoleoylglycerol there are ions corresponding to $[RCO - 2]^+$ and $[RCO - 3]^+$. The $[RCO + 74]^+$ ions retain the glycerol backbone and are possibly formed by the loss of a long-chain ketene, R′CH=C=O, from the $[M - RCOO]^+$ ions. Another series of ion occurs at m/e $[RCO + 128 + 14n]$ ($n = 0,1,2, \ldots$). The relative intensity of the first member of the series, $[RCO + 128]^+$, may be as high as 100 for saturated triacylglycerol, but it decreases with increasing unsaturation in the molecule. The increment of 128 mass units corresponds to $C_6H_8O_3$ and includes the glycerol backbone plus the first three carbons of another acyl chain. There is also an ion corresponding to $[RCO + 115]^+$ in the spectra of saturated triacylglycerols.

In a few cases, mass spectrometry has been used for the identification of unknown triacylglycerols. Sprecher *et al.* (1965), for example, used mass spectrometry along with other techniques to identify an unusual triacylglycerol isolated from the seed oil of *Sapium sebiferum*. Although the major fatty acids appeared to be oleic, linoleic, linolenic, and 2,4-decadienoic acids, the mass spectrum indicated that the molecular weight was higher than that expected from an earlier study (Maier and Holman, 1964). With the aid of IR, MS, and chemical evidence, Sprecher *et al.* (1965) were then able to determine that the missing fragment is 8-hydroxy-5,6-octadienoic acid. The structure of the intact triacylglycerol was deduced from the mass spectra of the original and hydrogenated samples.

The major component of the natural sample has a glycerol backbone with linolenic acid esterified at one primary position and linoleic acid esterified at the secondary position. The remaining primary position is esterified to 8-hydroxy-5,6-octadienoic acid, which in turn is esterified via the ω-hydroxyl group to 2,4-decadienoic acid. In spite of this unusual structure, the mass spectrum is very similar to that found for normal triacylglycerols. The spectrum of the hydrogenated product contains $[RCO]^+$ type fragments for decanoic acid, stearic acid, and the esterified moiety composed of the eight- and ten-carbon acyl groups, ($C_9H_{19}COOC_7H_{14}CO$), as well as $[RCO + 74]^+$ type fragments corresponding to the latter two moieties above. An $[RCO + 74]^+$ type fragment for decanoic acid does not occur because this acid is not connected directly to the glycerol backbone. The occurrence of the $[M - RCOO]^+$ ions that result from loss of the acyloxy groups corresponding to stearic acid and the diester moieties is also consistent with the noted structure. Finally, the location of the diester moiety at one of the primary positions of glycerol was established by the presence of a fragment resulting from the loss of C_9H_{19}-$COOC_7H_{14}COOCH_2$. The mass spectrum of the original material has similar features, indicating that linolenate also occupied one of the primary positions, leaving linoleic and oleic acids to occupy the 2 position of glycerol.

Varanasi *et al.* (1973) used mass spectrometry to investigate the diisovaleroyl-long-chain acylglycerols obtained from the mandibular tissue of the porpoise. The mass spectrometric approach was particularly valuable in this case because an earlier attempt to carry out stereospecific analysis (Malins and Varanasi, 1972) was hindered by the resistance of the highly branched compounds to porcine pancreatic lipase. The mass spectrum of the natural diisovaleroyl-long-chain acylglycerols was compared with the spectrum of synthetic 1,2(2,3)-diisovaleroyl-myristoylglycerol. The intensity ratio, R, for ions formed by loss of short-chain acyloxymethylene to that of ions formed by loss of long-chain acyloxymethylene is greater for the natural sample ($R = 18.3$) than for the synthetic sample ($R = 0.5$), and, since acyloxymethylene is lost only from the 1,3 positions, it was concluded that the natural sample is composed almost exclusively of 1,3-diisovaleroyl-2-long-chain acylglycerols. Similarly, Blomberg (1974) determined by GC/MS that the diisovaleroyl-long-chain acylglycerols obtained from the head oil of the North Atlantic pilot whale are predominantly 1,3-diisovaleroyl-2-long-chain acylglycerols. In this case, the ion ratio, R, for the natural sample of diisovaleroyl-isopentadecanoylglycerol was found to be $R = 15$–20. These results are consistent with the finding of other workers that short-chain fatty acids are located at the *sn*-3 position of natural triacylglycerols (Kuksis and Breckenridge, 1968).

Two mass spectrometric approaches have been used to analyze triacylglycerol mixtures. Triacylglycerols may be admitted to the mass spectrometer ion source directly (Hites, 1970, 1975) or by GLC (Murata and Takahashi, 1973).

By means of the first approach (Hites, 1970), it is possible to determine the molecular weight distribution of natural triacylglycerol mixtures. A sample (100–500 μg) is admitted into the mass spectrometer via a direct insertion probe and heated isothermally at 289°C. When the system has equilibrated, three or more partial mass spectra (to include all $[M]^+$ and $[M - 18]^+$ ions) are obtained. The ion intensities are corrected for contributions due to the natural heavy isotopes (^{13}C, ^{2}H, ^{17}O, ^{18}O), and then the corrected intensities of $[M]^+$ and $[M - 18]^+$ are summed for each molecular weight. A further correction is necessary for the dependence of the evaporation rate on molecular weight and for the different molecular ion abundances observed for saturated and unsaturated triacylglycerols. Using this procedure, a good correlation was found between the experimental values and those calculated according to the 1,3-random 2-random hypothesis (i.e., assuming that only unsaturated fatty acids are esterified in the 2 position of glycerol) for the triacylglycerols from a number of vegetable oils: Kokum butter, cocoa butter, olive oil, peanut oil, cottonseed oil, corn oil, soybean oil, sunflower oil, and linseed oil.

The second analytical procedure combines gas chromatographic resolution with mass spectrometry (Murata and Takahashi, 1973). In addition to the molecular weight distribution for the mixture, this procedure allows one to determine the fatty acid composition for triacylglycerols having the same carbon number. The fatty acids are determined by means of the intensities of the respective $[M - RCOO]^+$, $[RCO]^+$, and $[RCO + 74]^+$ ions. Unfortunately, the procedure does not generally allow one to determine the fatty acid composition of triacylglycerols having a particular molecular weight, but this information could be obtained if the triacylglycerols were first resolved according to their degree of unsaturation by argentation TLC. A GC/MS determination would then provide the same information as could be obtained by preparative GLC or reversed-phase chromatography (TLC or LC) followed by fatty acid analysis. In order to improve the accuracy of the GC/MS procedure, however, it is necessary to calibrate the system with mixtures of known composition. An alternative procedure would be to reduce each of the fractions from argentation TLC with Pt/deuterium (deutrogenation) prior to GC/MS. Not only would this procedure remove the need for calibration factors for unsaturated compounds, but also the original degree of unsaturation for each fatty acid could still be determined from the *m/e* value of the respective $[RCO]^+$ or $[RCO + 74]^+$ ions. The *m/e* values would increase by 2 for each double bond originally present. The $[M - RCOOCH_2]^+$ fragments could also be more effectively used to determine the positional placement of the fatty acids. Except for enantiomers, this scheme could provide almost complete resolution of a natural mixture.

3.10.7. Other Modes of Ionization

Most of the mass spectrometry of lipids has been conducted using electron impact (EI) as the mode of ionization, but other techniques including chemical ionization (CI) (Munson, 1971) and field ionization (FI) and field desorption (FD) (Evans *et al.*, 1974) have been explored. The use of these other techniques has been compared for a number of biologically important compounds including tricaproin and trilaurin (Fales *et al.*, 1975). Electron impact spectra usually contain more diagnostically important ions but often have molecular ions of very low intensity. The EI spectra of triacylglycerols, for example, have molecular ion abundances that are $\leq 1\%$. For short-chain triacylglycerols such as tricaproin, $[M]^+$ and $[M + 1]^+$ type ions are prominent features in the spectra obtained by FI or CI (isobutane). However, for higher-molecular-weight triacylglycerols such as trilaurin, these ions are unfortunately not observed. The $[M - RCOO]^+$ type ion is the highest-mass ion found in these spectra. FD, however, leads to spectra of triacylglycerols consisting almost exclusively of $[M]^+$ and $[M + 1]^+$ ions. Evans *et al.* (1974) were thus able to determine the major triacylglycerol species (without regard to positional distribution) in olive and safflower oils, but some overlapping of species arises due to the presence of $[M + 2]^+$ type ions.

3.11. Other Techniques

3.11.1. Infrared Spectroscopy

Absorptions in the range of 600–5000 cm^{-1}, due to the stretching or deformation vibrations of molecules, may be used to detect or confirm the presence of certain functional groups within an unknown compound. The method generally requires at least a milligram of purified sample, but Fourier transform IR technology (Griffiths, 1975) provides sensitivities good enough to allow even GLC-IR coupling. For experimental and theoretical details of IR spectroscopy, books by the following authors may be consulted: Parikh (1974), Stewart (1970), Dyer (1965), and Bellamy (1958). The use of IR in the field of lipids has also been reviewed (Davenport, 1971; Freeman, 1968; Chapman, 1965; Baumann and Ulshöfer, 1968), and a partial list of diagnostic absorption frequencies is given in Table XIX.

Certain structural features that can be characterized by IR are difficult to determine in another way. IR, for example, was used to determine the geometric configuration of the alk-1-enyl double bond of natural plasmalogens (Norton *et al.*, 1962; Warner and Lands, 1963). The most significant evidence for a *cis* configuration is provided by the absorption singlet at 1667 cm^{-1} in the IR spectrum of natural 1-monoalk-1-enylglycerol, which is in agreement with the C=C stretching absorption found for a synthetic *cis* vinyl ether. In contrast, the *trans* isomer has a doublet in the 1660–1672 cm^{-1} region. The *cis* configuration of the alk-1-enylglycerol double bond is confirmed by the fact that the C—H out-of-plane deformation vibration is found at approximately 738 cm^{-1} like that of the *cis* standard; the corresponding absorption in *trans* vinyl ethers is found at 930 cm^{-1}. All these features disappear after hydrogenation, leaving a spectrum identical to that of the monoalkylglycerols.

Table XIX. Infrared Absorption Frequencies for Selected Functional Groups[a, b]

Functional group	Vibration mode	Frequency (cm^{-1})	Relative intensity
—C—O—C—alkyl ether	C—O str	1060–1150	vs
—CH=CH=O—C	C—O str	1230–1270	s
cis	C=C str	1660–1672 (singlet)	m
trans	C=C str	1660–1672 (doublet)	m
cis	C—H o-o-p def	~730	m
trans	C—H o-o-p def	~930	m
1-CH=CH—O—C			
(1-alk-1-enylglycerol)	C=C str	1667 (singlet)	
	C—H o-o-p def	738	
RCO—OR (ester)	C=O str	1730–1750	s
	C—O— str (acetate)	1230–1280	m
	C—O— str (long chain)	1150–1200	m

[a] Data taken from Kates (1972). See also Table XXV, Chapter 1.

[b] Abbreviations: str, stretching; def, deformation; o-o-p def, out-of-plane deformation; s, strong; m, medium; vs, very strong.

Raman spectroscopy can also be used to measure vibrational energy transitions (Parikh, 1974). When a sample is irradiated with monochromatic visible radiation, light may be scattered either without loss of energy (Rayleigh scattering) or with energy losses equal to the allowed vibrational transitions (Raman scattering). Bailey and Horvat (1972) used the Raman shifts of 1654, 1657, and 1655 cm^{-1} found, respectively, for the *cis* monoene, diene, and triene C=C stretching frequencies and the shifts of 1668.6 and 1670.7 cm^{-1} found for the *trans* monoenes and diene frequencies to determine the *cis/trans* isomer composition of number of vegetable oils.

3.11.2. Nuclear Magnetic Resonance Spectroscopy

Nuclear magnetic resonance (NMR) absorption spectroscopy is used to measure transitions that occur between the magnetic field-dependent energy levels associated with the nuclear spin. Although most attention has been focused on the magnetic resonance of protons (^{1}H), resonance measurements of other nuclei such as ^{31}P, ^{13}C, and ^{19}F are also useful. Nuclear magnetic resonance, for example, is the best procedure available for detecting and locating ^{13}C within a molecule.

Since the absorption energy of a particular nucleus depends on the "local" electronic and nuclear environment, the resonance energies are dependent on molecular structure. The resonance energy of a proton in a particular electronic environment relative to that of some internal standard (usually tetramethylsilane) is called the chemical shift. A partial listing of chemical shifts for protons in various functional groups is given in Table XX, but it should be noted that spin–spin coupling due to the interaction of neighboring nuclei can cause these resonances to be split into two or more components. Special techniques such as nuclear magnetic double resonance, deuterium substitution, derivative formation, and use of lanthanide shift reagents can also be used to evaluate complex spectra. For a more complete consideration of NMR spectroscopy, books by the following authors may be consulted: Parikh (1974), Jackman and Sternhill (1969), Chapman and Magnus (1966), and Dyer (1965). The uses of NMR for the analysis of lipids have been reviewed (Chapman, 1965), and a few examples are outlined below.

The presence of acetic acid in the hydrogenated triacylglycerols derived from *Euonymus verrucosus* (Kleiman *et al.*, 1967) was confirmed by the sharp resonance at 7.95τ due to the CH_3COO protons. NMR spectrometry was also used to determine the conformation of the 1,3-dioxanes and 1,3-dioxolanes formed by the acid-catalyzed cyclization of 1-*O*-*cis*-alk-1-enyl-*sn*-glycerol (Baumann *et al.*, 1972). The analyses of the NMR spectra were made on the basis of the differential shielding that occurs with axial and equatorial substituents arising from the magnetic anisotropy of the ring and other substituents and on the basis of magnitude of the vicinal coupling constants in relation to the dihedral angle between coupling protons.

An NMR technique using lanthanide shift reagents (Pfeffer, 1975) has been developed to determine the positional distribution of isovaleric acid in triacylglycerols (Wedmid and Litchfield, 1975). The method makes use of deuterated 1,1,1,2,2,2,3,3-heptafluoro-7,7-dimethyl-4,6-octadione ("fod") derivatives of europium III and praeseodymium III to produce different α-methyl proton signals for isovaleroyl chains in the 1,3 or 2 position of triacylglycerols. Longer acyl groups

Table XX. Chemical Shifts of Protons in Various Functional Groups[a]

Functional group	τ (ppm)[b]
$(CH_3)_4Si$ (internal reference)	10.00
$-CH_2-$ (cyclopropane)	9.78
CH_3-C- (saturated)	8.7–9.3
$-CH_2-$ (saturated)	8.52–8.80
$-CH_2-C-O-COR$	8.50
$CH_3-C=C-$	8.40–8.82
$-C-H$ (saturated)	8.35–8.60
$CH_3-C=O$	7.4–8.1
$-CH_2-CO$ (fatty acid esters)	7.8–7.9
$-CH_2-CO$ (ketones)	7.8–7.9
$-CH_2-O-$ (alcohol or ether)	6.3–6.6
$C=CH-O-$ (*cis* vinyl ether)	4.2
$C=CH-O-$(*trans* vinyl ether)	4.0
$CH_2=C$ (nonconjugated)	5.0–5.4
$H-C=C-$ (nonconjugated)	4.3–4.8
$CH_2=C$ (conjugated)	4.3–4.7
$H-C=C-$ (conjugated)	3.5–4.0
$R-O-H$[c]	4.8–7.0

[a] Data taken from Kates (1972) and Parikh (1974).
[b] $\tau = 10.00 - \delta$ and δ (parts per million) = (sample absorption frequency − reference absorption frequency)/(spectrometer frequency $\times 10^{-6}$).
[c] For concentrations less than 1 M in an inert solvent. In general, the chemical shift of these protons depends on concentration, temperature, and the presence of other exchangeable protons.

within the same molecule do not interfere with these signals. When this technique was applied to samples of melon, jaw, and blubber triacylglycerols from 13 cetacean genera, Wedmid and Litchfield (1976) found that isovaleric acid was esterified only at the 1,3-positions in either monoisovaleroyl or diisovaleroyl triacylglycerols. These findings agree with a mass spectrometric determination of the 1,3-diisovaleroyl-2-long-chain acylglycerols in *Tursiops* mandibular blubber fat (Varanasi *et al.*, 1973).

3.11.3. Optical Rotation and Stereospecific Analysis

The absolute stereochemical configuration for an unknown is determined by a comparison of its specific optical rotation (polarimetry; optical rotatory dispersion) with that of a standard having a known stereochemical configuration. For example, the isopropylidene derivative of a natural monoalkylglycerol (derived from phospholipid) was found to have the same optical rotation as that of the isopropylidene derivative of synthetic 1-alkyl-*sn*-glycerol (Hanahan, 1972). Similarly, when natural monoalk-1-enylglycerols were hydrogenated, they were found to have the same optical rotation as the natural monoalkylglycerols. Therefore, in both cases, the ether function is located at the *sn*-1 position of glycerol.

The method of stereospecific analysis which depends on the known stereospecificity of certain enzymes (see Chapter 4) has been used to determine the stereochemical placement of the fatty acids in diacylglycerols, triacylglycerols (Brockerhoff, 1971), and certain ether analogues (Wood and Snyder, 1969).

3.11.4. Calculation Methods

It is often instructive to compare the results obtained from a particular experimental procedure with those obtained by some simple calculation process. Depending on the information available, one of the following procedures may be used for triacylglycerols (Litchfield, 1972).

If the fatty acid composition for each of the *sn*-1, *sn*-2, and *sn*-3 positions is known from stereospecific analysis, we can use a 1-random 2-random 3-random calculation. This calculation is based on the assumption that the fatty acids in each of these three pools are combined with each other in random fashion. The mole percent, [XYZ], of any triacylglycerol (*sn*-XYZ) having fatty acids X, Y, and Z at positions 1, 2, and 3, respectively, is given by $[XYZ] = [X]\,[Y]\,[Z]/10^4$, where the quantities in square brackets represent mole percent.

If the fatty acid compositions for the 2 position and the combined 1 and 3 positions is known, we can use the 1,3-random 2-random procedure. The method of calculation is the same as before except that there are only two pools of fatty acids. Hence the mole percent of any fatty acid is the same for the 1 and 3 positions and $[XYZ] = [ZYX]$.

If only the total fatty acid composition is known, we can assume that the mole percent of any fatty acid is the same, regardless of position. The method of calculation is still the same except that now $[XYZ] = [XZY] = [YXZ] = [YZX] = [ZXY] = [ZYX]$, and the procedure is called the 1,2,3-random hypothesis.

Diacylglycerol species may be calculated in a similar way, except that the fatty acid pools are considered two at a time instead of three at a time. For example, the species composition of 1,3-diacylglycerols may be calculated using a 1-random 3-random calculation. The mole percent, [XZ], of any 1,3-diacylglycerol (*sn*-XZ) having fatty acids X and Z at positions 1 and 3, respectively, is given by the expression $[XZ] = [X]\,[Z]/10^2$. Similarly, 1,2- or 2,3-diacyl-*sn*-glycerol species can be calculated by 1-random 2-random and 2-random 3-random calculation, respectively, and, if enantiomers are not considered, $[XZ] = [ZX]$. If the fatty acid pools are considered to be completely random (such as in diacylglycerols derived from a randomized triacylglycerol sample), then the compositions of the 1,2-, 2,3-, and 1,3-diacylglycerols are equivalent.

3.12. Combined Techniques

For a mixture of triacylglycerols composed of N fatty acids, there are N^3 possible triacylglycerol species. The mixture is therefore very complex, and effective resolution requires the application of a number of techniques even if enantiomers are not considered. The number of significant species, however, may be much smaller if the fatty acids are not randomly distributed among all three positions of glycerol. Seed oils, for example, generally have only a very small proportion of their total saturated fatty acids situated at the *sn*-2 position of glycerol (Kuksis, 1972*a*; Litchfield, 1972). Natural samples may also be stereochemically asymmetrical. The butyric acid in butterfat is situated mostly in the *sn*-3 position (Breckenridge and Kuksis, 1967; Pitas *et al.*, 1967; Kuksis *et al.*, 1973; Barbano and Sherbon, 1975), and the long-chain fatty acids of peanut oil are located almost exclusively in the *sn*-3 position of glycerol (Myher *et al.*, 1977*b*).

For a mixture of diacylglycerols composed of N fatty acids, there are N^2 possible species for each isomer (1,2-, 2,3-, or 1,3-diacyl-*sn*-glycerol). Once again, however, a natural sample may be simplified by the fact that certain acids are restricted to a particular position on the glycerol backbone. For example, the diacylglycerols derived from phospholipids have a predominance of saturated fatty acids esterified at the *sn*-1 position and unsaturated fatty acids esterified at *sn*-2 position (Kuksis, 1972*a*). The diacylglycerols derived by partial hydrolysis of triacylglycerols are generally more complex.

In general, a complete analysis of di- and triacylglycerols requires the combination of a number of separatory techniques. Hammond (1969) has outlined a strategy that would result in the complete separation of the triacylglycerol species within a mixture that chromatographs as a single band on normal silica gel (e.g., hexane–diethyl ether–acetic acid, 80 : 20 : 1, v/v/v). The steps in this procedure are as follows:

1. The triacylglycerols are separated into fractions having the same degree of unsaturation by argentation TLC.
2. Each of the above fractions is then separated according to carbon number by preparative gas chromatography or reversed-phase chromatography (TLC or LC).
3. There are now a number of fractions, each of which consists of triacylglycerols having a particular carbon number and degree of unsaturation. It is then necessary to generate representative diacylglycerols from each by Grignard hydrolysis [1,2(2,3) and 1,3] or pancreatic lipase.

4a. If no distinction is made between enantiomeric pairs, then the analysis is complete once the 1,3-diacylglycerols are analyzed on the basis of carbon number and degree of unsaturation. This may be done either by argentation TLC combined with gas chromatography or reversed-phase chromatography or by gas chromatography using a polar liquid phase.

4b. If a complete stereochemical analysis is required, the 1,3-diacylglycerols from step 3 must first be preparatively separated by combined argentation TLC and preparative GLC or reversed-phase chromatography. Each of the separated species is then subjected to stereospecific analysis.

Although this scheme can be carried out with the use of known techniques, a considerable amount of work would be involved and no one has yet attempted the complete program. Some examples of the use of combined techniques are given below, but more detailed reviews of this subject can be found elsewhere (Kuksis, 1972*a*; Litchfield, 1972).

Wessels and Rajagopal (1969) used reversed-phase TLC to separate, according to carbon number, the triacylglycerol fractions obtained by argentation TLC. For direct quantitation of the argentation fractions, GLC is quicker and more reliable, but when used preparatively to prepare fractions of uniform degree of unsaturation and carbon number the procedure using combined argentation and reversed-phase TLC is competitive with the procedure using argentation TLC plus preparative GLC (Kuksis and Ludwig, 1966; Breckenridge and Kukšis, 1968; Bugaut and Bezard, 1973). Reversed-phase partition TLC was combined with GLC in order to separate the triacylglycerols of *Ephedra nevadensis* seed fat (Litchfield, 1968). The triacylglycerols were resolved according to partition number* by reversed-phase TLC and then into peaks having uniform carbon number and unsaturation by GLC. Numerous studies have also combined argentation TLC and analytical GLC (Culp *et al.*, 1965; Breckenridge and Kuksis, 1969; Breckenridge *et al.*, 1969; Bezard and Bugaut, 1972).

Excluding studies where only FAMEs were determined, the majority of the analyses of triacylglycerols have been made using argentation TLC in conjunction with quantitative fatty acid analysis (Kuksis, 1972*a*; Litchfield, 1972). Argentation TLC by itself is not necessarily a good test of nonrandomness. The distribution of fatty acids among the various argentation TLC bands was found to be almost

* Partition number = carbon number − 2 (number of double bonds).

Table XXI. Major Molecular Species of Peanut Oil[a]

Triacyl-*rac*-glycerol[b] Position			Mole %	
1	2	3	Reconstituted[c]	Calculated[d]
16:0	16:0	16:0	—	0.06
16:0	18:0	16:0	—	0.02
16:0	16:0	18:0	—	0.03
16:0	18:0	18:0	—	—
16:0	16:0	20:0		0.02
16:0	16:0	22:0		0.03
16:0	16:0	24:0		0.01
16:0	18:1	16:0	1.02	1.29
16:0	18:1	18:0	0.76	0.60
16:0	16:0	18:1		0.37
16:0	18:0	18:1		0.17
18:0	16:0	18:1		0.04
18:0	18:0	18:1		0.02
16:0	16:0	20:1		0.01
16:0	18:1	20:0		0.40
16:0	18:1	22:0		0.74
16:0	18:1	24:0		0.03
18:1	20:0	16:0		—
18:1	22:0	16:0		—
18:1	24:0	16:0		—
16:0	18:2	16:0	0.98	1.18
16:0	18:2	18:0	0.72	0.55
16:0	18:1	18:1	8.01	8.55
16:0	16:0	18:2	—	0.12
16:0	18:0	18:2	—	0.03

(*Continued*)

Table XXI—Continued

Triacyl-*rac*-glycerol[b] position			Mole %	
1	2	3	Reconstituted[c]	Calculated[d]
18:0	18:1	18:1	2.3	1.85
18:0	16:0	18:2	—	0.03
18:0	18:0	18:2	—	0.01
18:1	18:0	18:1	—	0.26
18:1	16:0	18:1	—	0.56
16:0	18:1	20:1		0.29
16:0	20:1	18:1		
16:0	18:2	20:0		0.37
16:0	18:2	22:0		0.67
16:0	18:2	24:0		0.30
18:1	18:1	20:0		1.05
18:1	18:1	22:0		1.79
18:1	18:1	24:0		0.80
18:1	20:0	18:1		0.01
18:1	22:0	18:1		0.02
18:1	24:0	18:1		0.01
16:0	18:2	18:1	7.62	7.77
16:0	18:1	18:2	2.96	2.84
18:0	18:2	18:1	1.70	1.69
18:0	18:1	18:2	0.77	0.69
18:1	18:1	18:1	16.3	13.03
18:1	16:0	18:2	—	0.43
18:1	18:0	18:2	—	0.19
18:1	18:2	20:0		0.95
18:1	18:2	22:0		1.63
18:1	18:2	24:0		0.45
18:2	18:1	20:0		0.55
18:2	18:1	22:0		0.98
18:2	18:1	24:0		0.44
16:0	18:2	18:2	2.82	2.58
18:0	18:2	18:2	0.56	0.63
18:1	18:2	18:1	11.8	11.85
18:1	18:1	18:2	9.1	9.89
18:2	16:0	18:2	—	0.06
18:2	18:0	18:2	—	0.03
18:1	18:2	20:1		0.81
18:1	20:1	18:2		—
18:1	18:2	18:2	6.6	8.99
18:2	18:1	18:2	1.9	1.50
18:2	18:2	20:1		0.39
18:2	20:1	18:2		—
18:2	18:2	18:2	1.4	1.36

[a] Data from Myher *et al.* (1977*b*).
[b] Triacylglycerols identified by fatty acids in positions 1, 2, and 3, respectively.
[c] Reconstituted from GLC resolution of the diacylglycerols generated by Grignard hydrolysis.
[d] By 1-random 2-random 3-random calculation using data from stereospecific analysis and adding values of the respective enantiomers.

identical for both natural peanut oil (PNO) and randomized peanut oil (PNOR) even though stereospecific analysis showed that the natural oil was highly asymmetrical (Myher *et al.*, 1977*b*). Blank *et al.* (1965) and Jurriens and Kroesen (1965) combined argentation TLC with lipase hydrolysis to the triacylglycerols from a number of seed oils and from rat liver and were able to demonstrate quantitative differences from those expected on a strictly random basis. Later, Roehm and Privett (1970) used the same technique to study the changes taking place in the structure of soybean triacylglycerols during maturation. Similarly, Weber (1973) combined argentation TLC with stereospecific analysis to study the structural changes occurring in corn kernel triacylglycerols during maturation. Breckenridge and Kuksis (1968) carried out a very exhaustive analysis of the short- and medium-chain-length triacylglycerols of bovine milk fat. The combined application of argentation TLC, preparative and analytical GLC, plus fatty acid analysis led to the resolution of 38 triacylglycerol species which accounted for about 77% of the short-chain triacylglycerols.

Myher *et al.* (1977*b*) have compared the results obtained by several methods for PNO and PNOR. The major molecular species of PNO are given in Table XXI. The values in the first column are reconstituted from the GLC resolution on Silar 5CP of the diacylglycerols generated by Grignard hydrolysis. The triacylglycerols are reconstituted by first assuming that the 1,3-diacylglycerols are randomly combined with the fatty acids in the 2 position. Then, by trial and error, these values are corrected in accordance with GLC analysis of the 1,2(2,3)-diacylglycerols. Since these values and the values calculated by a 1-random 2-random 3-random calculation are in good agreement (Table XXI), the calculation can be used with confidence to generate a more complete composition table. The randomness of the PNOR sample was confirmed by the fact that the profiles obtained for the 1,2(2,3)- and 1,3-diacylglycerols were identical.

Tetraacylglycerols (estolides) are possible if a hydroxy acid esterified to glycerol is itself esterified to a fatty acid via the hydroxy group (Phillips and Smith, 1970; Kleiman *et al.*, 1972), and a number of procedures have been used to derive the structure of these complex glycerolipids. Sprecher *et al.* (1965) used mass spectrometry to demonstrate that the estolide groupings in *Sapium sebiferum* oil were located at the 1 or 3 position of glycerol. Phillips and Smith (1970) used pancreatic lipase hydrolysis to show that the hydroxy acids in *Monnina emarginata* oil were also esterified principally (>97%) to the 1 or 3 position of glycerol. Christie (1969) applied the method of stereospecific analysis to the argentation TLC fractions obtained for the tetraesters of *Sapium sebiferum* seed oil and found that the estolide fatty acids were located entirely in the *sn*-3 position of glycerol.

A combination of techniques can also be used for the analysis of diacylglycerols and the corresponding ether analogues. Argentation TLC followed by GLC and fatty acid analysis has been used to separate diacylglycerol species according to carbon number and degree of unsaturation (Renkonen, 1967*b*; Kuksis and Marai, 1967; Kuksis, 1972*a*). This procedure provides an almost complete resolution of diacylglycerols derived from phospholipids, but, in general, reverse isomers and enantiomers are left unresolved. Replacement of the analytical GLC step with preparative GLC (Kuksis and Marai, 1967) or reverse-phase partition TLC (Renkonen, 1967*b*) would allow a further resolution using the method of stereospecific analysis.

3.13. Conclusions

As pointed out in the last section, it is possible to outline schemes that can be used for the complete analysis of di- and triacylglycerol mixtures. Such comprehensive routines, however, are impractical for two reasons: the time required is generally prohibitive and the sample size is often too small. There is therefore a great need for methods that require a minimum of sample prefractionation. The technique of GC/MS shows promise in that it provides the investigator with detailed structural information and requires relatively small samples.

Acknowledgments

Investigations by the author and his collaborators were supported by the Medical Research Council of Canada, Ottawa, Ontario, and the Ontario Heart Foundation, Toronto, Ontario.

3.14. References

Aasen, A. J., Lauer, W. M., and Holman, R. T., 1970, Mass spectrometry of triglycerides. II. Specifically deuterated triglycerides and elucidation of fragmentation mechanisms, *Lipids* **5**:869.

Aitzetmüller, K., 1975, The liquid chromatography of lipids: A critical review, *J. Chromatogr.* **113**:231.

Anonymous, 1967, IUPAC-IUB Commission of Biochemical Nomenclature. The nomenclature of lipids, *Biochemistry* **105**:897.

Arpino, P., Baldwin, M. A., and McLafferty, F. W., 1974*a*, Liquid chromatography–mass spectrometry. II. Continuous monitoring. *Biomed. Mass Spectrom.* **1**:80.

Arpino, P. J., Dawkins, B. G., and McLafferty, F. W., 1974*b*, A liquid chromatography/mass spectrometry system providing continuous monitoring with nanogram sensitivity, *J. Chromatogr. Sci.* **12**:574.

Assmann, G., Krauss, R. M., Frederickson, D. S., and Levy, R. I., 1973, Positional specificity of triglyceride lipases in post heparin plasma, *J. Biol. Chem.* **248**:7148.

Bailey, G. F., and Horvat, R. J., 1972, Raman spectroscopic analysis of the *cis*/*trans* isomer composition of edible vegetable oils, *J. Am. Oil Chem. Soc.* **49**:494.

Barbano, D. M., and Sherbon, J. W., 1975, Stereospecific analysis of high melting triglycerides of bovine milk fat and their biosynthetic origin, *J. Dairy Sci.* **58**:1.

Barber, M., Merren, T. O., and Kelly, W., 1964, The mass spectrometry of large molecules. I. The triglycerides of straight chain fatty acids, *Tetrahedron Lett.* **18**:1063.

Barber, M., Chapman, J. R., and Wolstenholme, W. A., 1968, Lipid analysis by coupled mass spectrometry–gas chromatography (MS-GLC). 1. Diglycerides, *J. Mass. Spectrometry Ion Phys.* **1**:98.

Barrett, C. B., Dallas, M. S. J., and Padley, F. B., 1962, The separation of glycerides by thin-layer chromatography on silica impregnated with silver nitrate, *Chem. Ind.* (*London*) **1962**:1050.

Baumann, W. J., and Ulshöfer, H. W., 1968, Characteristic absorption bands and frequency shifts in the infrared spectra of naturally-occurring long-chain ethers, esters and ether esters of glycerol and various diols, *Chem. Phys. Lipids* **2**:114.

Baumann, W. J., Madson, T. H., and Weseman, B. J., 1972, "Plasmalogen-type" cyclic acetals: Formation and conformation of the 1,3-dioxanes and 1,3-dioxolanes from 1-*O*-*cis*-alk-1-enyl-*sn*-glycerols, *J. Lipid Res.* **13**:640.

Bellamy, L. J., 1958, *The Infrared Spectra of Complex Molecules*, 2nd ed., Methuen, London.

Bezard, J., and Bugaut, M., 1972, The component triglycerides of rat adipose tissue. I. As studied after fractionation into classes by silver ion–thin-layer chromatography, *J. Chromatogr. Sci.* **10**:451.

Biemann, K., 1962, *Mass Spectrometry: Organic Chemical Applications*, McGraw-Hill, New York.

Blank, M. L., and Snyder, F., 1975, Quantitative aspects of thin-layer chromatography in the analysis of phosphorus-free lipids, in: *Analysis of Lipids and Lipoproteins* (E. G. Perkins, ed.), pp. 63–69, American Oil Chemists' Society, Champaign, Ill.

Blank, M. L., Verdino, B., and Privett, O. S., 1965, Determination of triglyceride structure via silver nitrate–TLC, *J. Am. Oil Chem. Soc.* **42:**87.

Blank, M. L., Kasama, K., and Snyder, F., 1972, Isolation and identification of an alkyldiacylglycerol containing isovaleric acid, *J. Lipid Res.* **13:**390.

Blank, M. L., Cress, E. A., Piantadosi, C., and Snyder, F., 1975, A method for the quantitative determination of glycerolipids containing *O*-alkyl and *O*-alk-1-enyl moieties, *Biochim. Biophys. Acta* **380:**208.

Blankenhorn, D. H., Rouser, G., and Weimer, T. J., 1961, A method for the estimation of blood glycerides employing Florisil, *J. Lipid Res.* **2:**281.

Bligh, E. G., and Dyer, W. J., 1959, A rapid method of total lipid extraction and purification, *Can. J. Biochem. Physiol.* **37:**911.

Blomberg, J., 1974, Unusual lipids. II. Head oil of the North Atlantic pilot whale, *Globicephala melaena malaena, Lipids* **9:**461.

Breckenridge, W. C., and Kuksis, A., 1967, Specific distribution of short chain fatty acids in bovine milk fat, *Proc. Can. Fed. Biol. Soc.* **10:**156.

Breckenridge, W. C., and Kuksis, A., 1968, Structure of bovine milk fat triglycerides. I. Short and medium chain lengths, *Lipids* **3:**291.

Breckenridge, W. C., and Kuksis, A., 1969, Structure of bovine milk fat triglycerides. II. Long chain lengths, *Lipids* **4:**197.

Breckenridge, W. C., Marai, L., and Kuksis, A., 1969, Triglyceride structure of human milk fat, *Can. J. Biochem.* **47:**761.

Brockerhoff, H., 1971, Stereospecific analysis of triglycerides, *Lipids* **6:**942.

Budzikiewicz, H., Djerassi, C., and Williams, D. H., 1967, *Mass Spectrometry of Organic Compounds,* Holden-Day, San Francisco.

Bugaut, M., and Bezard, J., 1973, The component triglycerides of rat adipose tissue. II. As studied after fractionation of classes into groups by gas liquid chromatography, *J. Chromatogr. Sci.* **11:**36.

Burlingame, A. L., Kimble, B. J., and Derrick, P. J., 1976, Mass spectrometry, *Anal. Chem.* **48:**368R.

Burton, R. M., 1974, Lipid extraction and separation procedures, in: *Fundamentals of Lipid Chemistry* (R. M. Burton and F. C. Guerra, eds.), pp. 11–45, BI-Science Publications Division, Webster Groves, Mo.

Carroll, D. I., Dzidic, I., Stillwell, R. N., Haegele, K. D., and Horning, E. C., 1975, Atmospheric pressure ionization mass spectrometry: Corona discharge ion source for use in lipid chromatograph–mass spectrometer–computer analytical system, *Anal. Chem.* **47:**2369.

Carroll, K. K., 1976, Column chromatography of neutral glycerides and fatty acids, in: *Lipid Chromatographic Analysis*, Vol. 1 (G. V. Marinetti, ed.), pp. 205–237, Marcel Dekker, New York.

Casparrini, G., Horning, M. G., and Horning, E. C., 1968, Gas chromatographic study of phosphatidylserines, *Anal. Lett.* **1:**481.

Chapman, D., 1965, *The Structure of Lipids by Spectroscopic and X-ray Techniques*, Wiley, New York.

Chapman, D., and Magnus, P. D., 1966, *Introduction to Practical High Resolution NMR Spectroscopy*, Academic Press, New York.

Christie, W. W., 1969, The glyceride structure of *Sapium sebiferum* seed oil, *Biochim, Biophys. Acta* **187:**1.

Corey, E. J., and Venkateswarlu, A., 1972, Protection of hydroxyl groups as tert-butyldimethylsilyl derivatives, *J. Am. Chem. Soc.* **94:**6190.

Cram, S. P., and Juvet, R. S., Jr., 1976, Gas chromatography, *Anal. Chem.* **48:**411R.

Culp, T. W., Harlow, R. D., Litchfield, D., and Reiser, R., 1965, Analysis of triglycerides by consecutive chromatographic techniques. II. Ucuhaba kernel fat, *J. Am. Oil Chem. Soc.* **42:**974.

Curstedt, T., 1974, Mass spectra of trimethylsilyl ethers of ^{2}H-labelled mono- and diglycerides, *Biochim. Biophys. Acta* **360:**12.

Curstedt, T., and Sjövall, J., 1974, Analysis of molecular species of ^{2}H-labelled phosphatidylcholines by liquid-gel chromatography and gas chromatography–mass spectrometry, *Biochim. Biophys. Acta* **360:**24.

Davenport, J. B., 1971, Infrared spectroscopy of lipids, in: *Biochemistry and Methodology of Lipids* (A. R. Johnson and J. B. Davenport, eds.), pp. 231–242, Wiley-Interscience, New York.

De Stefano, J. J., and Kirkland, J. J., 1975*a*, Preparative high-performance liquid chromatography, Part I, *Anal. Chem.* **47:**1103A.

De Stefano, J. J., and Kirkland, J. J., 1975*b*, Preparative high-performance liquid chromatography, Part II, *Anal. Chem.* **47:**1193A.

Dyer, J. R., 1965, *Applications of Absorption Spectroscopy of Organic Compounds,* Prentice-Hall, Englewood Cliffs, N.J.

Ellingboe, J. E., Nyström, J. F., and Sjövall, J., 1968, A versatile lipophilic Sephadex derivative for "reversed-phase" chromatography, *Biochim. Biophys. Acta* **152:**803.

Ellingboe, J. E., Nyström, J. F., and Sjövall, J., 1970, Liquid-gel chromatography on lipophilic-hydrophobic Sephadex derivatives, *J. Lipid Res.* **11:**266.

Evans, N., Games, D. E., Harwood, J. L., and Jackson, A. H., 1974, Field-desorption mass spectrometry of triglycerides and phospholipids, *Biochem. Soc. Trans.* **2:**1091.

Fales, H. M., Milne, G. W. A., Winkler, H. U., Beckey, H. D., Damico, J. N., and Barron, R., 1975, Comparison of mass spectra of some biologically important compounds as obtained by various ionization techniques, *Anal. Chem.* **47:**207.

Ferrell, W. J., Radloff, J. F., and Jackiw, A. B., 1969, Quantitative analysis of free and bound fatty aldehydes: Optimum conditions for *p*-nitrophenyl-hydrazone formation, *Lipids* **4:**278.

Fiorti, J. A., Kanuk, M. J., and Sims, R. J., 1969, Gas chromatography of epoxyglycerides, *J. Chromatogr. Sci.* **7:**448.

Folch, J., Lees, M., and Sloane-Stanley, G. H., 1957, A simple method for the isolation and purification of lipid extracts from brain tissue, *J. Biol. Chem.* **191:**833.

Freeman, N. K., 1968, Applications of infrared absorption spectroscopy in the analysis of lipids, *J. Am. Oil Chem. Soc.* **45:**798.

Gelman, R. A., and Gilbertson, J. R., 1969, The quantitative and qualitative analysis of alkyl α-glycerol ethers as alkoxy acetaldehydes, *Anal. Biochem.* **31:**463.

Gray, G. M., 1969, The preparation and assay of long-chain fatty aldehydes, in: *Methods in Enzymology* (J. M. Lowenstein, ed.), pp. 678–684, Academic Press, New York.

Griffiths, P. R., 1975, *Chemical Infrared Fourier Transform Spectroscopy,* Wiley-Interscience, New York.

Gunstone, F. D., and Padley, F. B., 1965, Glyceride studies. Part III. The component glycerides of five seed oils containing linolenic acid, *J. Am. Oil. Chem. Soc.* **42:**957.

Haken, J. K., 1974, Influence of reference substances on retention behavior of homologous compounds on stationary phases of increasing polar character, *J. Chromatogr.* **99:**329.

Hallgren, B., and Larsson, S., 1962, The glyceryl ethers in the liver oils of elasmobranch fish, *J. Lipid Res.* **3:**31.

Hallgren, B., and Ställberg, G., 1967, Methoxy-substituted glycerol ethers isolated from Greenland shark liver oil, *Acta Chem. Scand.* **21:**1519.

Hallgren, B., Niklasson, A., Ställberg, G., and Thorin, H., 1974, On the occurrence of 1-*O*-(2-methoxy-alkyl)glycerols and 1-*O*-phytanylglycerol in marine animals, *Acta Chem. Scand.* **B28:**1035.

Hamming, M. C., and Foster, N. G., 1972, *Interpretation of Mass Spectra of Organic Compounds,* Academic Press, New York.

Hammond, E. G., 1969, The resolution of complex triglyceride mixtures, *Lipids* **4:**246.

Hanahan, D. J., 1972, Ether-linked lipids: Chemistry and methods of measurement, in: *Ether Lipids, Chemistry and Biology* (F. Snyder, ed.), pp. 25–50, Academic Press, New York.

Hanahan, D. J., Ekholm, J., and Jackson, C. M., 1963, Studies on the structure of glyceryl ether phospholipids of bovine erythrocytes, *Biochemistry* **2:**630.

Hasegawa, K., and Suzuki, T., 1973, Determination of molecular species of ovolecithin using gas chromatography–mass spectrometry, *Lipids* **8:**631.

Hasegawa, K., and Suzuki, T., 1975, Examination of acetolysis products of phosphatidylcholine by gas chromatography–mass spectrometry, *Lipids* **10:**667.

Hites, R. A., 1970, Quantitative analysis of triglyceride mixtures by mass spectrometry, *Anal. Chem.* **42:**1736.

Hites, R. A., 1975, Mass spectrometry of triglycerides, in: *Methods of Biochemical Analysis*, Vol. XXXV (J. M. Lowenstein, ed.), pp. 348–359, Wiley-Interscience, New York.

Holla, K. S., Horrocks, L. A., and Cornwell, D. G., 1964, Improved determination of glycerol and fatty acids in glycerides and ethanolamine phosphatides by gas–liquid chromatography, *J. Lipid Res.* **5**:263.

Horning, E. C., Carroll, D. I., Dzidic, I., Haegele, K. D., Horning, M. G., and Stillwell, R. N., 1974, Atmospheric pressure ionization (API) mass spectrometry: Solvent-mediated ionization of samples introduced in solution and in a liquid chromatograph effluent stream, *J. Chromatogr. Sci.* **12**:725.

Horning, M. G., Casparrini, G., and Horning, E. C., 1969, The use of gas phase analytical methods for the analysis of phospholipids, *J. Chromatogr. Sci.* **7**:267.

Horning, M. G., Murakami, S., and Horning, E. C., 1971, Analyses of phospholipids, ceramides and cerebrosides by gas chromatography–mass spectrometry, *Am. J. Clin. Nutr.* **24**:1086.

Horrocks, L. A., and Cornwell, D. G., 1962, The simultaneous determination of glycerol and fatty acids in glycerides by gas–liquid chromatography, *J. Lipid Res.* **3**:165.

Huebner, V. R., 1959, Preliminary studies on the analysis of mono- and di-glycerides by GLPC, *J. Am. Oil Chem. Soc.* **36**:262.

Jackman, L. M., and Sternhill, S., 1969, *Applications to NMR Spectroscopy in Organic Chemistry*, Pergamon Press, New York.

James, A. T., and Martin, A. J. P., 1952, Gas–liquid partition chromatography: The separation and microestimation of volatile fatty acids from formic acid to dodecanoic acid, *Biochem. J.* **50**:679.

James, A. T., and Martin, A. J. P., 1956, Gas–liquid chromatography: The separation and identification of the methyl esters of saturated and unsaturated acids from formic acid to *n*-octadecanoic acid, *Biochem. J.* **63**:144.

Johnson, C. B., and Holman, R. T., 1966, Mass spectrometry of lipids. II. Monoglycerides, their diacetyl derivatives and their trimethylsilyl ethers, *Lipids* **1**:371.

Jost, U., 1974, 1-Alkyl-2,3-diacyl-*sn*-glycerol, the major lipid in the harderian gland of rabbits, *Hoppe-Seyler's Z. Physiol. Chem.* **355**:422.

Joustra, M., Söderquist, B., and Fisher, L., 1967, Gel filtration in organic solvents, *J. Chromatogr.* **28**:21.

Jurriens, G., and Kroesen, A. C. J., 1965, Determination of glyceride composition of several solid and liquid fats, *J. Am. Oil Chem. Soc.* **42**:9.

Kasama, K., Rainey, W. T., and Snyder, F., 1973, Chemical identification and enzymic synthesis of a newly discovered lipid class—hydroxyalkylglycerols, *Arch. Biochim. Biophys.* **154**:648.

Kates, M., 1972, *Techniques of Lipidology: Isolation, Analysis and Identification of Lipids*, American Elsevier, New York.

Kelly, R. W., and Taylor, P. L., 1976, Analysis of steroids and prostaglandins by gas phase methods, *Anal. Chem.* **48**:465.

Kirkland, J. J., 1972, High-performance liquid chromatography with porous silica microspheres, *J. Chromatogr. Sci.* **10**:593.

Kleiman, R., Miller, R. W., Earle, F. R., and Wolff, I. A., 1967, (s)-1,2-Diacyl-3-acetins: Optically active triglycerides from *Euonymus verrucosus* seed oil, *Lipids* **2**:473.

Kleiman, R., Spencer, G. F., Earle, F. R., Nieschlag, H. J., and Barclay, A. S., 1972, Tetra-acid triglycerides containing a new hydroxy eicosadienoyl moiety in *Lequerella auriculata* seed oil, *Lipids* **7**:660.

Kuksis, A., 1967, Gas chromatography of neutral glycerides, in: *Lipid Chromatographic Analysis*, Vol. 1 (G. V. Marinetti, ed.), pp. 239–337, Marcel Dekker, New York.

Kuksis, A., 1971, Gas–liquid chromatographic fractionation of natural diglycerides on organo-silicone–polyester liquid phases, *Can. J. Biochem.* **49**:1245.

Kuksis, A., 1972*a*, New developments in the determination of structure of glycerides and phosphoglycerides, in: *Progress in the Chemistry of Fats and Other Lipids*, Vol. 12 (R. T. Holman, ed.), pp. 1–163, Pergamon Press, New York.

Kuksis, A., 1972*b*, Gas–liquid chromatographic fractionation of natural diglycerides on stabilized polyester liquid phases, *J. Chromatogr. Sci.* **10**:53.

Kuksis, A., 1973, Progress in the analysis of lipids. XIII. Gas chromatography, Part 5, *Fette Seifen Anstrichm.* **75**:517.

Kuksis, A., 1975, Gas liquid chromatography of neutral lipids, in: *Analysis of Lipids and Lipoproteins* (E. G. Perkins, ed.), pp. 26–62, American Oil Chemists' Society, Champaign, Ill.

Kuksis, A., 1976, Gas chromatography of neutral acylglycerols, in: *Lipid Chromatographic Analysis*, Vol. 1 (G. V. Marinetti, ed.), 2nd ed., pp. 215–337, Marcel Dekker, New York.

Kuksis, A., and Breckenridge, W. C., 1968, Triglyceride composition of milk fats, in: *The Symposium: Dairy Lipids and Lipid Metabolism* (M. F. Brink and D. Kritchevsky, eds.), pp. 28–98, Avi Publishing Co., Westport, Conn.

Kuksis, A., and Ludwig, J., 1966, Fractionation of triglyceride mixtures by preparative gas chromatography, *Lipids* **1**:202.

Kuksis, A., and Marai, L., 1967, Determination of the complete structure of natural lecithins, *Lipids* **2**:217.

Kuksis, A., and Myher, J. J., 1976, Analysis of subsets of molecular species of glycerophospholipids, in: *Lipids*, Vol. 1 (R. Paoletti, G. Porcellati, and G. Jacini, eds.), pp. 23–38, Raven Press, New York.

Kuksis, A., Marai, L., and Gornall, D. A., 1967, Direct gas chromatographic examination of total lipid extracts, *J. Lipid Res.* **8**:352.

Kuksis, A., Marai, L., Breckenridge, W. C., Gornall, D. A., and Stachnyk, O., 1968*a*, Molecular species of lecithins of some functionally distinct rat tissues, *Can. J. Physiol. Pharmacol.* **46**:511.

Kuksis, A., Breckenridge, W. C., Marai, L., and Stachnyk, O., 1968*b*, Quantitative gas chromatography in the structural characterization of glyceryl phosphatides, *J. Am. Oil Chem. Soc.* **45**:537.

Kuksis, A., Marai, L., and Myher, J. J., 1973, Triglyceride structure of milk fats, *J. Am. Oil Chem. Soc.* **50**:193.

Kuksis, A., Myher, J. J., Marai, L., and Geher, K., 1975*a*, Determination of plasma lipid profiles by automated gas chromatography and computerized data analysis, *J. Chromatogr. Sci.* **13**:423.

Kuksis, A., Myher, J. J., Marai, L., Yeung, S. K. F., Steiman, I., and Mookerjea, S., 1975*b*, Distribution of newly formed fatty acids among glycerolipids of isolated perfused rat liver, *Can. J. Biochem.* **53**:509.

Kuntz, F., 1973, Separation of "neutral" lipids, particularly of all classes of partial glycerides by one-dimensional thin-layer chromatography, *Biochim. Biophys. Acta* **296**:331.

Landowne, R. A., and Lipski, S. R., 1961, A simple method for distinguishing between unsaturated and branched fatty acid isomers by gas chromatography, *Biochim. Biophys. Acta* **47**:589.

Lauer, W. M., Aasen, A. J., Graff, G., and Holman, R. T., 1970, Mass spectrometry of triglycerides. I. Structural effects, *Lipids* **5**:861.

Lindqvist, B., Sjögren, I., and Nordin, R., 1974, Preparative fractionation of triglyceride mixtures according to acyl carbon number, using hydroxyalkoxypropyl Sephadex, *J. Lipid Res.* **15**:65.

Litchfield, C., 1968, Triglyceride analysis by consecutive liquid-liquid partition and gas–liquid chromatography. *Ephedra nevadensis* seed fat, *Lipids* **3**:170.

Litchfield, C., 1972, *Analysis of Triglycerides*, Academic Press, New York.

Litchfield, C., Harlow, R. D., and Reiser, R., 1967, Gas–liquid chromatography of triglyceride mixtures containing both odd and even carbon number fatty acids, *Lipids* **2**:363.

Mahadevan, V., 1976, Thin-layer chromatography of neutral glycerides and fatty acids, in: *Lipid Chromatographic Analysis*, Vol. 3 (G. V. Marinetti, ed.), 2nd ed., pp. 777–789, Marcel Dekker, New York.

Maier, R., and Holman, R. T., 1964, Naturally occurring triglycerides possessing optical activity in the glycerol moiety, *Biochemistry* **3**:270.

Malins, D. C., and Varanasi, U., 1972, Isovaleric acid in acoustic tissues of porpoises: Triacylglycerols resistant to porcine pancreatic lipase, in: *Proceedings of XIX Colloquium, Protides of Biological Fluids* (Peeters, ed.), pp. 127–129, Pergamon, Oxford.

Mangold, H. K., 1969, Aliphatic lipids, in: *Thin-Layer Chromatography* (E. Stahl, ed.), pp. 363–421, Springer-Verlag, New York.

McCloskey, J. A., 1969, Mass spectrometry of lipids and steroids, in: *Methods in Enzymology*, Vol. XIV (J. M. Lowenstein, ed.), pp. 382–450, Academic Press, New York.

McFadden, W. H., 1973, *Techniques of Combined Gas Chromatography/Mass Spectrometry: Applications in Organic Analysis*, Wiley-Interscience, New York.

McLafferty, F. W., 1973, *Interpretation of Mass Spectra*, W. A. Benjamin, Reading, Mass.

McLafferty, F. W., Knutti, R., Venkataraghavan, R., Arpino, P. J., and Dawkins, B. G., 1975, Continuous mass spectrometric monitoring of a liquid chromatograph with subnanogram sensitivity using an on-line computer, *Anal. Chem.* **47**:1503.

McReynolds, W. O., 1970, Characterization of some liquid phases, *J. Chromatogr. Sci.* **8**:685.

Mikolajczak, K. L., and Smith, C. R., Jr., 1967, Optically active trihydroxy acids of *Chamaepeuce* seed oils, *Lipids* **2**:261.

Morley, N. H., Kuksis, A., Buchnea, D., and Myher, J. J., 1975, Hydrolysis of diacylglycerols by lipoprotein lipase, *J. Biol. Chem.* **250**:3414.

Morris, L. J., 1966, Separations of lipids by silver ion chromatography, *J. Lipid Res.* **7**:717.

Morrison, A., Barratt, M. D., and Aneja, R., 1970, Mass spectrometry of some deuterated 1,3-distearins, *Chem. Phys. Lipids* **4**:47.

Munson, B., 1971, Chemical ionization mass spectrometry, *Anal. Chem.* **43**:28A.

Muramatsu, T., and Schmid, H. H., 1972, 1-*O*-2′-Hydroxylalkyl and 1-*O*-2′-ketoalkyl glycerols, *Chem. Phys. Lipids* **9**:123.

Murata, T., and Takahashi, S., 1973, Analysis of triglyceride mixtures by gas chromatography–mass spectrometry, *Anal. Chem.* **45**:1816.

Myher, J. J., and Kuksis, A., 1974, Gas chromatographic resolution of homologous monoacyl and monoalkyl glycerols, *Lipids* **9**:382.

Myher, J. J., and Kuksis, A., 1975, Improved resolution of natural diacylglycerols by gas–liquid chromatography on polar siloxanes, *J. Chromatogr. Sci.* **13**:138.

Myher, J. J., Marai, L., and Kuksis, A., 1974, Identification of monoacyl- and monoalkylglycerols by gas–liquid chromatography–mass spectrometry using polar siloxane liquid phases, *J. Lipid Res.* **15**:586.

Myher, J. J., Marai, L., Yeung, S. K. F., and Kuksis, A., 1977*a*, A micromethod for determination of molecular species of glycerophospholipids in lipoproteins, *Proc. Can. Fed. Biol. Soc.* **20**:133.

Myher, J. J., Marai, L., Kuksis, A., and Krichevsky, D., 1977*b*, Acylglycerol structure of peanut oils of different atherogenic potential, *Lipids* (in press).

Nelson, G. J., 1975, Isolation and purification of lipids from animal tissues, in: *Analysis of Lipids and Lipoproteins* (E. G. Perkins, ed.), American Oil Chemists' Society, Champaign, Ill.

Nickel, E. C., and Privett, O. S., 1967, Fractionation of triglycerides by reversed-phase partition chromatography, *Sep. Sci.* **2**:307.

Nichols, P. L., Jr., 1952, Coordination of silver ion with methyl esters of oleic and elaidic acids, *J. Am. Chem. Soc.* **74**:1091.

Norton, W. T., Gottfried, E. L., and Rapport, M. M., 1962, The structure of plasmalogens. VI. Configuration of the double bond in the α,β-unsaturated ether linkage of phosphatidylcholine, *J. Lipid Res.* **3**:456.

Nutter, L. J., and Privett, O. S., 1968, An improved method for the quantitative analysis of lipid classes via thin-layer chromatography employing charring and densitometry, *J. Chromatogr.* **35**:519.

Nyström, E., and Sjövall, J., 1975, Chromatography on lipophilic Sephadex, in: *Methods in Enzymology*, Vol. XXXV (J. M. Lowenstein, ed.), Academic Press, New York.

O'Brien, J. F., and Klopfenstein, W. E., 1971, Gas–liquid chromatographic analysis of diglycerides, *Chem. Phys. Lipids* **6**:1.

Odham, G., and Stenhagen, E., 1972, Complex lipids, in: *Biochemical Applications of Mass Spectrometry* (G. R. Waller, ed.), pp. 229–249, Wiley, New York.

Parikh, V. M., 1974, *Absorption Spectroscopy of Organic Molecules*, Addison-Wesley, Reading, Mass.

Parodi, P. W., 1975, Detection of acetodiacylglycerols in milk fat lipids by thin-layer chromatography, *J. Chromatogr.* **111**:223.

Perkins, E. G., 1975, Gas chromatography–mass spectrometry of lipids, in: *Analysis of Lipids and Lipoproteins* (E. G. Perkins, ed.), pp. 183–203, American Oil Chemists's Society, Champaign, Ill.

Perkins, E. G., and Johnston, P. V., 1969, Pyrolysis–gas chromatography of phosphoglycerides: A mass spectral study of the products, *Lipids* **4**:301.

Pfeffer, P. E., 1975, NMR of lipids: The use of chemical shift reagents, in: *Analysis of Lipids and Lipoproteins* (E. G. Perkins, ed.), pp. 153–169, American Oil Chemists' Society, Champaign, Ill.

Phillipou, G., Bigham, D. A., and Seamark, R. F., 1975, Subnanogram detection of *t*-butyldimethylsilyl fatty acid esters by mass fragmentography, *Lipids* **10**:714.

Phillips, B. E., and Smith, C. R., Jr., 1970, Glycerides of *Monnina emarginata* seed oil, *Biochim. Biophys. Acta* **218**:71.

Pitas, R. E., Sampugna, J., and Jensen, R. G., 1967, Triglyceride structure of cow's milk fat. I. Preliminary observations in the fatty acid composition of positions 1, 2 and 3, *J. Dairy Sci.* **50**:1332.

Pollack, J. D., Clark, D. S., and Somerson, N. L., 1971, Four-directional-development thin-layer chromatography of lipids using trimethyl borate, *J. Lipid Res.* **12**:563.

Powell, R. G., Kleiman, R., and Smith, C. R., Jr., 1969, New sources of 9-D-Hydroxy-*cis*-12-octadecenoic acid, *Lipids* **4**:450.

Privett, O. S., and Erdahl, W. L., 1975, Liquid chromatography of lipids, in: *Analysis of Lipids and Lipoproteins* (E. G. Perkins, ed.), pp. 123–137, American Oil Chemists' Society, Champaign, Ill.

Privett, O. S., Blank, M. L., and Romanus, O., 1963, Isolation analysis of tissue fatty acids by ultra-micro-ozonolysis in conjunction with thin-layer chromatography and gas liquid chromatography, *J. Lipid Res.* **4**:260.

Privett, O. S., Blank, M. L., Codding, D. W., and Nickell, E. C., 1965, Lipid analysis by quantitative thin-layer chromatography, *J. Am. Oil Chem. Soc.* **42**:381.

Radford, T., and DeJongh, D. C., 1972, Carbohydrates, in: *Biochemical Applications of Mass Spectrometry* (G. R. Waller, ed.), pp. 313–350, Wiley, New York.

Radin, N. S., 1969, Preparation of lipid extracts, in: *Methods in Enzymology*, Vol. XIV (J. M. Lowenstein, ed.), pp. 245–254, Academic Press, New York.

Ramachandran, S., Sprecher, H. W., and Cornwell, D. G., 1968, Studies on the preparation and analysis of glyceryl ether derivatives and the isolation and reductive ozonolysis of unsaturated glyceryl ethers, *Lipids* **3**:511.

Renkonen, O., 1965, Individual molecular species of different phospholipid classes. II. A method of analysis, *J. Am. Oil Chem. Soc.* **42**:299.

Renkonen, O., 1966, Individual molecular species of phospholipids. III. Molecular species of ox brain lecithins, *Biochim. Biophys. Acta* **125**:288.

Renkonen, O., 1967*a*, Individual molecular species of phospholipids. IV. Gas–liquid chromatography of different types of diglyceride acetates derived from ox brain lecithins, *Biochim. Biophys. Acta* **137**:575.

Renkonen, O., 1967*b*, The analysis of individual molecular species of polar lipids, in: *Advances in Lipid Research* (R. Paoletti and D. Kritchevsky, eds.), pp. 329–351, Academic Press, New York.

Renkonen, O., 1968, Individual molecular species of phospholipids. VII. Analysis of lecithins containing ten to twelve double bonds, *Lipids* **3**:191.

Renkonen, O., 1971, Thin-layer chromatographic analysis of subclasses and molecular species of polar lipids, in: *Progress in Thin-Layer Chromatography and Related Methods*, Vol. II (A. Niederwieser and G. Pataki, eds.), pp. 143–182, Ann Arbor Science Publishers, Ann Arbor, Mich.

Renkonen, O., and Luukkonen, A., 1976, Thin-layer chromatography of phospholipids and glycolipids, in: *Lipid Chromatographic Analysis*, Vol. 1 (G. V. Marinetti, ed.), 2nd ed., pp. 1–58, Marcel Dekker, New York.

Roberts, R. N., 1967, Gas chromatography of inositol and glycerol, in: *Lipid Chromatographic Analysis*, Vol. 1 (G. V. Marinetti, ed.), pp. 447–463, Marcel Dekker, New York.

Rock, C. O., and Snyder, F., 1975, Metabolic inter-relations of hydroxy-substituted ether-linked glycerolipids in the pink portion of the rabbit Harderian gland, *Arch. Biochim. Biophys.* **171**:631.

Roehm, J. N., and Privett, O. S., 1970, Changes in the structure of soybean triglycerides during maturation, *Lipids* **5**:353.

Rouser, G., 1973, Quantitative liquid column and thin-layer chromatography of lipids and other water insoluble substances, elution selectivity principles and a graphic method for pattern analysis of chromatographic data, *J. Chromatogr. Sci.* **11**:60.

Rouser, G., Kritchevsky, G., and Yamamoto, A., 1967, Column chromatographic and associated procedures for separation and determination of phosphatides and glycolipids, in: *Lipid Chromatographic Analysis* (G. V. Marinetti, ed.), pp. 99–162, Marcel Dekker, New York.

Ryhage, R., and Stenhagen, E., 1960, Mass spectrometry in lipid research, *J. Lipid Res.* **1**:361.

Saito, K., and Gamo, M., 1973, The distribution of diol waxes on preen glands of some birds. III. The occurrence of 1,2-diols, *Comp. Biochem. Physiol.* **45B**:603.

Satouchi, K., and Saito, K., 1976, Studies on trimethylsilyl derivatives of 1-alkyl-2-acylglycerols by gas–liquid chromatography mass spectrometry, *Biomed. Mass Spectrometry* **3**:122.

Schlenk, W., Jr., 1965, Synthesis and analysis of optically active triglycerides, *J. Am. Oil Chem. Soc.* **42**:945.

Schmid, H. H. O., Bandi, P. C., Mangold, H. K., and Baumann, W. J., 1969, Alkoxylipids. V. The isomeric monounsaturated substituents of neutral alkoxylipids and triglycerides of ratfish liver, *Biochim. Biophys. Acta* **187**:208.

Schmid, H. H. O., Bandi, P. C., and Kwei, L. S., 1975, Analysis and quantification of ether lipids by chromatographic methods, *J. Chromatogr. Sci.* **13**:478.

Schomberg, G., Dielmann, R., Husmann, H., and Weeke, F., 1976, Gas chromatographic analysis with glass capillary columns, *J. Chromatogr.* **122**:55.

Schwartz, D. P., Weihrauch, J. L., and Burgwald, L. H., 1969, A periodic acid column procedure for the oxidation of vic-glycols, epoxides and α-hydroxy acids at the micromole level, *Anal. Chem.* **41**:984.

Scott, R. P. W., Scott, C. G., Munroe, M., and Hess, J., Jr., 1974, Interface for on-line liquid chromatography–mass spectroscopy analysis, *J. Chromatogr.* **99**:395.

Sjövall, J., Nyström, E., and Haahti, E., 1968, Liquid chromatography on lipophilic Sephadex: Column and detection techniques, in: *Advances in Chromatography*, Vol. 6 (J. C. Giddings and R. A. Keller, eds.), pp. 119–170, Marcel Dekker, New York.

Skipski, V. P., and Barclay, M., 1969, Thin-layer chromatography of lipids, in: *Methods in Enzymology*, Vol. XIV (J. Lowenstein, ed.), pp. 530–598, Academic Press, New York.

Snyder, F., 1973, Thin-layer chromatographic behavior of glycerolipid analogs containing ether, ester, hydroxyl, and ketone groupings, *J. Chromatogr.* **82**:7.

Snyder, F., 1976, Chromatographic analysis of alkyl and alk-1-enyl ether lipids and their derivatives, in: *Lipid Chromatographic Analysis*, Vol. 1 (G. V. Marinetti, ed.), 2nd ed., pp. 111–148, Marcel Dekker, New York.

Snyder, F., and Blank, M. L., 1969, Relationships of chain lengths and double bond locations in *O*-alkyl, *O*-alk-1-enyl, acyl and fatty alcohol moieties in preputial glands of mice, *Arch. Biochim. Biophys.* **130**:101.

Snyder, F., Rainey, W. T., Jr., Blank, M. L., and Christie, W. H., 1970, The source of oxygen in the ether bond of glycerolipids, *J. Biol. Chem.* **245**:5853.

Snyder, F., Blank, M. L., and Wykle, R. L., 1971, The enzymic synthesis of ethanolamine plasmalogens, *J. Biol. Chem.* **246**:3639.

Sommer, L. H., 1965, *Stereochemistry, Mechanism and Silicon*, McGraw-Hill, New York.

Sprecher, H. W., Maier, R., Barber, M., and Holman, R. T., 1965, Structure of an optically active allene-containing tetraester triglyceride isolated from the seed oil of *Sapium sebiferum*, *Biochemistry* **4**:1856.

Stahl, E., ed., 1969, *Thin-Layer Chromatography—A Laboratory Handbook*, 2nd ed., Springer-Verlag, New York.

Stewart, J. E., 1970, *Infrared Spectroscopy: Experimental Methods and Techniques*, Marcel Dekker, New York.

Su, K. L., Baumann, W. J., Madson, T. H., and Schmid, H. H. O., 1974, Long-chain cyclic acetals of glycerols: Metabolism of the stereomeric 1,3-dioxanes and 1,3-dioxolanes in the myelinating rat brain, *J. Lipid Res.* **15**:39.

Sun, K. K., and Holman, R. T., 1968, Mass spectrometry of lipid molecules, *J. Am. Oil Chem. Soc.* **45**:810.

Thomas, A. E., III, Scharoun, J. E., and Ralston, H., 1965, Quantitative estimation of isomeric monoglycerides by thin-layer chromatography, *J. Am. Oil Chem. Soc.* **42**:789.

Thompson, G. A., and Kapoulas, V. M., 1969, Preparation and assay of glyceryl ethers, in: *Methods in Enzymology*, Vol. XIV (J. M. Lowenstein, ed.), pp. 668–678, Academic Press, New York.

Thompson, G. A., and Lee, P., 1965, Studies of the α-glyceryl ether lipids occurring in molluscan tissues, *Biochim. Biophys. Acta* **98**:151.

Van Golde, L. M. G., and Van Deenan, L. L. M., 1966, The effect of dietary fat on the molecular species of lecithin from rat liver, *Biochim. Biophys. Acta* **125**:496.

Van Golde, L. M. G., Pietersen, W. A., and Van Deenen, L. L. M., 1968, Alternations in the molecular species of rat liver lecithin by corn oil feeding to essential fatty acid deficient rats as a function of time, *Biochim. Biophys. Acta* **152**:84.

Varanasi, U., Everitt, M., and Malins, D. C., 1973, The isomeric composition of diisovaleroylglycerides: A specificity for the biosynthesis of the 1,3-diisovaleroyl structures, *Int. J. Biochem.* **4**:373.

Viswanathan, C. V., 1974, Chromatographic analysis of alkoxy-lipids, *J. Chromatogr.* **98**:129.

Warner, H. R., and Lands, W. E. M., 1963, The configuration of the double bond in naturally occurring alkenyl ethers, *J. Am. Chem. Soc.* **85**:60.

Watts, R., and Dils, R., 1969, Isomerization of mono- and diglyceride trimethylsilyl ethers, *Chem. Phys. Lipids* **3**:168.

Weber, E. J., 1973, Changes in structure of triglycerides from maturing kernels of corn, *Lipids* **8**:295.

Wedmid, Y., and Litchfield, C., 1975, Positional analysis of isovaleroyl triglycerides using proton magnetic resonance with $Eu(fod)_3$ and $Pr(fod)_3$ shift reagents. I. Model compounds, *Lipids* **10**:145.

Wedmid, Y., and Litchfield, C., 1976, Positional analysis of isovaleroyl triglycerides using proton magnetic resonance with $Eu(fod)_3$ and $Pr(fod)_3$ shift reagents. II. Cetacean triglycerides, *Lipids* **11**:189.

Wessels, H., and Rajagopal, N. S., 1969, Die DC-Trennung von isomeren und natürlichen triglycerid-Gemischen, *Fette Seifen Anstrichm.* **71**:543.

Witter, R. F., and Whitner, V. S., 1972, Determination of serum triglycerides, in: *Blood Lipids and Lipoproteins* (G. Nelson, ed.), pp. 181–272, Wiley-Interscience, New York.

Wood, R., 1967, GLC and TLC analysis of isopropylidene derivatives of isomeric polyhydroxy acids derived from positional and geometrical isomers of unsaturated fatty acids, *Lipids* **2**:199.

Wood, R., 1968, Gas–liquid chromatographic analysis of long-chain fatty alcohols, *J. Gas Chromatogr.* **6**:94.

Wood, R., and Snyder, F., 1966, Gas–liquid chromatographic analysis of long-chain isomeric glyceryl monoethers, *Lipids* **1**:62.

Wood, R., and Snyder, F., 1967*a*, Chemical and physical properties of isomeric glyceryl monoethers, *Lipids* **2**:89.

Wood, R., and Snyder, F., 1967*b*, Characterization and identification of glyceryl ether diesters present in tumor cells, *J. Lipid Res.* **8**:494.

Wood, R., and Snyder, F., 1968, Quantitative determination of alk-1-enyl and alkyl-glyceryl ethers in neutral lipids and phospholipids, *Lipids* **3**:129.

Wood, R., and Snyder, F., 1969, Tumor lipids: Metabolic relationships derived from structural analysis of acyl, alkyl, and alk-1-enyl moieties of neutral glycerides and phosphoglycerides, *Arch. Biochim. Biophys.* **131**:478.

Wood, R. C., Raju, P. K., and Reiser, R., 1965, Gas–liquid chromatographic analysis of monoglycerides and their trimethylsilyl ether derivatives, *J. Am. Oil Chem. Soc.* **42**:161.

Wood, R. C., Piantadosi, C., and Snyder, F., 1969*a*, Quantitative analysis and comparison of the physical properties of *O*-alkyl and *S*-alkyl monoethers of glycerol, *J. Lipid Res.* **10**:370.

Wood, R., Baumann, W. J., Snyder, F., and Mangold, H. K., 1969*b*, Gas–liquid chromatography of dialkyl, alkylacyl, and diacyl derivatives of glycerol, *J. Lipid Res.* **10**:128.

Yeung, S. K. F., and Kuksis, A., 1974, Molecular species of ethanolamine phosphatides of dog and pig kidney, *Can. J. Biochem.* **52**:830.

Yurkowski, M., and Brockerhoff, H., 1966, Fatty acid distribution of triglycerides determined by deacylation with methyl magnesium bromide, *Biochim. Biophys. Acta* **125**:55.

Chapter 4

Stereospecific Analysis of Triacylglycerols

W. Carl Breckenridge

4.1. Introduction

Natural triacylglycerols consist of complex mixtures of fatty acids esterified to specific hydroxyl groups of the glycerol molecule. Although glycerol has a plane of symmetry, asymmetrical triacylglycerols are formed when the primary hydroxyl groups are esterified with different fatty acids. Furthermore, the primary hydroxyl groups can be distinguished by stereospecific enzymes.

Although it had been realized that different fatty acids at positions 1 and 3 of the glycerol molecule would produce asymmetrical triacylglycerols, physical methods initially could not demonstrate significant optical rotation in natural triacylglycerols because of small differences in the chain length of the fatty acids. With the use of pancreatic lipase, which released fatty acids at both the primary positions, theories were developed which generally did not consider differences in fatty acid composition at the primary positions. With the development of procedures for the independent analysis of the fatty acids at positions 1 and 3 of the glycerol molecule, it has been found that most natural triacylglycerols possess some asymmetry. Although enzymatic methods can define selective associations of fatty acids with each position of the triacylglycerol molecule, the exact molecular structure can be determined only following the isolation of specific groups of triacylglycerols by complementary chromatographic techniques. The results to date on total mixtures have served to dispel various random hypotheses for triacylglycerol structure and have stimulated considerable research into biosynthetic mechanisms of triacylglycerols. The biological significance of a specific triacylglycerol structure in tissues has remained largely obscure.

The present discussion will be limited to a detailed review of methods for enzymatic stereochemical analyses followed by a summary of data obtained for the stereospecific distribution of fatty acids in natural triacylglycerols. The reader is

W. Carl Breckenridge • Departments of Clinical Biochemistry and Medicine, University of Toronto, Toronto, Ontario, M5S 2J5, Canada.

referred to earlier reviews for background covering this information (Brockerhoff, 1971; Christie, 1973; Kuksis, 1972; Litchfield, 1972).

4.2. Physical and Chemical Methods for the Determination of Stereochemical Distribution of Fatty Acids in Triacylglycerols

Although asymmetrical triacylglycerols can be defined by the Fischer convention, a form of stereospecific numbering proposed by Hirschman (1960) is now used generally.* Stereospecific numbering of the carbon atoms of glycerol uses the Fischer projection,† which places the secondary hydroxyl group to the left and numbers the glycerol carbon atoms 1 to 3 from top to bottom. Thus L-glycerol-3-phosphate or D-glycerol-1-phosphate is equivalent to *sn*-glycerol-3-phosphate, and 1,2-dioleoyl-3-palmitoyl L-glycerol would be equivalent to 1,2-dioleoyl-3-palmitoyl-*sn*-glycerol.

Various approaches employing physical, chemical, and enzymatic methods have been used for assessing asymmetry in triacylglycerols. The physical and chemical approaches have limitations since some selective fractionation of the natural mixture is necessary to demonstrate asymmetry. Triacylglycerols with fatty acids differing greatly in chain length show measurable optical rotation (Schlenk, 1965), and optically active monoacetyl triacylglycerols have been identified (Kleiman *et al.*, 1966) in *Euonymus verrucosus* seed oils where the acetyl group is located specifically in position 3. However, the small differences in optical rotation of most asymmetrical natural triacylglycerols can be detected only by the highly sensitive techniques of optical rotatory dispersion and circular dichroism, which allow measurements in the vacuum UV region. Using these techniques, it was observed that a saturated triacylglycerol with the chain length greater at position 1 than at position 3 possessed a negative optical rotation (Gronowitz *et al.*, 1975). When the chain length was greater at position 3, the rotation was positive. Thus in a series of 1,2-diacyl-3-myristoyl-*sn*-glycerols the optical rotation decreased as the acyl chain length in position 1 increased. The optical rotation of 1,2-dilauroyl-3-myristoyl-*sn*-glycerol could be detected. If a complex mixture of triacylglycerols possessed a constant pattern of chain length at positions 1 and 3, it would be possible to demonstrate asymmetry, as was observed in monoaceto triacylglycerols. However, in complex long-chain triacylglycerols, considerable fractionation would be required. Assume that a saturated class of triacylglycerols could be isolated containing myristic (M), palmitic (P), and stearic (S) acids and molecular species of SPP and SMM. If SPP contained all stearic acid at position 1 but SMM contained all stearic acid at position 3, the optical rotations of these species would be opposing rotations based on the data of Gronowitz *et al.* (1975) and would tend to cancel in the total rotation. While optical rotation could be measured if specific triacyl-

* The IUPAC-IUB Commission on Biochemical Nomenclature formed a subcommittee to study the nomenclature of lipids. In 1967, the committee proposed a trial use of stereospecific number as described in the *European Journal of Biochemistry* **2**:127.

† In the Fischer convention for the planar projection of glycerol described in detail in Chapter 5, the molecule is oriented so that the carbon chain is vertical and projecting away from the viewer while hydrogen and hydroxyl groups of C_2 project toward the viewer. The model is then flattened into a two-dimensional plane.

glycerols could be isolated, the data would indicate only the presence or absence of asymmetry and would provide little data about the location of the fatty acids in the glycerol molecule.

Recently, proton magnetic resonance (PMR) spectroscopy in combination with a chiral shift reagent has been used to resolve the PMR signals from ester groups at the center of chirality of triacylglycerols (Bus *et al.*, 1976). This method allowed a measurement of enantiomeric purity of synthetic triacylglycerols and confirmed the presence of acetic acid in position 3 for the seed oil triacylglycerols of *Euonymus alatus* (Kleiman *et al.*, 1966) and the presence of butyric acid in position 3 for bovine butterfat triacylglycerols (Breckenridge and Kuksis, 1968). The chiral shift reagents form complexes with the free electron pairs in the ester groups of the triacylglycerols. In the presence of the shift reagent, the enantiomers form diastereoisomer associates which have different PMR spectra. Using tris(3-heptafluorobutyryl-d-camphorato)europium(III), it was possible to separate the acetyl proton signals of the enantiomers of 1-acetyl-2,3-distearoyl-*rac*-glycerol. The 1-acetyl-2,3-distearoyl-*sn*-glycerol isomer possessed a higher δ value than 1,2-distearoyl-3-acetyl-*sn*-glycerol. On addition of 1-acetyl-2,3-distearoyl-*rac*-glycerol to the seed oil of *Euonymus alatus*, a second signal appeared with a higher δ value, indicating that the natural triacylglycerol contained the acetyl group at position 3. A similar phenomenon was observed with the butyryl PMR triplet when 1-butyryl-2,3-distearoyl-*sn*-glycerol was added to a hydrogenated fraction of butterfat containing triacylglycerols rich in butyric acid. The procedure may be useful for assessing asymmetry of purified fractions from natural triacylglycerols if suitable spectra can be obtained for small differences in fatty acid chain length at positions 1 and 3. The spectra will be more complex and the differences in δ values much smaller than those noted for triacylglycerols with two to six carbon atoms in primary positions.

A combined enzymatic and chemical method has been developed by Morris (1965*a*, *b*). Diacylglycerols released by pancreatic lipase were converted to trimethylsilyl (TMS) ethers. The resulting trisubstituted glycerols were separated by argentation thin-layer chromatography. Starting with a triacylglycerol such as monooleoyldipalmitin, dipalmitin and monopalmitoylmonoolein were formed by lipase hydrolysis provided that oleic acid was in the primary positions of the glycerol molecule. If the palmitic and oleic were each selectively associated with one of the primary positions, the resulting preparations of dipalmitin or monopalmitoylmonoolein silyl ethers, separated by argentation chromatography, would show optical rotation. Morris (1965*b*) checked several species of triacylglycerols from plant and animal sources and found most to be optically active. However, the diacylglycerol TMS ethers from natural triacylglycerols never possessed as much optical rotation as the equivalent synthetic diacylglycerol TMS ethers. It was concluded that this could have been due to a predominance of one enantiomer, incomplete purity of the triacylglycerol species, or decomposition of the TMS ether. Tertiarybutyldimethylsilyl ethers would be more stable for chromatographic purposes than the TMS derivatives, while the Grignard degradation would eliminate any specificity due to lipase hydrolysis. The procedure works effectively only on isolated species of triacylglycerols in which the resulting diacylglycerols can be resolved. If trisaturated species are involved, argentation chromatography would have to be replaced by high-pressure liquid–liquid chromatography.

4.3. Enzymatic Methods for Determination of Stereochemical Distribution of Fatty Acids in Triacylglycerols

The enzymatic procedures have been used by many laboratories to study the association of fatty acids with each position of the glycerol molecule. Even in natural oils, where equal amounts of a fatty acid may occur in both primary positions, it is necessary to isolate selective groups of triacylglycerols before it can be shown that the mixture is not racemic. The combined techniques of argentation TLC, gas–liquid chromatography, and stereospecific analysis allow an extensive assessment of the molecular species of triacylglycerols (Hammond, 1969).

The enzymatic methods rely on the ability of stereospecific enzymes to differentiate between diacylglycerols or phosphorylated derivatives. In a method developed by Lands *et al.* (1966), a stereospecific diacylglycerol kinase is used to selectively phosphorylate 1,2-diacyl-*sn*-glycerols from a mixture of 1,2- and 2,3-diacyl-*sn*-glycerols produced by pancreatic lipolysis of triacylglycerols. In the methods developed by Brockerhoff (1965*a*, 1967), 1,2- and 2,3-diacyl-*sn*-glycerols are generated from pancreatic lipase or Grignard reagent. Preparation of phosphatidylphenol derivatives allows digestion with phospholipase A_2, which selectively attacks the 1,2-diacyl-*sn*-glycerol-3-phosphoryl phenols, liberating fatty acids characteristic of position 2 and lysophosphatidylphenols characteristic of position 1. Position 3 is obtained by difference. A separate procedure (Brockerhoff, 1967) uses the 1,3-diacylglycerols produced by the Grignard reagent. Following conversion to the phosphatide, phospholipase A_2 will yield free fatty acids characteristic of position 1 and lysophosphatides characteristic of position 3. Although this method gives a direct estimate of position 3, it is complicated by acyl migration during the generation of diacylglycerols.

4.3.1. Generation of Diacylglycerols

The most important step in enzymatic methods for stereospecific analyses is the generation of diacylglycerols which are representative of the fatty acid composition of each position in the original triacylglycerol.

4.3.1.1. Pancreatic Lipase

The crude lipase of pancreas extracts or a semipurified preparation (Lands *et al.*, 1966) exhibits a high degree of specificity for hydrolysis of acyl groups associated with the primary positions of glycerol. It was used exclusively in early studies to generate 1,2- and 2,3-diacyl-*sn*-glycerols from natural or synthetic triacylglycerols and for accurate assessment of fatty acids associated with position 2 of triacylglycerols. While maximum yields (25% of total glyceride glycerol) of diacylglycerols are produced after short periods (1–2 min) of digestion (Luddy *et al.*, 1964), yields will usually be of the order of 10–15%.

For representative generation of 1,2- and 2,3-diacyl-*sn*-glycerols, the lipase must not show selective hydrolysis of specific ester bonds. Early studies (Tattrie *et al.*, 1958) established that palmitic and oleic acids were released at equal rates from

1,2-dipalmityl-3-oleoyl-*sn*-glycerol, but it has been claimed (Anderson *et al.*, 1967) that, depending on the methodology used, pancreatic lipase may not give a random generation of diacylglycerols from complex triacylglycerol mixtures. The enzyme is also extremely inactive toward ester bonds containing fatty acids with olefinic groups at atoms less than carbon 9 from the carboxyl group (Brockerhoff, 1968; Kleiman *et al.*, 1970) and toward some long-chain saturated fatty acids (Bottino *et al.*, 1967). Thus the enzyme cannot be used in stereospecific analyses of marine oils (Bottino *et al.*, 1967) which contain large amounts of long-chain acids esterified to the primary hydroxyl groups. Soluble fatty acid esters of short chain length are usually attacked more rapidly (Brockerhoff, 1969). While triacylglycerols containing butyric acid are attacked more rapidly than those containing only long-chain fatty acids (Jensen *et al.*, 1964), there is no preferential release of butyric acid within the molecule. The problem of intermolecular specificity can be overcome by using hexane or ether as a diluent to give more uniform particle size in the two-phase emulsification (Brockerhoff, 1965*b*; Breckenridge and Kuksis, 1968; Marai *et al.*, 1969).

In most studies, it has been assumed that there is an equal production of 1,2- and 2,3-diacyl-*sn*-glycerols since a calculation of the diacylglycerol composition (Brockerhoff and Yurkowski, 1966), based on the fatty acid composition of position 2 and the total triacylglycerols, agreed with the composition of the diacylglycerol products. Studies with synthetic triacylglycerols (Table I) have indicated that the proportions of 1,2- and 2,3-diacyl-*sn*-glycerols produced by pancreatic lipolysis may not be equal when different fatty acids are associated with positions 1 and 3. Thus 1-palmitoyl-2-oleoyl-3-linoleoyl-*sn*-glycerol gave an excess yield (Morley *et al.*, 1974) of 1,2- over 2,3-diacyl-*sn*-glycerol, while the reverse was the case with the enantiomer. The difference in yields may be less pronounced in natural mixtures, but diacylglycerol composition should always be compared with the calculated composition. It is not necessarily imperative that equal amounts of 1,2- and 2,3-diacyl-*sn*-glycerols be formed from triacylglycerols since the subsequent steps will allow an assessment of the fatty acids at each position. However, the 1,2- and 2,3-diacyl-*sn*-glycerols must accurately reflect the proportions of the fatty acids at the respective positions.

Procedure: Triacylglycerol (50–100 mg) is dissolved in 0.2 ml of hexane or ether (optional) and added to 2.0 ml tris HCl buffer (1 M, pH 8.0) containing $CaCl_2$

Table I. Relative Yields of Diacylglycerols during Pancreatic Lipolysis[a]

Substrate	Isomer[b] 1,2	2,3
sn-Glycerol-1-palmitate-2-oleate-3-linoleate	61	39
sn-Glycerol-1-linoleate-2-oleate-3-palmitate	44	56
sn-Glycerol-1-oleate-2,3-dipalmitate	35	65
sn-Glycerol-1,2-dipalmitate-3-oleate	67	33

[a] Data from Morley *et al.* (1974).
[b] Isomers determined by gas–liquid chromatography of intact diglycerides.

(0.1 M), bile salts (0.02%), an indicator such as bromophenol blue, and 50 mg pancreatic acetone powder (steapsin). After homogenization, the mixture is incubated with shaking at 37°C. The reaction is stopped by adding ethanol when approximately 30% of the triacylglycerols have been removed. The lipids are extracted with ether and applied to thin layers of silica gel G containing 6% by weight boric acid and developed in hexane–ether (1 : 1, v/v). The 1,2- and 2,3-diacyl-*sn*-glycerols (R_f 0.4) are resolved from any 1,3-diacyl-*sn*-glycerols (R_f 0.5), fatty acids (R_f 0.7), and triacyl (R_f 0.9) and monoacyl (R_f 0.1) glycerols. They are located by spraying the plate with dichlorofluorescein (0.05% in methanol). Following collection, the diacylglycerols are eluted with ether. The ether extract is diluted with $\frac{1}{3}$ vol of petroleum ether, washed with water, dried over anhydrous sodium sulfate, and evaporated under nitrogen at room temperature. The diacylglycerols should be derivatized immediately to prevent acyl migration.

4.3.1.2. Grignard Reagent

Yurkowski and Brockerhoff (1966) used methyl magnesium bromide to effect a random degradation of triacylglycerols to 1,2-, 2,3-, and 1,3-diacyl-*sn*-glycerols. The reagent appeared to show little preference for different fatty acid ester bonds. In subsequent modifications, ethyl magnesium bromide (Wood and Snyder, 1969; Christie and Moore, 1969*a*) has been used to give better resolution of tertiary alcohols from 1,3-diacyl-*sn*-glycerols (Brockerhoff, 1971).

The reaction is stopped by the addition of acetic acid, which is subsequently neutralized by sodium bicarbonate. The exposure to acid conditions should be as short as possible to prevent acyl migration. The Grignard reagent also appears to cause acyl migration since the 1,3-diacyl-*sn*-glycerols are contaminated (5–10%) with isomerized 1,2- and 2,3-diacyl-*sn*-glycerols (Brockerhoff, 1967) when the 1,3-diacyl-*sn*-glycerols are used in the stereospecific analysis. In modifications (Wood and Snyder, 1969; Christie and Moore, 1969*a*), the 1,2- and 2,3-diacyl-*sn*-glycerols were employed for the derivatization and phospholipase A_2 digestion. Although the use of the 1,3-diacyl-*sn*-glycerols allows a direct estimate of fatty acids at position 3, the contamination due to isomerization of 1,2-diacyl-*sn*-glycerols negates the results.

If the Grignard reagent is used in analysis of milk triacylglycerols containing short- and long-chain fatty acids, great care must be taken in the isolation of 1,3-diacyl-*sn*-glycerols containing short-chain fatty acids which may overlap with 1,2- and 2,3-diacyl-*sn*-glycerols containing only long-chain fatty acids. The original procedure has been scaled down to allow a stereospecific analysis on 10 mg of triglyceride (Christie and Moore, 1969*a*; Wood and Snyder, 1969).

Procedure: A solution (1 ml) of ethyl magnesium bromide (0.5 M) in anhydrous diethyl ether is added to the triacylglycerols (10–40 mg in 1 ml anhydrous diethyl ether). A white precipitate indicates a lack of anhydrous conditions. (Ether can be dried over molecular sieve 4A.) After agitation for 1 min, glacial acetic acid (0.05 ml) is added, followed by water (2 ml). The ethereal layer is removed and washed with dilute $NaHCO_3$ and water, dried over anhydrous Na_2SO_4, and evaporated under nitrogen. The diacylglycerols can be isolated as described previously by TLC on silica gel impregnated with borate as described in Section 4.3.1.1. The tertiary alcohols (R_f 0.70) migrate between the di- and triacylglycerols.

4.3.2. Stereospecific Selection of Diacylglycerols or Their Derivatives

The major approach to stereospecific selection involves the conversion of the diacylglycerols to a phospholipid which will serve as a substrate for phospholipase A_2, an enzyme stereospecific for 1,2-diacyl-*sn*-glycerol-3-phospholipids. Diglyceride kinase can also be used to convert 1,2-diacyl-*sn*-glycerols to 1,2-diacyl-*sn*-glycerol-3-phosphate.

4.3.2.1. Derivatization of Diacylglycerols

The diacylglycerols are converted to suitable substrates for phospholipase A_2 digestion by condensation with phenyl dichlorophosphate in the presence of pyridine as a catalyst (Brockerhoff, 1965*a*). The diacylglycerols should be added slowly at low temperatures to an excess of phenyl dichlorophosphate since there are two potential reactive groups on the phenol phosphate which could react with separate diacylglycerols. There has been no evidence of acyl migration or isomerization of the diacylglycerols during the derivatization. Excess phenol phosphate can be removed by the addition of water and partitioned by the procedure of Bligh and Dyer (1959). Following the formation of a salt, the phosphatidylphenols can be purified by TLC; however, if care has been taken in the purification of the diacylglycerol, this step is not essential.

Procedure: The diglycerides (2–20 mg in 0.5 ml of ether) are added slowly with cooling to 0.05 ml of phenyl dichlorophosphate which is dissolved in chloroform (1 ml) and pyridine (0.5 ml). After standing for 4 hr, pyridine (0.5 ml) and water (1 ml) are added with cooling. One hour later, chloroform (6 ml), methanol (6 ml), and water (5 ml) are added and the mixture is agitated. The chloroform layer is recovered and washed with dilute HCl (3×2 ml) to remove pyridine and water (2×10 ml). If purification is desired, the phosphatidylphenols can be isolated by TLC on silica gel G using a developing solution of chloroform–methanol–3% aqueous ammonia (63 : 30 : 7, v/v/v) where they have an R_f of 0.90.

4.3.2.2. Stereospecific Hydrolysis with Phospholipase A_2

Phospholipase A_2 from several snake venoms is now used routinely to hydrolyze fatty acids from position 2 of 1,2-diacyl-*sn*-glycerol-3-phospholipid. The specificity for position 2 is maintained regardless of the fatty acids present at position 1 or 2 (Bird *et al.*, 1965), but most snake venom phospholipases show selectivity in a more rapid rate of release of saturated fatty acids from position 2 of diacyl phosphoglycerides (Nutter and Privett, 1966). Of six venoms studied to date, only *Ophiophagus hannah* shows no specificity in the rate of release of different fatty acids. Although *Crotalus atrox* has slight specificity, it was used successfully in early studies by Brockerhoff (1965*a*) and Breckenridge and Kuksis (1968) since the digestions were carried to completion. Brockerhoff (1967) also found that 1,3-diacyl-2-phosphophenyl-*sn*-glycerols were digested by phospholipase A_2, resulting in the liberation of fatty acids from position 1, with the lysophosphatide representing position 3.

The original procedures (Brockerhoff, 1965*a*) used triethylamine bicarbonate as a buffer. Since the pH was adjusted by the addition of carbon dioxide, it was important to use a closed vessel to prevent loss of CO_2 and a change in pH. Tris HCl buffer can be readily substituted for triethylamine (Christie and Moore, 1969*a*). Phospholipase A_2 has a pH optimum of 7.5 and also requires the presence of calcium ions. The digestion can be carried out in ether saturated with the buffer (Christie and Moore, 1969*a*) or in a two-phase system (Brockerhoff, 1965*a*; Breckenridge and Kuksis, 1968; Wood and Snyder, 1969). The latter allows autodigestion and extraction of endogenous lipids in the phospholipase preparation before addition of the phosphatidylphenols. If long incubation times are required for samples containing polyunsaturated fatty acids, the system should be maintained in a nitrogen atmosphere.

Procedure: The phosphatidylphenols in diethyl ether (3 ml) are mixed with 5 ml tris buffer (0.1 M, pH 7.2), $CaCl_2$ (0.001 M), and *Ophiophagus hannah* venom (0.5 mg). After shaking for 16 hr, acetic acid (0.02 ml) is added and the mixture is extracted by the method of Bligh and Dyer (1959). If problems are encountered in the recovery of lysophosphatidylphenols, isobutanol (10 ml) can be added and the mixture dried with a rotary evaporator. The residue is taken up in chloroform–methanol (2 : 1, v/v) and subjected to TLC on silica gel G (Christie, 1975). An initial development with hexane–ether–formic acid (50 : 50 : 1, v/v/v) causes the migration of the free fatty acids with an R_f of 0.80. Under these conditions, the lyso and residual (2,3-) phosphatidylphenols remain at the origin. A second development with chloroform–methanol–14M ammonia (90 : 8 : 2, v/v/v) for two-thirds of the dimension of the plate results in the resolution of 2,3- and lysophosphatidylphenols. The 1,3-phosphatidylphenols are treated in a similar fashion except that there are only free fatty acids and lysophosphatidylphenols remaining after the digestion.

4.3.2.3. Diacylglycerol Kinase

The method of Lands *et al.* (1966) provides an alternative approach to stereospecific analysis since 1,2-diacyl-*sn*-glycerols are selectively phosphorylated. Diacylglycerol kinase must be purified from *Escherichia coli.* Although the specificity of this enzyme for atypical fatty acids has not been thoroughly tested, the enzyme was specific for phosphorylation of 1,2-diacyl-*sn*-glycerols. No specificity was demonstrated for the phosphorylation of 1,2-diacyl-*sn*-glycerols, derived from pig liver lecithins, provided that the reaction was taken to completion. The 1,2-diacyl-*sn*-glycerols and phosphatidic acid must be representative of the respective positions since they are used in the indirect calculations for the composition of positions 1 and 3. Discrepancies may also occur because of the cumulative error.

Procedure: The method of Lands *et al.* (1966) uses diacylglycerol kinase prepared by the method of Pieringer and Kunnes (1965). The incubation contains diacylglycerol (200–500 μg), bile salts (10 μl of a 20% solution), sodium phosphate buffer, pH 7.95 (50 μl of a 0.5 M solution), ATP (100 μl of 0.05 M solution), $MgCl_2$

(50 μl of a 1.0 M solution), and diacylglycerol kinase (1 mg). After 1 hr at 37°C, dilute HLC is added and the lipids are extracted with 2 ml of chloroform–methanol (2 : 1, v/v) followed by chloroform (1.3 ml). Triethyl amine is added, the solvent is evaporated, and the residue is taken up in chloroform–methanol (2 : 1, v/v). The phosphatidic acid is resolved from diacylglycerols by TLC first in 80% diethyl ether in petroleum ether followed by chloroform–methanol–formic acid (100 : 10 : 5, v/v/v).

4.3.3. Calculations and Checks on Accuracy

In order to ensure subsequent verification and allow further comparison, the results of the analyses should always be presented as the percent of fatty acid of each individual position. The results may also be expressed as a percent of the fatty acid in the total triacylglycerol. However, the latter alone gives no indication of the contribution of a particular fatty acid to the total pattern of each position.

A scheme of the best approach to stereospecific analysis is listed in Table II. In

Table II. Recommended Procedures for Stereospecific Analyses of Triacylglycerols

I. Chemical and enzymatic manipulations	a. Deacylation with ethyl magnesium bromide and isolation of 1,2-, 2,3-, and 1,3-diacyl-*sn*-glycerols
	b. Derivatization to phosphatidylphenols and digestion with phospholipase A_2
	c. Isolation of 2,3-phosphatidylphenols from (b) and digestion with pancreatic lipase
	d. Pancreatic lipase hydrolysis of triacylglycerols
II. Calculation of results	a. Position 1:
	i. Lysophosphatidylphenol from phospholipase A_2 digestion of 1,2-phosphatidylphenols
	ii. Free fatty acids from phospholipase A_2 digestion of 1,3-phosphatidylphenols
	b. Position 2:
	i. Free fatty acids from phospholipase digestion of 1,2-phosphatidylphenols
	ii. Monoacylglycerols from pancreatic lipase digestion of triacylglycerols
	c. Position 3:
	i. 3 × triacylglycerols − (position 1 + position 2)
	ii. 2 × 2,3-phosphatidylphenols − position 2
	iii. Free fatty acids liberated by pancreatic lipase hydrolysis of 2,3-phosphatidylphenols
	iv. Lysophosphatidylphenol remaining after phospholipase A_2 digestion of 1,3-phosphatidylphenols
III. Checks	a. 1,2- and 2,3-diacyl-*sn*-glycerols $= \dfrac{3 \times \text{triacylglycerols} + \text{monoacylglycerols}}{4}$
	b. Position 3: 2,3-phosphatidyl phenols $= \dfrac{\text{position 2} + \text{position 3}}{2}$

all procedures, it is important to obtain a random generation of 1,2-diacyl-*sn*-glycerols. This can be checked by calculation of the composition on the basis of the total triacylglycerol plus the monoacylglycerols formed by pancreatic lipase (Table II). This check is limited by the inability of pancreatic lipase to liberate long-chain fatty acids. In the Brockerhoff procedure (1965*a*) using the 1,2-diacyl-*sn*-glycerols, the composition of position 1 is derived directly from the composition of the lysophosphatidylphenols liberated by phospholipase A_2 digestion of the mixture of 1,2- and 2,3-phosphatidylphenols. The composition of the lysophosphatidylphenols could be affected by nonrepresentative production of diacylglycerols or by selective digestion of the phosphatidylphenols if the digestion is not complete. Position 1 can also be determined from the free fatty acids released by phospholipase A_2 digestion of the 1,3-diacyl-2-phosphophenyl-*sn*-glycerols. Since evidence (Brockerhoff, 1967) suggests that 1,3-diacyl-phosphatidylphenols are also formed by the isomerization of 1,2- or 2,3-diacyl-*sn*-glycerols, the calculation of position 1 by this method alone is not recommended.

Position 2 can be estimated from the two monoacylglycerols produced by pancreatic lipase hydrolysis of the triacylglycerols or from the free fatty acids released from phospholipase A_2 digestion of the 1,2-diacyl-*sn*-phosphatidylphenols. Wood and Snyder (1969) have recommended the elimination of pancreatic lipase hydrolysis for analysis of position 2.

Position 3 can be calculated by subtracting the sum of positions 1 and 2 from three times the fatty acid composition of the total triacylglycerols. Using the 1,3-diacyl-*sn*-glycerols produced by the Grignard reagent, position 3 can be calculated directly by analyzing the lysophosphatidylphenols produced by phospholipase A_2 hydrolysis of the 1,3-phosphatidylphenol. This estimate is again subject to error due to any acyl migration. Position 3 could also be ascertained by digestion of residual 2,3-diacylphosphatidylphenols with pancreatic lipase as proposed by Brockerhoff (1973). The accuracy of the estimates for positions 2 and 3 can be checked by comparing the sum of these estimates with the 2,3-phosphatidylphenols remaining after phospholipase A_2 digestion, provided that the digestion has gone to completion (Wood and Snyder, 1969). The various estimates for a given position should differ by no more than 10% for major components.

Using the diglyceride kinase method (Lands *et al.*, 1966), position 2 is obtained from the 2-monoacylglycerols. Position 1 is obtained by subtracting the monoacylglycerol composition from two times the phosphatidic acid composition. Position 3 is obtained by subtracting the phosphatidic acid composition from three times triacylglycerol composition. Thus indirect estimates are obtained for both primary positions, and no cross-checks are possible.

4.3.4. Studies with Radioactive Acylglycerols

Stereochemical analyses can also be applied to the study of the incorporation of radioactive fatty acids or glycerol into triacylglycerols. In such studies, it is extremely important to ascertain the completeness of digestion and recovery of the radioactive phosphatidylphenols which may be minor components of carrier materials. If free fatty acids are used, the recovery at each step must be ascertained, since the number of counts in the free fatty acids and lyso and residual phosphatidylphenols will be used to assess the relative incorporation of the fatty acid

Table III. Correction of Recoveries during Stereochemical Analysis of Radioactive Tracer Experiments[a]

	Experiment[b]			
	1		2	
Lipid	^3H	^{14}C	^3H	^{14}C
		(dpm)		
Total diacylglycerols	82,263	41,302	166,724	39,065
Phospholipase A_2 products				
Fatty acids		11,836	51	9,958
Residual phosphatidylphenols	40,247	12,953	52,061	9,076
Lysophosphatidylphenols	18,092	8,698	10,950	3,708
sn-2,3-DG/*sn*-1,2-DG[c]	69.0/31.0	42.7/57.3	81.1/18.9	52.4/47.6
Normalized *sn*-2,3-DG/*sn*-1,2-DG	74.9/25.1	50.0/50.0	79.5/20.5	50.0/50.0

[a] Data from Morley *et al.* (1977).
[b] ^3H-Glycerol-labeled diacylglycerols were formed from glycerol-labeled triolein by the action of lipoprotein lipase; ^{14}C-fatty-acid-labeled 1.2- and 2.3-diacyl-*sn*-glycerols were formed from labeled [^{14}C-oleic acid]triolein by the action of pancreatic lipase and were added as an internal standard.
[c] Refers to 1,2- and 2,3-diacyl-*sn*-glycerols.

into each position. Christie (1975) has described a procedure using an internal standard triacylglycerol of an odd-chain fatty acid. The fatty acid analyses are completed by radio GLC, and the radioactive fatty acid is expressed in terms of specific activity related to the internal standard. This approach has been used to assess the specificity of milk fat synthesis (Christie, 1974).

Similar procedures can be used to assess the stereospecific hydrolysis of triacylglycerols by lipoprotein lipase (Table III). In this case, radioactive [^3H-glycerol]triolein was subjected to hydrolysis by lipoprotein lipase. After isolation of the diacylglycerols, 1,2- and 2,3-diacyl-*sn*-glycerols generated by the action of pancreatic lipase from [^{14}C-oleyl]triolein were added. The amount of ^{14}C radioactivity recovered at each step, as a function of the theoretical amount added, is a direct indication of the recovery of the ^3H-glycerol products. Thus any loss of radioactivity through poor extraction or removal of material from silica gel can be detected as shown in the second last line of Table III, and corrections can be made, as shown in the last line, by adjusting the amount of lysophosphatidylphenols and residual phosphatidylphenols to 50 : 50 for ^{14}C.

4.4. Positional Distribution of Fatty Acids in Natural Triacylglycerols

The following section documents the positional composition of triacylglycerols from various biosynthetic origins. Tabulation of data has been selective in some cases where multiple analyses were done with essentially similar results. Particular attention has been given to studies involving preliminary fractionation of the triacylglycerol mixtures from specific tissues, since this gives a more accurate picture of the molecular structure of the triacylglycerols and the specificities involved in the biosynthetic mechanisms.

4.4.1. Animal Fats

Virtually all animal fats studied show some differences in the composition of fatty acids in positions 1 and 3. The differences are very marked in most tissues involved in synthesis and secretion of triacylglycerols or relatively small in some types of depot fat.

4.4.1.1. Milk

The triacylglycerols formed by the mammary gland result from extremely active biosynthetic mechanisms using fatty acids originating from the bloodstream or from *de novo* synthesis (Smith and Abraham, 1975). In most milk triacylglycerols studied to date, the distribution of fatty acids between positions 1 and 3 has been extremely different. Initial studies by Pitas *et al.* (1967) on total bovine milk fat and by Breckenridge and Kuksis (1968) on a molecular distillate of butteroil indicated that short-chain fatty acids were selectively associated with position 3 (Table IV). Inherent problems occurred with total milk fat mixtures since Jensen *et al.* (1964) had previously shown that triglycerides containing short-chain fatty acids were hydrolyzed more readily than those containing only long-chain fatty acids. However, in the molecular distillate used by Breckenridge and Kuksis, over 90% of the triacylglycerols were comprised of two long-chain fatty acids ($>C_{12}$) and one short-chain fatty acid ($<C_{12}$). After lipase hydrolysis, a mixture of diacylglycerols containing either two long-chain fatty acids or one long- and one short-chain fatty acid was obtained. After derivatization and digestion with phospholipase A_2, it was found that all the short-chain fatty acids were located in the residual phosphatidylphenol. By converting the residual phenols to acetates, it was demonstrated that only diacylglycerols containing one short-chain and one long-chain fatty acid remained. Further studies indicated that butyric acid was still preferentially associated with position 3 in rearranged butterfat, but this appeared to be due to a lack of complete equilibration of the fatty acids during the interesterification process (Kuksis *et al.*, 1973).

Barbano and Sherbon (1975) found that the stereochemical distribution of fatty acids in bovine milk fat differs for high-melting (long-chain) and low-melting (short-chain) triacylglycerols (Table IV). In low-melting fractions, specific placement of short-chain fatty acids at position 3 was confirmed, as was selective esterification of palmitate at position 1. However, in the high-melting triglycerides, palmitic acid was selectively esterified to position 2. It was suggested by these investigators that 2-monoglycerides originating from plasma triacylglycerol might be a substrate for the formation of high-melting milk fat triacylglycerols. In contrast, Marai *et al.* (1969) noted no extensive differences in the relative positioning of palmitic acid between positions 1 and 2 of 1,2-diacyl-*sn*-glycerols from short- and long-chain triacylglycerols of sheep and goat milk (Table V). In both types of triacylglycerols, palmitic acid was preferentially located in position 1. The short-chain fatty acids were also associated with position 3, as had been noted for bovine milk fat. The positional specificity of long-chain fatty acids was also less marked when labeled fatty acids

Table IV. Positional Distribution of Fatty Acids in Bovine Milk Fat and Butteroil

Source	Position[a]	Fatty acid[b] (mole %)												
		4:0	6:0	8:0	10:0	12:0	14:0	15:0	16:0	17:0	18:0	18:1	18:2	Others[f]
Milk fat														
Total[c]	1	5.0	3.0	0.9	2.5	3.1	10.5		35.9		14.5	20.6	1.2	
	2	2.9	4.8	2.3	6.1	6.0	20.4		32.8		6.4	13.7	2.5	
	3	43.3	10.8	2.2	3.6	3.5	7.1		10.1		4.0	14.9	0.5	
Low melting[d]	1					1.5	13.3		34.3		14.9	26.0	0.7	
	2		0.7	2.5	7.4	7.6	19.4		28.3		3.8	18.1	2.7	
	3	9.0	7.3	4.7	8.9	7.5	4.5		17.3		3.7	22.5	5.2	
High melting[d]	1						9.6		43.5		38.3	2.1		
	2						13.6		64.0		14.6	1.3		
	3						0.8		35.1		53.3	4.3		
Butteroil[e]														
Short chain	1				0.9	3.1	10.8	2.3	41.1	3.3	14.8	19.8	1.0	2.9
	2			0.9	4.3	6.5	22.8	4.4	37.4	1.3	3.5	11.8	1.2	5.9
	3	53.9	24.3	5.1	5.3	−0.3	−0.6	−1.3	4.9	−0.7	1.8	7.4	0.8	0.4

[a] In this and all subsequent tables, the positions 1, 2, and 3 refer to the positions 1, 2, and 3 of *sn*-glycerol.
[b] In this and all subsequent tables, fatty acids are identified by total number of carbon atoms and double bonds.
[c] Pitas *et al.* (1967).
[d] Barbano and Sherbon (1975). Various triacylglycerol fractions were obtained by recrystallization from cold acetone.
[e] Breckenridge and Kuksis (1968). The triacylglycerols contain two long-chain fatty acids (C_{12}–C_{18}) and one short-chain fatty acid (C_4–C_{10}) obtained by molecular distillation of butteroil.
[f] Comprises 14:1, 16:1, and 20:1 fatty acids. Distributions were parallel.

Table V. Positional Distribution of Fatty Acids in Triacylglycerols in Animal Milks

Source	Position	Fatty acid (mole %)														
		4:0	6:0	8:0	10:0	12:0	14:0	15:0	16:0	16:1	17:0	18:0	18:1	18:2	18:3	20:1
Goat milk[a]																
Short chain	1			2.3	5.0	[illegible].2	9.5	1.5	41.3	4.0	1.6	11.9	14.3	0.8		1.2
	2			2.4	11.1	[illegible].7	21.4	2.8	29.2	2.0	0.5	5.2	14.3	2.6		1.1
	3	30.6	24.6	9.4	10.3	[illegible].2	0.0	0.2	3.3	0.0		0.9	13.4	2.6		3.4
Long chain	1			1.2	2.0	[illegible].2	7.5	2.2	45.4	2.1	1.0	17.9	17.5			0.6
	2			0.3	3.6	[illegible].8	19.5	2.7	37.6	2.0	0.5	7.1	17.6	2.5		1.7
	3			1.2	18.6	[illegible].1	4.5	2.6	3.1	2.8	2.4	13.1	44.1	6.8		2.8
Sheep milk[a]																
Short chain	1			0.9	2.3	[illegible].3	11.4	2.2	38.4	2.1	1.4	14.8	16.4	2.9	3.9	
	2			2.7	8.2	[illegible].6	19.2	2.7	21.8	1.9	0.6	10.7	17.3	4.5	2.0	0.5
	3	24.0	21.6	7.8	12.0	[illegible].0	2.7	1.7	3.4	0.2	0.4	3.9	11.6	3.7	1.6	1.0
Long chain	1			0.5	0.7	[illegible].2	5.5	3.0	37.7	2.2	1.9	22.8	20.7	2.6	0.8	
	2			1.4	2.7	[illegible].1	16.3	5.3	25.5	2.4	1.2	14.3	21.8	3.9	1.5	
	3		1.5	2.0	8.6	[illegible].1	7.0	1.0	1.3	2.3	0.3	13.3	42.7	7.3	6.7	
Pig milk[b]	1						2.4		21.8	6.6		6.9	49.6	11.3	1.4	
	2						6.8		57.6	11.2		1.1	13.9	8.4	1.0	
	3						3.7		15.4	10.4		5.5	51.7	11.5	1.8	

[a] Marai *et al.* (1969). Short- and long-chain triacylglycerols were separated by TLC.
[b] Christie and Moore (1970*b*).

Table VI. Effect of Feeding Protected Unsaturated Fats on Structure of High-Molecular-Weight Milk Fat Triacylglycerols

		Fatty acid (mole %)					
Source	Position	<14:0	14:0	16:0	18:0	18:1	18:2
Goat[a]							
Control	1	7.7	5.7	31.0	9.7	18.8	0.7
	2	18.1	12.8	26.5	6.5	18.0	2.1
	3	26.6	6.1	5.7	9.2	27.8	4.9
Supplement[b]	1	19.7	16.3	17.4	6.5	15.9	18.9
	2	17.8	6.8	24.4	13.2	23.0	15.9
	3	25.9	4.2	10.1	6.1	21.7	13.5
Bovine[c]							
Control	1	8	9	29	12	26	5
	2	6	10	30	13	28	4
	3	41	4	2	12	23	5
Supplement[b]	1	16	5	20	19	22	11
	2	12	10	22	12	20	15
	3	26	4	3	10	22	25

[a] Mills *et al.* (1976).
[b] Animals were fed protected seed oils rich in linoleic acid. Protected seed oils have been coated with formylated casein to prevent hydrogenation of unsaturated fatty acids by the rumen bacteria.
[c] Approximate results recalculated from Mills *et al.* (1976).

were incorporated into goat milk triacylglycerols by the esterifying enzymes present in fresh milk (Christie, 1974).

Recently it has been observed that the composition of ruminant milk fats was altered by the feeding of protected seed oils* (Mills *et al.*, 1976). The unsaturated fatty acids were absorbed unaltered, with the result that linoleic acid levels reached 20% in the milk triacylglycerols. In bovine milk fat (Table VI), the linoleic acid was preferentially esterified to position 3. By contrast, in goats fed the protected oils, the linoleic acid was present in similar quantities in all positions of the milk triacylglycerols, indicating clear species differences in the esterification mechanisms.

In pig milk triacylglycerols (Christie and Moore, 1970*b*) varying in unsaturation (Tables VII and VIII), palmitic acid was selectively esterified to position 2. Although stearic, oleic, and linoleic acids were enriched in the primary positions, there appeared to be no selectivity between positions 1 and 3.

The data from milk fat triacylglycerols indicate that the mechanism of biosynthesis consists of complex enzyme specificities which direct the selective esterification of fatty acids. Regardless of the pathways involved, it is clear that the synthesis of di- and triacylglycerols is highly specific, placing short-chain fatty acids largely at position 3 in the final esterification step. This may assure a supply of long-chain diacylglycerols for phospholipid synthesis in the cell.

* Protected seed oils or diets have been coated with formylated casein, which is resistant to the rumen enzymes. As a result, the unsaturated fatty acids are not hydrogenated by the rumen microflora. In the intestine, the formylated casein is broken down and the unsaturated fatty acids are released for absorption and incorporation into the animal fat.

Table VII. Positional Distribution of Fatty Acids in Dienoic Triacylglycerols from Pig Tissues[a]

		Fatty acid (mole %)				
Source	Position	14:0	16:0	16:1	18:0	18:1
Inner back fat	1	0.8	3.7	3.1	13.6	78.8
	2	3.8	74.6	4.5	3.0	14.1
	3	−0.1		2.3	1.1	96.7
Liver	1	1.4	50.1	3.5	9.6	34.3
	2	0.8	16.2	6.2	1.8	74.7
	3	0.6	12.6	1.1	8.7	75.2
Blood	1	2.4	28.8	3.3	12.6	52.9
	2	2.5	44.2	5.7	3.5	44.1
	3	−0.4	7.7	3.0	3.4	86.3
Milk	1	2.5	10.5	7.4	4.8	74.8
	2	7.2	67.9	11.0	0.6	13.3
	3	0.5	6.8	16.7	2.7	73.3

[a] Data from Christie and Moore (1970*b*).

Table VIII. Positional Distribution of Fatty Acids in Trienoic Triacylglycerols from Pig Tissues[a]

		Fatty acid (mole %)					
Source	Position	14:0	16:0	16:1	18:0	18:1	18:2
Inner back fat	1		5.5	2.1	15.9	49.3	27.2
	2	4.3	67.7	3.6	3.4	10.8	10.2
	3	0.2	0.3	1.2	2.9	34.7	60.7
Liver	1	1.9	49.0	3.1	14.0	23.2	8.1
	2	0.4	9.4	3.3	1.5	39.3	45.8
	3	0.1	13.0	1.4	4.6	36.2	43.6
Blood	1	2.0	33.4	6.3	14.0	32.7	11.6
	2	2.8	33.0	5.1	2.8	31.1	25.2
	3		2.0	2.7	0.3	38.8	56.2
Milk	1	1.5	11.0	6.5	3.8	42.1	35.1
	2	6.2	57.3	10.5	1.3	8.8	15.9
	3	0.4	6.1	9.4	3.0	39.1	42.0

[a] Data from Christie and Moore (1970*b*).

4.4.1.2. Liver

Liver triacylglycerols are produced by an extremely active biosynthetic system responsible for the formation of 1,2-diacyl-*sn*-glycerol precursors for both triacylglycerols and diacylglycerol phospholipids. In liver triacylglycerols from several animals (Table IX), position 1 contains predominantly palmitic, stearic, and oleic acids, while position 2 contains oleic and linoleic acids. Oleic acid and any long-

Table IX. Positional Distribution of Fatty Acids in Liver Triacylglycerols

Source	Position	Fatty acid (mole %)													
		14:0	16:0	16:1	17:1	18:0	18:1	18:2	18:3 and 20:1	20:3	20:4	20:5	22:5	22:6	Other
Rat[a]	1	1.0	63.0			5.1	18.3	13.5							
	2	1.2	8.7	3.9		0.9	39.0	42.0							
	3	1.5	10.5	5.7		0.3	33.0	19.8							
Rat[b]	1	1.8	57.2	4.2	0.3	8.5	21.9	4.4							1.5
	2	1.0	12.7	2.9	0.5	3.5	58.3	19.2		0.5	0.9				1.1
	3	0.4	12.1	3.9	0.6	7.8	60.9	11.3		1.1	1.2				1.8
Rat[c]	1		73.1	3.8		5.0	12.7	4.0							
	2		5.5	3.2		0.7	46.0	41.6	1.3		1.0	0.1	0.1	0.6	
	3		14.0	3.3		3.6	47.7	23.6	1.4		2.0	0.8	1.0	2.7	
Pig[d]	1		41.3	1.6		17.1	23.8	9.8	1.3		2.5	0.5			
	2		18.3	4.1		1.5	40.9	25.0	0.9		3.6	0.6	1.9	1.2	
	3		13.0	1.8		9.0	45.1	12.6	0.6		7.7	2.2	3.2	1.8	
Sheep[e]	1	2.9	46.8	2.1	0.5	16.5	21.0	1.0	0.4	0.2	0.3	0.7			7.2
	2	2.4	13.7	2.6	1.1	17.6	55.2	8.2	1.3	1.5	0.2	1.6			4.2
	3	1.6	15.7	2.5	0.8	12.5	54.0	4.9	0.7	1.0	1.3	2.5			2.4
Rabbit[f]	1	2.3	32.8	2.2		8.1	24.7	29.9							
	2	1.7	14.2	1.7		1.0	28.1	53.3							
	3	0.5	16.9			3.8	35.4	43.4							

[a] Data from Slakey and Lands (1968).
[b] Data from Wood and Harlow (1969).
[c] Data from Akesson (1969).
[d] Data from Hunter *et al.* (1973).
[e] Data from Christie and Moore (1971).
[f] Data from Christie *et al.* (1974). Animals were on a diet containing safflower oil.

chain fatty acids are enriched in position 3. Considerable variation has been noted for rat liver triacylglycerols depending on the method of stereospecific analysis. Using the diglyceride kinase method, Slakey and Lands (1968) found much higher levels of linoleic acid in position 1 relative to positions 2 and 3 than Wood and Harlow (1969) and Akesson (1969), who used the Grignard reagent and phospholipase A_2 digestion. By contrast, the latter two groups of investigators found considerably more stearic acid in position 3 than Slakey and Lands (1968) did. These differences point out the extreme care that must be exercised to assure that enzyme selectivities do not occur during the hydrolytic steps. Liver triacylglycerols from pig, sheep, and rabbit contain relatively more palmitic acid in positions 2 and 3 than those from rat. In contrast to most species where stearic acid is enriched at position 1, sheep liver triacylglycerols contain large quantities of stearic acid in all three positions.

The molecular structure of the triacylglycerol is still poorly defined when no preliminary fractionation is completed prior to the stereospecific analyses. The necessary fractionations, however, have been completed for rat (Akesson, 1969) and pig (Christie and Moore, 1970*b*) liver triacylglycerols. The results indicate that the relative positional specificity noted for the total triacylglycerols is generally maintained for the various classes of unsaturation. In rat liver (Table X), the distribution of palmitic acid among the three positions was very similar to the total for the triacylglycerol groups of the classification: saturated, monoenoic, monoenoic; saturated, monoenoic, dienoic; and saturated, dienoic, dienoic. A similar trend was noted in pig liver (Tables VII and VIII). When two residues of palmitic acid occurred per molecule, these residues were found essentially in positions 1 and 3. Thus, in rat liver, the major components are 1-palmitoyl-2-,3-dioleoyl-*sn*-glycerol, 1 palmitoyl-2-oleoyl-3-linoleoyl-*sn*-glycerol, and 1-palmitoyl-2,3-dilinoleoyl-*sn*-glycerol. In each case, the enantiomer accounts for approximately 20% of the above structures.

Table X. Positional Distribution of Fatty Acids in Major Classes of Rat Liver Triacylglycerols[a]

		Fatty acid (mole %)				
Class[b]	Position	16:0	16:1	18:0	18:1	18:2
SSM	1	88.9		5.3	5.8	
	2	20.9	6.5	3.5	69.1	
	3	73.9	0.5	8.1	17.5	
SMM	1	80.5	1.3	5.0	13.2	
	2	3.8	2.7	0.8	92.7	
	3	16.8	7.7	2.0	73.5	
SMD	1	79.9	2.4	2.8	9.7	5.2
	2	3.3	1.3	0.4	37.7	57.3
	3	10.7	1.9	0.6	49.3	37.5
SDD	1	72.4		2.6	2.7	22.3
	2	4.0		0.5	3.5	92.0
	3	12.2		0.7	0.1	87.1

[a] From the data of Akesson (1969).
[b] S, M, and D represent saturated, monoenoic, and dienoic fatty acids.

Table XI. Stereospecific Analyses of Triacylglycerols Synthesized during Fat Absorption

Source	Type of feeding	Position	Fatty acid (mole %)											
			10:0	12:0	14:0	14:1	15:0	16:0	16:1	18:0	18:1	18:2	18:3	20:4
Rat mucosa[a]	Butteroil fat	1		3.3	9.0	1.3	1.5	44.0	1.5	15.6	22.3	1.5		
		2	1.4	4.7	26.9	3.1	2.3	35.9	2.4	3.5	16.3	3.1		0.3
		3	2.8	7.9	16.0	1.3	1.0	36.2	3.0	4.3	22.9	3.8		0.6
	Free fatty acids	1		1.5	8.8	1.3	4.6	39.7	6.4	13.8	18.5	3.6		0.9
		2		1.2	10.6	1.7	1.6	36.1	3.6	3.9	28.1	10.6		2.1
		3		3.9	12.4	2.4	0.5	22.6	5.3	6.3	25.7	11.9		6.0
Sheep[b] thoracic lymph	Control diet	1			3			14	4	44	23	4	4	
		2			3			45	4	17	18	8	3	
		3			2			10	6	43	20	10	4	
	Supplemented diet	1			1			8	2	48	14	25	2	
		2			1			26	3	18	12	24	2	
		3			2			8	5	28	20	36	2	

[a] Data from Breckenridge *et al.* (1976). The butteroil fraction was obtained from a molecular distillate of butteroil and contained two long-chain fatty acids and one short-chain fatty acid. It was fed by stomach tube as such or after hydrolysis to the free fatty acids.

[b] Data from Mills *et al.* (1976). The control diet contained lucerne and oats. The supplement was 200 g per day formaldehyde-treated safflower oil casein (2 : 1 w/w).

4.4.1.3. Intestine

The intestine represents one of the most complex systems with regard to triacylglycerol synthesis since the phosphatidic acid and 2-monoacylglycerol pathways contribute to lymph triacylglycerols. It has also been suggested that 1-monoacylglycerols may be acylated to triacylglycerols. Even with this level of complexity, a limited number of analyses of mucosal or lymph triacylglycerols have indicated that some selectivity occurs during the esterification process. After feeding butteroil, it was noted (Table XI) that the position 2 of the mucosal triacylglycerol was essentially similar to position 2 of the absorbed fat (Breckenridge *et al.*, 1976). However, stearic acid was present in higher concentrations in position 1 as opposed to position 3, while lauric and myristic acids were more readily esterified to position 3. When an identical free fatty acid mixture was fed to the rat, forcing the reesterification to proceed via the phosphatidic acid pathway, again stearic acid was selectively associated with position 1 while myristic, lauric, linoleic, and arachidonic acids were preferentially located in position 3.

Studies by Mills *et al.* (1976) on lymph triacylglycerols from sheep fed control or protected diets, rich in linoleic acid (Table XI), have also shown that palmitic and oleic acids are enriched in position 1 compared to position 3, while the reverse is true for linoleic acid. With protected diets, linoleate increased in all positions, but the proportional increase was greatest at position 3, while there was a marked increase of stearic acid in position 1. Thus 2-monoacylglycerols are reesterified in the intestine in a manner that indicates that positions 1 and 3 are not equivalent during the reacylation process.

4.4.1.4. Depot Fat

The depot fat is a major storage site in the body. Both the phosphatidic acid and monoacylglycerol pathways have been claimed (Polheim *et al.*, 1973) to be present in the adipocyte. The latter, combined with the lipase systems in the tissue, could affect random interesterification at positions 1 and 3. However, the asymmetrical distributions noted below do not support this concept.

In examples of the canine, rodent, and ruminant families (Table XII), there is a general distribution of palmitic and oleic acids in all three positions, while stearic acid is preferentially esterified to position 1, except for beef depot fat where it is located in position 3. There is a selective association of linoleic acid to position 2 in the cat and horse depot fat and position 3 for rabbit and beef depot fat.

In pig outer back fat (Table XIII), palmitic acid is preferentially esterified with position 2, while stearic and linoleic acids are primarily associated with positions 1 and 3, respectively, and oleic acid is present in large quantities in both positions 2 and 3. Further studies indicated (Christie and Moore, 1970*a*) that changes in the distribution of fatty acids in triaclyglycerols of pig depot fat occurred with changes in the total fatty acid composition. In position 1, palmitic acid was constant but stearic acid decreased, and the concentration of unsaturates increased as the amount of oleic acid in the total triacylglycerols increased. In position 2, stearic acid was constant while palmitic acid decreased when unsaturated acids increased. Very little change occurred in the concentration of saturated acids in position 3 when unsaturated acids increased in the total triacylglycerols. In contrast to pig depot fat, that of sheep

Table XII. Positional Distribution of Fatty Acids in Depot Fat Triacylglycerols of Various Mammals[a]

Mammal	Position	Fatty acid (mole %)						
		14:0	16:0	16:1	18:0	18:1	18:2	18:3
Cat	1	3	27	4	20	36	5	1
	2	5	20	5	5	49	11	<1
	3	2	20	3	14	46	8	1
Dog	1	2	26	4	22	36	5	<1
	2	9	27	7	4	35	12	<1
	3	1	16	4	14	57	7	<1
Rat	1	2	32	5	9	32	15	1
	2	1	10	4	1	37	45	1
	3	2	27	5	7	37	17	1
Rabbit	1	3	34	9	6	25	14	2
	2	6	25	12	1	26	23	5
	3	1	24	7	3	35	22	5
Beef	1	4	41	6	17	20	4	1
	2	9	17	6	9	41	5	1
	3	1	22	6	24	37	5	1
Horse	1	3	39	7	6	27	5	11
	2	7	9	10	1	29	17	25
	3	3	30	6	7	37	5	11

[a] From Brockerhoff *et al.* (1966).

contained low amounts of palmitic acid at position 2 but very large quantities of stearic acid at positions 1 and 3 (Christie and Moore, 1971).

Nutritional studies with rabbits have shown (Christie *et al.*, 1974) that the relative positional distribution of various fatty acids in triacylglycerols of adipose tissue remained constant even though the total concentration of a particular fatty acid in the depot fat could be altered considerably by diet, suggesting a common effect on all molecular species.

The depot fats of several marine oils from an extensive series of analyses (Brockerhoff *et al.*, 1968) are shown in Table XIV. In fish, palmitic acid and oleic acid are located in positions 1 and 2. Long-chain monoenoic fatty acids are preferentially esterified in positions 1 and 3, while long-chain polyunsaturated fatty acids are in position 2. In invertebrates, polyenoic fatty acids are enriched in position 2, saturated acids in position 1, and longer-chain acids in position 3. On the basis of his results, Litchfield (1968) has claimed that the distribution of 22:6 and 22:5 can be predicted by proportionality equations and the total fatty acid composition. In sea mammals, shorter fatty acids accumulate in position 2, while polyenoic fatty acids occupy position 3.

4.4.1.5. *Other Organs and Tissues*

Christie and Moore have carried out extensive comparisons of triacylglycerol structures in various tissues from pig (1970*b*) and sheep (1971). In pig kidney, heart,

Table XIII. Positional Distribution of Fatty Acids in Triacylglycerols from Various Pig Tissues[a]

		Fatty acid (mole %)										
Tissue	Position	14:0	16:0	16:1	18:0	18:1	18:2	18:3	20:1	20:2	20:3	20:4
Kidney	1	1.4	14.2	1.6	38.3	37.9	4.7		1.9			
	2	3.8	71.4	3.7	3.8	13.4	3.9					
	3	0.2	6.5	2.8	12.2	59.7	14.2	1.5	1.1	0.3	0.9	0.6
Outer back fat	1	0.9	9.5	2.4	29.5	51.3	6.4					
	2	4.1	72.3	4.8	2.1	13.4	3.3					
	3	−0.2	0.4	1.5	7.4	72.7	18.2					
Heart	1	1.5	12.7	3.1	26.9	50.3	5.5					
	2	3.1	64.8	4.6	5.0	19.0	3.5					
	3	−1.3	−0.1	0.4	15.8	73.8	11.4					
Adrenals	1	1.4	11.9	1.9	40.2	40.3	4.3					
	2	4.0	70.5	4.4	4.5	13.4	3.2					
	3	0.3	8.2	1.2	13.5	63.6	13.8					

[a] From Christie and Moore (1970*b*).

Table XIV. Positional Distribution of Fatty Acids in Triacylglycerols of Marine Oils[a]

		Fatty acid (mole %)											
Animal	Position	14:0	16:0	16:1	18:0	18:1	18:2	20:1	22:1	20:5	22:5	22:6	Other
Herring	1	6	12	13	1	16	3	25	14	3	1	1	
	2	10	17	10	1	10	3	6	5	18	3	13	
	3	4	7	5	1	8	1	20	50	4	1	1	
Mackerel	1	6	15	11	3	21	2	8	18	5	1	2	
	2	10	21	6	1	9	1	5	5	12	3	20	
	3	2	5	4	2	21	2	19	24	10	1	5	
Skate	1	2	19	12	5	30	1	12	8	4	1	5	
	2	3	15	7	1	9	1	8	5	6	7	37	
	3	1	6	6	1	28	2	19	11	11	2	11	
Cod	1	6	15	14	6	28	2	12	6	2	1	1	
	2	8	16	12	1	9	2	7	5	12	3	20	
	3	4	7	14	1	23	2	17	7	13	1	6	
Squid	1	4	28	8	4	21	4	6	3	12	1	4	
	2	2	2	4	1	7	1	7	5	28	2	38	
	3	4	12	9	2	26	1	17	13	5	1	10	
Periwinkle	1	2	13	6	4	27	5	18	9	6	1	4	
	2	9	17	9	1	12	5	13	3	16	2	7	
	3	3	4	10	1	21	5	16	9	8	2	9	
Lobster	1	3	13	10	3	22	2	11	7	8	1	3	
	2	4	12	7	1	17	2	10	2	13	1	15	
	3	4	13	10	3	25	2	12	8	8	1	5	
Harbor seal	1	4	11	15	1	29	1	18	8	3	2	3	
	2	11	13	30	1	30	3	3	1	1	1	1	
	3	1	4	14	1	26	1	16	7	8	6	10	
Sea whale	1	3	13	3	4	14	1	33	10	3	1	6	6
	2	12	6	12	1	29	5	10	2	5	1	3	10
	3	4	6	2	2	7	1	28	16	6	3	16	4

[a] From Brockerhoff *et al.* (1968).

and adrenals (Table XIII), over 70% of the palmitic acid was found at position 2, with much of the rest at position 1. In contrast, stearic acid was located principally at position 1, with most of the remainder at position 3. Linoleic acid was located at position 3, and large quantities of oleic acid were associated with position 3, followed by positions 1 and 2, respectively. In most sheep tissues (Christie and Moore, 1971), palmitic was selectively esterified to position 1, while stearic was equally distributed between positions 1 and 3 and oleic was selectively esterified at positions 2 and 3. Clearly, intraspecies differences occur.

Egg yolk triacylglycerols are synthesized with a very high degree of specificity (Table XV) since 90% of palmitic acid is associated with position 1 in dienoic and trienoic triacylglycerols. However, the distribution of oleic acid varied with the type of molecular species. In monoenoic species it was primarily in position 2, but in trienoic species it was primarily in position 3. Thus common diacylglycerols were not used for acylation at position 3. The major molecular species are 1-palmityl 2,3-

Table XV. Positional Distribution of Fatty Acids in Selected Fractions of Egg Yolk Triacylglycerols[a]

Fraction[b]	Position	Fatty acid (mole %)					
		14:0	16:0	16:1	18:0	18:1	18:2
SSM	1	1.4	86.6	0.8	8.7	2.5	
	2	0.5	14.4	3.3	12.4	69.4	
	3	1.1	31.9	1.6	41.9	23.5	
SMM	1	0.5	83.4	1.7	2.9	11.5	
	2		0.9	3.4	0.7	95.0	
	3	0.1	7.5	7.8	4.8	79.8	
SMD	1	0.6	82.7	5.3	3.4	6.7	1.3
	2		0.2	0.8		14.9	84.1
	3	0.3	7.7	4.1	3.2	72.9	11.8

[a] From the data of Christie and Moore (1970*c*).
[b] Fractions were resolved from the total triacylglycerol by argentation TLC. S, M, and D represent saturated, monoenoic, and dienoic fatty acids, respectively.

dioleoyl-*sn*-glycerol and the enantiomer and 1-palmityl-2-linoleyl-3-oleoyl-*sn*-glycerol (Gornall and Kuksis, 1971).

4.4.2. Human Tissues

The triacylglycerols from human heart, adipose tissue, and aortic lesions all showed a similar pattern of fatty acid distribution (Table XVI). There were marked enrichments of palmitic and linoleic acids in positions 1 and 2, respectively, while oleic acid was enriched in positions 2 and 3. Human liver triacylglycerols had a fatty acid distribution similar to triacylglycerols from other mammalian livers (Table IX). However, milk fat triacylglycerols had a distribution distinct from that of other milk fats, with shorter-chain fatty acids selectively esterified to position 3, myristic and palmitic acid enriched in position 2, stearic acid in position 1, and oleic and linoleic acids in both positions 1 and 3.

Human plasma triacylglycerols have been investigated in view of the atherogenicity of plasma lipoproteins and defective lipoprotein clearance in some forms of hyperlipoproteinemia. In plasma triacylglycerols of normal subjects, palmitic and stearic acids are preferentially associated with position 1, while oleic and linoleic are each distributed about equally between positions 2 and 3 (Table XVII). In hypertriglyceridemic subjects (Type IV), there was a significant decrease in the amount of oleic and linoleic at position 3 (Parijs *et al.*, 1976). It has not been determined whether there was a selective loss of certain molecular species or a general decrease in the fatty acids of these positions in all triacylglycerols. However, the data suggest that the type of triacylglycerol synthesized or catabolized in hypertriglyceridemic subjects may be different from that of normal subjects.

Other studies by Gordon *et al.* (1975) have indicated that the positional distributions of fatty acids in triacylglycerols of very-low-density lipoprotein (VLDL) show greater amounts of palmitic and linoleic acids in positions 2 and 1, respectively, when compared to low-density lipoprotein (LDL). Differences in the positional

Table XVI. Positional Distribution of Fatty Acids of Triacylglycerols from Various Human Tissues

Source	Position	Fatty acid (mole %)													
		10:0	12:0	14:0	14:1	15:0	16:0	16:1	17:0	18:0	18:1	18:2	18:3 and 20:0	20:1	20:4
Milk fat[a]	1	0.2	1.3	3.2	0.3	0.4	16.1	3.6	0.9	15.0	46.1	11.0	0.4	1.5	
	2	0.2	2.1	7.3	0.2	1.0	58.2	4.7	0.8	3.3	12.7	7.3	0.6	0.7	0.9
	3	1.8	6.1	7.1	1.4	0.6	6.2	7.3	0.6	2.1	49.7	14.7	1.6	0.5	0.3
Heart muscle[b]	1			4.4	0.2	0.6	48.3	4.4		11.0	24.4	3.8	1.5	1.0	
	2			6.0	1.4	0.3	10.1	14.5	0.3	1.6	54.6	10.1	0.7		
	3			3.4	0.2		20.8	6.9	0.9	5.7	57.8	2.6	0.1	1.7	
Adipose[b]	1			3.9	0.1	0.7	42.3	3.4	1.0	14.7	27.2	3.8	1.9	1.0	
	2			6.0	1.3	0.4	10.4	12.0	0.2	2.0	54.7	11.3	1.1	0.2	
	3			3.9	0.7	0.4	19.0	6.2	0.3	5.8	56.7	3.8		2.1	
Aortic lesions[b]	1			2.7			53.8	3.7		10.9	24.4	3.6	0.3	0.6	
	2			3.4			15.0	7.7		2.1	51.2	15.7	1.2	0.7	1.7
	3			3.2			11.0	4.8		5.9	57.9	9.8	1.2	1.7	1.9
Liver[b]	1			1.1		0.2	60.8	3.2	1.2	11.5	18.1	2.7	0.1	0.6	0.5
	2			1.4		0.2	9.1	3.3	0.6	2.7	64.0	14.5	1.0	0.2	2.4
	3			0.8		0.2	14.7	1.0		8.0	69.4	4.1	0.1	0.7	1.0

[a] Data from Breckenridge *et al.* (1969).
[b] Christie *et al.* (1971).

Table XVII. Positional Distribution of Fatty Acids in Triacylglycerols of Human Plasma and Lipoprotein Fractions from Normal and Hyperlipoproteinemic Subjects

Source	Position	Fatty acid (mole %)												
		14:0	15:0	16:0	16:1	17:0	18:0	18:1	18:2	18:3	20:1	20:2	20:4	22:0
Plasma[a]														
Normal	1	2.1	0.7	55.0	7.0	1.0	6.3	19.0	6.2	0.8	0.4	0.2	0.8	0.2
	2	1.3	0.2	12.6	6.7	0.3	2.9	46.0	25.6	1.1	0.3	0.1	1.9	0.2
	3	2.8	0.6	13.6	4.9	0.4	3.0	51.7	18.2	2.5	0.4	0.3	0.8	0.3
Type IV HLP	1	2.5	0.6	49.9	6.4	1.1	8.9	21.2	5.0	1.2	0.7	0.3	1.1	0.3
	2	2.2	0.2	19.8	8.3	0.3	3.6	42.9	16.9	1.7	0.4	0.3	1.8	0.4
	3	4.2	0.1	19.7	6.6	0.3	5.0	43.7	14.3	2.7	0.6	0.2	1.0	0.5
Lipoproteins[b]														
Normal VLDL	1	0.3		66.5	0.9		1.5	26.6	3.5				0.7	
	2	0.5		7.7	3.9		0.3	60.6	25.5	0.1			1.4	
	3	0.1		25.1	0.9		1.5	58.9	10.9					
Normal LDL	1	1.2	0.9	61.3	2.9	1.3	6.3	13.0	13.1					
	2	1.0	0.3	16.3	4.6	0.8		45.9	30.3	0.5			0.3	
	3	−0.4	−0.6	18.1	−2.1	−1.2	−0.5	72.5	13.9					
Type IIa HLP[b] VLDL	1	3.9	0.4	75.7	2.9		6.9	7.8	1.4	0.3			0.7	
	2	1.5	1.9	10.6	3.2			55.0	25.9				1.9	
	3	−2.7	0.4	−8.3	5.9	0.9	0.9	76.4	21.0	0.6			0.4	
Type IIa HLP[b] LDL	1	1.7		64.4	6.0		5.4	9.5	13.0					
	2	0.7	0.2	11.4	8.4	0.2	0.1	44.7	33.7	0.6				
	3	0.9	0.4	4.6	1.2	0.4	4.4	84.7	3.4					

[a] Data from Parijs *et al.* (1976). Normal refers to subjects possessing normal lipid values and no clinical signs of disease; type IV HLP refers to patients possessing the clinical and plasma lipid characteristics of type IV hyperlipoproteinemia.

[b] Data from Gordon *et al.* (1975). VLDL and LDL refer to very-low-density ($d \leqslant 1.006$) and low density ($d = 1.006–1.063$) lipoproteins; type IIa subjects possessed clinical and plasma lipid characteristics of familial type II hyperlipoproteinemia.

distributions of fatty acids were also noted between lipoprotein fractions from normal subjects and from one subject with elevated LDL (Table XVII). Since LDL is believed to be a catabolic product of VLDL (Eisenberg and Levy, 1976), the differences may be due to selective hydrolysis of lipoprotein triacylglycerols by lipoprotein lipase, which is specific for position 1 (Morley and Kuksis, 1972) in the initial hydrolytic step.

4.4.3. Interorgan and Interspecies Comparisons

It is apparent that triacylglycerol structures from various tissues can be grouped into three classes: A, those originating from storage and depot fat, B, those originating from secretory tissues (liver, intestine, milk fat), and C, those in a transport form (plasma). The positional distributions of fatty acids in group A are generally consistent for many tissues from the same animal, but many vary between animals, such as depot fat from pig (Christie and Moore, 1970*b*) and sheep (Christie and Moore, 1971). On the other hand, in group B, liver triacylglycerols tend to be consistent among several species. The positional distribution in milk triacylglycerols varies depending on the composition of the milk fat, while intestinal triacylglycerols are influenced by the type of lipid being absorbed. Group C fatty acids are represented only in plasma and are under continuous metabolic alteration during transport.

The reasons for the differences in selective distribution between the three glycerol hydroxyl groups in different tissues are not immediately obvious. It is possible that group B may represent the true specificities of the synthetic process, whereas the storage form and transport forms may be the product of synthesis from different pools. It is also possible that small but very active metabolic pools are masked in total tissue analysis. The analyses of Christie and Moore (1970*b*) on pig tissue triacylglycerols constitute a good example (Tables VII and VIII) of the different tissue specificities in the same animal. Palmitic acid was primarily in position 1 for liver, but in position 2 for milk and depot fat, while the positional distribution for blood triacylglycerols was intermediate between the two groups. An analysis of the major classes of unsaturated triacylglycerols in these tissues showed that the distribution of palmitic acid noted in the total triacylglycerols was maintained in back fat and milk for dienoic species containing one saturated and two monoenoic fatty acids, or trienoic species containing one saturated, one monoenoic, and one dienoic fatty acid. It was not maintained for blood and liver. Similarly, the total positional specificity was not maintained in molecular species of egg yolk (Table XV).

The biochemical mechanisms governing the synthesis of triacylglycerols are not yet clear, but it has been suggested that the triacylglycerol structure of rat liver may conform to a 1-random 2-random 3-random hypothesis (Slakey and Lands, 1968; Akesson, 1969). However, these calculations combine the fatty acids into groups of saturates, monoenes, dienes, etc. Any generalization of the composition will tend to favor an approximation of the observations to random theories. The noncorrelative specificity or 1-random 2-random 3-random hypothesis is, in fact, extremely specific. It suggests that enzymes have selectivity for the fatty acid substrate but not for the lysophosphatidic acid or diacylglycerol substrate. This type of specificity is difficult to envisage since the physicochemical properties of the fatty acid are still apparent whether it is esterified to a glycerol backbone or an acyl CoA. It would seem more plausible that for phospholipid and triacylglycerol synthesis the selective acylation process is relative to the integrated requirements of the tissue for diacylglycerols. In this regard, Hill *et al.* (1968) have suggested that the biosynthesis of liver triacylglycerols involves the synthesis of specific 1,2-diacyl-*sn*-glycerols followed by a random acylation of fatty acids at position 3.

4.4.4. Plant Oils

The triacylglycerols of plant seed oils which are rich in oleic, linoleic, and linolenic acids have a relatively low degree of asymmetry on the basis of stereochemical analysis of the total triglyceride mixture. In the oils of maize, soybean, linseed, olive, and wheat (Table XVIII), position 2 is almost exclusively occupied by unsaturated fatty acids, while saturated as well as unsaturated fatty acids occur in approximately similar quantities in positions 1 and 3. In cocoa butter, oleic acid is preferentially located in position 2. The distribution of oleic acid in the primary position is asymmetrical, with approximately one and one-half times as much in position 1 compared to position 3.

However, isolated triacylglycerol classes must be analyzed before it can be assumed that the structures are racemic. In such studies of monounsaturated triacylglycerols of cocoa butter (Sampugna and Jensen, 1968), it was found that most structures were racemic. In contrast to the unsaturated plant oils, a saturated fat, coconut oil, shows a high degree of asymmetry (Table XVIII). While dodecanoic acid was preferentially located in position 2, with lower but similar amounts in positions 1 and 3, octanoic acid was preferentially esterified to position 3, while myristate and palmitate were preferentially located in position 1.

Plant oils containing erucic acid also exhibited considerable selectivity in the placement of fatty acids (Table XIX). In rapeseed, much greater quantities of erucic acid were found in position 3 as opposed to position 1. Although other unsaturated fatty acids were preferentially esterified to position 2, the quantities of oleic and linoleic acids were greater in position 1 than in 3. In peanut oil, arachidic and behenic acids were preferentially esterified to position 3, while twice as much linoleic was found in position 1 compared to position 3. This specificity was not due to the analytical methodology since interesterification of the fat resulted in a random placement of these acids in all positions. Ohlson *et al.* (1975) have demonstrated that the fatty acids of various Cruciferae seed oils are nonrandomly distributed among all three positions. In general, oleic and linoleic were the major fatty acids in position 2, while saturated fatty acids were preferentially located in position 1, with erucic acid in positions 1 and 3. With increasing unsaturation of the oil, there was a preferential esterification of oleic and linoleic acids for position 1 as opposed to position 3.

Several vegetable oils contain atypical fatty acids. In all cases, these components have been found entirely in position 3 or enriched in this position relative to the other positions. The allenic estolide in *Sapium sebiferum* (Christie, 1973) was found entirely in position 3 (Table XX). Twice as much S-coriolic acid (13-L-hydroxy-*cis*-9,*trans*-11-octadecadienoic) of *Monnino emarginata* was found (Phillips and Smith, 1972) in position 3 compared to position 1.

Table XVIII. Positional Distribution of Fatty Acids in Triacylglycerols of Seed Oils

Source	Position	Fatty acid (mole %)											
		6:0	8:0	10:0	12:0	14:0	16:0	16:1	18:0	18:1	18:2	18:3	20:0
Coconut[a]	1	1.1	4.2	4.2	38.5	29.2	15.6		3.4	3.7			
	2	0.3	2.1	4.5	77.8	8.1	1.4		0.5	2.9	1.5		
	3	2.6	32.2	13.3	37.8	8.0	0.7		0.5	3.4	1.4		
Cocoa butter[b]	1						34.0	0.6	50.4	12.3	1.3		1.0
	2						1.7	0.2	2.1	87.4	8.6		
	3						36.5	0.3	52.8	8.6	0.4		2.3
Linseed[b]	1						10.1	0.2	5.6	15.3	15.6	53.2	
	2						1.6	0.1	0.7	16.3	21.3	59.8	
	3						6.0	0.3	4.0	17.0	13.2	59.4	
Soybean[b]	1						13.8		5.9	22.9	48.4	9.1	
	2						0.9		0.3	21.5	69.7	7.1	
	3						13.1		5.6	28.0	45.2	8.4	
Maize[b]	1						17.9	0.3	3.2	27.5	49.8	1.2	
	2						2.3	0.1	0.2	26.5	70.3	0.7	
	3						13.5	0.1	2.8	30.6	51.6	1.0	
Olive[b]	1						13.1	0.9	2.6	71.8	9.8	0.6	
	2						1.4	0.7		82.9	14.0	0.8	
	3						16.9	0.8	4.2	73.9	5.1	1.3	
Wheat flour[c]	1						23.9	1.0	2.3	14.2	53.9	4.7	
	2						3.6	0.3	0.4	15.0	75.7	5.0	
	3						20.6	1.0	2.0	12.2	59.7	4.7	

[a] Yeung and Kuksis (unpublished).
[b] Brockerhoff and Yurkowski (1966).
[c] Arunga and Morrison (1971).

Table XIX. *Positional Distribution of Fatty Acids in Plant Triacylglycerols Containing Long-Chain Fatty Acids*

Source	Position	Fatty acid (mole %)										
		16:0	16:1	18:0	18:1	18:2	18:3	20:0	20:1	22:0	22:1	24:0
Peanut[a]	1	13.6	0.3	4.6	59.2	18.5		0.7	1.1	1.3		0.7
	2	1.6	0.1	0.3	58.5	38.6			0.3	0.2		0.5
	3	11.0	0.3	5.1	57.3	10.0		4.0	2.7	5.7		2.8
Peanut[b]	1	19.9	0.1	3.5	47.5	26.1		0.3	0.7	0.1		
	2	2.2	0.1	1.0	50.8	46.2				0.1		
	3	12.8	0.2	3.7	54.0	11.3		4.0	2.9	7.3		3.3
Peanut, interesterified[b]	1	11.8	0.1	2.7	50.8	27.8		1.4	0.9	2.6		1.1
	2	11.9	0.1	3.3	50.9	26.8		1.2	1.2	2.6		0.8
	3	11.9	0.1	2.1	50.3	27.9		1.4	1.2	2.6		1.6
Rapeseed[a]	1	4.1	0.3	2.2	23.1	11.1	6.4		16.4	1.4	34.9	
	2	0.6	0.2		37.3	36.1	20.3		2.0		3.6	
	3	4.3	0.3	3.0	16.6	4.0	2.6		17.3	1.2	51.0	

[a] Brockerhoff and Yurkowski (1966).
[b] Myher *et al.* (1977).

Table XX. Positional Distribution of Fatty Acids in Seed Oil Triacylglycerols Containing Atypical Fatty Acids

Source	Position	Fatty acid (mole %)									
		10:2[c]	14:0	16:0	16:1	18:0	18:1	18:2	18:3	20:0	18:3 conj.[d]
Monnina emarginata[a]	1		0.4		11.1	6.2	33.0	31.0	2.4	0.4	15.3
	2		0.1		0.9	0.4	35.0	63.0	1.3	1.3	0.6
	3		0.2		4.5	3.1	25.0	27.0			39.0
Sapium sebiferum[b]											
Normal triacylglycerols	1			12.6		4.6	16.1	22.4	44.9		
	2			1.0			16.9	33.3	48.8		
	3			4.1		0.8	6.9	17.2	71.0		
Estolide fraction	1			13.5		3.9	14.8	26.7	41.1		
	2			1.8		0.6	17.6	33.9	46.1		
	3	100									

[a] Data from Phillips and Smith (1972).
[b] Data from Christie (1969).
[c] Produced from estolide.
[d] Produced from S-coriolic acid.

Thus, as a general rule, seed oils containing common fatty acids show a selective placement of unsaturated fatty acids at position 2 and only small differences in the composition of positions 1 and 3. When long-chain ($>C_{20}$) or anomalous fatty acids are present, they are selectively enriched in position 3.

4.4.5. Yeast and Microorganisms

Several stereochemical analyses have been completed on yeast and bacterial triacylglycerols (Table XXI). The distribution of fatty acids in triacylglycerols of *Lipomyces lipoferus* was very asymmetrical (Haley and Jack, 1974). Whereas twice as much palmitic acid was located in position 3 compared to position 1, oleic acid was preferentially located in position 2 followed by 1 and 3. Since *Lipomyces lipoferus* readily formed large amounts of triacylglycerols *de novo*, the data were consistent with the idea that specific synthesis of di- and triacylglycerols was occurring with specificities quite different from phosphoglyceride synthesis.

In *Mycobacterium bovis* BCG (Walker *et al.*, 1970), position 1 contained principally palmitic, stearic, and oleic acids, while position 2 contained essentially palmitic and stearic acids (Table XXI). Long-chain fatty acids ($>C_{20}$) were preferentially located in position 3. In *Mycobacterium smegmatis*, oleic acid was the major component in position 1, while oleic, stearic, and behenic acids were located in position 3. In such triacylglycerols, it is important to establish that the long-chain fatty acids are not accumulating in the residual phosphatidyl phenols due to resistance to phospholipase hydrolysis.

4.5. Concluding Remarks

The methodology for the stereochemical analysis of triacylglycerols has been well developed and applied to triacylglycerols from many sources. The Grignard deacylation gives representative 1,2- and 2,3-diacyl-*sn*-glycerols reproducibly for subsequent derivatization and phospholipase A_2 hydrolysis. The 1,3-diacyl-*sn*-glycerols produced by the same procedure can be used for a direct assessment of the fatty acid composition of position 3 but are slightly inaccurate since they are contaminated by significant amounts of isomerized 1,2- and 2,3-diacyl-*sn*-glycerols. If atypical fatty acids react with the ethyl magnesium bromide, diacylglycerols can be produced by pancreatic lipase hydrolysis, but a careful assessment of the intermediate diacylglycerols must be made to assure that no selective hydrolysis has occurred. Although lipoprotein (Morley and Kuksis, 1972) and hepatic (Akesson *et al.*, 1976) lipase show preferential release of fatty acids at position 1, the specificity is not sufficiently complete to use the enzymes as an alternative approach to stereospecific analyses at the present time. Cross-checks on the fatty acid composition are possible for each position. In addition, labeled substrates have been included in the procedure to assess recoveries and completeness of the digestion at each step.

The stereochemical analyses of triacylglycerols from many animal sources have shown that there are specific fatty acid distributions in each position of the glycerol

Table XXI. Positional Distribution of Fatty Acids in Triacylglycerols of Yeast and Bacteria

Source	Position	Fatty acid (mole %)														
		14:0	16:0	16:1	17:0	17:1	18:0	18:1	18:2	19Br	20:0	22:0	24:0	24:1	26:0	Others[c]
Mycobacterium smegmatis[a]	1	1.0	8.0	8.7	0.2	1.5	6.5	60.4		7.4	1.1	0.6	1.1	0.6		1.5
	2	6.6	57.3	13.1	1.7	0.8	5.9	9.3		1.2	0.4	0.3	0.6	0.4		1.9
	3	1.4	7.0	7.0	0.4	0.1	15.5	17.9		6.1	6.9	7.2	18.1	6.5		3.7
Mycobacterium bovis BCG[a]	1	0.7	10.5	2.1	2.5	1.9	26.2	35.7		4.4	4.9	0.9	1.4		4.0	3.9
	2	4.5	5.0	1.0	6.9	0.1	20.1	1.1		0.3	1.8	0.9	0.9		1.4	3.5
Lipomyces lipoferus[b]	1	2.7	13.9	8.0			4.1	61.0	9.7	9.7						1.2
	2		1.2	1.9				87.7	8.8							0.4
	3	6.3	29.3	12.8			8.7	36.7	5.8							

[a] Data from Walker *et al.* (1970).
[b] Data from Haley and Jack (1974).
[c] Fatty acids each accounting for less than 1% of total fatty acids.

molecule which are characteristic for the tissue or, in some cases, for the animal. The triacylglycerols of tissues secreting these lipids show a high degree of asymmetry. Such triacylglycerol structures may be influenced by the specific diacylglycerol molecular species formed by the tissue for phospholipid synthesis since atypical fatty acids, which are present in triacylglycerols but are not found in phospholipids of the tissue, are usually enriched in position 3 of the triacylglycerols. In view of such specificities, it is difficult to define simple rules for the structure of triacylglycerols. It is more important to determine the role of such specific structures on the metabolism and function of triacylglycerols.

4.6. References

Akesson, B., 1969, Composition of rat liver triacylglycerols and diacylglycerols, *Eur. J. Biochem.* **9**:463.

Akesson, B., Gronowitz, S., and Herstof, B., 1976, Stereospecificity of hepatic lipase, *FEBS Lett.* **71**:241.

Anderson, R. E., Bottino, N. R., and Reiser, R., 1967, Pancreatic lipase as a source of diglycerides for the stereospecific analysis of triglycerides, *Lipids* **2**:410.

Arunga, R. O., and Morrison, W. R., 1971, The structural analysis of wheat flour glycerolipids, *Lipids* **6**:768.

Barbano, D. M., and Sherbon, J. W., 1975, Stereospecific analysis of high melting triglycerides of bovine milk fat and their biosynthetic origin, *J. Dairy Sci.* **58**:1.

Bligh, E. G., and Dyer, W. J., 1959, A rapid method of total lipid extraction and purification, *Can. J. Biochem. Physiol.* **37**:911.

Bird, P. R., De Haas, H., Heemskirk, C. H., and Van Deenen, L. L. M., 1965, Synthetic lecithins containing one short chain fatty acid and their breakdown by phospholipase A, *Biochim. Biophys. Acta* **98**:566.

Bottino, N. R., Vandenburg, G. A., and Reiser, R., 1967, Resistance of certain long chain polyunsaturated fatty acids of marine oils to pancreatic lipase hydrolysis, *Lipids* **2**:489.

Breckenridge, W. C., and Kuksis, A., 1968, Specific distribution of short chain fatty acids in molecular distillates of bovine milk fat, *J. Lipid Res.* **9**:388.

Breckenridge, W. C., Marai, L., and Kuksis, A., 1969, Triglyceride structure of human milk fat, *Can. J. Biochem.* **47**:761.

Breckenridge, W. C., Yeung, S. K. F., and Kuksis, A., 1976, Biosynthesis of triacylglycerols by rat intestinal mucosa *in vivo*, *Can. J. Biochem.* **54**:145.

Brockerhoff, H., 1965*a*, A stereospecific analysis of triglycerides, *J. Lipid Res.* **6**:10.

Brockerhoff, H., 1965*b*, Stereospecific analysis of triglycerides: An analysis of human depot fat, *Arch. Biochem. Biophys.* **110**:586.

Brockerhoff, H., 1967, Stereospecific analysis of triglycerides: An alternative method, *J. Lipid Res.* **8**:167.

Brockerhoff, H., 1968, Substrate specificity of pancreatic lipase, *Biochim. Biophys. Acta* **159**:296.

Brockerhoff, H., 1969, Action of pancreatic lipase on emulsions of water-soluble esters, *Arch Biochem. Biophys.* **134**:366.

Brockerhoff, H., 1970, Substrate specificity of pancreatic lipase: Influence of the structure of fatty acids on the reactivity of esters, *Biochim. Biophys. Acta* **212**:92.

Brockerhoff, H., 1971, Stereospecific analysis of triglycerides, *Lipids* **6**:942.

Brockerhoff, H., 1973, Determination of fatty acids in position 3 of triglycerides, *Lipids* **8**:439.

Brockerhoff, H., and Yurkowski, M., 1966, Stereospecific analyses of several vegetable fats, *J. Lipid Res.* **6**:62.

Brockerhoff, H., Hoyle, R. J., and Wolmark, N., 1966, Positional distribution of fatty acids in triglycerides of animal depot fats, *Biochim. Biophys. Acta* **116**:67.

Brockerhoff, H., Hoyle, R. J., Hwang, P. C., and Litchfield, C., 1968, Positional distribution of fatty acids in depot triglycerides of aquatic animals, *Lipids* **3**:24.

Bus, J., Lock, C. M., and Groenewegen, A., 1976, Determination of enantiomeric purity of glycerides with a chiral PMR shift reagent, *Chem. Phys. Lipids* **16:**123.

Christie, W. W., 1969, The glyceride structure of *Sapium sebiferum* seed oil, *Biochim. Biophys. Acta* **187:**1.

Christie, W. W., 1973, *Lipid Analysis*, Pergamon Press, Oxford.

Christie, W. W., 1974, Biosynthesis of triglycerides in freshly secreted milk from goats, *Lipids* **9:**876.

Christie, W. W., 1975, Structural analysis of triglycerides containing isotopically-labelled fatty acids, *J. Chromatogr. Sci.* **13:**411.

Christie, W. W., and Moore, J. H., 1969*a*, A semimicro method for the stereospecific analysis of triglycerides, *Biochim. Biophys. Acta* **176:**445.

Christie, W. W., and Moore, J. H., 1969*b*, The effect of dietary copper on the structure and physical properties of adipose tissue triglycerides in pigs, *Lipids* **4:**345.

Christie, W. W., and Moore, J. H., 1970*a*, The variation of triglyceride structure with fatty acid composition in pig adipose tissue, *Lipids* **5:**921.

Christie, W. W., and Moore, J. H., 1970*b*, A comparison of the structures of triglycerides from various pig tissues, *Biochim. Biophys. Acta* **210:**46.

Christie, W. W., and Moore, J. H., 1970*c*, The structure of egg yolk triglycerides, *Biochim. Biophys. Acta* **218:**83.

Christie, W. W., and Moore, J. H., 1971, Structures of triglycerides isolated from various sheep tissues, *J. Sci. Food Agr.* **22:**120.

Christie, W. W., Moore, J. H., Lorimer, A. R., and Lawrie, T. D. V., 1971, The structures of triglycerides from atherosclerotic plaques and other human tissues, *Lipids* **6:**854.

Christie, W. W., Moore, J. H., and Gottenbos, J. J., 1974, Effect of dietary saturated fatty acids and linoleic acid upon the structures of triglycerides in rabbit tissues, *Lipids* **9:**201.

Eisenberg, S., and Levy, R. T., 1976, Lipoprotein metabolism, in: *Advances in Lipid Research*, Vol. 13 (R. Paoletti and D. Kritchevsky, eds.), pp. 1–80, Academic Press, New York.

Gordon, D. T., Pitas, R. E., and Jensen, R. G., 1975, Effects of diet and type IIa hyperlipoproteinemia upon structure of triacylglycerols and phosphatidyl cholines from human plasma lipoproteins, *Lipids* **10:**270.

Gornall, D. A., and Kuksis, A., 1971, Molecular species of glycerophosphatides and triacylglycerols of egg yolk lipoproteins, *Can. J. Biochem.* **49:**51.

Gronowitz, S., Herslof, B., Ohlson, R., and Toregard, B., 1975, ORD and CD studies of saturated glycerides, *Chem. Phys. Lipids* **14:**174.

Haley, J. E., and Jack, R. C., 1974, Stereospecific analysis of triacylglycerols and major phosphoglycerides from *Lipomyces lipoferus, Lipids* **9:**679.

Hammond, E. G., 1969, The resolution of complex triglyceride mixtures, *Lipids* **4:**246.

Hill, E. E., Lands, W. E. M., and Slakey, P. M., 1968, The incorporation of ^{14}C-glycerol into different species of diglycerides and triglycerides, *Lipids* **3:**411.

Hirschman, H. J., 1960, The nature of substrate asymmetry in stereoselective reactions, *J. Biol. Chem.* **235:**2762.

Hunter, M. L., Christie, W. W., and Moore, J. H., 1973, The structures of the principal glycerolipids of pig liver, *Lipids* **8:**65.

IUPAC-IUB Commission on Biochemical Nomenclature, 1967, Nomenclature of lipids, *Eur. J. Biochem.* **2:**127.

Jensen, R. G., Sampugna, J., and Pereira, R. L., 1964, Intermolecular specificity of pancreatic lipase and the structural analysis of milk fat triglycerides, *J. Dairy Sci.* **47:**1.

Kleiman, R., Miller, R. W., Earle, F. R., and Wolfe, I. A., 1966, Optically active aceto-triglycerides of oil from *Euonymus verrucosus* seeds, *Lipids* **1:**1286.

Kleiman, R., Earle, F. R., Tallent, W. H., and Wolfe, I. A., 1970, Retarded hydrolysis by pancreatic lipase of seed oils with *trans*-3-unsaturation, *Lipids* **5:**513.

Kuksis, A., 1972, Newer developments in determination of structure of glycerides and phosphoglycerides, in: *Progress in the Chemistry of Fats and Other Lipids*, pp. 1–163, Pergamon Press, Oxford.

Kuksis, A., Marai, L., and Myher, J. J., 1973, Triglyceride structure of milk fats, *J. Am. Oil Chem. Soc.* **50:**193.

Lands, W. E. M., Pieringer, R. A., Slakey, P. M., and Zschocke, A., 1966, A micromethod for the stereospecific determination of triglyceride structure, *Lipids* **1:**444.

Litchfield, C., 1968, Predicting the positional distribution of docosahexanoic and docosapentaenoic acids in aquatic animal triglycerides, *Lipids* **3**:417.

Litchfield, C., 1972, *Analysis of Triglycerides*, Academic Press, New York.

Luddy, F. E., Barford, R. A., Herb, S. F., Magidman, P., and Riemenschneider, R. W., 1964, Pancreatic lipase hydrolysis of triglycerides by a semimicro technique, *J. Am. Oil Chem. Soc.* **41**:693.

Marai, L., Breckenridge, W. C., and Kuksis, A., 1969, Specific distribution of fatty acids in the milk fat triglycerides of goat and sheep, *Lipids* **4**:562.

Mills, S. C., Cook, L. J., Scott, T. W., and Nestel, P. J., 1976, Effect of dietary fat supplementation on the composition and positional distribution of fatty acids in ruminant and porcine glycerides, *Lipids* **11**:49.

Morley, N. H., and Kuksis, A., 1972, Positional specificity of lipoprotein lipase, *J. Biol. Chem.* **247**:6389.

Morley, N. H., Kuksis, A., and Buchnea, D., 1974, Hydrolysis of synthetic triacylglycerols by pancreatic and lipoprotein lipase, *Lipids* **9**:481.

Morley, N. H., Kuksis, A., Hoffman, A. G. D., and Kakis, G., 1977, Preferential *in vivo* accumulation of *sn*-2,3-diacylglycerols in post heparin plasma of rats, *Can. J. Biochem.* **55** (in press).

Morris, L. J., 1965*a*, A synthetic optically active trialiphatic triglyceride and a method for the detection of optical activity in natural asymmetric triglycerides, *Biochem. Biophys. Res. Commun.* **18**:495.

Morris, L. J., 1965*b*, The detection of optical activity in natural asymmetric triglycerides, *Biochem. Biophys. Res. Commun.* **20**:340.

Myher, J. J., Marai, L., Kuksis, A., and Kritchevsky, D., 1977, Acylglycerol structure of peanut oils of different atherogenic potential, *Lipids* **12** (in press).

Nutter, L. J., and Privett, O. S., 1966, Phospholipase A properties of several snake venom preparations, *Lipids* **1**:258.

Ohlson, R., Podlaha, O., and Toregard, B., 1975, Stereospecific analysis of some Cruciferae species, *Lipids* **10**:732.

Parijs, J., DeWeerdt, G. A., Beke, R., and Barbier, F., 1976, Stereospecific distribution of fatty acids in human plasma triglycerides, *Clin. Chim. Acta* **66**:43.

Phillips, B. E., and Smith, C. R., 1972, Stereospecific analysis of triglycerides from *Monnina emarginata* seed oil, *Lipids* **7**:215.

Pieringer, R. A., and Kunnes, R. S., 1965, The biosynthesis of phosphatidic acid and lysophosphatidic acid glycerides by phosphokinase pathways in *E. coli*, *J. Biol. Chem.* **240**:2833.

Pitas, R. E., Sampugna, J., and Jensen, R. G., 1967, Triglyceride structure of cow's milk fat. I. Preliminary observations on the fatty acid composition of positions 1, 2 and 3, *J. Dairy Sci.* **50**:1332.

Polheim, D., David, J. S. K., Schultz, F. M., Wylie, M. B., and Johnston, J. M., 1973, Regulation of triglyceride biosynthesis in adipose and intestinal tissue, *J. Lipid Res.* **14**:415.

Sampugna, J., and Jensen, R. G., 1968, Analysis of triglycerides with *Geotrichum condidum* lipase, *Lipids* **3**:519.

Schlenk, W., 1965, Optical activity of triglycerides, *J. Am. Oil Chem. Soc.* **42**:945.

Slakey, P. M., and Lands, W. E. M., 1968, The structure of rat liver triglycerides, *Lipids* **3**:30.

Smith, S., and Abraham, S., 1975, The composition and biosynthesis of milk fat, in: *Advances in Lipid Research*, Vol. 13 (R. Paoletti and D. Kritchevsky, eds.), pp. 195–239, Academic Press, New York.

Tattrie, N. H., Bailey, R. A., and Kates, M., 1958, The action of pancreatic lipases on stereoisomeric triglycerides, *Arch. Biochem. Biophys.* **78**:319.

Walker, R. W., Barakat, H., and Hung, J. G. C., 1970, The positional distribution of fatty acids in phospholipids and triglycerides of *Mycobacterium smegmatis* and *M. bovis* BCG, *Lipids* **5**:684.

Wood, R., and Harlow, R. D., 1969, Structural studies of neutral glycerides and phosphoglycerides of rat liver, *Arch. Biochem. Biophys.* **131**:495.

Wood, R., and Snyder, F., 1969, Tumor lipids: Metabolic relationships derived from structural analyses of acyl, alkyl, and alk-1-enyl moieties of neutral glycerides and phosphoglycerides, *Arch. Biochem. Biophys.* **131**:478.

Yurkowski, M., and Brockerhoff, H., 1966, Fatty acid distribution of triglycerides determined by deacylation with methyl magnesium bromide, *Biochim. Biophys. Acta* **125**:55.

Chapter 5

Stereospecific Synthesis of Enantiomeric Acylglycerols

Dmytro Buchnea

5.1. Introduction

The natural triacylglycerols are mixtures of glycerol esters containing one, two, or three different fatty acids per molecule. By means of stereospecific analysis of the mixed triacylglycerols, it has been possible to demonstrate that they are made up largely of asymmetrical or enantiomeric molecules. The isolation and identification of individual acylglycerols from such mixtures have proved extremely difficult because of the large variety of fatty acids represented and the great similarity in the physicochemical properties of their glycerol esters. With modern chromatographic methods, it has been possible to separate and identify natural triacylglycerols on the basis of their fatty acid composition, but a resolution of the positional isomers and enantiomers has not been achieved. Stereospecific chemical synthesis therefore remains the only means of obtaining pure enantiomeric triacylglycerols for physicochemical and metabolic studies.

Long before the enantiomeric nature of natural acylglycerols became recognized, chemists had endeavored to obtain optically active acylglycerols. Abderhalden and Eichwald (1914, 1915) attempted to prepare optically active acylglycerols by direct resolution of synthetic racemates. Bergmann *et al.* (1921), Bergmann and Sabetay (1924), and Grün and Limpächer (1927) tried to resolve the chiral esters of acylglycerols from racemic intermediates containing acidic or basic functional groups. None of these methods yielded optically active acylglycerols. The synthesis of pure optically active acylglycerols without recourse to chiral resolution of racemates became possible with the preparation of 1,2- and 2,3-isopropylidene-*sn*-glycerols by Fischer and Baer (1937) and Baer and Fischer (1939*a*). These compounds now constitute the key substances for the synthesis of stereospecifically substituted *sn*-glycerol derivatives, as well as for the structural correlation of the acyl-*sn*-glycerol enantiomers to the *sn*-1- and *sn*-3-glyceraldehydes, the conventional stereochemical reference compounds.

Dmytro Buchnea • Banting and Best Department of Medical Research, University of Toronto, Toronto, Ontario, M5G 1L6, Canada.

A wide variety of methods have been proposed for the preparation of synthetic acylglycerols and related compounds. These have been discussed at length in many excellent reviews, including those of Malkin and Bevan (1957), Hartman (1958), Mattson and Volpenhein (1962), van Deenen and de Haas (1964), Baer (1963, 1965), Slotboom and Bonsen (1970), Shvets (1971), Jensen and Gordon (1972), Jensen (1972), and Jensen and Pitas (1976). In the present discussion, only subjects related to synthesis of enantiomeric normal-chain acylglycerols have been considered. In order to provide a critical account, emphasis has been placed on those techniques which the author himself has pioneered or has employed in his own laboratory.

5.2. *General Procedures for Specification of Absolute Configuration of Enantiomeric sn-Glycerol Derivatives*

Several different methods have been used for specifying the absolute configuration of synthetic triacylglycerols. All of the configurational assignments rest upon the pioneering work of Baer and Fischer (1939*a*). The original D/L system placed an α-monoacylglycerol in the same category as the glyceraldehyde into which it could be transformed by oxidation without any alteration or removal of substituents. Steric representation of the glycerol molecule is then possible in a Fischer projection formula (Fig. 1, left side). Equivalent but easier to visualize is the structure shown in the center of Fig. 1. The chiral C-2 is in the plane of the paper, the broken line pointing away from the observer (below the plane of the paper), and wedges pointing toward the observer (above the plane of the paper). This nomenclature was later extended by Baer and Buchnea (1959) to accommodate compounds that could not be named under the original rule, such as diacylphosphatidylethanolamine. However, an optical classification of the triacylglycerols could not be made by an extension of this system, and confusion arose between the two primary positions of the acylglycerol molecules.

This difficulty is avoided in the R/S system (Cahn *et al.*, 1956, 1966), which identifies the substituents on the chiral carbon of glycerol by the clockwise (R) or counterclockwise (S) order of their priorities. Under the rules of the R/S system, the priority of substituents on the C-2 atom of a triacylglycerol molecule is such that the oxygen atom has the highest rank and the hydrogen atom the lowest rank (Fig. 1, right side). The decision on the second and third rank rests with the chain length and substitution of the fatty acids esterified to the primary positions, the long-chain saturated acids being ranked over the unsaturated and short-chain acids. As a result,

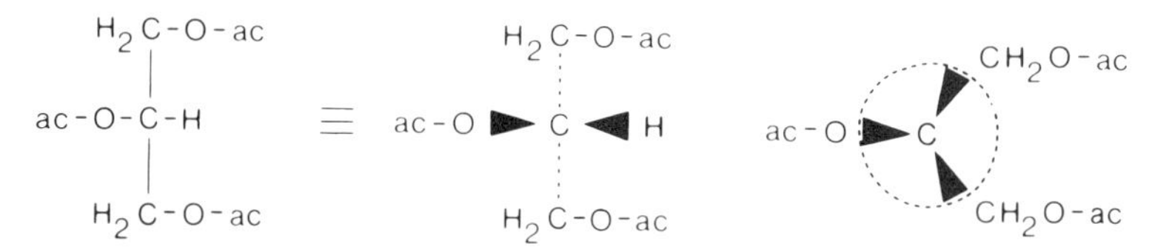

Fig. 1. Antipode of a triacylglycerol. Illustration of the Cahn–Ingold–Prelog nomenclature, developed from the Emil Fischer projection formula.

the description of a simple synthetic substitution of fatty acids at the primary positions of the glycerol molecule requires frequent changes of the configurational prefixes, which is undesirable. Moreover, the R/S system, like the old D/L system, does not account in a consistent manner for the stereospecificity of the acylglycerols toward phospholipase A_2 when tested in the form of appropriate phosphorylated derivatives (Brockerhoff, 1965). Finally, a nonrandom distribution of fatty acids in natural or synthetic enantiomeric triacylglycerols cannot be systematically correlated by reference to either the R and S or the D and L configuration.

All of these problems are avoided if the stereochemistry of the acylglycerols is described by stereospecific numbering (*sn* system) as proposed by Hirschmann (1960). This system recognizes the fact that the two primary carbinol groups of the glycerol molecules are not interchangeable in their reaction with asymmetrical structures, which includes nearly all biochemical processes (Ogston, 1948). The method used to decide which carbinol group is to receive the lower number is a general one and is based on the priorities of the R/S system of Cahn *et al.* (1956, 1966). Thus, if the secondary hydroxyl group is shown to the left of C-2 in the Fischer projection (Fig. 1), the carbon atom above C-2 is called C-1, and the one below is called C-3. The use of the stereospecific numbering is indicated by the prefix "*sn*" before the stem name of the acylglycerol. One disadvantage of the *sn* system of specifying configurations of acylglycerols is the inability to express chilarity in the usual manner by configurational prefixes. However, the fact that C-1 and C-3 lie across a plane of symmetry of glycerol should be sufficient to show that the appropriate pairs of acylglycerols are optical antipodes. In the discussion of the acylglycerol syntheses in this chapter, only the *sn* system of nomenclature is employed. This system has been adopted and recommended by the IUPAC-IUB Commission on Nomenclature of Glycerolipids (Anonymous, 1967).

5.3. Sources and Practical Methods for Preparation of Enantiomeric sn-Glycerol Derivatives

Since the two primary hydroxyl groups of glycerol are not identical, they must be distinguished during chemical synthesis of enantiomeric acylglycerols. This distinction is best achieved if the glycerol is prepared *de novo* in the form of a derivative with specifically protected hydroxyl groups. Only a few natural products are practically suitable for this purpose.

5.3.1. D-Mannitol and L-Mannitol

The naturally occurring D-mannitol is commercially available and presently provides the most significant starting material for stereospecific synthesis of optically active *sn*-glycerol derivatives. With D-mannitol as the starting material, Fischer and Baer (1937) and Baer and Fischer (1939*a*) laid the foundation for the classic or unambiguous stereospecific synthesis of acylglycerols. D-Mannitol was converted into 1,2:5,6-diisopropylidene-D-mannitol, which by oxidative cleavage with lead tetraacetate afforded two molecules of 1,2-isopropylidene-*sn*-glyceraldehyde. A catalytic reduction of the aldehyde group resulted in 1,2-isopropylidene-*sn*-glycerol.

The 2,3-isopropylidene-*sn*-glycerol can be prepared by the same route of synthesis from L-mannitol. Since L-mannitol is not found in nature, it must be prepared synthetically. Kiliani (1887) and Fischer (1890) first obtained L-mannitol by reduction of the dilactone of L-mannaric acid and L-mannose with sodium amalgam. Later, Baer and Fischer (1939*a*) started with L-arabinose, which they prepared according to the method of Anderson and Sands (1929) from mesquite gum. Using the method of Kiliani (1922, 1925), from L-arabinose they obtained L-mannonic lactone, which was then reduced to L-mannitol. Since L-mannose is now readily synthesized from L-arabinose by the nitromethane route described by Sowden

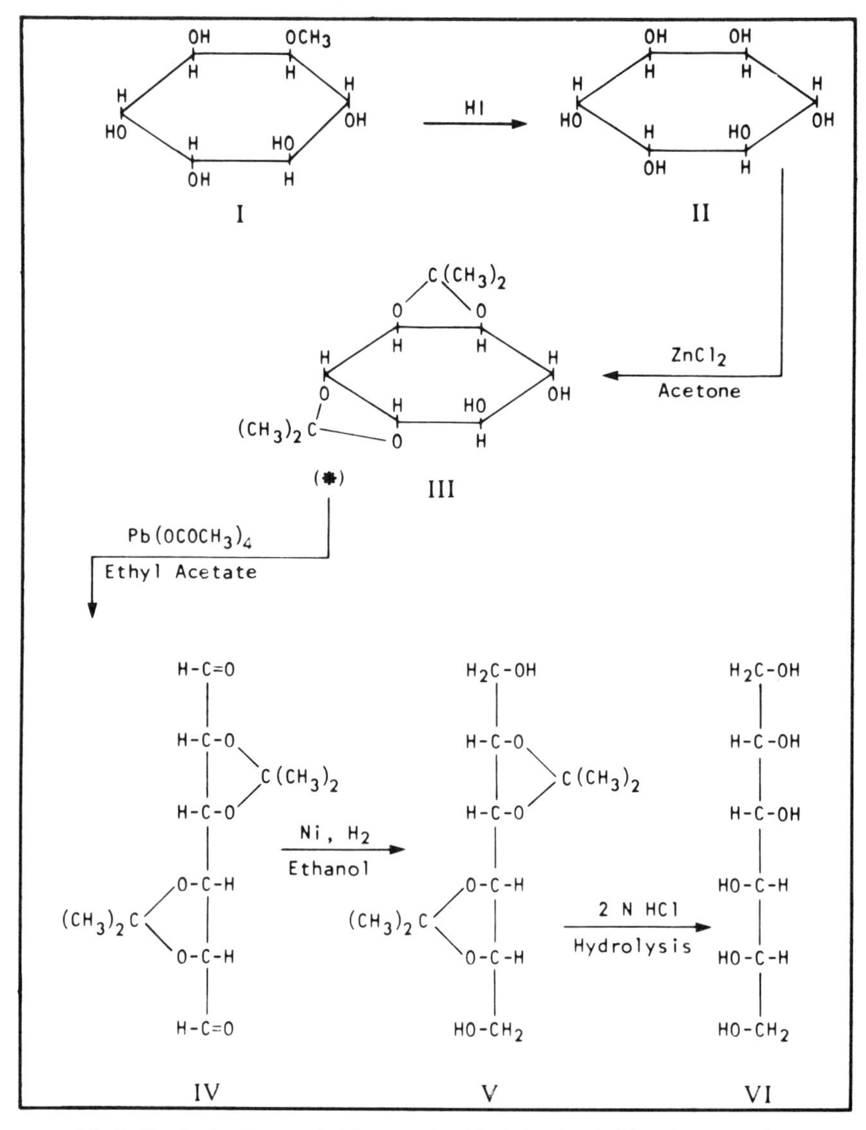

Fig. 2. Synthesis of L-mannitol from quebrachitol via L-inositol (Buchnea, 1971).

(1962), L-mannitol can be obtained conveniently from L-mannose by reduction with sodium borohydride as described by Wright and Tove (1967).

Alternatively, L-mannitol may be synthesized from quebrachitol via L-inositol as described by Angyal and MacDonald (1952), Angyal *et al.* (1953), and Angyal and Hoskinson (1962). For this synthesis, these workers adapted the method used for synthesis of D-inositol as described by Ballou and Fischer (1953). Buchnea (1971) has simplified the preparation of L-mannitol described by Angyal and Hoskinson (1962) and has adapted it for large-scale preparation. Figure 2 illustrates the modified method of preparation of L-mannitol from quebrachitol (Buchnea, 1971).

Diisopropylidene-L-inositol (0.21 mole) is suspended in 500 ml of absolute ethyl acetate in a 1-liter Erlenmeyer flask equipped with a magnetic stirrer and immersed in a water bath maintained at 60°C. Lead tetraacetate (0.25 mol) is added portion by portion to the agitated suspension over a period of 30 min, after which time the reaction mixture is checked with potassium iodide test paper for the excess of lead tetraacetate. After the oxidative cleavage is complete, the mixture, while still hot, is filtered by suction using filter aid (Kieselguhr). The solid cake of lead acetate is washed three times with 70-ml portions of hot absolute ethyl acetate, and the filtrate is then concentrated by distillation under reduced pressure. The residue, 1,6-dialdehyde (IV), is redissolved in 400 ml of absolute ethanol, and 100 g of freshly prepared catalyst (Raney-nickel) is added to the solution. The catalytic reduction is carried out in an atmosphere of pure hydrogen until the consumption of hydrogen ceases (about 3 hr) with an uptake of about 9 liters of hydrogen (calc. 9.5 liters of hydrogen). After the reduction is complete, the catalyst is removed by centrifugation and is washed twice with 100-ml portions of hot ethanol. The combined supernatants are evaporated by distillation under reduced pressure, and L-mannitol (27 g, 70% of theory based on diisopropylidene-L-inositol) is subsequently recovered after removal of isopropylidene protective groups by hydrolysis as described by Angyal and Hoskinson (1962). Since the specific rotation of L-mannitol in water is very small, it is measured in the presence of sodium borate decahydrate, when it exhibits a strong negative rotation (Baer and Fischer, 1939*a*). A mixture of 1 part L-mannitol and 2 parts sodium borate decahydrate in aqueous equilibrium solution gives a specific rotation of $[\alpha]_D^{25} - 30.5°$ (5% L-mannitol and 10% $Na_2B_4O_7 \cdot 10H_2O$), and melting point 166–167.5°C. A specific rotation of $[\alpha]_D^{25} - 29°$ (c,10) and melting point 166°C has been reported by Angyal and Hoskinson (1962).

5.3.2. Preparation of 1,2:5,6-Diisopropylidene-D- and -L-mannitols

The original method for the preparation of 1,2:5,6-diisopropylidene-D- and -L-mannitols using zinc chloride as catalyst was developed by Fischer and Taube (1927). This method was later studied by Fischer and Baer (1930) and Baer and Fischer (1939*a*, *b*). Baer (1945) published an improved version of this method which gave about 55% yield for both enantiomers. The author (Buchnea, unpublished results, 1965) has further improved the method of Baer (1945) to obtain yields up to 70% of theoretical for the 1,2:5,6-diisopropylidene-D- and L-mannitols.

Zinc chloride (powder) is dried for 12 hr at 110–120°C in an oven, and then, while still warm, is added to absolute acetone and stirred under anhydrous conditions until the zinc–acetone solution reaches room temperature (20–25°C). The turbidity

($ZnCO_3$) is quickly filtered off with suction using dry filter aid (Kieselguhr). The clear zinc–acetone solution is added to a predried (over P_2O_5, 24 hr) D- or L-mannitol and the suspension is stirred under anhydrous conditions at room temperature until the D- and L-mannitols are completely dissolved (5–10 hr). Subsequently, 1,2:5,6-diisopropylidene-D- and L-mannitols are recovered as described by Baer (1945, 1952). The products are recrystallized from a hot benzene–petroleum ether (b.p. 60–80°C) mixture (1:1, v/v) and filtered through a hot filter (a filter funnel with a hot water jacket).

5.3.3. Preparation of 1,2- and 2,3-Isopropylidene-sn-glycerols

The 1,2:5,6-diisopropylidene-D- and L-mannitols are converted into the 1,2- and 2,3-isopropylidene-*sn*-glycerols according to the methods of Baer (1945, 1952). The 1,2:5,6-diisopropylidene-D- and L-mannitols are split by glycol fission with lead tetraacetate in absolute ethyl acetate at room temperature. The resulting 1,2- and 2,3-isopropylidene-*sn*-glyceraldehydes are catalytically reduced in the presence of hydrogen at 40–50°C with freshly prepared Raney-nickel catalyst to 1,2- and 2,3-isopropylidene-*sn*-glycerol, respectively. The catalyst is removed by filtration with suction using a filter aid (Kieselguhr) and is washed with ethyl acetate. During the filtration the catalyst must remain covered with solvent to prevent it from igniting in contact with air. After filtration the catalyst is inactivated by 2 N hydrochloric acid while still wet. The 1,2- and 2,3-isopropylidene-*sn*-glycerols are recovered from filtrate by distillation in vacuum.

Lok *et al.* (1976) have developed a method for the synthesis of chiral derivatives of *sn*-glycerol by using D- and L-serines as starting material, thus circumventing the

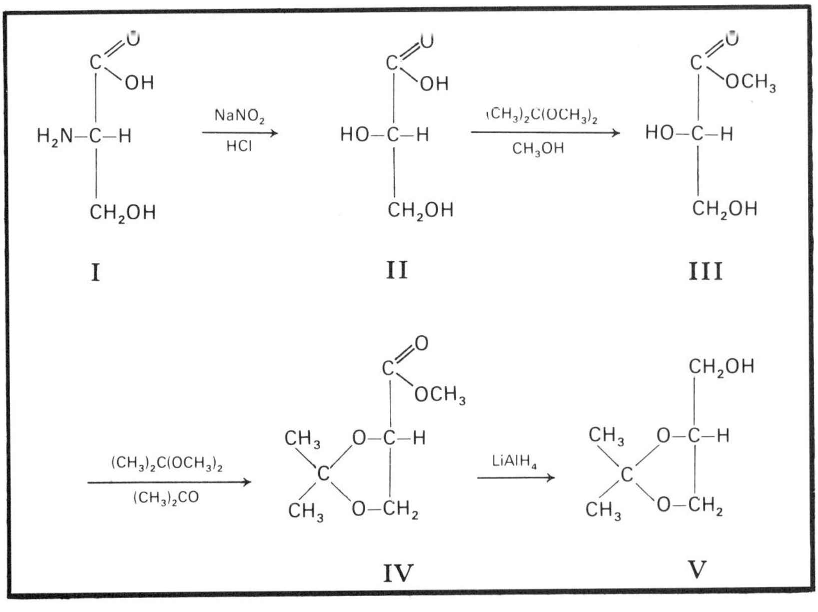

Fig. 3. Synthesis of 1,2- and/or 2,3-isopropylidene-*sn*-glycerol from D- and/or L-serine (Lok *et al.*, 1976).

use of D- and L-mannitols. The conversion of D- and L-serines into the corresponding 1,2- and 2,3-isopropylidene-*sn*-glycerols is shown in Fig. 3. L-Serine (I) is deaminated with sodium nitrate in dilute hydrochloric acid, resulting in L-glyceric acid (II). Without isolation, this is esterified with a mixture of 2,2-dimethroxypropane and methanol, giving III. Compound III is then acetonated with a mixture of acetone and 2,2-dimethroxypropane, resulting in 2,3-isopropylidene-L-glycerate (IV). Finally, reduction of the carboxylic group is accomplished with lithium aluminum hydride to yield 2,3-isopropylidene-*sn*-glycerol (V). By the same sequence of reactions, D-serine yields 1,2-isopropylidene-*sn*-glycerol.

The duplication of the preparation of 1,2- and 2,3-isopropylidene-*sn*-glycerol from D- and L-serines by chemical synthesis is of great importance because it confirms the structure and configuration of the isopropylidene-*sn*-glycerols derived from D- and L-mannitols. However, it cannot compete with D- and L-mannitols as a source of enantiomers since it is too expensive (at least 40 times more expensive than D-mannitol). Therefore, D-mannitol is likely to remain as the most convenient starting material for the preparation of chiral *sn*-glycerol derivatives. Moreover, Sections 5.4.1 and 5.4.2 discuss the conversion of 1,2-isopropylidene-*sn*-glycerol into 2,3-isopropylidene-*sn*-glycerol and the inversion of 3-benzyl-*sn*-glycerol into 1-benzyl-*sn*-glycerol, which leads to the preparation of both enantiomers of *sn*-glycerol derivatives from D-mannitol, which occurs abundantly in nature.

5.3.4. Preparation of 1- and 3-Benzyl-sn-glycerols

The first optically active 1- and 3-benzyl-*sn*-glycerols were synthesized by Sowden and Fischer (1941). Howe and Malkin (1951) improved the procedure for the preparation of racemic benzyl-glycerol. Buchnea (1967*a*) adapted the method of Kaufmann and Förster (1960) for the preparation of optically active 1,2-isopropylidene-3-benzyl- and 1-benzyl-2,3-isopropylidene-*sn*-glycerols by replacing the sodium metal and organic solvent (benzene or toluene) with 50% aqueous solution of sodium hydroxide. The complete synthesis of the enantiomeric benzyl-*sn*-glycerols is carried out as follows: An appropriate amount (0.5 mole) of pure 1,2- or 2,3-isopropylidene-*sn*-glycerol is added in a thin stream from a dropping funnel to a stirred 50% aqueous sodium hydroxide solution (1 mole NaOH) at 50°C over a period of 10–15 min, followed by a drop-by-drop addition of benzyl chloride (0.8 mole) over a period of 90 min. During this process, the reaction temperature is slowly raised from 50°C to about 95°C in order to induce gentle refluxing of the reaction mixture. The stirring and refluxing are continued for about 10 hr. The reaction mixture is then cooled in an ice bath, and the solid sodium chloride is dissolved with ice-cold distilled water. The reaction product, 1,2-isopropylidene-3-benzyl- or 1-benzyl-2,3-isopropylidene-*sn*-glycerol, is extracted four times with diethyl ether. The diethyl ether extracts are combined and washed twice with distilled water and dried with sodium or magnesium sulfate. The solvent, diethyl ether, is then evaporated by distillation under reduced pressure, and the 1,2-isopropylidene-3-benzyl- or 1-benzyl-2,3-isopropylidene-*sn*-glycerol is distilled in high vacuum. Finally, the isopropylidene protective groups are removed by hydrolysis with 10% acetic acid, after which the aqueous solution is evaporated by distillation under reduced pressure, and the 1- and 3-benzyl-*sn*-glycerols are recovered by distillation in high vacuum as described by Sowden and Fischer (1941) and Howe and Malkin (1951).

5.3.5. Preparation of Enantiomeric Alkyl- and Alkenyl-sn-glycerols

The homologous alkyl- and alkenyl-*sn*-glycerols occur in nature exclusively as the *sn*-1-glycerol derivatives (Baer and Fischer, 1941).

Optically active synthetic 1- and 3-alkyl-*sn*-glycerols were first prepared from the corresponding isopropylidene-*sn*-glycerols. These compounds can serve as starting material for the preparation of enantiomeric alkyl-diacyl-*sn*-glycerols. Baer and Fischer (1941, 1947) and Baer *et al.* (1944) obtained the optically active 1-octadecyl-*sn*-glycerol (batyl alcohol), 1-octadecenyl-*sn*-glycerol (selachyl alcohol), and 1-hexadecyl-*sn*-glycerol (chimyl alcohol) and their enantiomers by reacting the appropriate sodium isopropylidene-*sn*-glyceroxide with the alkyl or alkenyl chloride. Baumann and Mangold (1964) have obtained improved yields and purity of the 1- and 3-alkyl or alkenyl-*sn*-glycerol ethers by reacting the potassium 1,2- or 2,3-isopropylidene-*sn*-glyceroxide with alkyl- or alkenylmethanesulfonate. Buchnea (unpublished results, 1975, 1976) has synthesized 3-octadecyl-*sn*-glycerol ether according to the method of Baumann and Mangold (1964) and has compared its physical properties to those of the natural 1-octadecyl-*sn*-glycerol (batyl alcohol). Both ethers had the same melting points (71–72°C) as reported by Baer (1963). However, the specific rotation of the synthetic 3-alkyl-*sn*-glycerol ether was $[\alpha]_D^{25}$ –

Table I. Melting Points and Specific Rotations of Alkyl- and Alkenyl-sn-glycerol and Dialkyl- and Mixed Alkyl-alkenyl-sn-glycerol Ethers

	Substance	Melting point (°C)	Specific rotation $[\alpha]_D^{25}$ in $CHCl^3$ (deg)
A.[a]	1-Octadecyl-*sn*-glycerol	71–72	+4.0 (8.1%)
	3-Octadecyl-*sn*-glycerol	71–72	–2.3 (8.0%)
	1-Hexadecyl-*sn*-glycerol	62.5–63.5	+3.0 (11.0%)
	3-Hexadecyl-*sn*-glycerol	63–64	–2.2 (8.4%)
	1-Octadecenyl-*sn*-glycerol	17.6–19.0	–4.5 (in substance)
B.[b]	2,3-Dioctadecenyl-*sn*-glycerol	—	+7.4
	2-Octadecyl-3-octadecenyl-1-trityl-*sn*-glycerol	—	–4.3
	2-Octadecyl-3-octadecenyl-*sn*-glycerol	—	+6.7
	2-Octadecenyl-3-octadecyl-1-trityl-*sn*-glycerol	—	–4.2
	2-Octadecenyl-3-octadecyl-*sn*-glycerol	—	–6.9
	2-Hexadecanoyl-3-octadecyl-1-trityl-*sn*-glycerol	35	–7.6
	2-Hexadecanoyl-3-octadecyl-*sn*-glycerol	67.5–68.5	+0.43
	1,2-Dioctadecyl-*sn*-glycerol	53.5–54.5	–6.9
	1,2-Dihexadecanoyl-*sn*-glycerol	68–69	–2.9
	1,2-Dioctadecanoyl-*sn*-glycerol	76–77	–2.8

[a] Baer and Fischer (1941, 1947) and Baer *et al.* (1944).
[b] Palameta and Kates (1966).

1.2° ± 0.1°, and that of the natural 1-alkyl-*sn*-glycerol ether was $[\alpha]_D^{25}$ + 1.2° ± 0.1°; both rotations were taken in chloroform solution (c,10). The optical rotations and melting points of the natural and synthetic alkyl- and alkenyl-*sn*-glycerol ethers reported by Baer (1963) are summarized in Table IA. Palameta and Kates (1966) reported the synthesis of dialkenyl- and mixed alkyl-alkenyl-*sn*-glycerol ethers by alkylation of 3-triphenylmethyl-*sn*-glycerol with 1-bromo-*cis*-9-octadecene and potassium hydroxide in boiling benzene, followed by acid hydrolysis of the triphenylmethyl protecting group. These dialkenyl- and mixed alkyl-alkenyl-*sn*-glycerols are listed also in Table IB.

Åkesson *et al.* (1976) successfully used the inversion method of Lands and Zschocke (1965) to invert an aliphatic 3-alkyl-*sn*-glycerol ether into 1-alkyl-*sn*-glycerol ether (see Section 5.4.2).

Davies *et al.* (1934) and Gupta and Kummerow (1959) have synthesized the optically inactive 2-octadecyl-*sn*-glycerol ethers by reaction of 2-potassium-1,3-benzylidene-*sn*-glycerol with octadecyl-*p*-toluenesulfonate in absolute benzene. Paltauf and Spener (1968), Paltauf (1971), and Paltauf and Johnston (1971) have employed the methods of Baumann and Mangold (1964) for the synthesis of other long-chain glycerol ethers as model substances for the investigation of intestinal absorption of monoacyl-*sn*-glycerols. Baumann (1972) has described the above-mentioned method for the preparation of an extensive series of optically active glycol alkylenediol ethers.

5.4. Methods of Configurational Conversion and Inversion of sn-Glycerol Derivatives

In Section 5.3.1 of this chapter, various synthetic methods are described for the preparation of the naturally inaccessible L-mannitol, which is the essential starting material for the preparation of 2,3-isopropylidene-*sn*-glycerol and 1-benzyl-*sn*-glycerol. This synthesis, however, requires the preparation of numerous intermediates, which is laborious. To circumvent this difficulty, methods have been developed for the configurational conversion of 1,2-isopropylidene-*sn*-glycerol into 2,3-isopropylidene-*sn*-glycerol and for the inversion of 3-benzyl-*sn*-glycerol into 1-benzyl-*sn*-glycerol.

Fischer and Brauns (1914) were the first to demonstrate that exchanging the position of the hydroxyl and amide groups of dextrorotatory isopropylmalonamidic acid resulted in the formation of a levorotatory antipode with no change in the magnitude of the optical rotation. This successful configurational conversion is still considered to be an experimental confirmation of the spatial concept of the chiral carbon atom as postulated by van't Hoff and Le Bell (Fischer and Baer, 1937).

5.4.1. Conversion of 1,2-Isopropylidene-sn-glycerol into 2,3-Isopropylidene-sn-glycerol

Fischer and Baer (1937) first reported the configurational conversion of 1,2-isopropylidene- into 2,3-isopropylidene-*sn*-glycerol. The experimental details were later published by Baer and Fischer (1945*b*). The transformation proceeds through

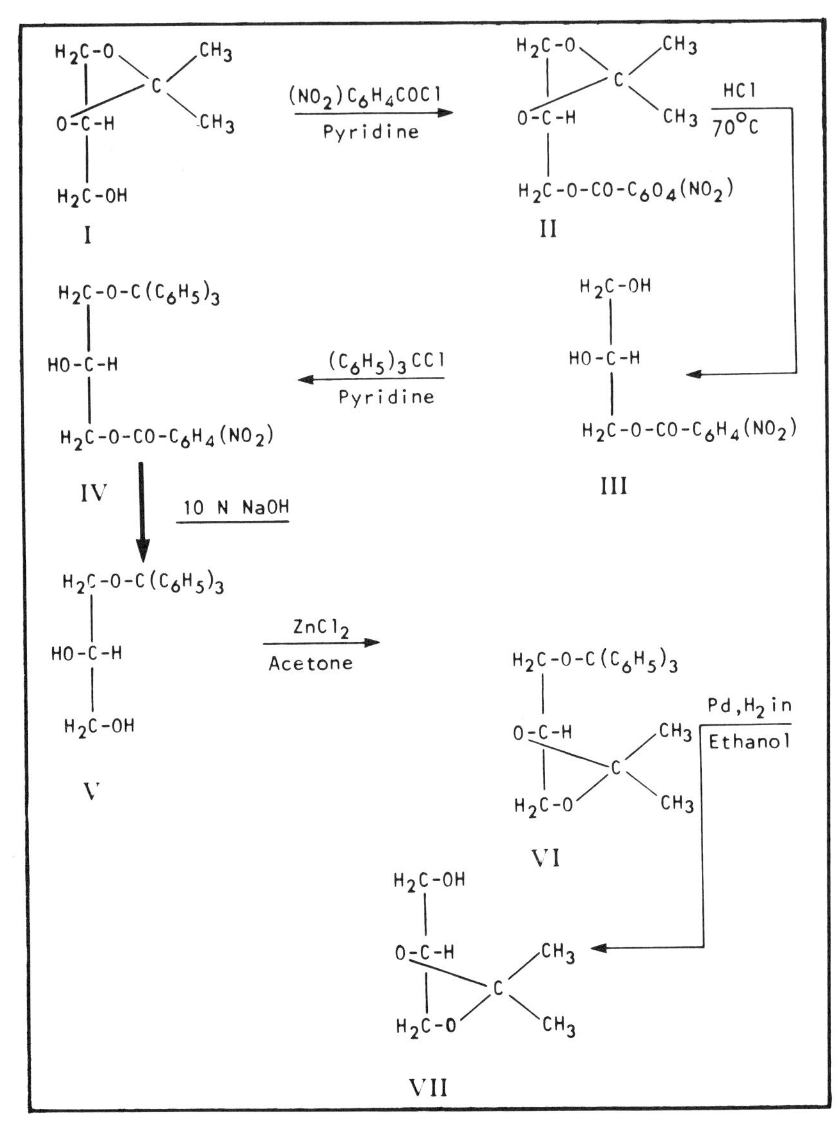

Fig. 4. Conversion of 1,2-isopropylidene-*sn*-glycerol into 2,3-isopropylidene-*sn*-glycerol (Baer and Fischer, 1945*b*).

these six steps, as shown in Fig. 4: 1,2-Isopropylidene-*sn*-glycerol (I) is nitrobenzoated in the *sn*-3 position (II). Removal of isopropylidene group with 0.5 N hydrochloric acid at 80°C results in 3-nitrobenzoyl-*sn*-glycerol (III). Tritylation of III gives 1-triphenylmethyl-3-nitrobenzoyl-*sn*-glycerol (IV). Saponification of IV with sodium hydroxide in ethanol yields 1-triphenylmethyl-*sn*-glycerol (V), which on acetonation with zinc chloride and acetone produces 1-triphenylmethyl-2,3-isopropylidene-*sn*-glycerol (VI). The removal of the triphenylmethyl group by catalytic hydrogenolysis with hydrogen in presence of palladium black as a catalyst results in 2,3-isopropylidene-*sn*-glycerol (VII).

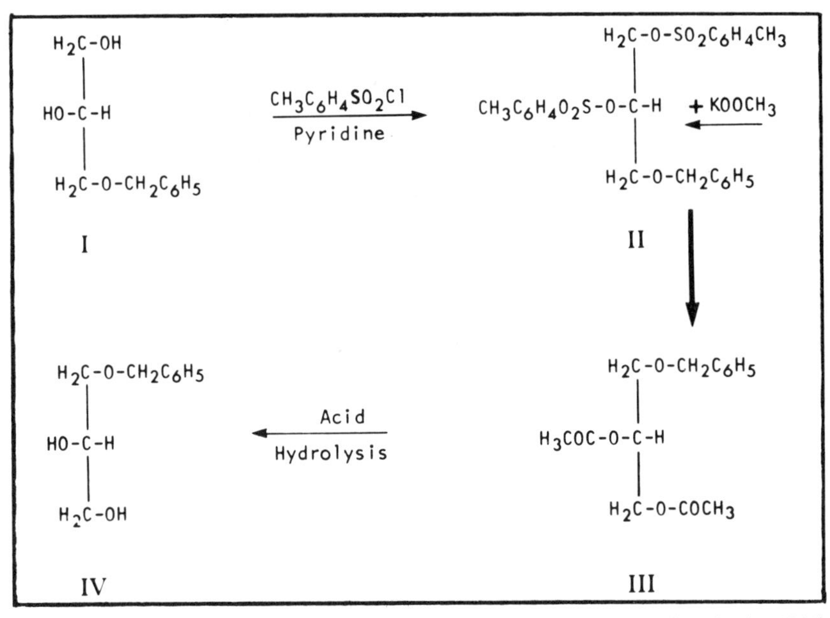

Fig. 5. Inversion of 3-benzyl-*sn*-glycerol into 1-benzyl-*sn*-glycerol (Lands and Zschocke, 1965).

5.4.2. *Inversion of 3-Benzyl-sn-glycerol into 1-Benzyl-sn-glycerol*

Lands and Zschocke (1965) have reported an elegant and practical method for the inversion of 3-benzyl-*sn*-glycerol into its enantiomer 1-benzyl-*sn*-glycerol by a nucleophilic displacement with potassium acetate. This "Walden inversion" proceeds as shown in Fig. 5. 3-Benzyl-*sn*-glycerol (I) is tosylated in *sn*-1 and *sn*-2 positions with *p*-toluenesulfonyl chloride in pyridine. The resulting 1,2-ditosyl-3-benzyl-*sn*-glycerol (II) is attacked from the rear of chiral carbon-2 of the *sn*-glycerol moiety (II) with freshly fused potassium acetate, resulting in 1-benzyl-2,3-diacetoyl-*sn*-glycerol (III). A hydrolysis of the acetoyl groups affords 1-benzyl-*sn*-glycerol (IV) with no loss in optical rotation. The formulation of the mechanism and an analysis of the factors influencing competition between different groups are discussed in detail by Ingold (1973) and his school.

5.4.3. *Preparation of 1-Benzyl-sn-glycerol from* D-*Mannitol via 1,3:2,5:4,6-O-Trimethylene-*D-*mannitol*

Gigg and Gigg (1967*b*) have developed an alternative procedure for the synthesis of 1-benzyl-*sn*-glycerol from D-mannitol. D-Mannitol is converted into 1,3:2,5:4,6-*O*-trimethylene-D-mannitol as described by Schulz and Tollens (1896). As shown in Fig. 6, this compound is converted through 2,5-*O*-methylene-D-mannitol and 1,6-dibenzoyl-3,4-benzylidene-D-mannitol (I) as described by Ness *et al.* (1943) and Wickberg (1958) into the corresponding benzyl derivative by the action of benzyl chloride and sodium hydroxide. The benzylidene group is hydrolyzed with dilute hydrochloric acid, resulting in 1,6-di-*O*-benzyl-2,5-*O*-methylene-D-mannitol (II). An oxidative cleavage of compound II followed by

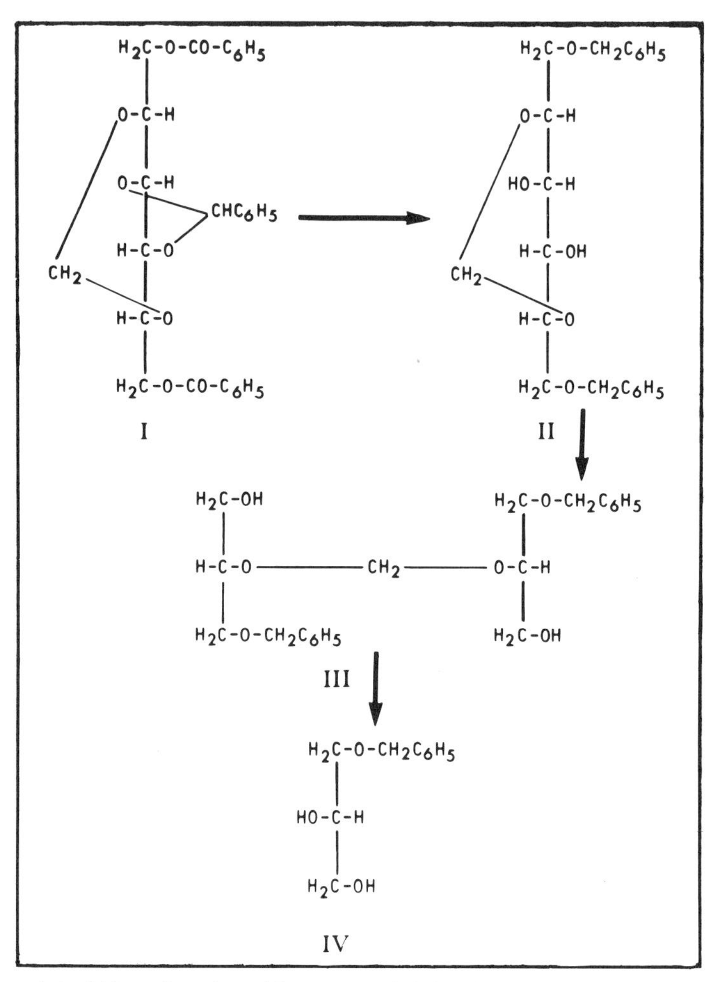

Fig. 6. Preparation of 1-benzyl-*sn*-glycerol from D-mannitol via 1,3:2, 5:4,6-*O*-trimethylene-D-mannitol (Gigg and Gigg, 1967*b*).

aldehyde reduction with lithium aluminum hydride produces methylene-bis-2-*sn*-(benzyl-*sn*-glycerol) (III), which on hydrolysis of methylene linkages with acid, yields 1-benzyl-*sn*-glycerol (IV).

5.5. *Selection of Protective Groups for Stereospecific Synthesis of Acyl-, Alkyl-, and Alkylacyl-sn-glycerols*

A stereospecific synthesis of *sn*-glycerol derivatives depends on a selective acylation and/or alkylation of the different hydroxyl groups of the glycerol molecule. This can be accomplished if the other hydroxyl functions are temporarily blocked by one or two different blocking groups, which can be selectively removed without displacing the acyl or alkyl groups previously introduced. Since racemic derivatives of *sn*-glycerol cannot be resolved, it is necessary to introduce at least one of the blocking groups during the initial preparation of the enantiomeric glycerol.

5.5.1. Isopropylidene Group

The isopropylidene moiety is a widely used protective group in polyol chemistry. Under normal conditions, the reaction with the acetone–zinc reagent is limited to the blocking of the vicinal hydroxyl groups of a polyol, as first demonstrated by Irvine *et al.* (1915) and later confirmed by Hibbert and Carter (1929). Isopropylidene was introduced as a protective group in acylglycerol synthesis by Fischer *et al.* (1920), resulting in the formation of racemic monoacylglycerols. An attempt to synthesize 1,2:5,6-diisopropylidene-D- and L-mannitols yielded only 2% of the products. The first successful syntheses of 1,2:5,6-diisopropylidene-D- and L-mannitols were accomplished by Fischer and Baer (1937).

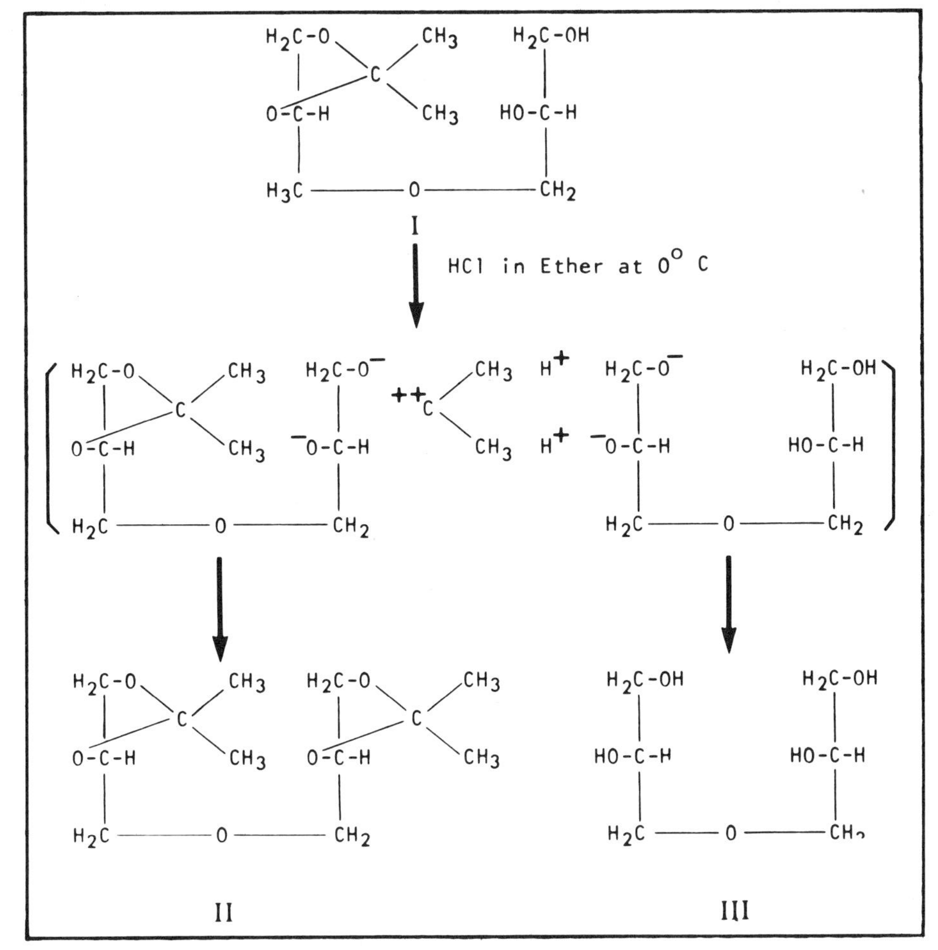

Fig. 7. Mechanism of disproportionation of 1,2-isopropylidene-*sn*-glycerol-3-*sn*-glycerol ether to bis(1,2-isopropylidene-*sn*-glycerol-3) ether and bis(3-*sn*-glycerol) ether (Buchnea, 1967*a*, *b*).

The 1,2:5,6-diisopropylidene-D- and L-mannitols were then converted into the corresponding optically active 1,2- and 2,3-isopropylidene-*sn*-glycerols, the key substances for the synthesis of all stereospecifically substituted *sn*-glycerol derivatives.

The isopropylidene protective group exposes the *sn*-1- or *sn*-3-hydroxyl function either for direct acylation and/or alkylation or for the selective introduction of other blocking groups, e.g., benzyl, nitrobenzoyl.

The optically active 1,2- and 2,3-isopropylidene-*sn*-glycerols and their derivatives are stable only in a very pure state and when stored at −10°C or lower temperatures. In the presence of traces of acid, racemization or loss of the protective groups takes place. Buchnea (1967*a*, *b*) has shown that the 1,2-isopropylidene-*sn*-glycerol-3-*sn*-glycerol ether (I) (Fig. 7) disproportionates to bis(1,2-isopropylidene-*sn*-glycerol-3) ether (II) and bis(3-*sn*-glycerol) ether (III) when stored in diethyl ether solution at 0°C in the presence of traces of dry HCl. The disproportionation can be prevented by storing the compounds in diethyl ether containing a few pellets of sodium hydroxide at −10°C. The isopropylidene group is best removed by refluxing (<100°C) the isopropylidene-*sn*-glycerol derivative in trimethylborate in the presence of boric acid (Hartman, 1959; Mattson and Volpenhein, 1962). After cooling, the reaction mixture is washed with cold water to effect hydrolysis of the trimethyl borate, which proceeds at room temperature without acyl migration.

Serdarevich (1967) studied the structure of racemic isopropylideneglycerols and their monoacyl esters by NMR spectroscopy and confirmed the five-membered ring structure of isopropylidene glycerol.

5.5.2. Benzylidene Group

For the synthesis of 2-monoacyl-*sn*-glycerol, it is necessary to protect the *sn*-1- and *sn*-3-hydroxyl functions of the glycerol molecule. This can be accomplished by using a benzylidene protective group. Hibbert and Carter (1929) were the first to demonstrate that the condensation of glycerol with benzaldehyde produces mainly 1,3-benzylidene-*sn*-glycerol. Bergmann and Carter (1930) and later Stimmel and King (1934) used 1,3-benzylidene-*sn*-glycerol for the synthesis of 2-monoacyl-*sn*-glycerols. Verkade and van Roon (1942) have extensively investigated the products of the condensation of glycerol and benzaldehyde and have found that 1,3-benzylidene-*sn*-glycerol appears in *cis* and *trans* forms. Baggett *et al.* (1961), Serdarevich and Carroll (1966), and Serdarevich (1967) have investigated the structure of 1,3-benzylidene-*sn*-glycerol by NMR spectroscopy and have confirmed the presence of a six-membered ring. Subsequently, Serdarevich (1967) has claimed that the isomer with the higher melting point (82.5–83.5°C) possesses the *cis* configuration, and the isomer with the lower melting point (63.5–64.5°C) possesses the *trans* conformation.

The benzylidene protective group is removed by first replacing it with triethyl or trimethyl borate, which can then be hydrolyzed with water at room temperature (Martin, 1953).

5.5.3. Triphenylmethyl (Trityl) Group

The triphenylmethyl or trityl group has also been extensively utilized as a protective group in polyol chemistry. Helferich and Sieber (1927, 1928) introduced it as a protective group in acylglycerol syntheses. The triphenylmethyl chloride reacts preferentially with the primary hydroxyl groups of glycerol. The trityl protective group has been widely used in acylglycerol syntheses by Verkade *et al.* (1953).

The trityl protective group is easily removed either by catalytic hydrogenolysis with hydrogen in presence of a catalyst (palladium black) or with HCl in ethyl ether or petroleum ether solution at 0° C.

Although the use of the trityl protective method of synthesis produces high yields, it cannot compete with the use of the benzylidene protective group, since detritylation leads to the formation of by-products such as triphenylmethyl halides or triphenyl carbinol which must be removed during wasteful crystallization. Buchnea and Baer (1960) have demonstrated, however, that the triphenylmethyl halides and triphenyl carbinol can be completely separated from any acylglycerols by silicic acid column chromatography with a nonpolar solvent, e.g. benzene. Subsequently, Buchnea (1971, 1974) has shown that the removal of the trityl protective groups (detritylation) and the chromatographic separation of the acylglycerols from the triphenyl carbinol can be accomplished in one operation by column chromatography on a silicic acid–boric acid mixture (10 : 1, w/w).

To avoid the use of the trityl protective group, Daubert and King (1938) employed the carbobenzoxy group as the protecting group, which on hydrogenolysis is converted into toluene. Unfortunatley, the intermediate carbobenzoxyglycerol decomposes on distillation.

5.5.4. Benzyl Group

A conversion of the enantiomeric isopropylidene glycerols into the corresponding benzyl ethers which are stable to acid hydrolysis allows the selective exposure of the *sn*-1,2- and the *sn*-2,3-hydroxyl functions to acylation. Sowden and Fischer (1941) introduced the benzyl moiety as a protective group for the synthesis of optically active monoacid diacyl-*sn*-glycerols, which can be further acylated to triacylglycerols. The blocking group is introduced by reacting the appropriate sodium or potassium isopropylidene-*sn*-glyceroxide with benzyl chloride. Since the 1-benzyl- and 3-benzyl-*sn*-glycerols can also be protected at the remaining primary positions with trityl groups prior to acylation, this method is suitable for the preparation of mixed-acid diacyl and triacyl-*sn*-glycerols (Buchnea and Baer, 1960). The benzyl protective group, however, can be removed only by catalytic hydrogenolysis with pure hydrogen in the presence of palladium as a catalyst. The use of this blocking group is limited to synthesis of saturated diacyl-*sn*-glycerols, although an unsaturated acid may be introduced during the completion of the triacylglycerol formation.

5.5.5. 2,2,2-Trichloroethoxycarbonyl Group

In order to allow the preparation of unsaturated 1,2- and 2,3-diacyl-*sn*-glycerols, Pfeiffer *et al.* (1968) employed the 2,2,2-trichloroethoxycarbonyl group to protect the free primary hydroxyl group. This group is introduced as the carbonyl chloride in the presence of pyridine and is retained without isomerization during the removal of the isopropylidene groups. The 2,2,2-trichloroethoxycarbonyl group is removed following acylation with zinc in acetic acid at room temperature in a few hours. The use of the 2,2,2-trichloroethoxycarbonyl blocking group was first described by Woodward (1966) and Woodward *et al.* (1966) during their synthesis of cephalosporin. Later, Windholz and Johnston (1967) studied its general application as a protective group in synthetic organic chemistry.

5.5.6. Other Protective Groups

There are other protective groups which have been used alone or in combination with others for the synthesis of diacid or mixed-acid diacyl-*sn*-glycerols. Cunningham and Gigg (1965) have employed carbonate and benzyl protective groups in their synthetic work. Gigg and Gigg (1967*a*) have used the tetrahydropyranyl group in combination with carbonate protective groups for the synthesis of enantiomeric unsaturated diacyl-*sn*-glycerols. They employed phosgene for the introduction of the carbonate group in combination with benzyl and triphenylmethyl as protecting groups for the synthesis of mixed-acid diacyl-*sn*-glycerols and plasmalogens.

Baer and Fischer (1945*b*) used alternately isopropylidene, *p*-nitrobenzoyl, and triphenylmethyl protective groups for the conversion of 1,2- into 2,3-isopropylidene-*sn*-glycerol. Buchnea and Baer (1960) employed benzyl and triphenylmethyl protective groups for the synthesis of mixed-acid diacyl-*sn*-glycerols.

5.6. Stereospecific Acylation of Enantiomeric sn-Glycerol Derivatives

The stereospecific synthesis of enantiomeric acyl-*sn*-glycerols consists of finding effective means of introducing acyl groups into *sn*-glycerol derivatives with suitably exposed hydroxyl functions. This requires appropriate reagents in the pure state and carefully controlled reaction conditions.

5.6.1. Selection of Acylating Reagents

Experience has shown that only the acid chlorides and acid anhydrides are suitable for an efficient and reliable introduction of acyl groups into glycerol molecules with selectively protected hydroxyl groups (Mattson and Volpenhein, 1962; Jensen, 1972). Both reagents must be freshly prepared and are preferably distilled, although washing with ice-cold water may be a satisfactory means of purification of polyunsaturated derivatives which isomerize during distillation.

5.6.1.1. Preparation of Acid Chlorides

The chlorides of the saturated fatty acids are generally prepared by reaction with thionyl chloride, which is an efficient and economical reagent. This method was extensively investigated by Fierz-David and Kuster (1939), who prepared the acid chlorides of some 18 different fatty acids. The chlorides of unsaturated fatty acids are best prepared with oxalyl chloride (Mattson and Volpenhein, 1962). The author (Buchnea, unpublished results) has routinely purified the thionyl chloride by successive distillation from quinoline and boiled linseed oil as described by Vogel

(1959), while the oxalyl chloride has been subjected to a simple initial distillation and a redistillation when it developed a slight yellow color. Both chlorinating reagents are poisonous and must be handled with care.

A fast, clean, and effective method of chlorination of fatty acids is as follows: 1 part fatty acid and either 1.2 parts oxalyl chloride or 1.8 parts thionyl chloride (mole ratios) are placed in a suitable round-bottom flask provided with a reflux condenser and a calcium chloride tube. The reaction mixture is gently refluxed at 68–70°C in an oil bath until the evolution of gas, HCl, CO_2, or SO_2, ceases (4–5 hr). The excess oxalyl or thionyl chloride is removed by distillation under anhydrous conditions at reduced pressure at a bath temperature of 70°C, and the acid chlorides are distilled in a vacuum of 0.005–0.001 mm Hg. To protect the mercury from reacting with traces of HCl or SO_2, a trap filled with NaOH pellets is connected between the acyl chloride receiver and the high-vacuum pump. The heat is applied only after the vacuum has been achieved. Once the vacuum is constant, the acid chlorides distill smoothly at constant temperatures. The yields are 85–90% of theory of very pure acid chloride, provided that the fatty acids are pure and the chlorinating agents are of high quality.

5.6.1.2. Preparation of Acid Anhydrides

The fatty acid anhydrides may be prepared via the mixed-acid anhydride intermediates using either acetyl chloride (Mattson *et al.*, 1964) or trifluoroacetic anhydride (Bourne *et al.*, 1949). The fatty acids are refluxed with an equal weight of the reagent for about 3 hr. The excess anhydride is removed by distillation and the mixed-acid anhydrides are converted into the fatty acid anhydrides by heating for 30 min to 150°C at 1–2 mm Hg pressure.

5.6.2. Synthesis of Enantiomeric 1- and 3-Monoacyl-sn-glycerols

The optically active monoacyl-*sn*-glycerols were first synthesized by Baer and Fischer (1945*a*) from D- and L-mannitols as shown in Fig. 8. They obtained four different saturated 1-monoacyl-*sn*-glycerols and fifteen 3-monoacyl-*sn*-glycerols. The physical properties of these acylglycerols are listed in Tables II, III, and IV. The synthesis is effected by dissolving the 1,2- or 2,3-isopropylidene-*sn*-glycerol (0.11 mole) in dry benzene and adding dry pyridine (0.2 mole). To the mixture is immediately added 0.11 mole of saturated or unsaturated acid chloride, and the reaction mixture is kept under anhydrous conditions at 40°C for 10 hr. Baer and Fischer (1945*a*) removed the protective isopropylidene groups by acid hydrolysis. The monoacyl-*sn*-glycerols with eight or fewer carbon atoms in the fatty acid radicals were freed of their protective groups by hydrolysis with 10% acetic acid, and the monoacyl-*sn*-glycerols containing more than eight carbon atoms were hydrolyzed with a mixture of concentrated hydrochloric acid and diethyl ether (1 : 1, v/v) at low temperature. The removal of the protective isopropylidene groups by acid hydrolysis unfortunately involves some acyl migration, and therefore, during the hydrolysis, the temperature and time must be very carefully controlled to minimize isomerization of the optically active monoacyl-*sn*-glycerols. A much more effective method of removing the isopropylidene groups involves refluxing with trimethyl

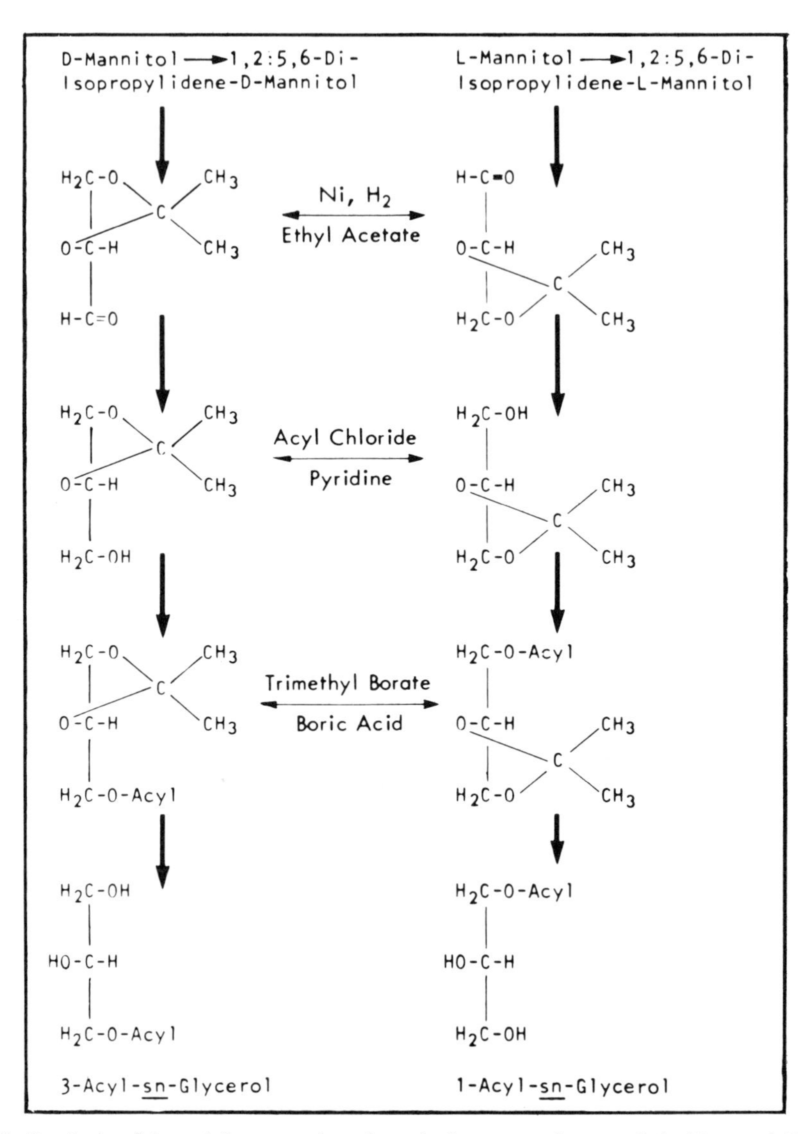

Fig. 8. Synthesis of 1- and 3-monoacyl-*sn*-glycerols from D- and L-mannitols (Baer and Fischer, 1945*a*).

borate in the presence of boric acid, which results in the replacement of the isopropylidene groups by trimethyl borate groups. The latter are readily removed by hydrolysis in water at room temperature without any acyl migration (Hartman, 1959; Mattson and Volpenhein, 1962). The improved procedure of preparation of the enantiomeric monoacyl-*sn*-glycerols was employed by Schlenk (1965*a*, *b*) to prepare the series of homologues shown in Table V along with their optical rotations. Gronowitz *et al.* (1975) have used the original procedure of Baer and Fischer (1945*a*) to prepare selected 3-monoacyl-*sn*-glycerols which they have employed in studies of optical rotatory dispersion and circular dichroism (Tables VI and VII).

Table II. Formulas, Yields, and Physical Data of 1,2-Isopropylidene-3-acyl-sn-glycerols[a, b]

Acid	Formula	Yield (%)	Boiling point °C (mm) or melting point °C	Refractive index		Density (d)	$[\alpha]_D$ (+)	
				n_D	°C		°C	deg (in substance)
Acetic	$C_8H_{14}O_4$	85.6	77 (6)	1.4258	23	1.070	25	1.95
Propionic	$C_9H_{16}O_4$	93.6	88–89 (6)	1.4270	24	1.046	21	4.66
Butyric	$C_{10}H_{18}O_4$	81.2	101–102 (6)	1.4279	25	1.024	22	5.13
n-Valeric	$C_{11}H_{20}O_4$	73.2	108–110 (6)	1.4297	26	1.005	26	5.65
Isovaleric	$C_{11}H_{20}O_4$	86.4	101–103 (6)	1.4270	27	1.000	27	5.74
Caproic	$C_{12}H_{22}O_4$	82.2	125–126 (7)	1.4340	24	0.994	18	5.50
Enanthic	$C_{13}H_{24}O_4$	60.0	139–140 (9)	1.4360	23	0.973	25	5.22
Caprylic	$C_{14}H_{26}O_4$	73.7	148–149 (7)	1.4376	22	0.973	26	5.08
Pelargonic	$C_{15}H_{26}O_4$	71.0	156–157 (6)	1.4390	24	0.965	23	5.22
Capric	$C_{16}H_{30}O_4$	81.1	163–164 (6)	1.4402	24	0.960	22	5.03
Hendecanoic	$C_{17}H_{32}O_4$	57.0	118–120 (2×10^{-2})	1.4415	25	0.952	21	5.14
Lauric	$C_{18}H_{34}O_4$	69.0	130–131 (2×10^{-3})	1.4407	28	0.943	28	5.06
Myristic	$C_{20}H_{38}O_4$	72.0	166–168 (2×10^{-2})	1.4448	27	0.350	24	4.79
Palmitic	$C_{22}H_{42}O_4$	89.0	33–34.5	1.4430	50	0.910	50	4.95
Stearic	$C_{24}H_{46}O_4$	87.0	41–42	1.5355	60	0.911	50	4.94

[a] Prepared from 1,2-isopropylidene-*sn*-glycerol, $[\alpha]_D + 13.8°$ (in substance). [b] From Baer and Fischer (1945*a*).

Table III. Formulas, Yields, and Physical Data of 1-Acyl-2,3-isopropylidene-sn-glycerols[a, b]

Acid	Formula	Yield (%)	Boiling point °C (mm)	Refractive index		Density (d)	$[\alpha]_D$ (−)	
				n_D	°C		°C	deg (in substance)
Acetic	$C_8H_{14}O_4$	87.0	79–80 (7)	1.4252	23	1.070	25	2.0
Propionic	$C_9H_{16}O_4$	80.0	88–89 (7)	1.4269	24	1.044	24	4.6
Butyric	$C_{10}H_{18}O_4$	81.0	98–99 (7)	1.4276	25	1.025	24	4.9
Caproic	$C_{12}H_{22}O_4$	75.0	119–121 (7)	1.4340	22	0.993	19	5.6

[a] Prepared from 2,3-isopropylidene-*sn*-glycerol $[\alpha]_D - 13.8°$ (in substance). [b] From Baer and Fischer (1945*a*).

Table IV. Formulas, Yields, and Physical Constants of 3-Monoacyl-sn-glycerols[a]

Acid	Formula	Yield (%)	Melting point °C	Refractive index n_D	Refractive index °C	$[\alpha]_D$ in dry pyridine (deg)	$[M]_D$ (deg)
Acetic	$C_5H_{10}O_4$	94.0	—	1.4500	22	10.5 (c, 5)	14.07
Propionic	$C_6H_{12}O_4$	95.0	—	1.4515	24	9.0 (c, 10)	13.32
Butyric	$C_7H_{14}O_4$	70.0	—				
		43.0	—	1.4500	24	8.3 (c, 10)	13.44
n-Valeric	$C_8H_{16}O_4$	77.5					
		61.5	—	1.4498	25	7.8 (c, 10)	13.73
Isovaleric	$C_8H_{16}O_4$	75.7					
		62.3	—	1.4470	25	7.3 (c, 10)	12.85
Caproic	$C_9H_{18}O_4$	83.0					
		47.5	7–9	1.4513	25	7.7 (c, 9)	14.63
Enanthic	$C_{10}H_{20}O_4$	75.0	—				
		49.0	14–15	1.4525	24	7.3 (c, 9.5)	14.89
Caprylic	$C_{11}H_{22}O_4$	73.0					
		57.0	28–30	1.4548	22	6.6 (c, 10.2)	14.38
Pelargonic	$C_{12}H_{24}O_4$	47.0	34–35	1.4548	24	6.27 (c, 9.9)	14.61
Capric	$C_{13}H_{26}O_4$	82.5	44	1.5000	40	5.55 (c, 10)	13.78
Hendecanoic	$C_{14}H_{28}O_4$	80.6	49–50	—	—	5.40 (c, 10)	14.04
Lauric	$C_{15}H_{30}O_4$	57.0	54–55	—	—	4.90 (c, 10)	13.43
Myristic	$C_{17}H_{34}O_4$	85.0	62–64	—	—	4.60 (c, 10)	13.89
Palmitic	$C_{19}H_{38}O_4$	73.0	71–72	—	—	4.37 (c, 7.8)	14.52
Stearic	$C_{21}H_{42}O_4$	56.0	76–77	—	—	3.58 (c, 12.3)	12.89

[a] From Baer and Fischer (1945*a*).

An alternative route for the synthesis of saturated and monounsaturated 1-monoacyl-*sn*-glycerols has been developed by Buchnea and Baer (1960) and Buchnea (1971). This method involves the preparation of 3-benzyl-*sn*-glycerol from D-mannitol, followed by tritylation in the *sn*-1 position and acylation in the *sn*-2 position. Removal of the trityl group with hydrogen chloride results in a simultaneous migration of the acyl group to the *sn*-1 position. The benzyl group is removed by catalytic reduction. The transformations are outlined in Fig. 9. The method is limited to the preparation of the saturated and monounsaturated fatty acid derivatives, since the single double bond can be protected by the reversible bromination without *trans* isomerization.

It has been demonstrated by several investigators (Crossley *et al.*, 1959; Serdarevich, 1967; Buchnea, 1967*c*) that 1- and 3-monoacyl-*sn*-glycerols isomerize under acidic, basic, or thermal conditions to an equilibrium mixture of 90% racemic monoacylglycerol and 10% 2-monoacylglycerol.* The isomerization rate appears to be strongly dependent on the size and nature of the acyl group. A knowledge of the rate of isomerization is important in evaluating the storage possibilities of the acylglycerols and in deciding on the timing of any subsequent acylations during the preparation of the diacyl- and triacyl-*sn*-glycerols, respectively, using 1- and 3-monoacyl-*sn*-glycerols as precursors.

* This author has demonstrated that another component of the equilibrium mixture is most likely bis(monoacylglycerol) ether (Buchnea, 1967*c*).

Table V. Rotatory Dispersion of 3-Monoacyl-sn-glycerols[a]

		$[\alpha]$ at nm (deg)							$[\alpha]_D$ in pyridine (deg)	
Acid	Percent in solvent	578	545	436	405/8	366	334	313	Found	Literature value[b]
Butyric	10 pyridine	−8.26	−9.39	−15.87	−18.94	−24.53	−29.97		−8.1	−8.3
Pivalic	None	+2.2	+2.8	+4.6	+5.2	+6.2	+7.2	+0.8		
	10 pyridine	−2.32	−3.2	−5.2	−6.2	−8.0			−2.2	
	10 benzene	+3.8	+4.3	+6.7	+7.7	+9.2	+10.6	+11.6		
	10 methanol	−5.2	−6.0	−10.4	−12.4	−19.7	−19.7			
	10 dioxane		−0.1	−0.5	−0.8	−1.6	−2.5	−3.4		
Lauric	10 pyridine	−5.2	−5.54	−9.46	−11.4	−14.7	−18.9	−22.4	−5.0	−4.9
	5 benzene	−1.1	−0.35	−0.65	−0.73	−1.73	−2.65	−4.7		
	10 methanol	−5.85	−6.58	−11.15	−13.7	−17.7	−22.5	−26.9		
Myristic	10 pyridine	−4.48	−5.38	−9.00	−10.62	−14.42	−18.42		−4.4	−4.5
	1.2 benzene	−1.2	−1.6	−4.0	−4.8	−6.3	−11.2			
	10 methanol	−5.85	−6.58	−11.15	−13.7	−17.7	−22.5	−26.9		
Palmitic	10 pyridine	−4.40	−4.83	−8.20	−9.85	−12.8	−15.8		−4.8	−4.37

[a] From Schlenk (1965*a*).
[b] Baer and Fischer (1945*a*).

Table VI. Physical Data and Rotational Values for Some 1,2-Isopropylidene-3-acyl-sn-glycerols[a]

Isopropylidene compound	Melting point (°C)	Boiling point (°C)	Literature value for melting and boiling points[b] (°C)	$[M]_{300}$ (deg)	$[M]_{250}$ (deg)	$[M]_{200}$ (deg)	$[\phi]_{max}$ (deg)
					(in hexane, 27°C)		
Acetyl		89–90/11 mm	77/6 mm	134	302	1308	−124
Butyryl		71–78/1 mm	101–102/2 mm	160	311	1337	−124
Decanoyl		157–160/3 mm	163–164/6 mm	165	325	1393	−130
Lauroyl		120–125/5 × 10^{-3}	130–131/2 × 10^{-3}				
Myristoyl	22–23		23–24	170	340	1418	−135
Palmitoyl	33–34		33–34.5	179	347	1429	−135
Stearoyl	42–43		41–42	178	355	1452	−126

[a] From Gronowitz *et al.* (1975). [b] Baer and Fischer (1945*a*).

Table VII. Physical Data and Rotational Values for Some 3-Acyl-sn-glycerols[a]

3-Acyl-*sn*-glycerol	Melting point (°C)	n_D^{τ}	Literature value[b]: Melting point (°C)	Literature value[b]: n_D^{τ}	$[\alpha]_D^{25}$	Literature value for $[\alpha]_D^{25}$ [b]	$[M]_{300}$ (deg)	$[M]_{250}$ (deg)	$[M]_{210}$ (deg)	$[M]_{200}$ (deg)	$[\phi]_{max}$ (deg)
					(in pyridine)			(in ethanol, 27°C)			
Acetyl		1.4498^{21}		1.4500[illegible]	−10.12 (c,4)	−10.5 (c,5)	−125	−216	−818	−1330	221
Butyryl		1.4493^{20}		1.4500[illegible]	−7.40 (c,2)	−8.3 (c,10)	−120	−214	−746	−1182	147
Decanoyl	42–43		44		−5.85 (c,2)	−5.55 (c,10)	−141	−262	−864	−1408	221
Lauroyl	53–54		54–55		−4.83 (c,3)	−4.90 (c,10)	−130	−238	−814	—	152
Myristoyl	61–62		62–64		−4.40 (c,10)	−4.60 (c,10)	−156	−270	−852	—	160
					−4.07 (c,3)						
Palmitoyl	68–69		71–72		−4.07 (c,8)	−4.37 (c,8)	−138	−238	−825	−1363	256
Stearoyl	74–75		76–77		−4.47 (c,12)	−3.58 (c,12)	−122	−245	−922	−1487	174

[a] From Gronowitz *et al.* (1975). [b] Baer and Fischer (1945*a*).

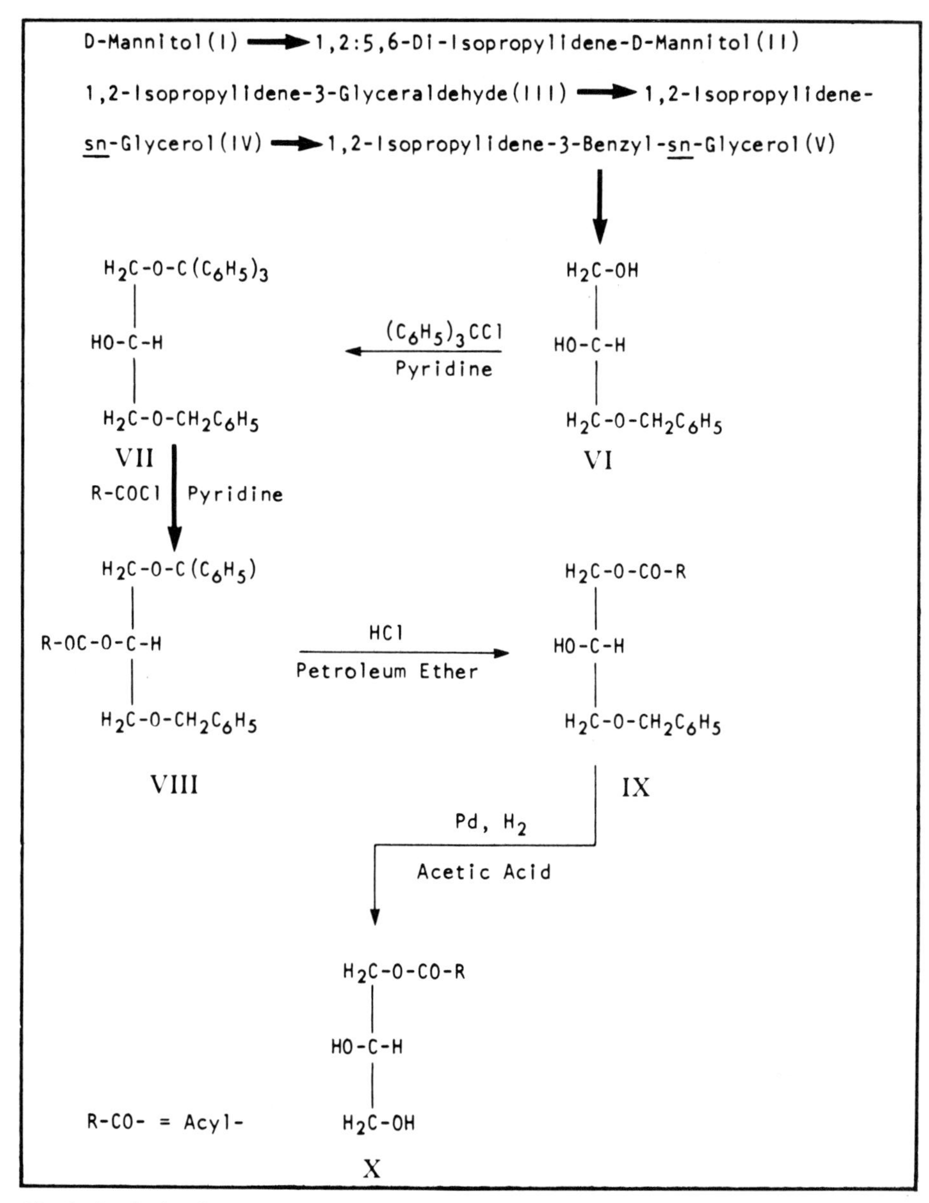

Fig. 9. Synthesis of 1-monoacyl-*sn*-glycerol from D-mannitol (Buchnea and Baer, 1960; Buchnea, 1971).

Gronowitz *et al.* (1975) investigated polarimetrically the racemization of 3-palmitoyl-*sn*-glycerol in three different solvents (chloroform, pyridine, and ethanol) and found that practically no racemization took place within 24 hr at room temperature or 50°C in ethanol solution. It is therefore apparent that the thermal racemization is unimportant as long as the further acylation of the optically active monoacyl-*sn*-glycerol enantiomers is carried out immediately after preparation of monoacyl-*sn*-glycerols. Furthermore, Gronowitz *et al.* (1975) found that the storage of 3-monostearoyl-*sn*-glycerol at −20°C for 1 month did not change the rotation value, while storage of 3-monoacetyl-*sn*-glycerol at room temperature for 1 month led to a loss of 75% of optical activity. The author (Buchnea, unpublished results) has found that his synthetic long-chain 1- and 3-monoacyl-*sn*-glycerol enantiomers (myristoyl-, palmitoyl-, stearoyl-, and arachidoyl-*sn*-glycerols) did not change their optical rotations when stored at −10°C for many years. Schlenk (1965*b*) has likewise reported that his preparations of optically active mono-, di-, and triacyl-*sn*-glycerols did not show any significant racemization after storage for 6–7 years.

5.6.3. Synthesis of 2-Monoacyl-sn-glycerols

Although the 2-monoacyl-*sn*-glycerols are not enantiomeric, their synthesis is mentioned in this chapter because these derivatives are key intermediates in several metabolic transformations of diacyl- and triacyl-*sn*-glycerols, and there is a great demand for pure synthetic standards. The saturated 2-monoacyl-*sn*-glycerols are prepared in good yield by acylation of 1,3-benzylidene-*sn*-glycerol with the acyl chloride in the presence of pyridine as described by Bergmann and Carter (1930), Stimmel and King (1934), and Daubert (1940). The original method is not suitable for the preparation of unsaturated 2-monoacyl-*sn*-glycerols because the protective benzylidene group could be removed only by catalytic hydrogenolysis with hydrogen in presence of palladium black as a catalyst.

5.6.4. Synthesis of Enantiomeric 1,2- and 2,3-Diacyl-sn-glycerols

The optically active saturated monoacid 1,2-diacyl-*sn*-glycerols were first prepared by Sowden and Fischer (1941). In this procedure, the free *sn*-3-hydroxyl group in the 1,2-isopropylidene-*sn*-glycerol is protected with a benzyl group. An acid hydrolysis of the protective isopropylidene group results in 3-benzyl-*sn*-glycerol. The liberated free hydroxyl groups are acylated with fatty acid chloride in the presence of pyridine to give 1,2-diacyl-3-benzyl-*sn*-glycerol. The benzyl group is removed by catalytic hydrogenation. The 2,3-diacyl-*sn*-glycerols are obtained if the starting material is 1-benzyl-*sn*-glycerol. Since the benzyl protective group can be removed only by catalytic hydrogenation, the method is limited to the synthesis of saturated monoacid diacyl-*sn*-glycerols.

Gronowitz *et al.* (1975) synthesized saturated monoacid 1,2-diacyl-*sn*-glycerol according to the method of Howe and Malkin (1951).

Baer and Buchnea (1958) have adopted the original procedure of Sowden and Fischer (1941) for the synthesis of 1,2- and 2,3-dioleoyl-*sn*-glycerols. In the modified procedure, the double bonds of the oleic acid radical are protected by bromine. After the removal of the protective benzyl group by catalytic reduction, the double bonds of the oleic acid residues are regenerated with zinc dust in diethyl ether, with a minimum amount of *trans* isomerization. Traces of *trans* isomer can be separated by crystallization from petroleum ether (b.p. 30–60°C).

Buchnea and Baer (1960) have further modified the benzyl ether procedure for the preparation of 1,2- and 2,3-diacyl-*sn*-glycerols to permit the synthesis of mixed-acid saturated and saturated monounsaturated diacyl-*sn*-glycerols. In the latter modification, the enantiomeric benzyl-*sn*-glycerols are first tritylated, which results in blocking of the remaining primary hydroxyl group in the benzyl-*sn*-glycerol. The tritylated benzyl ethers are then acylated at the free hydroxyl at the *sn*-2 position with a fatty acid chloride. The removal of the trityl group under acidic conditions results in the transfer of the fatty acid from the *sn*-2 position to the *sn*-1- or *sn*-3-hydroxyl group, depending on the enantiomeric benzyl ether selected. A subsequent acylation at the *sn*-2 position can then be carried out with another saturated or bromine-protected monounsaturated fatty acid chloride. The unsaturated acids are regenerated following removal of the benzyl group by catalytic hydrogenolysis. In view of the isomerization step, the fatty acids intended for the primary positions of the benzylglycerol molecules are introduced first into these glycerol molecules. Buchnea and Baer (1960) used this procedure to obtain the 1-stearoyl-2-oleoyl- and 1-oleoyl-2-stearoyl-*sn*-glycerols.

Gigg and Gigg (1967*a*) have described an alternate method for the preparation of mixed-acid 1,2- and 2,3-diacylglycerols. In this procedure, the 3-benzyl-*sn*-glycerol is converted to the 1,2-carbonate-3-benzyl-*sn*-glycerol by treatment with phosgene. The benzyl group is then removed by catalytic hydrogenolysis and the resulting 1,2-carbonate-*sn*-glycerol is converted into the 1,2-carbonate-3-tetrahydropyranyl-*sn*-glycerol by reacting with dihydropyran in the presence of an acid catalyst to give 1,2-carbonate-3-tetrahydropyranyl-*sn*-glycerol. A subsequent removal of the carbonate group with KOH yields 3-tetrahydropyranyl-*sn*-glycerol, which on acylation with saturated or unsaturated acid chloride in the presence of pyridine gives 1,2-diacyl-3-tetrahydropyranyl-*sn*-glycerol. The removal of the fatty acid substituent from the 1-*sn* position with pancreatic lipase yields 2-monoacyl-3-tetrahydropyranyl-*sn*-glycerol, which can be reacylated at the free *sn*-1 position with a different saturated or unsaturated fatty acid chloride in the presence of pyridine. The protective tetrahydropyranyl group is eventually removed with boric acid to yield the mixed-acid 1,2-diacyl-*sn*-glycerol. The mixed-acid 2,3-diacyl-*sn*-glycerol is obtained if the starting material is 1-benzyl-*sn*-glycerol (Gigg and Gigg, 1967*b*).

Pfeiffer *et al.* (1968) have reported the synthesis of optically active mono- and polyunsaturated monoacid 1,2- and 2,3-diacyl-*sn*-glycerols via the 1,2- and 2,3-carbonate-*sn*-glycerols. These workers employed 2,2,2-trichloroethoxycarbonyl groups to protect the primary hydroxyl groups. Following acylation, the protective groups were removed with zinc in acetic acid at 25°C with little or no isomerization.

A simplified synthesis of mixed-chain-length saturated and polyunsaturated acid 1,2- and 2,3-diacyl-*sn*-glycerols has been reported by Buchnea (1971, 1974). This procedure is based on the judicious use of the enantiomeric monoacyl-*sn*-glycerols as the starting materials. With an aim to duplicating the natural 1,2-diacyl-*sn*-glycerols and their enantiomers, the saturated 1- and 3-monoacyl-*sn*-glycerols were prepared as shown in the reaction schemes in Figs. 8 and 9. The monoacyl-*sn*-glycerols are dissolved in a mixture of anhydrous pyridine and anhydrous benzene (1 : 1). A solution of pure triphenylmethyl chloride (trityl chloride) in anhydrous benzene is then added with stirring under anhydrous conditions. The reaction mixture is kept at 40°C for 24 hr and the reaction products, the monoacyl-triphenylmethyl-*sn*-glycerols, are isolated, and acylated with the desired fatty acid chloride at the free *sn*-2-hydroxyl group of the glycerol molecule. The resulting mixed-acid 1,2- or 2,3-diacyl-triphenylmethyl-*sn*-glycerols are isolated and purified by chromatography on a silicic acid–boric acid (10 : 1, w/w) column, which also removes the protective trityl group. Buchnea (1971, 1974) has utilized this method in the preparation of an extensive series of diacyl-*sn*-glycerols, which are listed in Tables VIII and IX along with their physical properties.

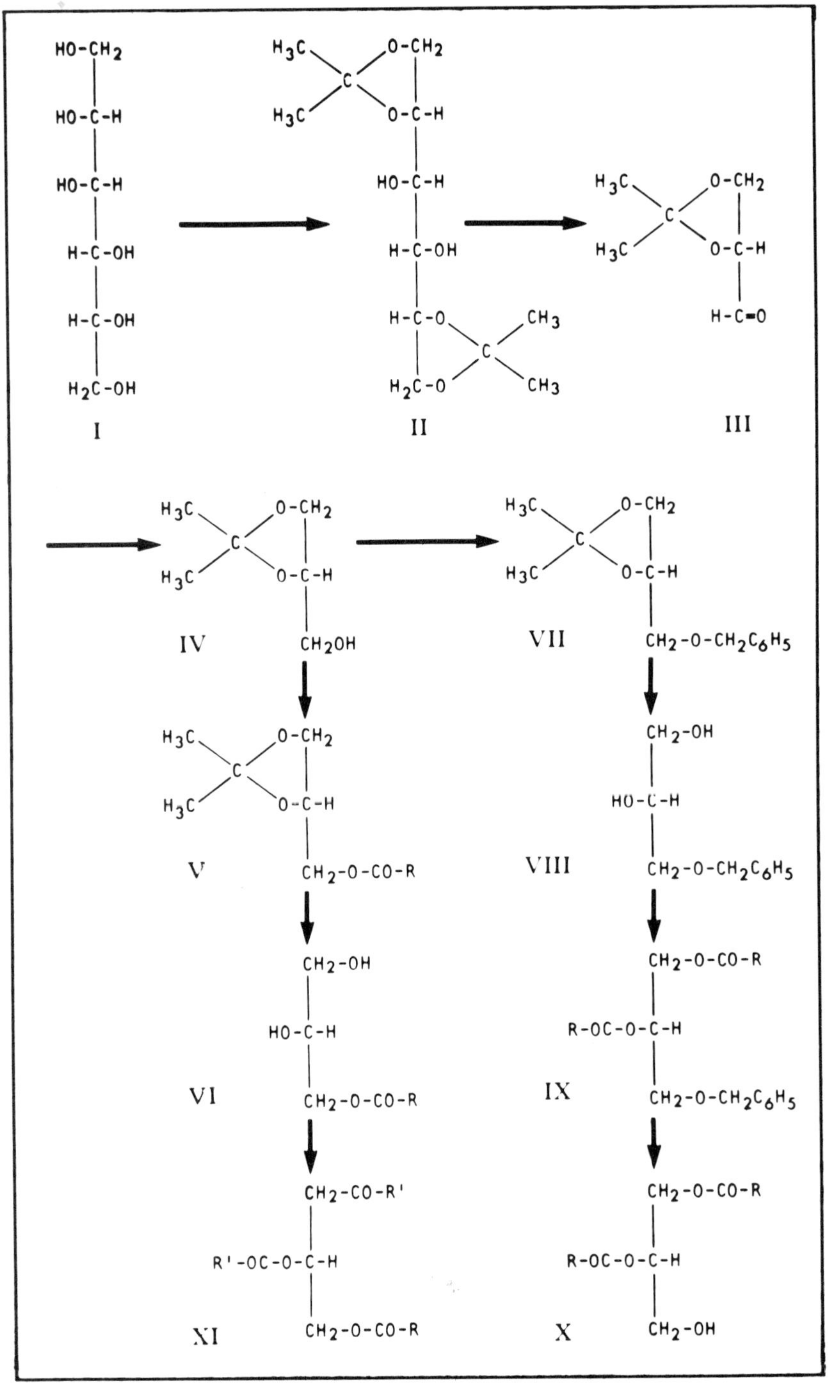

HO-CH_2 | HO-C-H | HO-C-H | H-C-OH | H-C-OH | H_2C-OH

I

H_3C, H_3C >C< O-CH_2 | O-C-H | HO-C-H | H-C-OH | H-C-O, H_2C-O >C< CH_3, CH_3

II

H_3C, H_3C >C< O-CH_2 | O-C-H | H-C=O

III

H_3C, H_3C >C< O-CH_2 | O-C-H | CH_2OH

IV

H_3C, H_3C >C< O-CH_2 | O-C-H | CH_2-O-CO-R

V

CH_2-OH | HO-C-H | CH_2-O-CO-R

VI

H_3C, H_3C >C< O-CH_2 | O-C-H | CH_2-O-$CH_2C_6H_5$

VII

CH_2-OH | HO-C-H | CH_2-O-$CH_2C_6H_5$

VIII

CH_2-O-CO-R | R-OC-O-C-H | CH_2-O-$CH_2C_6H_5$

IX

CH_2-O-CO-R | R-OC-O-C-H | CH_2-OH

X

CH_2-CO-R' | R'-OC-O-C-H | CH_2-O-CO-R

XI

Fig. 10. Synthesis of mono-, di-, and triacyl-*sn*-glycerols (Gronowitz *et al.*, 1975).

Table VIII. Formulas, Iodine Values, and Yields of Saturated and Monounsaturated Monoacid Diacyl-sn-glycerols and Mixed Saturated, Monounsaturated, and Polyunsaturated Diacid Diacyl-sn-glycerols[a]

Diacyl-*sn*-glycerol	Formula	Iodine value	Yield
sn-1,2-Distearoyl-[b]	$C_{39}H_{76}O_5$	—	70.0
sn-1,2-Dipalmitoyl-[b]	$C_{35}H_{68}O_5$	—	89.0
sn-1,2-Dioleoyl-[c]	$C_{39}H_{72}O_5$	81.0	95.0
sn-2,3-Dioleoyl-[c]	$C_{39}H_{72}O_5$	80.6	90.0
sn-1-Stearoyl-2-oleoyl-[d]	$C_{39}H_{74}O_5$	40.5	89.9
sn-1-Oleoyl-2-stearoyl-[d]	$C_{39}H_{74}O_5$	40.3	85.5
sn-1-Stearoyl-2-linoleoyl-	$C_{39}H_{72}O_5$	82.1	95.0
sn-1-Stearoyl-2-linolenoyl-	$C_{39}H_{70}O_5$	122.5	75.0
sn-2-Linoleoyl-3-stearoyl-	$C_{39}H_{72}O_5$	82.0	75.0
sn-2-Linolenoyl-3-oleoyl-	$C_{39}H_{68}O_5$	165.0	56.0
sn-1-Palmitoyl-2-oleoyl-	$C_{37}H_{70}O_5$	42.1	82.0
sn-1-Palmitoyl-2-linoleoyl-	$C_{37}H_{68}O_5$	83.8	79.0
sn-1-Palmitoyl-2-linolenoyl-	$C_{37}H_{66}O_5$	128.2	70.0
sn-2-Oleoyl-3-palmitoyl-	$C_{37}H_{70}O_5$	42.5	85.0
sn-2-Linoleoyl-3-palmitoyl-	$C_{37}H_{68}O_5$	83.5	78.0
sn-2-Linolenoyl-3-palmitoyl-	$C_{37}H_{66}O_5$	128.5	71.0

[a] From Buchnea (1971, 1974). [b] Sowden and Fischer (1941). [c] Baer and Buchnea (1958).
[d] Buchnea and Baer (1960).

Table IX. Specific Rotations, Refractive Indices, and Physical States at 20°C and Melting Points of Saturated and Monounsaturated Monoacid Diacyl-sn-glycerols and Mixed Saturated, Monounsaturated, and Polyunsaturated Diacid Diacyl-sn-glycerol Enantiomers[a]

Diacyl-*sn*-glycerol	Specific rotation in $CHCl_3$ (c, 10, deg)	Refractive index at 25°C	Physical state at 20°C
sn-1,2-Distearoyl-[b]	−2.7	—	(m.p. 74.5–75°C)
sn-1,2-Dipalmitoyl-[b]	−2.3	—	(m.p. 67–67.5°C)
sn-1,2-Dioleoyl-[c]	−2.8	1.4679	Oil
sn-2,3-Dioleoyl-[c]	+2.8	1.4679	Oil
sn-1-Stearoyl-2-oleoyl-[d]	−2.8	—	Paste
sn-1-Oleoyl-2-stearoyl-[d]	−2.8	—	Paste
sn-1-Stearoyl-2-linoleoyl-	−2.8	1.4710	Oil
sn-1-Stearoyl-2-linolenoyl-	−2.7	1.4702	Oil
sn-2-Linoleoyl-3-stearoyl-	+2.7	1.4710	Oil
sn-2-Linolenoyl-3-oleoyl-	+2.6	1.4800	Oil
sn-1-Palmitoyl-2-oleoyl-	+2.5[e]	—	Oil
sn-1-Palmitoyl-2-linoleoyl-	+2.5[e]	—	Oil
sn-1-Palmitoyl-2-linolenoyl-	+2.5[e]	—	Oil
sn-2-Oleoyl-3-palmitoyl-	−2.6[e]	—	Oil
sn-2-Linoleoyl-3-palmitoyl-	−2.4[e]	—	Oil
sn-2-Linolenoyl-3-palmitoyl-	−2.4[e]	—	Oil

[a] From Buchnea (1971, 1974). [b] Sowden and Fischer (1941). [c] Baer and Buchnea (1958).
[d] Buchnea and Baer (1960). [e] Specific rotation taken in substance.

Table X. Yields and Physical Constants of Some Diacyliodohydrin-sn-glycerols

Diacyliodohydrin-*sn*-glycerol	Specific rotation (deg)	Melting point (°C)	Yield (%)
sn-1-Stearoyl-2-oleoyl-3-iodohydrin-[a]	+3.1	17	80–90
sn-1-Iodohydrin-2-oleoyl-3-stearoyl-[a]	−4.0	—	—
sn-1-Oleoyl-2-stearoyl-3-iodohydrin-[a]	+3.2	25	70
sn-1-Stearoyl-2-lauroyl-3-iodohydrin-[a]	+3.3	35–36	—
sn-1-Palmitoyl-2-oleoyl-3-iodohydrin-[b]	+3.4	10.5–11	80
sn-1-Palmitoyl-2-linoleoyl-3-iodohydrin-[c]	+3.3	2	—

[a] de Haas and van Deenen (1961).
[b] Daemen *et al.* (1963).
[c] Daemen (1967).

An ingenious procedure for the preparation of mixed-acid 1,2-diacyl-*sn*-glycerols of well-defined composition is based on the acylation of 3-iodohydrin-*sn*-glycerols using a trityl protective group. The mixed-acid 1,2-diacyl-3-iodohydrin-*sn*-glycerols were first prepared by de Haas and van Deenen (1961), Daemen *et al.* (1963), Haverkate and van Deenen (1965), and Daemen (1967). Stoffel and Pruss (1969) extended this synthetic procedure to include the mixed-acid 1,2-diacyl-*sn*-glycerols containing linoleoyl, arachidonoyl, and phytanoyl radicals. Table X lists the enantiomeric 1,2-diacyl-3-iodohydrin-*sn*-glycerols thus far prepared, along with their physical properties. The protective iodohydrin group is removed by first replacing it with a phosphorylcholine moiety, which is then released by enzymatic hydrolysis with phospholipase C from *Bacillus cereus* (Haverkate and van Deenen, 1965). The intermediate formation of the 1,2-diacyl-*sn*-glycerophospholipid permits a verification of the distribution of the fatty acids by means of hydrolysis with phospholipase A_2 (*Crotalus adamanteus*), which releases the fatty acid from the *sn*-2 position. Since phospholipase C is not stereospecific, the iodohydrin method ought to be suitable also for the preparation of the 2,3-diacyl-*sn*-glycerols.

Under the synthesis of enantiomeric 1,2- and 2,3-diacyl-*sn*-glycerols may also be considered the preparation of 1-alkyl-2-acyl- and 1-alkenyl-2-acyl-*sn*-glycerols and their enantiomers. The alkyl-diacyl-*sn*-glycerols have been found in marine oils (André and Bloch, 1935). The 1,2-dialkyl-*sn*-glycerols occur in the form of natural alkylglycerol moieties in some glycerophospholipids. There is much interest in appropriate reference compounds. Stegerhock and Verkade (1956*a*) first prepared the racemic 1-alkyl-2-acylglycerols via the racemic 1-alkyl-3-tritylglycerol. Acylation of the free hydroxyl group in the *sn*-2 position gave racemic 1-alkyl-2-acyl-3-trityl-glycerol, which upon hydrogenolysis of the trityl group yielded racemic 1-alkyl-2-acylglycerol. Palameta and Kates (1966) used a similar procedure to prepare the optically active 1-trityl-2-acyl-3-alkyl-*sn*-glycerol from the corresponding 1-trityl-3-alkyl-*sn*-glycerol. However, a removal of the protective trityl group by hydrogen chloride led to acyl migration and formation of the 1-acyl-3-alkyl-*sn*-glycerol. The isomerization of the acyl group could have been prevented if the removal of the trityl group was effected by silicic acid/boric acid chromatography (Buchnea, 1971). By adopting the procedures developed by Buchnea (1971, 1974) for the synthesis of mixed-acid saturated and polyunsaturated diacyl-*sn*-glycerols (see above, this section), it is possible to synthesize both enantiomers of all alkylacyl- and alkenylacyl-*sn*-glycerols, using 1- and 3-alkyl- or 1- and 3-alkenyl-*sn*-glycerol ethers as starting materials. The latter compounds are easily prepared from 1,2- and 2,3-isopropylidene-*sn*-glycerols as described by Baumann and Mangold (1964).

The first preparations of long-chain 1,2-dialkyl-*sn*-glycerols were undertaken for establishing the structure of the naturally occurring glycerol diethers isolated from bacterial lipids. Kates *et al.* (1963) synthesized 1,2-dihexadecyl- and 1,2-dioctadecyl-*sn*-glycerols by alkylation of 3-benzyl-*sn*-glycerol with the corresponding long-chain alkyl bromides in boiling benzene using potassium hydroxide for condensation. Baumann and Mangold (1966) have described a synthesis applicable to the preparation of saturated and unsaturated, as well as mixed, dialkyl-*sn*-glycerol ethers, which is based on *sn*-2-alkylation of 1-alkyl-3-triphenylmethyl-*sn*-glycerol with alkyl methanesulfonates. Optically active 2,3-dialkyl-*sn*-glycerols can be prepared by a similar sequence of reactions starting with 3-alkyl- or 3-alkenyl-*sn*-glycerols (Baumann *et al.*, 1966; Palameta and Kates, 1966).

5.6.5. Synthesis of Enantiomeric 1,3-Diacyl-sn-glycerols

The enantiomeric 1,3-diacyl-*sn*-glycerols are of interest as intermediates in the synthesis of enantiomeric triacyl-*sn*-glycerols as well as reference compounds in physicochemical and metabolic investigations. The simplest route for the preparation of enantiomeric 1,3-diacyl-*sn*-glycerols is the isomerization of the 1,2- or 2,3-diacyl-*sn*-glycerols with HCl in diethyl ether solution at 0°C (Buchnea, 1967*c*).

Schlenk (1965*a*) has prepared a series of asymmetrical 1,3-diacyl-*sn*-glycerols as listed in Table XI, by employing the procedures of Baer and Fischer (1939*a*, 1945*a*).

Lok *et al.* (1976) have prepared asymmetrical 1,3-diacyl-*sn*-glycerols from an equimolar mixture of D-glycidol and fatty acid. The mixture was heated and stirred in the presence of tetraethylammonium bromide until all glycidol ester was used up. In the synthesis of 1,3-diacyl-*sn*-glycerols containing a short-chain fatty acid, the glycidol with shorter-chain fatty acid radical is preferable as starting material.

The syntheses of the enantiomeric 1-alkyl-3-acyl- and 1-acyl-3-alkyl-*sn*-glycerols are of interest in the preparation of enantiomeric alkyldiacyl-*sn*-glycerols. There are several reports discussing the appropriate methodology for their syntheses (Baumann, 1972; Palameta and Kates, 1966).

Table XI. Optical Rotatory Dispersion of 1,3-Diacyl-sn-glycerols[a]

sn-Diacylglycerols	Percent in solvent	[α] at nm (deg) 578	545	436	405/8	366	334	313
1-Butyroyl-3-myristoyl-	30 methanol	−0.42	−0.47	−0.80	−0.93	−1.23	−1.50	−1.69
	30 pyridine	−1.12	−0.17	−0.36	−0.45	−0.62	−0.84	−1.05
1-Isovaleroyl-3-myristoyl-	30 methanol	+0.07	+0.07	+0.11	+0.15	+0.20	+0.20	+0.37
1-Sorbinoyl-3-palmitoyl-	2.5 methanol	+1.84	+1.95	+2.93	+2.93	+5.08	+7.82	
	10 pyridine			+0.7	+1.5	+2.3		

[a] From Schlenk (1965*a*).

Table XII. *Rotatory Dispersion of Optically Active Triacylglycerols*[a]

sn-Triacylglycerols	Percent in solvent	[α] at nm (deg)								
		578	545	436	405/8	366	334	313	302	297
1,2-Divaleroyl-3-pivaloyl-	None	+1.35	+1.60	+2.70	+3.20	+4.12	+4.9	+5.6	+6.0	
	10 pyridine	+0.2	+0.22	+0.42	+0.48	+0.55				
	10 benzene	+0.30	+0.25	+0.15	+0.05	−0.05	−0.20	+0.85		
	10 methanol	+1.90	+2.20	+3.20	+3.70	+4.70	+6.00	+7.00	+7.65	+8.00
	10 dioxane	+1.04	+1.08	+2.00	+2.56	+3.48				
	10 ethyl acet.	+1.22	+1.42	+2.40	+2.85	+3.80				
	10 cyclohexane	+1.80	+2.06	+3.46	+4.20	−5.60	+6.8	+8.0	+8.8	
1,2-Dibutyroyl-3-lauroyl-	None	+0.05	+0.05	+1.11	+0.17					
	10 benzene	+1.10	+1.40		+2.20	+2.65				
1,2-Dibutyroyl-3-myristoyl-	None	+0.09	+0.10	+0.16	+0.18					
	30 benzene	+0.88	+0.94	+1.40	+1.66	+2.20				
1-Myristoyl-2,3-dibutyroyl-	None	−0.09	−0.13	−0.17						
	30 pyridine	−0.99	−1.12	−1.96	−2.36	−3.08				
	30 benzene	−0.79	−0.91	−1.62	−1.91	−2.57	−2.57	−2.74		
1-Myristoyl-2,3-diisovaleroyl-	None	−0.55	−0.64	−1.25	−1.39	−1.72				
	34 pyridine	−0.33	−0.42	−0.67	−0.90	−1.16				
	33 benzene	−0.41	−0.48	−0.92	−1.11	−1.53	−1.95			
1-Palmitoyl-2,3-disorbinoyl-	7.3 benzene	−6.85	−8.00							
1-Isovaleroyl-2,3-dimyristoyl-	30 benzene			−0.06	−0.12	−0.25	−0.43	−0.43		
1-Sorbinoyl-2,3-dipalmitoyl-	16.1 benzene	−5.57	−9.3	−17.9	−21.1	−30.95				
1,2-Dipalmitoyl-3-lauroyl-	32 benzene	+0.10	+0.10	+0.11	+0.10	+0.16	+0.16	+0.20	+0.23	
1-Lauroyl-2,3-dipalmitoyl-	21 benzene	−0.06	−0.07	−0.08	−0.11	−0.18				

[a] From Schlenk (1965*a*).

5.6.6. Synthesis of Enantiomeric Mixed-Acid Triacyl-sn-glycerols

The first enantiomeric triacyl-*sn*-glycerols were prepared by Baer and Fischer (1939*c*). They obtained the saturated diacid triacylglycerols by acylation of the enantiomeric 1- and 3-acyl-*sn*-glycerols with appropriate acid chlorides in the presence of pyridine or quinoline. This procedure was utilized by Schlenk (1962, 1965*a*, *b*) for the synthesis of several asymmetrical diacid triacyl-*sn*-glycerols. These triacylglycerols are listed in Tables XII and XIII, together with their piezoelectric properties. The enantiomeric monoacyl-*sn*-glycerols are now generally employed as starting materials for the preparation of enantiomeric saturated and unsaturated diacid triacyl-*sn*-glycerols. Thus Quinn *et al.* (1967) have prepared the 1,2-dioleoyl-3-palmitoyl-, 1-oleoyl-2,3-dipalmitoyl-, and 1,2-dipalmitoyl-3-oleoyl-*sn*-glycerols. Gronowitz *et al.* (1975) prepared a series of mixed-chain-length saturated diacid triacyl-*sn*-glycerols according to the method of Baer and Fischer (1939*c*) and investigated their optical activity (Tables XIV and XV).

Another route to enantiomeric diacid triacyl-*sn*-glycerols was described by Sowden and Fischer (1941). In this method, enantiomeric monoacid 1,2-diacyl- and 2,3-diacyl-*sn*-glycerols are acylated with different saturated or unsaturated fatty acid chlorides to yield diacid triacyl-*sn*-glycerols. Tattrie *et al.* (1958) employed this procedure for the synthesis of enantiomeric diacid triacyl-*sn*-glycerols for the investigation of the positional specificity of pancreatic lipase.

Quinn *et al.* (1967) and Jensen *et al.* (1970) acylated mixed-acid 1,3-diacyl-*sn*-glycerol in the *sn*-2 position with an appropriate fatty acid chloride in the presence of pyridine to yield mixed-triacid triacyl-*sn*-glycerols.

Lok *et al.* (1976) have synthesized several mixed-diacid and mixed-triacid triacyl-*sn*-glycerols from D-glycidol ester via 1,3-diacyl-*sn*-glycerols. The 1,3-diacyl-*sn*-glycerol was acylated with an appropriate fatty acid chloride in the presence of pyridine in hexane. The reaction mixture was refluxed until the reaction was complete. The resulting triacyl-*sn*-glycerols were then analyzed with pancreatic lipase for the purity of the *sn*-2 fatty acid composition of the triacyl-*sn*-glycerols.

Table XIII. Piezoelectric Investigation of Acylglycerols[a]

Acylglycerol	Form	Piezo effect
Triacyl-*sn*-glycerols		
1-Lauroyl-2,3-dipalmitoyl-	Antipodes	Yes
	Racemate	No
1-Stearoyl-2,3-dipalmitoyl-	Antipodes	Yes
	Racemate	No
1-Palmitoyl-2-oleoyl-3-stearoyl-	Antipodes	Yes
	Racemate	No
Diacyl-*sn*-glycerols		
1-Isovaleroyl-3-myristoyl-	Antipodes	Yes
	Racemate	No
1-Stearoyl-3-oleoyl-	Antipodes	Yes
	Racemate	No

[a] From Schlenk (1965*a*).

Table XIV. Physical Data and Rotational Values for Some Triacylglycerols[a]

Compound[b]	Melting point (°C)	n_D^T	$[M]_D^{25}$ in benzene (deg)	$[M]_{300}^{27}$ in benzene (deg)	$[M]_{300}^{27}$ in hexane (deg)	$[M]_{235}^{27}$ in hexane (deg)
sn-AcAcB		1.4332^{20}	+4.7	+30	+25	+83
sn-AcAcD		1.4407^{20}	+7.1	+39	+15	+55
sn-AcAcM		1.4461^{20}	+8.8	+57	+26	+77
sn-AcAcP	26–27		+9.1	+57	+27	+83
sn-AcAcSt	39–40		+10.2	+61	+26	+97
sn-BBM		1.4470^{20}		+27		
sn-HHM		1.4498^{21}		+12		
sn-OcOcM		1.4526^{20}		+5.7		
sn-DDM	26–26.5			+2.9		
sn-LaLaM	39.5–40			+0.8		
sn-PPM	54.5–55.5			−1.5		
sn-StStM	60–61			−1.4		
sn-LaLaB		1.4508^{20}		−26	−2.0	
sn-LaLaP	42.5–43			+2.0		
sn-LaLaSt	43.5–44.5			+1.8		
sn-MMAc	38–38.5			−51	−20	−76
sn-StStAc	52.5–53.5			−60	−28	−90
						Optical yield
sn-MAcAc		1.4477^{21}		−43		75%
sn-MHH		1.4505^{21}		−11		90%
sn-MDD	26–27			−2.6		90%
sn-PAcAc	24–25				−15	55%
sn-StLaLa	43–43.5			−1.7		94%

[a] From Gronowitz *et al.* (1975).
[b] Abbreviations for fatty acids: Ac, acetic; B, butyric; D, decanoic; H, hexanoic; La, lauric; M, myristic; Oc, octanoic; P, palmitic; St, stearic.

The asymmetrical diacid triacyl-*sn*-glycerols are conveniently prepared by appropriate acylation of 1- or 3-monoacyl-*sn*-glycerols or monoacid 1,2- or 2,3-diacyl-*sn*-glycerols. However, the synthesis of triacid triacyl-*sn*-glycerols requires mixed-acid diacyl-*sn*-glycerols as starting materials. The utilization of mixed-acid

Table XV. Stereospecific Analysis of Three Optically Active Triacylglycerols[a]

Compound[b]	Position 1 (%)	Position 2 (%)	Position 3 (%)
sn-PPM	96.3 P	98.1 P	0.5 P
	3.7 M	1.9 M	99.5 M
sn-DDM	96.4 D	98.6 D	4.8 D
	2.0 La	1.1 La	—
	1.6 M	0.4 M	95.2 M
sn-LaLaP	99.0 La	99.5 La	5.1 La
	0.2 M	—	0.7 La
	0.8 P	0.4 P	94.2 P

[a] From Gronowitz *et al.* (1975).
[b] Abbreviations; P, palmitic; M, myristic; D, decanoic; La, lauric.

1,2- and 2,3-diacyl-*sn*-glycerols (Buchnea, 1971, 1974) and mixed-acid 1,3-diacyl-*sn*-glycerols (Lok *et al.*, 1976) as starting materials allows the preparation of all varieties of stereospecifically substituted triacid triacyl-*sn*-glycerols.

5.7. *Synthesis of Enantiomeric Isotope-Labeled Acyl- and Alkylacyl-sn-glycerols*

Isotope-labeled mono-, di-, and triacyl-*sn*-glycerols of relatively high specific activity have been prepared as substrates for metabolic studies. The labeled atoms may be located in the fatty chain, in the glycerol moiety, or in both moieties. In most instances, the radioactive or stable isotope-labeled compounds can be synthesized by standard methods, although appropriate measures must be taken not to contaminate the equipment and environment with radioactivity. The 1-alkyl-2,3-diacyl-*sn*-glycerols are naturally occurring lipids (Baumann, 1972) which also have been prepared in synthetic radioactive form to facilitate simple enzymatic and metabolic studies.

5.7.1. *Synthesis of Enantiomeric Isotope-Labeled Acyl-sn-glycerols*

The preparation of the enantiomeric acyl- and alkylacyl-*sn*-glycerols containing isotope-labeled fatty acids is performed using the general methods described for the synthesis of the corresponding enantiomeric unlabeled compounds. In view of the expense of the isotope-labeled acids, the preparations are usually carried out on a smaller scale and involve more complete recoveries of intermediates, products, and unreacted starting materials. For the purposes of various metabolic studies, Buchnea (1973, unpublished results) has prepared the 1- and 3-[1′-^{14}C]palmitoyl-*sn*-glycerols using standard methods in amounts of several grams and with specific activities ranging in several thousands of cpm/mg. The trityl derivatives of the radioactive palmitoyl-*sn*-glycerols were acylated with oleoyl and linoleoyl chloride to give the 1-[1′-^{14}C]palmitoyl-2-oleoyl- and 1-[1′-^{14}C]palmitoyl-2-linoleoyl-*sn*-glycerols and their enantiomers, respectively. These compounds have been utilized by Morley *et al.* (1974) in studies of the positional specificity of lipoprotein lipase. The radioactive diacyl-*sn*-glycerols were further acylated with appropriate fatty acid chlorides to produce mixed-acid triacyl-*sn*-glycerols with the radioactive palmitic acid in either the *sn*-1 or the *sn*-3 position. These compounds have been used in the determination of the positional specificity of both pancreatic and lipoprotein lipase (Morley *et al.*, 1975).

Morley *et al.* (1975) have prepared the 1-palmitoyl-2-linoleoyl-3-(CD_3-)-palmitoyl-*sn*-glycerol by reacting the 1,2-diacyl-*sn*-glycerol with the chloride of deuterated palmitic acid. A Grignard degradation of the triacyl-*sn*-glycerol resulted in the formation of a racemic mixture of 1,2- and 2,3-diacyl-*sn*-glycerols, of which only the 2,3 isomer contained the deuterated palmitic acid. The enantiomeric triacylglycerol and diacylglycerols were employed in studies of the positional specificity of the lipases.

The general methods of acyl-*sn*-glycerol synthesis may also be employed for the preparation of acylglycerols containing labeled atoms in more than one position of the glycerol molecule, if such are required (Fig. 11).

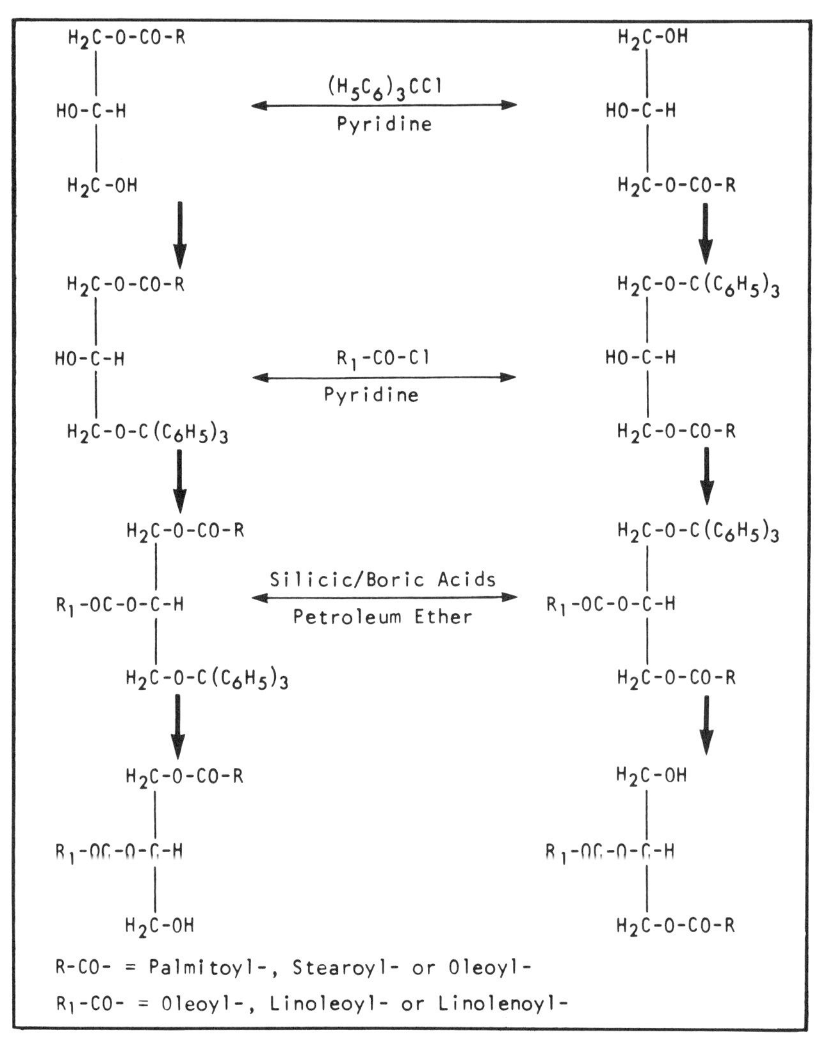

Fig. 11. Syntheses of mixed fatty acid saturated and unsaturated diacyl-*sn*-glycerol enantiomers (Buchnea, 1971, 1974).

The preparation of acylglycerols stereospecifically labeled in the glycerol moiety, however, is much more difficult, as it requires the chemical synthesis of the glycerol molecule. Karnovsky and Wolff (1960) prepared the *sn*-[1-^{14}C]- and the *sn*-[3-^{14}C]glycerol trioleates by acylation of the corresponding glycerols. The enantiomeric glycerols were obtained from D- and L-serine-3-^{14}C, respectively, by deamination of the acids and reduction of the resulting glyceric acid esters. The radioactive D-serine and L-serine were resolved from a racemic mixture prepared by a chemical synthesis with methanol-^{14}C.

5.7.2. Synthesis of Enantiomeric Isotope-Labeled Alkylacyl-sn-glycerols

The synthesis of enantiomeric isotope-labeled alkylacylglycerols and alkenylacylglycerols is of interest as a source of acylglycerol derivatives of increased metabolic stability and restricted positional accessibility to lipases. These compounds may be used as markers in both biosynthetic and degradative studies. The enantiomeric monoalkyl or monoalkenyl-*sn*-glycerols containing the labeled atoms in the fatty acid chains are readily prepared by reacting the 1,2- or 2,3-isopropylidene-*sn*-glycerols with the radioactive alkyl or alkenyl methanesulfonate as described by Baumann and Mangold (1964). Åkesson *et al.* (1976) have prepared the 1-*O*-tetradecyl-*sn*-glycerol from the 3-tetradecyl-*sn*-glycerol by inversion of configuration at C-2 by the method of Lands and Zschocke (1965). Åkesson *et al.* (1976) prepared the 1,2-[^{3}H]dioleoyl-3-tetradecyl-*sn*-glycerol and the 1-tetradecyl-2,3-[^{14}C]dioleoyl-*sn*-glycerol by acylation of the glyceryl ethers with oleoyl chloride. Paltauf (1971) has prepared the 1-[9,10-$^3H_2$]octadecenyl-*sn*-glycerol and its *sn*-3 isomer by catalytic tritiation of the corresponding octadecenyl derivatives. The enantiomeric octadecenyl-*sn*-glycerols were prepared as described by Chacko and Hanahan (1968). Paltauf *et al.* (1974) prepared the 1-*O*-[9,10-$^3H_2$]octadecyl-2,3-dioctadecenoyl-*sn*-glycerol and the 1,2-dioctadecenoyl-3-[1-^{14}C]octadecyl-*sn*-glycerol by acylation of the glyceryl ethers with the appropriate radioactive fatty acid chloride. Paltauf and Wagner (1976) have synthesized the 1-[1′-^{14}C]octadecenoyl-2,3-dioctadecenyl-*sn*-glycerol and the 1,2-dioctadecenyl-3-[9,10-$^3H_2$]octadecenoyl-*sn*-glycerol by the method of Baumann and Mangold (1966), starting from 3-octadecenyl-*sn*-glycerol and 1-octadecenyl-*sn*-glycerol.

The enantiomeric 1-alkyl-3-acyl- and 1-acyl-3-alkyl-*sn*-glycerols containing radioactive fatty acid chains may be synthesized by isomerization of the appropriate 1-alkyl-2-acyl- and 3-alkyl-2-acyl-*sn*-glycerols in ethereal HCl solution at low temperature.

5.8. Methods of Isolation and Purification of Synthetic Acyl-sn-glycerols

Even though the various starting materials and reagents employed in stereospecific syntheses of triacylglycerols are of the highest quality, impurities and by-products are generated during the synthetic transformations, and they must be removed after the completion of each step. It is also necessary to remove any unreacted reagents and starting materials. In the past, great reliance was put upon distillation, solvent extraction, and crystallization as methods of isolation and purification of the intermediates and final products. More recently, chromatographic methods have been employed with increasing frequency, especially for the isolation and purification of the higher-molecular-weight products.

5.8.1. Selection of Solvents

The nature of the solvents employed for the extraction, crystallization, and chromatographic purification of the acylglycerols is very important as it can readily destroy the intermediate or final product. As a general rule, the solvents must be inert and dry. Depending on the exact nature of the acylglycerol molecules, the polarity of the solvents ranges from pure petroleum ether (b.p. 30–60°C) to diethyl ether. Solvents which generate acids (e.g., chloroform) must be avoided for storage and prolonged chromatographic runs. When necessary, the organic phase is dried with anhydrous sodium or magnesium sulfate. Traces of acids can be conveniently removed from the organic phase by washing with a saturated solution of sodium bicarbonate, the bicarbonate being removed with water.

5.8.2. Distillation and Crystallization

Although most acylglycerols are stable to distillation in high vacuum, this method of isolation and purification is impractical when applied to the relatively small amounts of material usually being prepared. Distillation, however, is excellently suited to the isolation and purification of the lower-molecular-weight starting materials and certain intermediates. The purification may be efficiently carried out by fractional distillation in high vacuum.

The crystallization of acylglycerols is limited largely to the saturated derivatives. Appropriate solvents are petroleum ether (b.p. 30–60°C) alone or in mixture with diethyl ether, acetone, methanol, chloroform, etc. The unsaturated acylglycerols may be forced out of solution by lowering the temperature, but this practice severely limits the degree of purification attained.

5.8.3. Chromatography

The most effective method for the isolation and purification of the mono-, di-, and triacylglycerols is adsorption chromatography. However, the adsorbent must be carefully selected, especially for the handling of the partial acylglycerols, which may be subject to acyl migration. Aluminum oxide, which has been successfully employed for the separation and purification of glycerophospholipids (Hanahan and Jayko, 1952), leads to acyl migration in mono- and diacyglycerols, as well as hydrolysis of ester bonds and isomerization of double bonds in acylglycerols. Certain preparations of aluminum oxide of controlled polarity and alkalinity, however, may be less harmful than others. Some isomerization of partial acylglycerols also takes place during silicic acid chromatography (Borgström, 1952), but this can usually be prevented by including boric acid in the adsorbent mixture (Thomas *et al.*, 1965). Appropriate polarities of the eluting solvents are prepared by mixing petroleum ether (b.p. 30–60°C) with increasing proportions of diethyl ether (95:5 to 50:50 v/v, respectively). The silicic acid/boric acid column must be prepared with nonpolar solvents (e.g., petroleum ether) and must be eluted with solvents that do not dissolve boric acid.

Of special interest is the removal of the triphenylmethyl (trityl) protective group from the acylglycerol molecules during chromatography on silicic acid/boric acid columns, as first described by Buchnea (Buchnea and Baer, 1960; Buchnea, 1971, 1974). The removal of the trityl protective group takes place without isomerization of the acyl groups. Since it may be required that the trityl group be removed from glycerolipid derivatives of various degrees of polarities, the polarity of the eluting solvent, the temperature of elution, and the rate of flow of the solvent must be empirically adjusted. Allowance may also have to be made for any variability in the polarity and moisture content of the silicic acid/boric acid mixture. Buchnea (1974) has found that suitable solvent mixtures for the removal of trityl groups from diacylglycerols are petroleum ether and petroleum ether–diethyl ether mixture (95:5, v/v).

It is best to test the detritylation system with suitable reference compounds before committing it to the workup of previously untested material.

TLC with boric acid-treated silica gel (Thomas *et al.*, 1965) is widely used for the isolation of mono- and diacylglycerols without isomerization. This method, however, removes only one triphenylmethyl group from 1,3-di-(triphenylmethyl)-2-acyl-*sn*-glycerols (Buchnea, unpublished results, 1965).

Baer (1974) has questioned the nature of the form of boric acid in these adsorbents following heating at elevated temperatures during activation. He has suggested that all boric acid may be converted into metaboric and tetraboric acid.

Whatever the actual form of boric acid after activation at 115°C, the silicic acid/boric acid plates and columns are effective means of minimizing the isomerization of the acylglycerols during adsorption chromatography and are useful reagents for removing the protective triphenylmethyl groups from acylglycerols.

Conventional column and thin-layer chromatography are extensively utilized in the course of syntheses of acylglycerols, while gas–liquid chromatography may also be useful.

The general methodology of acylglycerol separation is discussed in great detail by Kates (1972), Jensen and Pitas (1976), and Myher (Chapter 3, this volume).

5.9. Determination of Structure and Absolute Configuration of Synthetic and Natural Acyl-sn-glycerols

The determination of structure and absolute configuration of synthetic acylglycerols serves to verify the identity of the final product at the end of a lengthy series of difficult transformations. A determination of structure and absolute configuration of a natural acylglycerol isolated in pure form allows a comparison of its structure to that of a synthetic acylglycerol prepared to duplicate the natural one. Chemical, enzymatic, and physical methods have all been employed for this purpose, and the more successful procedures are summarized in the following sections.

5.9.1. Chemical and Enzymatic Methods

The various synthetic methods described in this chapter allow the duplication of any natural acylglycerol structure, but it may be difficult to prove that a true synthetic analogue has been obtained, because the structure of the natural acylglycerols is not known. Despite provisional evidence for the enantiomeric nature of natural triacylglycerols (Fischer and Baer, 1941), there have been claims of the existence of various random distributions of fatty acids in natural triacylglycerols (Vander Wal, 1960; Coleman and Fulton, 1961; Gunstone, 1967).

5.9.1.1. Chemical Proof

The assessment of the exact structure and configuration of natural acylglycerols became possible with the synthesis of the optically active 1,2- and 2,3-isopropylidene-*sn*-glycerols (Fischer and Baer, 1937). These compounds became the key substances in synthesis of all optically active *sn*-glycerol derivatives. By virtue of the known stereochemical relationship to *sn*-1- and *sn*-3-glyceraldehyde, it also has been possible to establish unambiguously methods for proving the structure and configuration of naturally occurring acylglycerols and glycerophospholipids.

It is well known that natural triacylglycerols are complex mixtures of glycerol esters of different fatty acids. Pure individual acylglycerols have seldom been isolated, and enantiomers have never been resolved. Most observations on the triacylglycerol structure of natural fats have therefore been made on mixtures of many molecular species. The introduction of modern methods of acylglycerol analysis in combination with stereospecific enzymatic hydrolysis has demonstrated that natural fats are largely asymmetrical. The publications of Schwartz and Carter (1954) and Hirschmann (1960) have emphasized the fact that regardless of the fatty acid substitution the *sn*-1 and *sn*-3 positions of glycerol molecules are not interchangeable.

5.9.1.2. *Enzymatic Proof*

The stereospecific distribution of the fatty acids in the natural and synthetic triacylglycerols can be readily determined by the method proposed by Brockerhoff (1965, 1967). This analysis is based on the positional and enantiomeric specificity of phospholipase A_2. The free monoacylglycerols or diacylglycerols generated randomly from triacylglycerols are converted into *sn*-1,2(2,3)-diacylphosphatidylphenols, of which only the *sn*-1,2 isomers serve as substrates for the enzyme. Since the procedure does not indicate the molecular association of the fatty acids except in very simple species, exact identities of enantiomers are not usually obtained, but the overall degree of randomness or lack of it can always be demonstrated.

Lands *et al.* (1966) have shown that the diacylglycerol kinase from *Escherichia coli* selectively catalyzes the phosphorylation of the 1,2-diacyl-*sn*-glycerols but not of the 2,3-diacyl-*sn*-glycerols. In combination with a method for random generation of 1,2(2,3)-diacyl-*sn*-glycerols, this method also can be used for the assessment of the positional distribution of fatty acids in synthetic and natural triacylglycerols. The stereospecific analysis of natural triacyl-*sn*-glycerols has been discussed in great detail in this volume by Breckenridge (Chapter 4). The stereospecific analyses of natural triacylglycerols have largely discredited the various random distributions of fatty acids postulated for natural triacylglycerols by earlier investigators (Vander Wal, 1960; Coleman and Fulton, 1961; Gunstone, 1967).

5.9.1.3. *Assessment of Enzyme Specificity by Synthetic Acyl-sn-glycerols*

The synthetic enantiomers of acyl-*sn*-glycerols can serve as stereospecific probes of the positional specificity of lipases and acyltransferases and hence of the chiral nature of their natural substrates and reaction products. Kennedy and Weiss (1956) demonstrated that the transfer of phosphorylcholine from the cytidine nucleotide complex was stimulated by addition of 1,2-diacyl-*sn*-glycerol, resulting in the biosynthesis of phosphatidylcholine. Such a synthesis did not take place with the enantiomeric 2,3-diacyl-*sn*-glycerols (Weiss *et al.*, 1960). However, the acyltransferases involved in the biosynthesis of triacylglycerols appeared to catalyze the acylation of both 1,2- and 2,3-dioleoyl-*sn*-glycerol (Weiss *et al.*, 1960). More recently, O'Doherty *et al.* (1972) have employed mixed-acid 1,2- and 2,3-diacyl-*sn*-glycerols for quantitation of the relative activity of the *sn*-1- and *sn*-3-acyltransferases in the microsomes of rat liver and intestine. The hepatic enzyme, if a single entity, showed a definite preference for the acylation of the 1,2-diacyl-*sn*-glycerols. The intestinal enzyme esterified much more of the 2,3-diacyl-*sn*-glycerols than did the hepatic enzyme.

Enantiomeric triacylglycerols and alkyldiacylglycerols have been employed in studies of the positional specificity of lipases. Tattrie *et al.* (1958) used enantiomeric diacid triacylglycerols to demonstrate that pancreatic lipase shows no preference for one primary ester group over the other, but hydrolyzes both at the same rate. Karnovsky and Wolff (1960) examined the lipolysis of *sn*-[1-^{14}C]glycerol trioleate and *sn*-[3-^{14}C]glycerol trioleate by pancreatic and lipoprotein lipase and also concluded that both of these enzymes were not stereospecific. Morley *et al.* (1974), however, observed that from synthetic 1-palmitoyl-2-oleoyl-3-[1-^{14}C]linoleoyl-*sn*-glycerol and from 1-linoleoyl-2-oleoyl-3-[1-^{14}C]palmitoyl-*sn*-glycerol lipoprotein lipase released the fatty acids from the *sn*-1 position in preference to those of the *sn*-3 position, although pancreatic lipase showed no differentiation between the two primary positions. A similar positional specificity for lipoprotein lipase was demonstrated by Paltauf *et al.* (1974) and Åkesson *et al.* (1976), who used synthetic enantiomers of 1- and 3-alkyldiacyl-*sn*-glycerols. Åkesson *et al.* (1976) showed a comparable positional specificity for hepatic lipase using similar substrates. According to Paltauf *et al.* (1974), lingual lipase released the fatty acids from the *sn*-3 position in preference to those in the *sn*-1 position when tested with the enantiomeric alkyldiacyl-*sn*-glycerols. Interestingly, the presence of an alkyl group in the *sn*-2 position abolished the positional specificity of the lipoprotein lipase (Paltauf and Wagner, 1976). Morley *et al.* (1975) have also assessed the stereospecificity of lipoprotein lipase toward diacyl-*sn*-glycerols using radioactive mixed-acid 1,2- and 2,3-diacyl-*sn*-glycerols. Although a preferential release of the fatty acids from the *sn*-1 position was found, the effect was not large enough to account for the accumulation of the 2,3-diacyl-*sn*-glycerols during lipoprotein lipase hydrolysis of triacylglycerols.

5.9.2. Physical Methods

It is obvious that the triacid and the asymmetrical diacid triacylglycerols must exist in enantiomeric forms due to the chiral nature of C-2 of glycerol and the presence of two different fatty acids in the primary positions. As a result, it would be expected that differences would be seen in the optical rotation of the chemically indistinguishable forms of synthetic triacylglycerols as well as in pure triacylglycerol species isolated from natural fats. Although these differences are small, they can nevertheless be demonstrated experimentally using sensitive measuring techniques. In addition to differences in optical rotation, the enantiomeric acylglycerols have been shown to be different in the piezoelectricity and X-ray diffraction patterns of their crystals, as well as in the NMR or PMR spectra of certain metal complexes.

5.9.2.1. *Polarimetry*

In classical polarimetry, the specific rotation $[\alpha]_D$ is usually given for the strong yellow band of sodium light, called the D-line, corresponding to 589 nm. Because of the great similarity in the molecules of the common homologous fatty acid substituents in the *sn*-1 and *sn*-3 positions of the enantiomeric triacylglycerols, the differences recorded by the polarimetric readings are very small. The difficulty of detecting optical activity in enantiomerically pure triacyl-*sn*-glycerols of the common long-chain fatty acids was first pointed out by Sowden and Fischer (1941). Schlenk (1962, 1965*a*, *b*) synthesized ten different enantiomeric triacyl-*sn*-glycerols and demonstrated that several of them have small but measurable optical rotations (Table XII). However, the majority of long-chain triacyl-*sn*-glycerols failed to show optical rotation even in the ultraviolet regions, which are more sensitive. On the basis of the results obtained, Schlenk (1965*a*) proposed that if in a saturated triacyl-*sn*-glycerol the chain length at the *sn*-1 position is greater than that of the *sn*-3 position, the absolute configuration of the triacyl-*sn*-glycerol is "S," and the rotation is negative. In the opposite instance, the absolute configuration is "R," and the rotation is positive. These correlations have been recently verified by Gronowitz *et al.* (1975) as explained in Section 5.9.2.2. Schlenk (1965*a*,*b*) introduced the term "crypto-activity" to describe optical activity that is expected but not experimentally detected. Morris (1965*a*, *b*) has demonstrated that a higher optical activity may be recorded by simple polarimetry for the asymmetrical natural triacylglycerols following their degradation to diacylglycerols. Relatively pure natural asymmetrical triacylglycerols are subjected to a brief random digestion with pancreatic lipase, and the mixed-acid, 1,2- and 2,3-diacyl-*sn*-glycerols released are isolated and converted to trimethylsilyl ethers. Each pair of the diacylglycerol silyl ethers is separated by preparative TLC on the basis of unsaturation of the fatty acids in positions *sn*-1 and *sn*-3, respectively, and the optical rotations are determined. All natural asymmetrical triacylglycerols so examined were found to be optically active. In this connection, it is pertinent to note that Baer and Mahadevan (1959) had shown that the optical rotations measured with the D-line of sodium light vary with the concentration (5–80%) of the diacylglycerols (1,2-didecanoyl-*sn*-glycerol) and the nature of the solvent. They found that the solvents fall into two distinct groups: those in which the specific rotation of the diacylglycerols remains fairly constant over the whole concentration range, and those in which it undergoes considerable change from negative values for low concentrations to positive values for high concentrations. Zero values are observed at the approximate concentrations of 15% for tetrachloromethane, 40% for chloroform, and 43% for dichloromethane.

Occasionally, natural long-chain triacylglycerols have exhibited unusually high optical rotations on conventional polarimetry, but these have been shown to be due to the presence of optically active fatty acids. Thus the seed oil of the Chinese tallow tree *Sapium sebiferum* contains 8-hydroxy-5,6-octadienoic acid, which is esterified to a *trans*-2,*cis*-4-decadienoic acid. The optical activity is due to the allenic acid (Maier and Holman, 1964; Sprecher *et al.*, 1965), although it has been subsequently shown that the estolide moiety is specifically attached to the *sn*-3 position of the triacylglycerol molecule (Christie and Moore, 1969).

5.9.2.2. *ORD and CD Spectrometry*

Gronowitz *et al.* (1975) have studied in detail the more sensitive optical rotatory dispersion and circular dichroism spectra of enantiomeric acylglycerols and their derivatives. The acylglycerols investigated by Gronowitz *et al.* (1975) are listed in Table XIV. Figure 12 shows the ORD spectra of 1,2-isopropylidene-3-palmitoyl-*sn*-glycerol, 3-palmitoyl-*sn*-glycerol, 1,2-dipalmitoyl-*sn*-glycerol, and of the pseudo-enantiomeric 1,2-diacetyl-3-myristoyl- and 1,2-dimyristoyl-3-acetyl-*sn*-glycerols, as obtained by Gronowitz *et al.* (1975). The corresponding CD measurements are given in Fig. 13. The 1,2-isopropylidene-3-palmitoyl-*sn*-glycerol gives a plain ORD curve with a positive rotational sign (in hexane). However, the CD measurements indicate a "hidden" weak Cotton effect centered at 213–215 nm. The sign of this effect is negative, and, according to Gronowitz *et al.* (1975), is most probably due to the

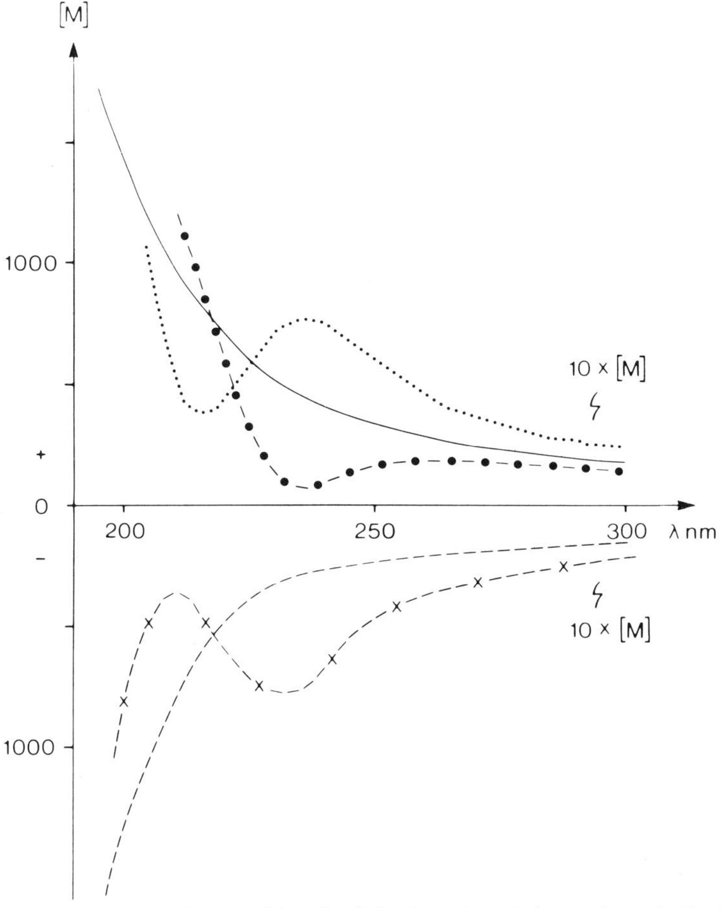

Fig. 12. ORD curves of 1,2-isopropylidene-3-palmitoyl-*sn*-glycerol (——, hexane), 3-palmitoyl-*sn*-glycerol (– – – –, ethanol), 1,2-dipalmitoyl-*sn*-glycerol (–●–●–, hexane–ether 1 : 1), 1,2-diacetyl-3-myristoyl-*sn*-glycerol (· · · ·, hexane), and 1,2-dimyristoyl-3-acetyl-*sn*-glycerol (× – – – – ×, hexane) (Gronowitz *et al.*, 1975).

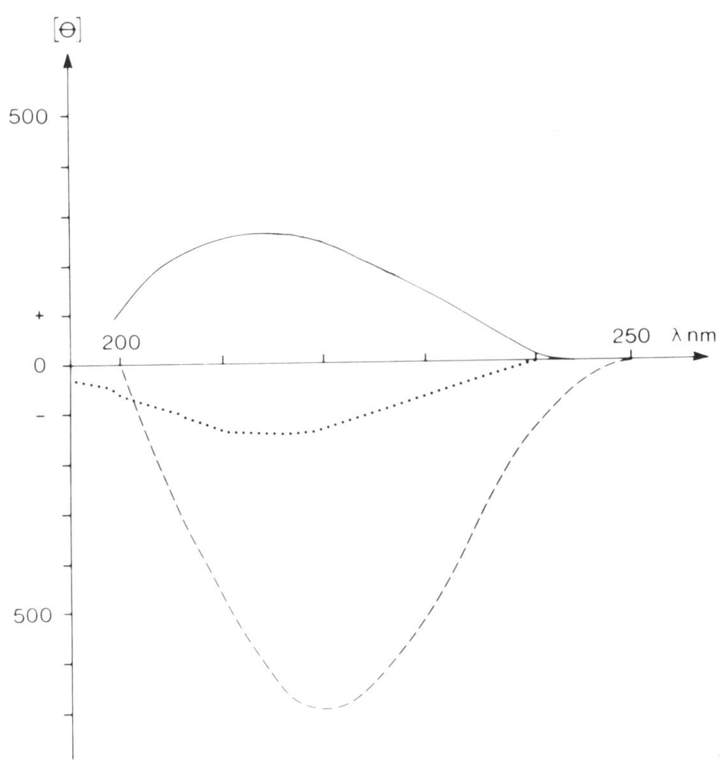

Fig. 13. CD curves of 1,2-isopropylidene-3-palmitoyl-*sn*-glycerol (····, hexane), 3-palmitoyl-*sn*-glycerol (——, ethanol), and 1,2-dipalmitoyl-*sn*-glycerol (----, hexane–ether 1 : 1) (Gronowitz *et al.*, 1975).

n → π* transition of the ester chromophore (for review, consult Crabbe, 1965). This Cotton effect could not be observed in ethanol solution. Thus it appears that the sign of the plain curve is not determined by the sign of this Cotton effect, but by a strong positive rotational effect in the vacuum-UV region. Gronowitz *et al.* (1975) also observed that the molecular rotation at 200 nm $[M]_{200}^{25}$ (in hexane) increased linearly with increasing chain length (Fig. 14).

The removal of the protective isopropylidene group leads to an ORD curve with a negative rotational sign (Fig. 12). The curve, however, is not completely plain. With decreasing wavelength, the rotation values decrease slowly until 230 nm and then decrease more sharply. This form is due to the existence of a weak Cotton effect at 215 nm (in ethanol), which is observed in the CD curve (Fig. 13), and a much stronger effect with negative signs in the vacuum-UV region. Gronowitz *et al.* (1975) found that both Cotton effects change signs upon removal of the protecting isopropylidene group. For 3-butyryl-*sn*-glycerol, the beginning of a negative effect beyond the positive one is observed in hexafluoroacetone–trihydrate ($[\theta]_{218} = +17$, $[\theta]_{190} = -15$). However, this Cotton effect appears to be too weak to be responsible for the strongly negative ORD curve and is most probably a conformational type of effect originating from n → π* transition of the ester chromophore (Gronowitz *et al.*, 1975). In contrast to the 1,2-isopropylidene-3-acyl-*sn*-glycerols, the molecular

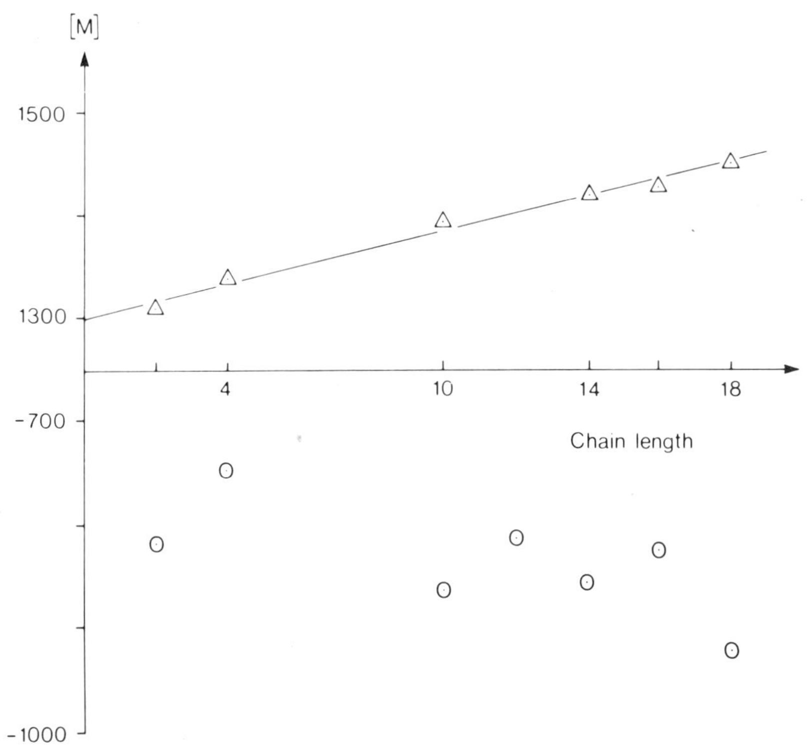

Fig. 14. Molecular rotation of 1,2-isopropylidene-3-acyl-*sn*-glycerols (△, hexane, 200 nm) and 3-acyl-*sn*-glycerols (⊙, ethanol, 210 nm) plotted against the chain length of the acyl group (Gronowitz *et al.*, 1975).

rotations of the 3-acyl-*sn*-glycerols at 210 nm show no linear correlation with the chain length, and only a slight decrease in the molecular rotation values is observed (Fig. 14).

The ORD curve for 1,2-dipalmitoyl-*sn*-glycerol (Fig. 12) indicates the beginning of a negative Cotton effect at 250 nm which is strongly overlapped by a positive effect at shorter wavelengths. The Cotton effect can be clearly recognized in the CD curve (Fig. 13) because of the much higher amplitude in diacyl-*sn*-glycerols than in monoacyl-*sn*-glycerols. Gronowitz *et al.* (1975) expected the sign of the Cotton effect of a 1,2-diacyl-*sn*-glycerol to be opposite to that of the corresponding 3-monoacyl-*sn*-glycerol.

Gronowitz *et al.* (1975) have prepared 1,2-diacyl-3-acetyl-*sn*-glycerols which can be regarded as "pseudoantipodes" of 1,2-diacetyl-3-acyl-*sn*-glycerols with respect to the terminal positions. The ORD and CD curves should therefore nearly be mirror images. This is illustrated with the 1,2-diacetyl-3-myristoyl- and 1,2-dimyristoyl-3-acetyl-*sn*-glycerols for the ORD (Fig. 12) and CD (Fig. 15) curves, respectively. This is also evident from the rotational values in hexane and benzene given in Table XIV for the two pseudoantipodal compounds. It is thus obvious that the length of the acyl chain in the *sn*-2 position has only a marginal influence on the rotational values. A comparison of the molecular rotations of five 1,2-diacetyl-3-acyl-*sn*-glycerols shows that the rotation increases linearly with the number of

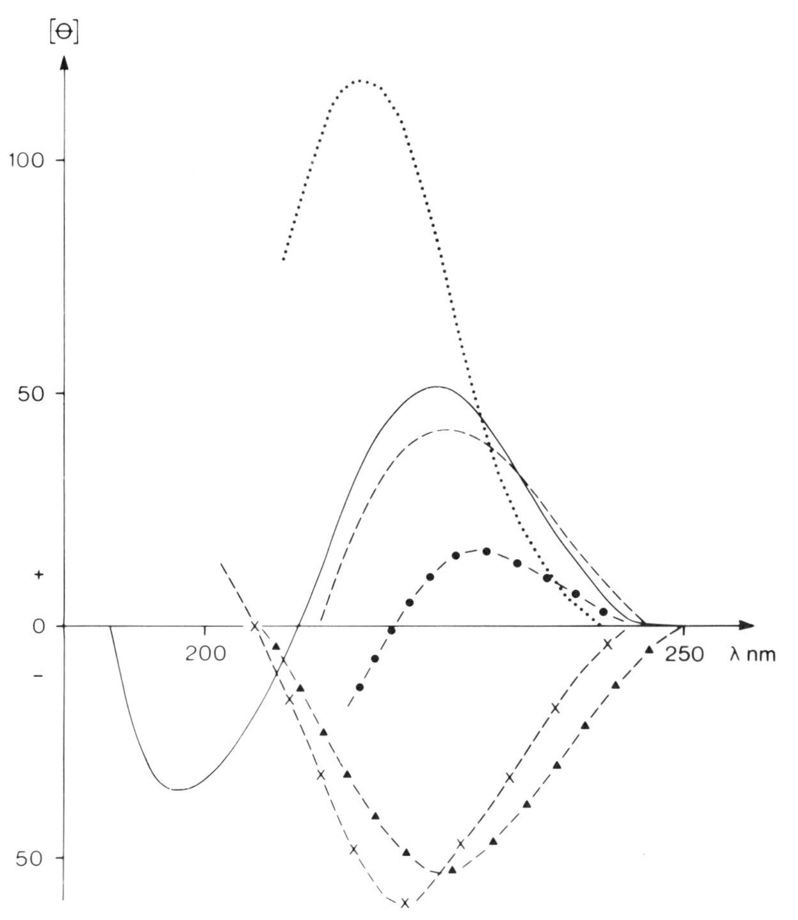

Fig. 15. CD curves of 1,2-diacetyl-3-myristoyl-*sn*-glycerol (——, hexane; – – – –, ethanol; · · · ·, hexafluoroacetonetrihydrate; –●–●–●–, acetonitrile) and 1,2-dimyristoyl-*sn*-glycerol (––▲––▲––, –×– – – –×–, hexane) (Gronowitz *et al.*, 1975).

carbon atoms in the acyl group (Fig. 16). The ORD curve of a 1,2-diacetyl-3-acyl-*sn*-glycerol in the region between 190 and 300 nm is an anomalous curve of the general type shown in Fig. 12 for 1,2-diacetyl-3-myristoyl-*sn*-glycerol. It shows a positive Cotton effect with a peak at 235 nm and a trough at 217 nm. In the CD curve, this is reflected in a maximum at 223 nm (Fig. 15). Depending on the solvent used, a negative maximum may be observed at 198 nm (hexane) as shown in Fig. 15. However, Gronowitz *et al.* (1975) concluded from an examination of the ORD curves at low wavelengths that this negative Cotton effect must be followed by another positive effect at still shorter wavelength. According to these findings, it appears that in triacyl-*sn*-glycerols the 223-nm Cotton effects and vacuum-UV effects have the same sign, in contrast to mono- and diacyl-*sn*-glycerols, where the signs are opposite.

In order to examine the effect of substitution of increasingly similar acyl groups into the acylglycerol molecule on the ORD and CD curves, Gronowitz *et al.* (1975)

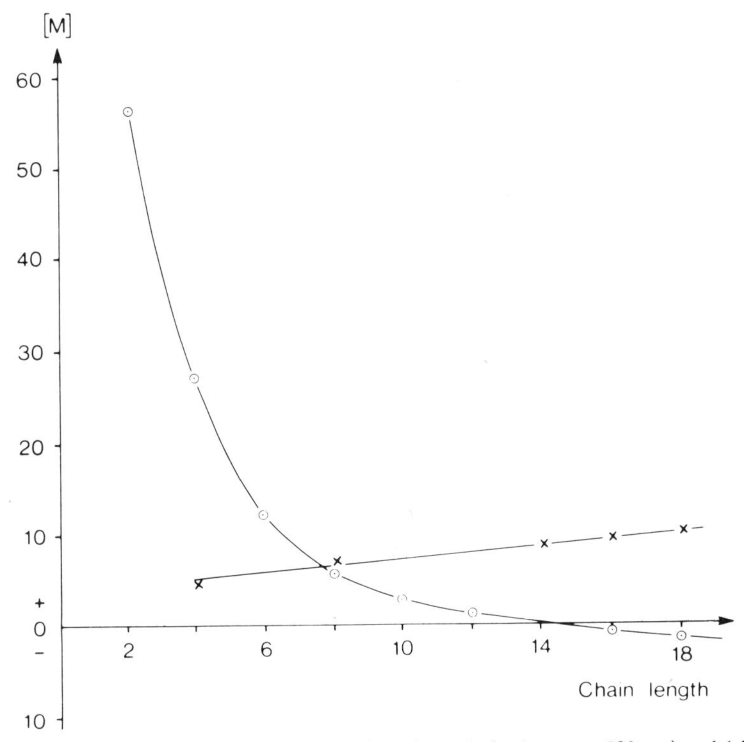

Fig. 16. Molecular rotation of 1,2-diacetyl-3-acyl-*sn*-glycerols (×, benzene, 589 nm) and 1,2-diacyl-3-myristoyl-*sn*-glycerols (O, benzene, 300 nm) plotted against the chain length of the acyl group (Gronowitz *et al.*, 1975).

studied a series of synthetic 1,2-diacyl-3-myristoyl-*sn*-glycerols. Unfortunately, these compounds (with the exception of the acetyl derivatives) could not be measured down to 200 nm because these absorption phenomena depend on the use of relatively high concentrations which could not be experimentally obtained. The ORD curves could be obtained in benzene up to 280–290 nm, but in this particular region only plain curves were obtained. The molecular rotations recorded in the benzene solutions of these compounds are given in Table XIV and in Fig. 16. It is obvious that the optical rotation decreases when the acyl group in the *sn*-1 position gets more similar in chain length to the myristoyl group in the *sn*-3 position. However, even small differences in chain lengths, such as those present in 1,2-dilauroyl-3-myristoyl-*sn*-glycerol, were measurable. Gronowitz *et al.* (1975) confirmed the theory of Schlenk (1965*a*) that in a saturated triacyl-*sn*-glycerol with the greater chain length in the *sn*-1 position ("S" configuration) the rotation is negative. In the opposite case ("R" configuration), the rotation is positive. For the acetyl-*sn*-glycerol derivatives, which allowed measurements down to 200 nm or below, Gronowitz *et al.* (1975) obtained a negative Cotton effect at 223 nm for the "S" isomers and a corresponding positive Cotton effect for the "R" isomers.

Gronowitz *et al.* (1975) also undertook limited studies on the solvent effect on the CD curves in the hope of obtaining higher rotational values. The highest rotations were obtained with benzene and chloroform, which, however, were not transparent at lower wavelengths. Hexane–fluoroacetone–trihydrate had a high transparency and high rotational values, but solubility problems arose with long-chain triacyl-*sn*-glycerols. Hexane alone also had high transparency; however, the rotational values were small compared to those obtained with other solvents. Gronowitz *et al.* (1975) observed a strict linearity with concentration of 1,2-diacyl-*sn*-glycerols in the range 0.1–4.0 g/100 ml of solvent, in general agreement with the concentration dependence demonstrated by Baer and Mahadevan (1959) for 1,2-dioleoyl-*sn*-glycerol (5–80 g/100 ml solvent).

5.9.2.3. Nuclear Magnetic Resonance Spectroscopy

High-frequency NMR has proved valuable in the solution of many problems in lipid chemistry. It distinguishes hydrogen atoms in different environments, both qualitatively and quantitatively (O'Connor, 1961; Hopkins, 1965; Chapman, 1965*a*). Recently much progress has been made in the analysis of enantiomeric acylglycerols by the use of signal shift reagents. Thus Bus *et al.* (1976) used tri(3-heptafluorobutyryl-d-camphorato)europium(III), Eu(hfbc)$_3$, to determine the optical purity of enantiomeric mixtures of tri-, di, and monoacyl-*sn*-glycerols with various fatty acid chain lengths in the proton magnetic resonance spectra. Synthetic model enantiomers of progressively increasing complexity were used to assign the PMR signals. The signal separation in the enantiomers became progressively more difficult as the chain length difference between fatty acids in the *sn*-1 and *sn*-3 positions of glycerol became smaller. The signal of the enantiomeric shift difference ($\Delta\Delta\delta$) of the terminal acyl CH_3 group in 1-acyl-2,3-distearoyl-*sn*-glycerol vs. its enantiomer remained the same in the series where the acyl group was alternately hexanoyl, butyryl, and propionyl, but it was reversed for acetoyl. The greatest separation of proton signals of enantiomeric triacyl-*sn*-glycerols was obtained when a large difference existed between the fatty acids in the *sn*-1 and *sn*-3 positions. Addition of more shift reagent increased the ($\Delta\Delta\delta$) for the low-field methyl signal. The best resolution was obtained by measurements at 220 MHz.

Since the shift reagents are Lewis acids capable of causing isomerization of acyl groups, the study of partial acylglycerols with this method might be hazardous. Bus *et al.* (1976), however, found that Eu(hfbc)$_3$ does not cause a measurable isomerization of diacyl-*sn*-glycerols within 24 hr. Table XVI shows the chemical shift difference for the $CH_2C{=}O$ protons of the 1-benzoyl-3-stearoyl-*sn*-glycerol and its enantiomer after addition of the chiral shift reagents. Table XVI also shows the results obtained with 1-stearoyl-*sn*-glycerol, 3-stearoyl-*sn*-glycerol, and the mixture of the two enantiomers. At a chiral shift reagent/substrate ratio of 1.30, $\Delta\Delta\delta = 0.18$ for the $CH_2C{=}O$ protons at 220 MHz, and the triplets were completely separated. A racemic mixture to which 1-stearoyl-*sn*-glycerol was added gave an increase of the lower-field triplet, and the opposite happened after addition of 3-stearoyl-*sn*-glycerol to the racemic mixture. However, 1-stearoyl-*sn*-glycerol and 3-stearoyl-*sn*-glycerol in $CDCl_3$ at 220 MHz gave identical PMR spectra.

Buchnea (unpublished results, 1976) has examined four different enantiomers of acyl-*sn*-glycerols, including 1- and 3-stearoyl-*sn*-glycerols in $CDCl_3$ at 220 MHz, and has found that the NMR spectra of the corresponding enantiomers were identical. However, when 1- and 3-stearoyl-*sn*-glycerols were phosphocyclized with phenyl-

Table XVI. Enantiomeric Shift Differences ($\Delta\Delta\delta$ ppm) of Proton Signals from Various Acylglycerols in $CDCl_3$[a]

Compound	Substrate (μmole)	Molar ratio CSR/substrate	MHz	Resonance observed	$\Delta\Delta\delta/\Delta\delta$
1,2-Diacetoyl-3-stearoyl-	32.8	1.56	60	*sn*-1-$CH_3C{=}O$	0.10/3.82
			220	*sn*-2-$CH_3C{=}O$	0.04/4.08
1-Acetoyl-2,3-distearoyl-	26.1	3.36	60	$CH_3C{=}O$	0.04/5.93
		2.32	60	$CH_2C{=}O$	0.08/4.48
1-Propionyl-2,3-distearoyl-	66.6	1.84	60	$\mathit{CH_3}CH_2C{=}O$	0.04/2.57
1-Butyryl-2,3-distearoyl-	26.2	2.80	220	$CH_2C{=}O$	0.005/1.68
				$CH_2C{=}O$	0.01/1.78
				$CH_3(CH_2)_2C{=}O$	0.01/0.50
1-Hexanoyl-2,3-dipalmitoyl-	35.3	4.10	220	$CH_3(CH_2)_4C{=}O$	0.006/0.13
1-Benzoyl-3-stearoyl-	48.0	1.20	60	$CH_2C{=}O$	0.04/2.93
1,2-Dipalmitoyl-	38.7	0.45	100	$CH_2C{=}O$	0.04/1.07
		1.36	100	$CH_2C{=}O$	0.075/2.95
1-Stearoyl-	65.4	1.30	60	$CH_2C{=}O$	0.18/2.27

[a] From Bus *et al.* (1976).

phosphoryl dichloride to 1-stearoyl-*sn*-glycerol-2,3- and 3-stearoyl-*sn*-glycerol-1,2-cyclic (phenyl)phosphotriesters, the NMR spectra were different. It is hoped that these compounds will provide an opportunity to confirm and extend the findings of Bus *et al.* (1976).

5.9.2.4. X-Ray Diffraction

X-ray crystallography has been used extensively in the investigation of acylglycerol structure because of the polymorphic nature of their crystals (Chapman, 1965*a*). Although normal X-ray diffraction can distinguish between symmetrical and asymmetrical molecules, it gives identical patterns for enantiomeric acylglycerols. Schlenk (1965*a*, *b*) has developed, however, a routine which allows differentiation between configurational types of acylglycerols by X-ray diffraction using synthetic acylglycerols as reference standards. This is accomplished by comparing the X-ray diffraction patterns of the unknown to that of its racemate and to that of one of its enantiomers. If the unknown is a racemate, its pattern will be similar to that of the synthetic racemate. If the unknown is an enantiomer, its diffraction pattern will be similar to that of the synthetic enantiomer, but the nature of the configuration will remain to be determined. This is done by mixing the unknown with one of its synthetic enantiomers in a 1 : 1 ratio and determining the X-ray diffraction pattern of the mixture. If the configuration of the unknown is opposite to that of the synthetic enantiomer, the diffraction pattern will be that of a racemate. If the mixture has the same pattern as the synthetic enantiomer, the unknown has the same configuration as the synthetic enantiomer. This approach has been found to be valid for all enantiomeric triacyl-, 1,3-diacyl-, and 1-monoacyl-*sn*-glycerols so far examined.

5.9.2.5. Piezoelectric Effect

The piezoelectric effect is the electric charge which appears on the surface of the crystals (without symmetrical center) when they are exposed to a mechanical pressure applied at a definite angle to their axes. The mechanical pressure distorts the ionic structure of the crystals, forcing the positive and negative ions to appear on the opposite surfaces of the crystals. All crystals built up of the asymmetrical molecules of an enantiomer must show piezoelectric effect. Thus the enantiomeric triacyl-*sn*-glycerols are piezoelectric.

Schlenk (1962, 1965*a*, *b*) studied the problem of triacylglycerol chirality by means of the piezoelectricity and showed that racemic and enantiomeric mono-, di-, and triacyl-*sn*-glycerols can be distinguished by this technique. Only the enantiomeric crystals can produce the piezoelectric effect. No direct differentiation can be made between the chirality of the enantiomers. Although studies with mixtures of enantiomers modeled on the work of Schlenk (1965*a*) with X-ray diffraction should allow a differentiation between enantiomers, no such studies have yet been performed in the piezoelectrometer described by Bergmann (1957).

The piezoelectric effect may not become apparent in some tests because the crystals are placed without a definite orientation, and only a small portion of them may be in the optimal position having the polar axes in the orientation angle required by the applied pressure. Moreover, the effect is weakened or canceled when several crystals are compressed at the same time while their polar axes are in the opposite directions. Therefore, numerous measurements are necessary with small crystals to demonstrate unambiguously the presence or absence of a piezoelectric effect. Single measurements are not reproducible.

5.10. Summary and Conclusions

The chemical syntheses of enantiomeric acyl-*sn*-glycerols are based on the preparation of the optically active enantiomers of 1,2- and 2,3-isopropylidene-*sn*-glycerols. The initial synthesis of the saturated monoacid diacyl-*sn*-glycerol enantiomers in the early 1940s was followed by the preparation of the monoacid monounsaturated diacyl-*sn*-glycerol enantiomers in the late 1950s, and the mixed-acid saturated and monounsaturated diacyl-*sn*-glycerols in the 1960s. Ten years later, the synthesis of mixed-acid, saturated, and polyunsaturated diacyl-*sn*-glycerol enantiomers was accomplished. Today, it is possible to duplicate any naturally occurring enantiomeric forms of mono-, di-, and triacyl-*sn*-glycerols or their alkyl and alkenyl analogues by chemical synthesis. In all instances, the synthetic methods involve the successive building up of individual structural parts of the molecules of the desired mono-, di-, and trisubstituted-*sn*-glycerol enantiomers.

The success of these syntheses rests upon the development of appropriate synthetic techniques and the introduction of different protective groups which can be selectively removed during the building-up process of the asymmetrical molecule without racemization of enantiomers and without reduction and isomerization of double bonds. The greater reactivity of the *sn*-1- and *sn*-3-hydroxyl groups in comparison to the *sn*-2-hydroxyl group of the glycerol molecule presents difficulties during the preparation of 1,2- and/or 2,3-diacyl-*sn*-glycerols. The difficulties arise largely because of the resulting migration of the acyl group from the secondary to the primary positions of the glycerol molecule. This process may affect both yield and purity of the diacyl-*sn*-glycerols.

Of importance comparable to that of the improved synthetic techniques has been the development and adoption to acyl-*sn*-glycerol synthesis of new separation and purification methods of starting materials, intermediates, and final products. Preparative column and thin-layer chromatography have been essential in providing adequate quantities of pure intermediates at the various stages of the acyl-*sn*-glycerol preparation. The gas chromatography has been helpful in the determination of their molecular composition. Indispensable in the characterization of the acyl-*sn*-glycerols and their derivatives have also been the infrared and nuclear magnetic resonance spectroscopic methods. Finally, optical rotatory dispersion and circular dichroism techniques have played a major part in the detection and measurement of the optical activity of glycerides.

The need for a proof of the structure of the synthetic and natural acyl-*sn*-glycerols through enzymatic methods, including complete stereospecific analysis, must also be emphasized.

In the past, the preparation of synthetic analogues of natural diacyl- and triacyl-*sn*-glycerols has greatly facilitated biochemical studies of the anabolic and catabolic pathways involving glycerolipids. At the present time, acyl-*sn*-glycerols of defined chemical structure in the form of glycerophospholipids are used in studies of membrane structure and function. Since the composition and molecular association of fatty acids in the acyl-*sn*-glycerol moieties of the glycerophospholipids determine membrane fluidity and permeability, the availability of synthetic acyl-*sn*-glycerol moieties helps in assessing the significance of the natural acyl-*sn*-glycerol structure in membrane architecture and function.

ACKNOWLEDGMENTS

The author is indebted to Drs. G. H. de Haas, editor of *Chemistry and Physics of Lipids*, Amsterdam, Holland, S. Gronowitz and B. Herslöf, University of Lund, Lund, Sweden, and R. Ohlson and B. Toregard, Karlshamns Oljefabriker, Stockholm, Sweden, for permission to reproduce tables and figures of ORD and CD studies of saturated acylglycerols. He is personally indebted to Dr. B. Herslöf for providing him with a copy of his Ph.D. thesis. Gratefully acknowledged is the invaluable help of Dr. H. Schlenk, University of Minnesota, Minneapolis, Minnesota, for making available unpublished material of his late brother, W. Schlenk, Jr.

The studies by the author and his collaborators referred to in this chapter were supported by the Medical Research Council of Canada.

5.11. References

Abderhalden, E., and Eichwald, E., 1914, Versuch über die Darstellung optisch aktiver Fette. I. Synthese optisch-aktiver Halogenglycerine, *Berichte* **47:**1856.

Abderhalden, E., and Eichwald, E., 1915, Darstellung von optisch-aktiver Dibromopropionsäure, Konfiguration der optisch-aktiver Glycerin-Derivative: Ihre Beziehung zur Glycerinsäure, *Berichte* **48:**113.

Åkesson, B., Gronowitz, S., and Herslöf, B., 1976, Stereospecificity of hepatic lipase, *FEBS Lett.* **51**:241.

Anderson, E., and Sands, L., 1929, Crude mesquite gum β-L-aribinose, *Organic Syntheses* **8**:18.

André, E., and Bloch, A., 1935, Sur an nouveau groupe de lipides: Les ether–ester du glycerol (glyceryoxy-alcoyl diglycerides), *Bull. Soc. Chim. Fr.* **5(2)**:789.

Angyal, S. J., and Hoskinson, R. M., 1962, L-Mannitol from L-inositol, in: *Methods in Carbohydrate Chemistry*, Vol. 1, p. 87, Academic Press, New York.

Angyal, S. J., and MacDonald, C. G., 1952, Isopropylidene derivatives of inositol and quercitol: The structure of pinitol and quebrachitol, *J. Chem. Soc.* **1952**:686.

Angyal, S. J., MacDonald, C. G., and Matheson, N. K., 1953, The structure of the di-*O*-isopropylidene derivatives of (−)-inositol and pinitol, *J. Chem. Soc.* **1953**:3321.

Anonymous, 1967, The nomenclature of lipids, *J. Lipid Res.* **8**:523.

Baer, E., 1945, 1,2,5,6-Diacetone D-mannitol and 1,2,5,6-diacetone L-mannitol, *J. Am. Chem. Soc.* **67**:338.

Baer, E., 1952, L-α-*Glycerophosphoric Acid* (*Barium Salt*) *Biochemical Preparation*, Vol. 2 (E. G. Ball, ed.), p. 31, Wiley, New York.

Baer, E., 1963, The synthesis of phospholipids, in: *Progress in the Chemistry of Fats and Other Lipids*, Vol. 6 (R. T. Holman, W. O. Lundberg, T. Malkin, eds.), pp. 33–86, Pergamon Press, New York.

Baer, E., 1965, From the trioses to the synthesis of natural phospholipids: A research trail of forty years, *J. Am. Oil Chem. Soc.* **42**:257.

Baer, E., 1974, Alleged role of boric acid in detritylation of diacyl-triphenylmethyl-*sn*-glycerol ether by silicic acid, *Lipids* **9**:833.

Baer, E., and Buchnea, D., 1958, Synthesis of unsaturated α,β-glycerides. I. D-α,β-diolein and L-α,β-diolein, *J. Biol. Chem.* **230**:447.

Baer, E., and Buchnea, D., 1959, Synthesis of L-α-(dioleoyl)-cephalin; with a comment on the stereochemical designation of glycerolphospholipides, *J. Am. Chem. Soc.* **81**:1758.

Baer, E., and Fischer, H. O. L., 1939*a*, Studies on acetone-glyceraldehyde. VII. Preparation of 1-glyceraldehyde and 1(−)acetone glycerol, *J. Am. Chem. Soc.* **61**:761.

Baer, E., and Fischer, H. O. L., 1939*b*, Studies on acetone-glyceraldehyde. IV. Preparation of *d*(+)acetone glycerol, *J. Biol. Chem.* **128**:463.

Baer, E., and Fischer, H. O. L., 1939*c*, Studies on acetone-glyceraldehyde. V. Synthesis of optically active glycerides from *d*(+)-acetone-glycerol, *J. Biol. Chem.* **128**:475.

Baer, E., and Fischer, H. O. L., 1941, Studies on acetone-glyceraldehyde, and optically active glycerides. IX. Configuration of natural batyl, chimyl, and selachyl alcohols, *J. Biol. Chem.* **140**:397.

Baer, E., and Fischer, H. O. L., 1945*a*, Synthesis of a homologous series of optically active normal aliphatic α-monoglycerides (L-series), *J. Am. Chem. Soc.* **67**:2031.

Baer, E., and Fischer, H. O. L., 1945*b*, Conversion of *d*(+)acetone glycerol into its enantiomorph, *J. Am. Chem. Soc.* **67**:944.

Baer, E., and Fischer, H. O. L., 1947, Naturally occurring glycerol ethers. III. Selachyl alcohol and its geometrical isomer, *J. Biol. Chem.* **170**:337.

Baer, E., and Mahadevan, V., 1959, Synthesis of L-α-lecithins containing shorter chain fatty acids, water-soluble glycerol-phosphatides, *J. Am. Chem. Soc.* **81**:2494.

Baer, E., Rubin, L. J., and Fischer, H. O. L., 1944, Naturally occurring glycerol ethers. II. Synthesis of selachyl alcohol, *J. Biol. Chem.* **155**:447.

Baer, E., Buchnea, D., and Newcombe, A. G., 1956, Synthesis of unsaturated α-lecithins. I. L-α-(Dioleoyl)-lecithin, *J. Am. Chem. Soc.* **78**:232.

Baggett, N., Dobinson, B., Foster, A. B., Homer, J., and Thomas, L. F., 1961, Proton magnetic resonance studies of some derivatives of 5-hydroxyl-1,3-dioxane(1,3-methylidene-glycerol), *Chem. Ind.* (*London*) **1961**:106.

Ballou, C. E., and Fischer, H. O. L., 1953, Derivatives of D-manno-hexodialdose (6-aldo-D-mannose), *J. Am. Chem. Soc.* **75**:3673.

Baumann, W. J., 1972, The chemical syntheses of alkoxylipids in: *Ether Lipids* (F. Snyder, ed.), pp. 51–79, Academic Press, New York.

Baumann, W. J., and Mangold, H. K., 1964, Reaction of aliphatic methanesulfonates. I. Syntheses of long-chain glycerol-(1) ethers, *J. Org. Chem.* **29**:3055.

Baumann, W. J., and Mangold, H. K., 1966, Reaction of aliphatic methanesulfonates. II. Syntheses of long-chain di- and trialkyl glycerol ethers, *J. Org. Chem.* **31**:498.

Baumann, W. J., and Ulshöfer, H. W., 1968, Characteristic absorption bands of naturally-occurring long-chain ethers, esters and ether esters of glycerol and various diols. *Chem. Phys. Lipids* **2**:114.

Baumann, W. J., Mahadevan, V., and Mangold, H. K., 1966, Optisch aktive synthetische und naturliche *O*-Alkyl-glyceride, *Hoppe-Seyler's Physiol. Chem.* **347**:52.

Bergmann, L., 1957, Eine Apparatur zur Messung der Piezoelektrizitat. *Z. Instrumentenkd.* **65**:2.

Bergmann, M., and Carter, N. M., 1930, Synthese von *β*-Glyceriden, *Z. Physiol. Chem.* **191**:211.

Bergmann, M., and Sabetay, S., 1924, Über *α*-Monoglyceride Hochmolekularen Fettesäuren, *Z. Physiol. Chem.* **137**:47.

Bergmann, M., Brand, E., and Dreyer, F., 1921, Synthese von *α,β*-Diglyceriden und unsymmetrische Triglyceriden, *Berichte* **54**:936.

Bligh, E. G., and Dyer, W. J., 1959, A rapid method of total lipid extraction and purification, *Can. J. Biochem. Physiol.* **37**:911.

Borgström, B., 1952, Investigation on lipid separation methods: Separation of phospholipids from neutral fat and fatty acids, *Acta Physiol. Scand.* **25**:101.

Bourne, E. J., Stacey, M., Tatlow, T. C., and Tedder, J. M., 1949, Studies on trifluoroacetic acid. I. Trifluoroacetic anhydride as promoter of ester formation between hydroxycompounds and carboxylic acids, *J. Chem. Soc.* **1949**:2976.

Breckenridge, W. C., and Kuksis, A., 1968, Specific distribution of short-chain fatty acids in molecular distillates of bovine milk fat, *J. Lipid Res.***9**:388.

Breckenridge, W. C., and Kuksis, A., 1969, Structure of bovine milk fat triglycerides. II. Long-chain length, *Lipids* **4**:1.

Brockerhoff, H. J., 1965, Stereospecific analysis of triglycerides, *J. Lipid Res.* **6**:10.

Brockerhoff, H. J., 1967, Stereospecific analysis of triglycerides: An alternative method, *J. Lipid Res.* **8**:167.

Brockerhoff, H. J., Hoyle, R. J., and Wolmark, N., 1966, Positional distribution of fatty acids in triglycerides of animal depot fats, *Biochim. Biophys. Acta* **116**:67.

Brockerhoff, H. J., Hoyle, R. J., Hwang, P. C., and Litchfield, C., 1968, Positional distribution of fatty acids in depot triglycerides of aquatic animals, *Lipids* **3**:24.

Buchnea, D., 1967*a*, Synthesis and conversion of bis-glycerol ethers. III. Disproportionation of 1,2-isopropylidene-glycerol-3-glycerol ether to bis(1,2-isopropylidene-glycerol-3)-ether and bis-glycerol ether, *Chem. Phys. Lipids* **1**:177.

Buchnea, D., 1967*b*, Synthesis and conversion of bis-glycerol ethers. IV. Mechanism of disproportionation of 1,2-isopropylidene-3-glycerol-3-glycerol ether to bis(1,2-isopropylidene-glycerol-3)-ether and bis-glycerol ether, *Chem. Phys. Lipids* **6**:734.

Buchnea, D., 1967*c*, Acyl migration in glycerides. I. A bimolecular resonant ion complex as intermediate in acyl migration of monoglycerides, *Chem. Phys. Lipids* **1**:113.

Buchnea, D., 1971, Synthesis of C-18 mixed acid diacyl-*sn*-glycerol enantiomers, *Lipids* **6**:734.

Buchnea, D., 1974, Detritylation by silicic acid and boric acid column chromatography, *Lipids* **9**:55.

Buchnea, D., and Baer, E., 1960, Synthesis of enantiomeric mixed-acid *α,β*-diglycerides, *J. Lipid Res.* **1**:405.

Bus, J., Lok, C. M., and Groenewengen, A., 1976, Determination of enantiomeric purity of glycerides with PMR shift reagent, *Chem. Phys. Lipids* **16**:123.

Cahn, R. S., Ingold, C. K., and Prelog, V., 1956, The specification of asymmetric configuration in organic chemistry, *Experientia* **12**:81.

Cahn, R. S., Ingold, C. K., and Prelog, V., 1966, Specification of molecular chirality, *Angew. Chem. Int. Ed. Engl.* **5**:385.

Chacko, G. M., and Hanahan, D. J., 1968, Chemical synthesis of 1-*O*(D) and 3-*O*(L) glycerol monoethers, diethers and derivatives: Glycerides, monoester phospholipids and diester phospholipids, *Biochim. Biophys. Acta* **164**:252.

Chapman, D., 1965*a*, *The Structure of Lipids by Spectroscopy and X-Ray Techniques*, Methuen, London.

Chapman, D., 1965*b*, Infra-red spectroscopy of lipids, *J. Am. Oil Chem. Soc.* **42**:353.

Christie, W. W., and Moore, J. H., 1969, A semimicro method for the stereospecific analysis of triglycerides, *Biochim. Biophys. Acta* **176**:445.

Coleman, M. H., and Fulton, W. C., 1961, The structural investigation of natural fats by the partial hydrolysis technique, in: *Enzymes of Lipid Metabolism* (P. Desnuelle, ed.), pp. 127–137, Pergamon Press, New York.

Crabbe, P., 1965, An introduction to optical rotatory dispersion and circular dichroism in organic chemistry, in: *Optical Rotatory Dispersion and Circular Dichroism in Organic Chemistry* (G. Snatzke, ed.), Chap. 1, Heyde and Son, London.

Crossley, A., Freeman, I. P., Hudson, B. J. F., and Pierce, J. H., 1959, Acyl migration in diglycerides, *J. Chem. Soc.* **152:**760.

Cunningham, J., and Gigg, R., 1965, Glycerol 1,2-carbonate, *J. Chem. Soc.* **1965:**1553.

Daemen, F. J. M., 1967, A convenient synthesis of phosphatidylethanolamines, *Chem. Phys. Lipids* **1:**476.

Daemen, F. J. M., de Haas, G. H., and van Deenen, L. L. M., 1963, An improved synthesis of mixed acid L-α-phosphatidylethanolamine containing a polyunsaturated fatty acid, *Rec. Trav. Chim. Pays-Bas* **82:**487.

Daubert, B. F., 1940, Preparation of fatty acid β-monoglycerides, *J. Am. Chem. Soc.* **62:**1713.

Daubert, B. F., and King, C. G., 1938, The relative stability of aromatic and aliphatic monoglycerides, *J. Am. Chem. Soc.* **60:**3003.

Daubert, B. F., and King, C. G., 1941, Synthetic fatty acid glycerides of known constitution, *Chem. Rev.* **29:**269.

Davies, W. H., Heilborn, I. M., and Jones, W. E., 1934, The unsaponifiable matter from the oil of elasmobranch fish. X. The structure of batyl alcohol and the synthesis of α-octadecyl glycerol ether, *J. Chem. Soc.* **1934:**1232.

de Haas, G. H., and van Deenen, L. L. M., 1961, Synthesis of enantiomeric mixed-acid phosphatides, *Rec. Trav. Chim. Pays-Bas* **80:**951.

Fierz-David, H. E., and Kuster, W., 1939, Herstellung der Chloriden der Fattsäuren von der Propion- bis zur Nonadecansäure, *Helv. Chim. Acta* **22:**82.

Fischer, E., 1890, Synthese der Mannose und Levulose (1-Mannit), *Berichte* **23:**375.

Fischer, E., and Brauns, F., 1914, Verwandlung der *d*-Isopropylmalonaminsäure in den optischen Antipoden durch Vertauschung von Carboxyl- und Säureamide-Grupe, *Berichte* **47:**3181.

Fischer, E., Bergmann, M., and Barwind, E., 1920, Neue Synthese von α-Monoglyceriden, *Berichte* **53:**1589.

Fischer, H. O. L., and Baer, E., 1930, Über Hydrozinoderivative des Glycerinaldehyds und Dioxy-Acetons, *Berichte* **63:**1749.

Fischer, H. O. L., and Baer, E., 1937, Synthese optisch-aktiver Glyceride, *Naturwissenschaften* **25**(**36**)**:**588.

Fischer, H. O. L., and Baer, E., 1941, Preparation and properties of optically active derivatives of glycerols, *Chem. Rev.* **29:**287.

Fischer, H. O. L., and Taube, C., 1927, Über Acetonieren mit Zinkchlorid, *Berchite* **60:**485.

Fritz, J. S., and Schenk, E. H., 1959, Acid catalyzed acetylation of organic hydroxyl groups, *Anal. Chem.* **31:**1808.

Gigg, J., and Gigg, R., 1967*a*, Preparation of unsymmetrical diglycerides, *J. Chem. Soc.* **1967:**431.

Gigg, J., and Gigg, R., 1967*b*, 1-*O*-Benzyl-L-glycerol and D-(glycerol 1,2-carbonate), *J. Chem. Soc.* **1967:**1865.

Gronowitz, S., Herslöf, B., Ohlson, R., and Toregard, B., 1975, ORD and CD studies of saturated glycerides, *Chem. Phys. Lipids* **14:**174.

Grün, A., and Limpächer, R., 1927, Spaltung asymmetrischer Glyceride in die Antipoden. I. Über optisch aktiven Glycerid-Schwefelsäuren und die Thermolabilität des Drehungsvermögens ihrer Salze, *Berichte* **60:**255.

Gunstone, F. D., 1967, *An Introduction to the Chemistry and Biochemistry of Fatty Acids and Their Glycerides*, pp. 138–149, Chapman and Hall, London.

Gupta, S. C., and Kummerow, F. A., 1959, An improved procedure for preparing glycerol ethers, *J. Org. Chem.* **24:**409.

Hanahan, D. J., and Jayko, M. E., 1952, The isolation of dipalmitoleoyl-L-α-glycerylphosphorylcholine from yeast: A new route to (dipalmitoyl)-L-α-lecithin, *J. Am. Chem. Soc.* **74:**5070.

Hartman, L., 1958, Advances in the synthesis of glycerides of fatty acids, *Chem. Rev.* **58:**845.

Hartman, L., 1959, Hydrolysis of isopropylidene esters of fatty acids, *J. Chem. Soc.* **1959:**4134.

Haverkate, F., and van Deenen, L. L. M., 1965, Isolation and chemical characterization of phosphatidyl glycerol from spinach leaves, *Biochim. Biophys. Acta* **106:**78.

Helferich, B., and Sieber, H., 1927, Zur Synthese partiall acylierter Glyceride, *Z. Physiol. Chem.* **170:**31.

Helferich, B., and Sieber, H., 1928, Zur Synthese partiall acylierter Glyceride, *Z. Physiol. Chem.* **175:**311.

Hibbert, H., and Carter, N. M., 1929, Mechanism of organic reactions. I. Wandering of acyl groups in glycerol esters, *J. Am. Chem. Soc.* **51:**1601.

Hill, E. E., and Lands, W. E. M., 1970, Phospholipid metabolism, in: *Lipid Metabolism* (S. J. Wakil, ed.), pp. 185–277, Academic Press, New York.

Hirschmann, H., 1960, The nature of substrate asymmetry in stereoselective reactions, *J. Biol. Chem.* **235:**2762.

Hopkins, C. Y., 1965, Nuclear magnetic resonance of fatty acids and lipids, in: *Progress in the Chemistry of Fats and Other Lipids*, Vol. 8 (R. T. Holman, W. O. Lundberg, and T. Malkin, eds.), p. 213, Pergamon Press, New York.

Howe, R. J., and Malkin, T., 1951, An X-ray and thermal examination of glycerides. XI. The 1:2-diglycerides and further observations on 1:3-diglycerides, *J. Chem. Soc.* **1951:**2663.

Ingold, C. K., 1973, Kinetics of nucleophilic aliphatic substitution: Second-order and first-order reaction, in: *Organic Chemistry*, 3rd ed. (R. T. Morrison and R. N. Boyd, eds.), pp. 459–460, Allyn and Bacon, Boston.

Irvine, J. C., MacDonald, J., and Soutar, C. W., 1915, Condensation of acetone and benzylidene with glycerol: Preparation of α-methyl ether, *J. Chem. Soc.* **1915:**335.

Jensen, R. G., 1972, Synthetic glycerides, in: *Topics in Lipid Chemistry*, Vol. 3 (F. G. Gunstone, ed.), pp. 1–35, Wiley, New York.

Jensen, R. G., and Gordon, D. T., 1972, The synthesis of phosphoglycerides, *Lipids* **7:**621.

Jensen, R. G., and Pitas, R. E., 1976, Synthesis of some acyl-glycerols and phospholipids, in: *Advances in Lipid Research*, Vol. 14 (R. Paoletti and D. Kritchevsy, eds.), pp. 213–247, Academic Press, New York.

Jensen, R. G., Quinn, J. G., and Sampugna, J., 1970, Pancreatic lipolysis of enantiomeric triglycerides, *Lipids* **5:**580.

Karnovsky, M. L., and Hauser, G., 1960, The synthesis and metabolism of enantiomeric forms of glycerol-1-C^{14}, *J. Biol. Chem.* **226:**881.

Karnovsky, M. L., and Wolff, D., 1960, Studies on the stereospecificity of lipids, in: *Fifth International Conference on Biochemistry of Lipids* (G. Popjak, ed.), Pergamon Press, New York.

Kates, M., 1972, Techniques of lipidology, in: *Laboratory Techniques in Biochemistry and Molecular Biology*, Vol. 3, pp. 267–600, American Elsevier, New York.

Kates, M., Chan, T. H., and Stanacev, N. Z., 1963, Aliphatic diether analogs of glyceride-derived lipids. I. Synthesis of D-α,β-dialkyl glycerol ethers, *Biochemistry* **2:**394.

Kaufmann, H. P., and Förster, N., 1960, Über Fettsäurenester des Diglycerins, *Fette Seifen Anstrichm.* **62:**796.

Kennedy, E. P., and Weiss, S. B., 1956, The function of cytidine coenzymes in the biosynthesis of phospholipides, *J. Biol. Chem.* **222:**193.

Kiliani, H., 1887, Über das Doppellakon des Metazuckersäure, *Berichte* **20:**2710.

Kiliani, H., 1922, Darstellung von L-Mannonsäure und L-Glykonsäure, *Berichte* **55:**100.

Kiliani, H., 1925, Darstellung von L-Mannonsäure und L-Glykonsäure, *Berichte* **58:**2349.

Lands, W. E. M., and Zschocke, A., 1965, New synthesis of (*l*)-1-*O*-benzylglycerol, *J. Lipid Res.* **6:**324.

Lands, W. E. M., Pieringer, R. A., Slakey, S. P. M., and Zschocke, A., 1966, A micromethod for the stereospecific determination of triglyceride structure, *Lipids* **1:**444.

Lok, C. M., Ward, J. P., and van Dorp, D. A., 1976, The synthesis of chiral glycerides starting from D- and L-serine, *Chem. Phys. Lipids* **16:**115.

Maier, R., and Holman, R. T., 1964, Naturally occurring triglycerides possessing optical activity in the glycerol moiety, *Biochemistry* **3:**270.

Malkin, T., and Bevan, T. H., 1957, The synthesis of glycerides, in: *Progress in the Chemistry of Fats and Other Lipids*, Vol. 4 (R. T. Holman, W. O. Lundberg, and T. Malkin, eds.), pp. 63–77, Pergamon Press, New York.

Mangold, H. K., and Malins, D. C., 1960, Fractionation of fats, oil and waxes, *J. Am. Oil Chem. Soc.* **37:**383.

Mank, A. P. J., Ward, J. P., and van Dorp, D. A., 1976, A versatile, flexible synthesis of 1,3-diglycerides and triglycerides, *Chem. Phys. Lipids* **16:**107.
Martin, J. B., 1953, Preparation of saturated and unsaturated symmetrical monoglycerides, *J. Am. Chem. Soc.* **75:**5482.
Mattson, F. H., and Volpenhein, R. A., 1962, Synthesis and properties of glycerides, *J. Lipid Res.* **3:**281.
Mattson, F. H., Volpenhein, R. A., and Martin, J. B., 1964, Esterification of hydroxy compounds by fatty acid anhydride, *J. Lipid Res.* **5:**374.
Morley, N. H., Kuksis, A., and Buchnea, D., 1974, Hydrolysis of triacylglycerols by pancreatic and lipoprotein lipase, *Lipids* **9:**481.
Morley, N. H., Kuksis, A., Buchnea, D., and Myher, J. J., 1975, Hydrolysis of diacylglycerols by lipoprotein lipase, *J. Biol. Chem.* **250:**3414.
Morris, L. J., 1965*a*, The detection of optical activity in natural asymmetric triglycerides, *Biochem. Biophys. Res. Commun.* **20:**340.
Morris, L. J., 1965*b*, A synthetic optically active trialiphatic triglyceride and a method for the detection of optical activity in natural asymmetric triglycerides, *Biochem. Biophys. Res. Commun.* **18:**495.
Ness, A. T., Hann, R. M., and Hudson, C. S., 1943, The acetolysis of trimethylene-D-mannitol: 2,5-Methylene-D-mannitol, *J. Am. Chem. Soc.* **65:**2215.
O'Connor, R. T., 1961, Recent progress in application of infra-red absorption spectroscopy to lipid chemistry, *J. Am. Oil Chem. Soc.* **38:**648.
O'Doherty, P. J. A., Kuksis, A., and Buchnea, D., 1972, Enantiomeric diglycerides as stereospecific probes in triglyceride synthesis *in vitro, Can. J. Biochem.* **50:**881.
Ogston, A. G., 1948, Interpretation of experiments on metabolic processes using isotopic tracer elements, *Nature* (*London*) **18:**963.
Palameta, B., and Kates, M., 1966, Aliphatic diether analogs of glycerol-derived lipids. III. Synthesis of dialkenyl and mixed alkylalkenylglycerol ethers, *Biochemistry* **5:**618.
Paltauf, F., 1971, Metabolism of the enantiomeric 1-*O*-alkylglycerol ethers in the rat intestinal mucosa *in vivo*: Incorporation into 1-*O*-alkyl and 1-*O*-alk-1′-enyl glycerol lipids, *Biochim. Biophys. Acta* **239:**38.
Paltauf, F., and Johnston, J. M., 1971, The metabolism of enantiomeric 1-*O*-alkyl glycerols and 1,2- and 1,3-alkyl acyl glycerols in the intestinal mucosa, *Biochim. Biophys. Acta* **239:**47.
Paltauf, F., and Spener, F., 1968, An improved synthesis of 1,2-dialkyl glycerol ethers and synthesis of ^{14}C-labelled trialkyl glycerol ethers, *Chem. Phys. Lipids* **2:**168.
Paltauf, F., and Wagner, E., 1976, Stereospecificity of lipases, enzymatic hydrolysis of enantiomeric alkyldiacyl- and diacyl-alkylglycerols by lipoprotein lipase, *Biochim. Biophys. Acta* **431:**359.
Paltauf, F., Esfandi, F., and Holasek, A., 1974, Stereospecificity of lipases: enzymatic hydrolysis of enantiomeric alkyl diacyl-glycerols by lipoprotein lipase, lingual lipase and pancreatic lipase, *FEBS Lett.* **40:**119.
Pfeiffer, F. R., Cohen, S. R., Williams, K. R., and Weisbach, A., 1968, Glycerolipids. I. Synthesis of D and L mono- and polyunsaturated 1,2-diglycerides *via* glycerol carbonates, *Tetrahedron Lett.* **32:**3549.
Quinn, J. G., Sampugna, J., and Jensen, R. G., 1967, Synthesis of 100-gram quantities of highly purified mixed acid triglycerides, *J. Am. Oil Chem. Soc.* **44:**439.
Ryhage, R., and Stenhagen, E., 1960, Mass spectroscopy in lipid research, *J. Lipid Res.* **1:**361.
Schlenk, W., 1962, Optische Aktivität Bei Triglyceriden: Festschrift Karl Wuster 60 Gerbutstag, p. 105, *Chem. Abstr.* **57:**14930g.
Schlenk, W., 1965*a*, Synthesis and analysis of optically active triglycerides, *J. Am. Oil Chem. Soc.* **42:**945.
Schlenk, W., 1965*b*, Neuere Ergebnisse der Konfigurationsforschung, *Angew. Chem.* **77:**161.
Schmid, H. H. O., and Mangold, H. K., 1966, Neutrale Plasmalogene und Alkoxydiglyceride in menschlichem Depotfett, *Biochem. Z.* **346:**13.
Schmid, H. H. O., Baumann, W. J., and Mangold, H. K., 1967, The structure and configuration of "neutral plasmalogens," *J. Am. Chem. Soc.* **89:**4797.
Schulz, M., and Tollens, B., 1896, Über die Verbindungen der mehrwertigen Alkohole mit Formaldehyde, *Ann. Chem.* **289:**21.
Schwartz, P., and Carter, H. E., 1954, A nonenzymatic illustration of "citric acid type" asymmetry: The meso-carbon atom, *Proc. Natl. Acad. Sci.* (*USA*) **40:**499.

Serdarevich, B., 1967, Glyceride isomerization in lipid chemistry, *J. Am. Oil Chem. Soc.* **44:**381.

Serdarevich, B., and Carroll, K. K., 1966, Synthesis and characterization of 1- and 2-monoglycerol ethers of antesio fatty alcohols, and reinvestigation of benzylidene glycerol synthesis, *Can. J. Biochem.* **44:**743.

Shvets, V. I., 1971, Advances in the synthesis of glycerol phosphatide esters, *Russ. Chem. Rev.* **40:**330.

Slotboom, A. J., and Bonsen, P. P. M., 1970, Recent developments in the chemistry of phospholipids, *Chem. Phys. Lipids* **5:**301.

Sowden, J. C., 1962, α-L-Glucose and L-mannose from L-arabinose by the nitromethane synthesis, *Methods Carbohyd. Chem.* **1:**132–135.

Sowden, J. C., and Fischer, H. O. L., 1941, Optically active α-β-diglyceride, *J. Am. Chem. Soc.* **63:**3244.

Sprecher, H. W., Maier, R., and Holman, R. T., 1965, Structure of an optically active allene-containing tetraester triglyceride isolated from the seed oil of *Sapium sebiferum, Biochemistry* **4:**1856.

Stegerhock, L. J., and Verkade, P. E., 1956*a*, Ester derived from batyl alcohol, *Rec. Trav. Chim. Pays-Bas* **75:**143.

Stegerhock, L. J., and Verkade, P. E., 1956*b*, Phosphoric acid and derivatives. IV. Phosphoric acid derived from batyl alcohol, *Rec. Trav. Chim. Pays-Bas* **75:**467.

Stimmel, B. F., and King, C. G., 1934, Preparation and properties of α-monoglycerides, *J. Am. Chem. Soc.* **63:**3244.

Stoffel, W., and Pruss, H. D., 1969, Monolayer studies with synthetic saturated, mono- and polyunsaturated mixed 1,2-diglycerides, 1,2-diacylphosphatidylethanolamines and phosphatidylcholines at the air-water-interface, *Z. Physiol. Chem.* **350:**1385.

Tattrie, N. H., Bailey, R. A., and Kates, M., 1958, The action of pancreatic lipase on stereoisomeric triglycerides, *Archiv. Biochem. Biophys.* **78:**319.

Thomas, A. E., Scharoun, J. E., and Ralston, H., 1965, Quantitative estimation of isomeric monoglycerides by thin-layer chromatography, *J. Am. Oil Chem. Soc.* **42:**789.

van Deenen, L. L. M., and de Haas, G. H., 1964, The synthesis of phospholipids and some biochemical applications, in: *Advances in Lipid Research*, Vol. 2 (R. Paoletti and D. Kritchevsky, eds.), p. 167, Academic Press, New York.

Vander Wal, R. J., 1960, Calculation of distribution of the saturated and unsaturated acyl groups in fats from pancreatic lipase hydrolysis data, *J. Am. Oil Chem. Soc.* **37:**18.

Verkade, P. E., 1953, Synthesis of glycerides, *Chim. Ind.* (*Paris*) **69:**239.

Verkade, P. E., and van Roon, J. D., 1942, Über die Benzylidene-Glycerols, *Rec. Trav. Chim. Pays-Bas* **61:**831.

Vogel, A. I., 1959, *Textbook of Practical Organic Chemistry*, Third Ed., II-49, p. 189, Longmans, Spottiswood, Ballantyne and Co. Ltd., London and Colchester.

Weiss, S. B., Kennedy, E. P., and Kiyasu, J. Y., 1960, The enzymatic synthesis of triglycerides, *J. Biol. Chem.* **235:**40.

Wickberg, B., 1958, Synthesis of 1-glycerol D-galactopyranosides, *Acta Chem. Scand.* **12:**1187.

Windholz, T. B., and Johnston, D. B. R., 1967, Trichloroethoxycarbonyl: A generally applicable protecting group, *Tetrahedron Lett.* **27:**2555.

Woodward, R. B., 1966, Recent advances in the chemistry of natural products, *Science* **153:**487.

Woodward, R. B., Heusler, K., Gosteli, J., Naegeli, P., Oppolzer, W., Ramage, S., Raganathan, S., and Vorbugen, H., 1966, The total synthesis of cephalosporin C[1], *J. Am. Chem. Soc.* **88:**852.

Wright, W. R., and Tove, S. B., 1967, Metabolism of 1-palmitoyl dioleoyl and 3-palmitoyl dioleoyl glycerol by adipose tissue, *Biochim. Biophys. Acta* **137:**54.

Chapter 6

Metabolic Studies with Natural and Synthetic Fatty Acids and Enantiomeric Acylglycerols

Patrick J. A. O'Doherty

6.1. Introduction

Due to the concerted efforts of numerous investigators, the last 15 years has witnessed unprecedented advances in the elucidation of both the anabolism and catabolism of acylglycerols. The availability of synthetic substrates containing radioactive and stable isotope-labeled molecules, coupled with the development of chromatographic and spectroscopic methodologies, has heralded an era of intensified study of the metabolism of acylglycerols in the hope of increasing our understanding of the biological importance of the structure of these molecules. It is the purpose of this chapter to review the state of our current knowledge of metabolic studies with fatty acids and acylglycerols in animal tissues and to discuss some of the methods employed and difficulties encountered in these investigations.

6.2. Activation of Fatty Acids

The utilization of fatty acids in acylglycerol biosynthesis requires an initial activation to the CoA derivatives via acid : CoA ligases. Using the energy of the thiol ester bond, the acyl group is then able to participate in an acylation or esterification of a hydroxyl group of a suitable glycerol acceptor. This type of enzyme was first demonstrated in liver by Kornberg and Pricer (1953) and was subsequently shown to occur in most other tissues of various mammalian species (for review, see Brindley, 1973; Aas, 1971). Aas (1971) has determined the organ and subcellular distribution of the ATP-dependent fatty acid activating enzymes in the rat. When these results are correlated with earlier studies on isolated enzymes, the existence of five different ATP-dependent acyl CoA systems (I–V) can be recognized, with chain length

---

Patrick J. A. O'Doherty • G. F. Strong Laboratory, Department of Medicine, The University of British Columbia, Vancouver, British Columbia V5Z 1M9, Canada.

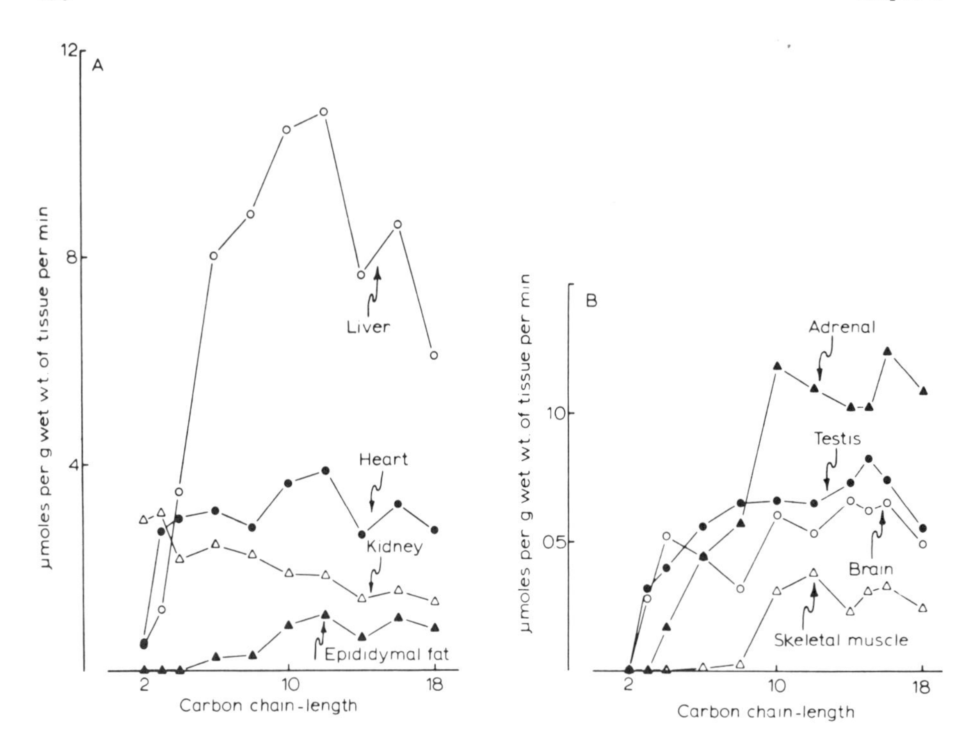

Fig. 1. Relative rates of activation of different fatty acids in several organs from the rat (Aas, 1971).

optima at C_2, C_4, C_7, C_{12}, and C_{16}, respectively. All tissues studied contain enzymes IV and V, while some tissues lack one or more of the enzymes I–III. Liver, kidney, and heart muscle apparently contain all five enzymes. Figure 1 shows the relative capacities for fatty acid activation in different organs from the rat. Substrates used were all the even-numbered straight-chain fatty acids from C_2 to C_{18}, the unsaturated C_{18} acids, propionate, and in some tissues C_{15}. In all tissues tested, the long-chain fatty acids were activated; oleate was activated as much as stearate, while linoleate and linolenate were activated at about half the rate of the former two acids. The epididymal fat tissue and skeletal muscle seem to lack the ability to activate short-chain fatty acids.

The subcellular distribution of fatty acid activating enzymes appears to be principally the same in all tissues. Short-chain fatty acids (enzymes I–III) are activated in the mitochondrial matrix, long-chain fatty acids (enzymes IV–V) in the outer membrane of the mitochondria and in the microsomes. The medium-chain acids are activated in all three places because of overlapping in enzyme specificity (Aas, 1971). Preparations obtained from pig, guinea pig, and cat intestinal mucosa have maximum activity toward laurate, myristate, and palmitate, respectively (Brindley, 1973).

The fatty acid specificity of activation may be one of the factors controlling the rate and route of utilization of fatty acids by acyltransferases. Unfortunately, this effect is lost in those *in vitro* comparisons of fatty acid specificity where the synthetic acyl CoA esters are employed instead of free fatty acids along with appropriate acyl

CoA generating systems. It is possible that some of the differences in the fatty acid utilization reported between *in vivo* and *in vitro* experiments are due to the absence of the differential activation of the free fatty acids in the latter systems.

6.3. Biosynthesis of Phosphatidic Acid

Phosphatidic acid occupies a pivotal position in lipid metabolism since it functions as a precursor in the biosynthesis of both higher acylglycerols and phospholipids. Since it is now established that phosphatidic acid can be synthesized via several metabolic pathways, an understanding of the biosynthesis of phosphatidic acid is imperative to our understanding of acylglycerol metabolism and its control.

Before discussing the role of *sn*-glycerol-3-phosphate as an acceptor of fatty acids and fatty acyl groups, it is relevant to discuss the synthesis of *sn*-glycerol-3-phosphate. In many animal tissues, as well as in plants and microorganisms, *sn*-glycerol-3-phosphate is derived from glucose. *sn*-Glycerol-3-phosphate may also originate from glycerol and ATP. This reaction is catalyzed by glycerol kinase, which has been purified 169-fold from rat liver by Bublitz and Kennedy (1954) and obtained in crystalline form by Wieland and Suyter (1958) and by Hayashi and Lin (1967) from pigeon liver and *Escherichia coli*, respectively. The enzyme does not exhibit an absolute specificity for glycerol as L-glyceraldehyde and dihydroxyacetone are phosphorylated at least equally as well. In tissues or organisms where both pathways for the formation of *sn*-glycerol-3-phosphate can take place, it is difficult to assess the relative contribution of each route. In many tissues, however, *sn*-glycerol-3-phosphate acting as a precursor for glycerolipids is probably derived mainly from glucose (Hübscher, 1970).

6.3.1. Acylation of sn-Glycerol-3-phosphate

It was Kornberg and Pricer (1953) who first showed that palmitic acid was incorporated into phosphatidic acid in a particulate fraction of guinea pig liver. From studies in rat liver, Weiss *et al.* (1960) suggested the involvement of *sn*-glycerol-3-phosphate as a precursor of triacylglycerols. In addition to liver, the phosphatidic acid or glycerol phosphate pathway has been found to occur in a wide variety of tissues and organisms: adipose tissue (Shapiro *et al.*, 1960; Steinberg *et al.*, 1961; Roncari and Hollenberg, 1967), brain (McMurray *et al.*, 1957; Hokin and Hokin, 1959; Martensson and Kanfer, 1968), mammary gland (McBride and Korn, 1964; Pynadath and Kumar, 1964; Dils and Clark, 1962), intestinal mucosa (Clark and Hübscher, 1960, 1961; Johnston, 1959; Coniglio and Cate, 1959), pancreas (Prottey and Hawthorne, 1967), diaphragm (Neptune *et al.*, 1963), aortic homogenates (Stein *et al.*, 1963), and spermatozoa (Terner and Korsh, 1962). Glycerol phosphate acyltransferase is not confined to the animal kingdom. It has also been found in higher plants by Cheniae (1965) and Sastry and Kates (1966), in yeast by Kuhn and Lynen (1965), and in bacteria by Ailhaud and Vagelos (1966) and Goldfine (1966).

The biosynthesis of phosphatidic acid has been studied for the most part in the endoplasmic reticulum in mammalian tissues (Van den Bosch *et al.*, 1972). As discussed by these authors, the elucidation of the exact pathway and mechanism of acylation of *sn*-glycerol-3-phosphate has been hindered by the fact that the product of acylation was phosphatidic acid, with little accumulation of the intermediate monoacylglycerol phosphate (Kornberg and Pricer, 1953; Lands and Hart, 1964; Stoffel *et al.*, 1967; Abou-Issa and Cleland, 1969; Possmayer *et al.*, 1969). Current evidence suggests that at least two distinct enzymes are involved in the biosynthesis of phosphatidic acid in mammalian tissues (Lands and Hart, 1964; Lamb and Fallon, 1970). Yamashita and Numa (1972) have reported a partially purified preparation of glycerol-3-phosphate acyltransferase from rat liver microsomes by techniques involving Triton X-100 treatment, Sepharose 2B filtration, and sucrose density centrifugation. Although only a fourfold purification was achieved, the enzyme preparation appeared to be free of 1-acylglycerol phosphate acyltransferase activity. This enzyme remained bound to membrane fragments which were separated from the glycerol phosphate acyltransferase during the sucrose gradient centrifugation step (Yamashita *et al.*, 1972). The partially purified glycerol phosphate acyltransferase required Ca^{2+} for activity and was stimulated by neutral phosphoglycerides. The product formed by the enzyme was identified as 1-acylglycerol phosphate, even when the incubation was carried out in the presence of both palmitoyl and linoleoyl CoA. Synthesis of phosphatidic acid was observed only after the addition of a sucrose gradient-prepared fraction containing the membrane-bound 1-acylglycerol phosphate acyltransferase, thus clearly establishing that two distinct enzymes are involved in the synthesis of phosphatidic acid (Yamashita *et al.*, 1972, 1975).

Examination of the acyl-donor specificity of partially purified glycerol phosphate acyltransferase revealed that palmitoyl CoA is the best acyl donor and that stearoyl CoA is utilized fairly effectively, as shown in Fig. 2. On the other hand, unsaturated fatty acyl CoA thioesters, such as oleoyl CoA, linoleoyl CoA, and arachidonyl CoA, were poor substrates (Yamashita and Numa, 1972).

Both saturated and unsaturated fatty acids can be used for the synthesis of phosphatidic acid in intact microsomes (Possmayer *et al.*, 1969; Husbands and

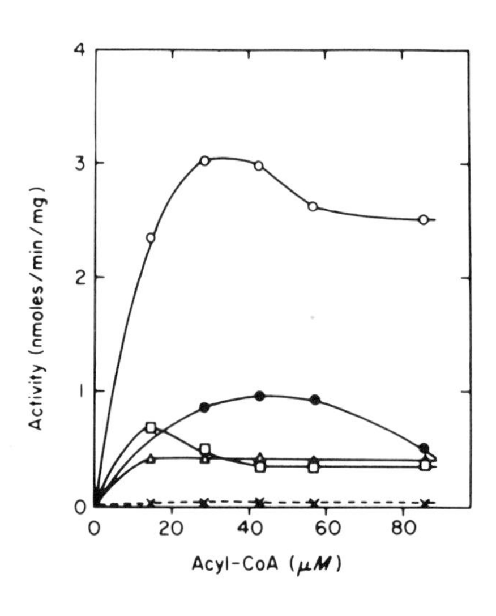

Fig. 2. Acyl-donor specificity of microsomal glycerol phosphate acyltransferase (Yamashita and Numa, 1972). ○—○, Palmitoyl CoA; ●—●, stearoyl CoA; □—□, oleoyl CoA; △—△, linoleoyl CoA; ×—×, arachidonyl CoA.

Lands, 1970; Hill *et al.*, 1968; Sanchez de Jimenez and Cleland, 1969; Yamashita and Numa, 1972; Daae, 1973). During purification of the glycerol phosphate acyltransferase by Yamashita *et al.* (1972), however, the enzyme lost its ability to utilize unsaturated fatty acyl CoA thioesters. It is possible that the purification procedure employed resulted in a change in the specificity of one microsomal glycerol phosphate acyltransferase or that a second enzyme with specificity for unsaturated acyl groups was lost during the purification procedure. The sucrose density gradient fractions of Yamashita *et al.* (1972) containing 1-acylglycerol phosphate acyltransferase effectively utilized both saturated and unsaturated fatty acids for phosphatidic acid synthesis with no apparent specificity. This is consistent with several earlier studies using total microsomes (Stoffel *et al.*, 1966; Hill and Lands, 1968; Barden and Cleland, 1969). In a kinetic study on the 1-acylglycerol phosphate acyltransferase of rat liver microsomes, Okuyama and Lands (1972) demonstrated that the selectivity of this enzyme was dependent on the incubation conditions employed. When the acylation was carried out at very low concentrations of 1-acylglycerol phosphate (below 10 μM), palmitate and arachidonate tended to be excluded from the 2 position of phosphatidic acid. In the experiments of Yamashita *et al.* (1972), the lysophosphatidic acid concentration was 7 μM, and the acylation of 2-acylglycerol phosphate proceeded at only 8–23% the rate of the 1-acyl isomer for the acyl CoAs tested, but saturated esters gave no higher rates of acyl transfer than unsaturated ones (Okuyama and Lands, 1972). A significant specificity for monoenoic and dienoic fatty acyl CoA thioesters, however, was found by Yamashita *et al.* (1973), as shown in Fig. 3. The most effective acyl donor was oleoyl CoA; linoleoyl CoA and palmitoyl CoA were also good substrates but were somewhat less effective. It is therefore clear that the selectivities observed during *in vitro* acylation of glycerol phosphate do not simply predict the ultimate distribution of saturated and

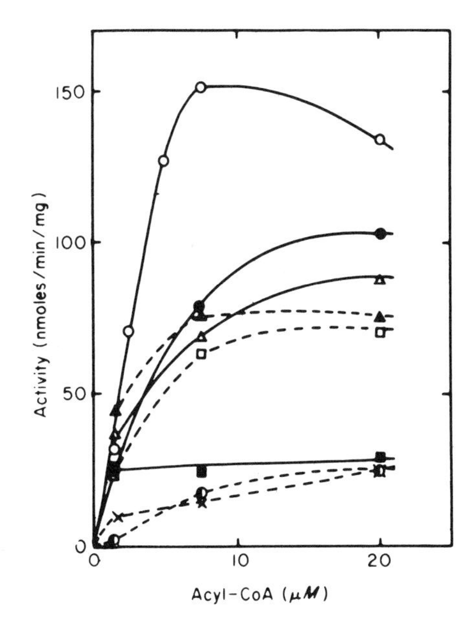

Fig. 3. Acyl-donor specificity of microsomal 1-acylglycerol phosphate acyltransferase (Yamashita *et al.*, 1973). O—O, Lauroyl CoA; ▲—▲, myristoyl CoA; □—□, palmitoyl CoA; ×—×, stearoyl CoA; △—△, palmitoleoyl CoA; O—O, oleoyl CoA; ●—●, linoleoyl CoA; ■—■, arachidonyl CoA.

unsaturated fatty acids in the end product. Factors such as pH (Lamb and Fallon, 1970), concentration of micellar and monomeric forms of acyl CoA (Zahler and Cleland, 1969), and isomeric monoacylglycerol phosphate concentration (Okuyama and Lands, 1972) are all now known to be involved. In addition, various treatments of microsomes (Yamashita and Numa, 1972; Okuyama and Lands, 1972) and protein concentrations (Lands and Hart, 1965) and whether or not mixtures of acyl CoA and/or monoacylglycerol phosphate are used (Okuyama and Lands, 1972) can affect the end result.

Akesson and co-workers (Akesson, 1970; Akesson *et al.*, 1970*a*, *b*) have reported on the *in vivo* synthesis of phosphatidic acid following the intraportal injection of labeled palmitate and linoleate. Even at very short time intervals following the administration of the labeled fatty acid, 90% of the palmitate was recovered at the 1 position of phosphatidic acid. On the other hand, 90% of the linoleate was found at position 2, indicating that *in vivo* the synthesis of phosphatidic acid occurred with a high degree of specificity. Analytical data obtained on the distribution of fatty acids in phosphatidic acid isolated from whole liver are in good agreement with this concept (Possmayer *et al.*, 1969; Akesson *et al.*, 1970*a*).

Detailed studies have also been carried out on the specificities of the mitochondrial acyltransferases. In contrast to microsomes, both whole mitochondria and mitochondrial outer membranes synthesize mainly monoacylglycerol phosphate (Daae and Bremer, 1970; Monroy *et al.*, 1972). Palmitate, supplied either as palmitoyl CoA or as palmitoyl-carnitine in the presence of carnitine palmitoyltransferase and CoA, is esterified at position 1 of the glycerol moiety (Daae, 1972; Monroy *et al.*, 1972). This preferential use of saturated fatty acids for the acylation of glycerol-3-phosphate is found for mitochondrial acyltransferases from different rat organs as well as from the livers of different species. Monroy *et al.* (1973) have reported a sixfold purification of the rat liver mitochondrial glycerol phosphate acyltransferase. The enzyme preparation was devoid of lysophosphatidate acyltransferase and produced essentially only 1-acylglycerol phosphate. As observed with intact mitochondria, a noticeable preference for saturated fatty acids was found, with the relative effectiveness shown in Table I. Linoleoyl CoA competed effectively with palmitoyl CoA in the acylation of 1-palmitoylglycerol phosphate by rat liver mitochondria (Monroy *et al.*, 1972). Thus the *in vitro* specificities of the mitochondrial acyltransferases are consistent with the asymmetrical distribution of fatty acids found in hepatic glycerolipids.

Some attention has been given recently to the possible mechanism(s) of glycerol phosphate acylation. The kinetics of appearance of 1-acylglycerol-2,3-phosphodiester has led to the suggestion that it may be a precursor of phosphatidic acid in rat liver mitochondria (Monroy *et al.*, 1972). O'Doherty *et al.* (1974*a*) showed that 1-acyl-*sn*-glycerol-2,3-phosphodiester could be used as a substrate for the synthesis of phosphatidic acid by rat liver microsomes and mitochondria. It was found that 1-palmitoyl-*sn*-glycerol-2,3-cyclic phosphodiester was utilized 3–4 times more readily than 2-palmitoyl-*sn*-glycerol-1,3-cyclic phosphodiester in the synthesis of phosphatidic acid. The 3-palmitoyl-*sn*-glycerol-1,2-cyclic phosphodiester was found to be inactive. These results are consistent with the hypothesis that acylglycerol cyclic phosphodiesters could serve as intermediates in glycerolipid metabolism. The formation of cyclic phosphodiester intermediates could possibly provide a mechanism whereby specific fatty acids are positionally placed on the glycerol molecule.

Table I. Substrate Specificity of Mitochondrial sn-Glycerol-3-phosphate Acyltransferase[a]

Acyl donor	Acyl CoA concentration (μmoles)	Total glycerol-3-phosphate incorporated[b] (nmoles)
None		0.00
Hexanoyl CoA	144	0.00
Octanoyl CoA	144	0.01
Decanoyl CoA	72	0.11
Myristoyl CoA	36	0.21
Palmitoyl CoA	72	0.55
Stearoyl CoA	38	0.19
Oleoyl CoA	10	0.04
Linoleoyl CoA	10	0.01

[a] From Monroy *et al.* (1973).
[b] Lysophosphatidic acid made up 97% of the product.

Mishkin and Turcotte (1974) have presented preliminary evidence that the Z protein of liver cytosol is superior to albumin in stimulating the esterification of *sn*-glycerol-3-phosphate in the presence of palmitoyl CoA and rat liver microsomes. These authors suggested that the Z protein may play a role in the positional specificity and fatty acid selectivity of the glycerol phosphate acyltransferases.

6.3.2. *Acylation of Dihydroxyacetone Phosphate*

Hajra and Agranoff (1968*a*) were the first to isolate and characterize acyldihydroxyacetone phosphate. In guinea pig liver mitochondria, Hajra *et al.* (1968) demonstrated that this lipid was present in small amounts and that it was rapidly labeled in the presence of ^{32}Pi or [^{32}P]ATP. An acyltransferase reaction was subsequently demonstrated in mitochondrial and microsomal fractions from guinea pig liver, giving rise to acyldihydroxyacetone phosphate from palmitoyl CoA and dihydroxyacetone phosphate (Hajra, 1968). When the rate of esterification was examined with different fatty acids, the highest rate was obtained with $C_{16:0}$ or $C_{17:0}$ in both subcellular fractions. In mitochondrial preparations, the rate of acylation of glycerol phosphate was 5–20% that of dihydroxyacetone phosphate, and glycerol phosphate did not compete with dihydroxyacetone phosphate in the acyltransferase reaction. This competition was observed, however, in the microsomal fraction.

Hajra and Agranoff (1968*b*) showed that the conversion of acyldihydroxyacetone phosphate to phosphatidic acid required two reactions. In the presence of NADP and a mitochondrial preparation from guinea pig liver, acyldihydroxyacetone phosphate was reduced to lysophosphatidic acid, which was then converted to phosphatidic acid if palmitoyl CoA was added to the assay system. These results suggested the occurrence of an alternate pathway for the biogenesis of phosphatidic acid.

Recent experiments have shown that the synthesis of acyldihydroxyacetone phosphate is not confined to liver but also occurs in other rat tissues including kidney, heart, testis, spleen, brain, and adipose tissue (La Belle and Hajra, 1972), as well as in yeast (Johnston and Paltauf, 1970). The formation of acyldihydroxyacetone by rat liver mitochondria was found to be 6 times more rapid than in microsomes. In contrast to an earlier report (Hajra and Agranoff, 1968*b*), in which the reduction of palmitoyl dihydroxyacetone phosphate was described as an exclusively mitochondrial event, both mitochondria and microsomes of different rat organs have been reported to reduce alkyl and acyl derivatives of dihydroxyacetone phosphate (La Belle and Hajra, 1972). Reduction of both derivatives was fairly specific, with rates at least 10–50 times higher with NADPH than with NADH. Reduction occurred specifically by transfer of the hydrogen from the B side at position 4 of the nicotinamide ring of NADPH.

6.3.3. Synthesis of Phosphatidic Acid from Monoacylglycerols and Diacylglycerols

An alternative series of reactions leading to the formation of phosphatidic acid from monoacylglycerols has been proposed. Paris and Clement (1965) provided evidence for the biosynthesis of 1-acyl-*sn*-glycerol-3-phosphate by rat intestinal mucosa following the absorption of a mixture of 1-[^{3}H]palmitoyl-*sn*-[^{14}C]glycerol and free fatty acids. The isotope ratios of both phosphatidic and lysophosphatidic acids suggested that direct phosphorylation of 1-palmitoyl-*sn*-glycerol had occurred, followed by acylation to produce phosphatidic acid. In a followup study, Paris and Clement (1969) showed that the *in vitro* incubation of monooleoylglycerol with ATP and fatty acids in the presence of subcellular fractions of intestinal cells resulted in the formation of lysophosphatidic acid. On the basis of these observations, they suggested that lysophosphatidic acid is an intermediate in the synthesis of phospholipids in the intestinal mucosa during fat absorption.

A monoacylglycerol kinase has been reported in brain by Pieringer and Hokin (1962). Using a deoxycholate-solubilized preparation from guinea pig brain, they showed that both 2-monoacylglycerol and D,L-1-monoacylglycerol were suitable substrates. The stereospecificity of the phosphorylation reaction is important if lysophosphatidic acid, after acylation to phosphatidic acid, is to serve as a precursor of glycerides and glycerophospholipids. The latter are known to be substituted derivatives of L-3-glycerol phosphate, and consequently the lysophosphatidic acid produced in the kinase reaction should also have this configuration. The exact physiological significance of the monoacylglycerol kinase reaction is undetermined.

Hokin and Hokin (1959) claimed the formation of phosphatidic acid from 1,2-diacyl-*sn*-glycerols in particulate fractions of brain. Phosphatidic acid synthesis appeared to be dependent on the addition of 1,2-diacyl-*sn*-glycerols, deoxycholate, ATP, and Mg^{2+}. Further studies (Hokin and Hokin, 1963) revealed the presence of diacylglycerol kinase in human erythrocyte ghosts. In erythrocytes, the synthesis of phosphatidic acid via this pathway was found to be 10–40 times greater than via the phosphorylation of monoacylglycerol, and 2500 times greater than the synthesis by acylation of *sn*-glycerol-3-phosphate. On the basis of this result, it was suggested (Hokin and Hokin, 1963) that diacylglycerol kinase played an important role in a phosphatidic acid cycle which was thought to be coupled to sodium transport. Diacylglycerol kinase activity has also been described in polymorphonuclear leukocytes (Sastry and Hokin, 1966), pancreas (Prottey and Hawthorne, 1967), and avian salt gland (Hokin and Hokin, 1964). Lapetina and Hawthorne (1971) have studied the enzyme in subcellular fractions of rat cerebral cortex. The enzyme was present in quite high levels in all subcellular fractions isolated from rat brain, with some enrichment observed in the cytosol and microsomal fractions. The rates of synthesis of phosphatidic acid from 1,2-diacyl-*sn*-glycerols and *sn*-glycerol-3-phosphate were compared and those of several reactions involved in the formation of phosphoinositides. It was concluded that diacylglycerol kinase represents an important route for the synthesis of phosphatidic acid and, consequently, inositol lipids. Table II gives the fatty acid composition of the product and unreacted substrate during the phosphorylation of mixed 1,2-diacyl-*sn*-glycerols with diacylglycerol kinase from *Escherichia coli* (Lands *et al.*, 1966*a*). The results show that the relative amounts of the component fatty acids in the substrate and product are essentially constant throughout the course of the reaction. The small differences in 20:4 ω6 and 22:6 ω3 could indicate a difference in phosphorylation rates, but when 45% of the diacylglycerol was reacted the phosphatidate composition was identical to that of the original diacylglycerols.

6.3.4. Relative Contributions of the Pathways of Phosphatidic Acid Synthesis

In recent years, much attention has been focused on studies to evaluate the relative contributions of the pathways of phosphatidic acid synthesis. The involvement of dihydroxyacetone phosphate as a precursor of higher acylglycerols

Table II. Fatty Acid Composition of Substrate and Product of Phosphorylation of Mixed 1,2-Diacyl-sn-glycerols by Diacylglycerol Kinase of Escherichia coli[a]

	0 min	15 min		30 min		45 min		60 min	
Fatty acid	DG[b]	DG	PA[b]	DG	PA	DG	PA	DG	PA
	(mole %)								
16:0	12	12	15	12	12	11	15	12	12
16:1	1	1	2	1	1	1	1	1	1
18:0	30	32	32	32	32	33	31	33	31
18:1	19	18	19	18	19	18	19	18	19
18:2	13	12	12	13	13	12	13	12	13
18:3 ω3	3.4	3.1	2.9	3.3	3.3	4.1	3.3	2.8	3.5
20:3 ω6	4.7	4.8	4.2	4.8	4.3	4.9	4.6	4.8	5.0
20:4 ω6	5.5	5.4	3.7	5.3	4.7	5.3	5.2	5.3	5.5
22:5 ω6	1.6	1.6	1.7	1.3	3.3	1.0	0.7	1.7	1.2
22:6 ω3	7.5	7.4	5.8	7.3	6.3	7.5	6.7	7.0	7.2
24:4 ω6	1.3	1.6	0.8	1.3	1.3	1.4	0.8	2.2	1.4
Extent of reaction	0	15%		19%		38%		45%	

[a] From Lands *et al.* (1966*a*).
[b] Abbreviations: DG, 1,2-diacyl-*sn*-glycerols; PA, phosphatidic acid.

has received close examination, and several laboratories have attempted to assess the relative contribution of both the glycerol phosphate and the dihydroxyacetone phosphate pathways to the synthesis of phosphatidic acid. Okuyama and Lands (1970) incubated rat liver slices with [1,3-^{14}C]- and [2-^{3}H]glycerol and determined the amount of both isotopes incorporated into phosphatidic acid. These authors theorized that if the glycerol was to be converted to a significant extent into dihydroxyacetone phosphate prior to incorporation into lipid, one would expect a relative loss of ^{3}H. It was found, however, that the ^{3}H/^{14}C ratio in the synthesized phosphatidic acid was increased rather than decreased when compared with the ^{3}H/^{14}C ratio of the substrates. Plackett and Rodwell (1970) reported similar findings in *Mycoplasma* strain Y grown on [2-^{3}H]glycerol and [1,3-^{14}C]glycerol. These workers claimed that *sn*-glycerol-3-phosphate, rather than dihydroxyacetone phosphate, is the preferred substrate for lipid biosynthesis. It was argued, however, by Agranoff and Hajra (1971) that in the experimental design of Okuyama and Lands (1970) and of Plackett and Rodwell (1970) the glycerol substrate must be phosphorylated and oxidized by enzymes not involved in the acyldihydroxyacetone phosphate pathway before its entry into this pathway. Agranoff and Hajra (1971) therefore studied the relative participation of both pathways of phosphatidic acid synthesis by measuring the incorporation of label from [^{3}H]NADH and [^{3}H]NADPH into C-2 of the glycerol moiety of phosphatidic acid. Their results suggested that the glycerol moiety of lipids derived from glucose, generally considered to be the usual source of lipid glycerol, is introduced at least partially via the acyldihydroxyacetone phosphate pathway in mouse liver. In Ehrlich ascites tumor cells, this pathway has been found to be a major route of synthesis (Snyder, 1972).

Manning and Brindley (1972) have also studied the relative rates of incorporation of [2-^{3}H]glycerol and [1-^{14}C]glycerol into lipids by rat liver slices. They argued that, provided the ^{3}H/^{14}C ratio in lipid is compared with that in glycerol phosphate, the values for the percentage of glycerol incorporated via the glycerol phosphate and dihydroxyacetone phosphate pathways can be obtained. By using this method, they claimed that 40–50% of the glycerol incorporated into lipid by rat liver slices proceeded via the glycerol phosphate pathway and 50–60% was incorporated via dihydroxyacetone phosphate. Pollock *et al.* (1975*a*) have compared the rates of phosphatidic acid synthesis from dihydroxyacetone phosphate via acyldihydroxyacetone phosphate or glycerol phosphate in homogenates of 13 tissues, most of them deficient in glycerol phosphate dehydrogenase. In all tissues examined, dihydroxyacetone phosphate entered phosphatidic acid more rapidly via acyldihydroxyacetone phosphate than via glycerol phosphate. Tissues with a relatively low rate of phosphatidic acid synthesis via glycerol phosphate showed no compensating increase in the rate of synthesis via acyldihydroxyacetone phosphate. The rates at which tissue homogenates synthesize phosphatidic acid from dihydroxyacetone phosphate via glycerol phosphate increase as glycerol phosphate dehydrogenase increases. Both glycerol phosphate dehydrogenase and glycerol phosphate : acyl CoA acyltransferase were more active than dihydroxyacetone phosphate : acyl CoA acyltransferase, indicating that all the tissues examined possessed an apparently greater capability to synthesize phosphatidic acid via glycerol phosphate than via acyldihydroxyacetone phosphate but did not express this potential.

It has been documented that the physiological glycerol phosphate concentration is much greater than the dihydroxyacetone phosphate concentration. In liver and in brain, the glycerol phosphate/dihydroxyacetone phosphate ratio is about 10 (Lowry *et al.*, 1964; Greenbaum *et al.*, 1971). Even in cells with very little glycerol phosphate dehydrogenase activity, such as Ehrlich ascites tumor cells, the glycerol phosphate/dihydroxyacetone phosphate ratio is also reported to be about 10 (Garfinkel and Hess, 1964). In addition, the glycerol phosphate concentration in a mammary gland tumor is as great as that in normal gland, yet the former has much less glycerol phosphate dehydrogenase (Rao and Abraham, 1973). Given the greater activity of glycerol phosphate acyltransferase than dihydroxyacetone phosphate acyltransferase in the tissues used and a glycerol phosphate concentration greater than that of dihydroxyacetone phosphate, it remains probable that *in vivo* the glycerol phosphate pathway to glycerololipids is more active than the acyldihydroxyacetone phosphate pathway (Pollock *et al.*, 1975*a*). On the other hand, two observations are consistent with a significant contribution by the acyldihydroxyacetone phosphate pathway to glycerolipid synthesis. First, most reductive, synthetic pathways use NADPH rather than NADH (Rao and Abraham, 1973). The 10^5-fold higher ratio of NADPH/NADP than NADH/NAD facilitates synthesis using NADPH (Krebs and Veech, 1970). The acyldihydroxyacetone phosphate pathway fits this generalization, but the glycerol phosphate pathway does not. Second, an assay of the glycerol phosphate pathway in liver slices implied that about one-half of the glycerolipid is synthesized from acyldihydroxyacetone phosphate (Manning and Brindley, 1972). This assay uses glycerol, and thus it does not include the normal branch point of the metabolic sequence. Glucose is the usual glyceride glycerol precursor (Hübscher, 1970), and its conversion to glycerol phosphate is subject to possible regulation by the intracellular NADH/NAD ratio. Thus the acyldihydroxyacetone phosphate pathway *in vivo* could be even more active than that reported. Rognstad *et al.* (1974) have claimed that isolated hepatocytes show less dependence on the acyldihydroxyacetone phosphate pathway than that reported for liver slices (Manning and Brindley, 1972). These two reports, however, are not readily comparable because of the different incubations used, but, even with isolated hepatocytes (Rognstad *et al.*, 1974), there is evidence for at least some lipid synthesis via the acyldihydroxyacetone phosphate pathway. It is possible that the use of variously labeled glucose precursors may permit quantitative evaluation of the alternative routes from glucose to lipid glycerol (Pollock *et al.*, 1975*b*).

6.4. Biosynthesis of Diacylglycerols

6.4.1. Synthesis of Diacylglycerols from Monoacylglycerols

It was Skipski *et al.* (1959) who first implied that 2-monoacylglycerols might be potential substrates in the resynthesis of triacylglycerols, while Savary *et al.* (1961) and Mattson and Volpenhein (1962, 1964) noted the remarkable similarity of position 2 of the fed and lymph triacylglycerols. An enzyme system with the ability to catalyze the conversion of monoacylglycerols to di- and triacylglycerols was first demonstrated by Clark and Hübscher (1960, 1961) in cell-free suspensions of rabbit small intestine. Their investigations showed that the addition of monoacylglycerols stimulated the incorporation of labeled fatty acids into glycerides. The reactions of the monoacylglycerol pathway have been studied most extensively in the mucosa of the small intestine of various animals (Clark and Hübscher, 1960, 1961; Senior and Isselbacher, 1962; Johnston and Brown, 1962; Ailhaud *et al.*, 1963; Gallo *et al.*, 1968; O'Doherty, 1974; O'Doherty and Kuksis, 1974*b*), but there is also evidence for its occurrence in kidney and pancreas (Hübscher, 1961), adipose tissue (Belfrage, 1964; Schultz and Johnston, 1971), the wall of the aorta (Stein *et al.*, 1963), Ehrlich ascites tumor cells (O'Doherty and Kuksis, 1974*b*), and the mammary glands of guinea pigs (McBride and Korn, 1964) and goats (Pynadath and Kumar, 1964; Dimick *et al.*, 1965).

The first enzyme in the monoacylglycerol pathway, monoacylglycerol acyltransferase (EC 2.3.1. type), was initially described with particulate subcellular fractions of rabbit small intestinal mucosa (Clark and Hübscher, 1960, 1961). The enzyme was also demonstrated in the intestinal mucosa of the rat (Senior and Isselbacher, 1962; Ailhaud *et al.*, 1963, 1964), hamster (Johnston and Brown, 1962; Brown and Johnston, 1963), cat (Brindley and Hübscher, 1965), chicken, pig (Bickerstaffe and Annison, 1969), and sheep (Leak and Cunningham, 1968; Bickerstaffe and Annison, 1969) and in hamster adipose tissue (Schultz and Johnston, 1971), pig kidney (Hübscher, 1961), and rabbit pancreas (Hübscher, 1961). Its activity has also been observed in the liver of rat (Hübscher, 1961; Snyder *et al.*, 1970; O'Doherty and Kuksis, 1974*b*) and pig (Sundler and Akesson, 1970), in several neoplastic tissues (Snyder *et al.*, 1970; O'Doherty and Kuksis, 1974*b*), in cultured fibroblasts (Lynch and Geyer, 1972; Bailey *et al.*, 1973), and in the fat body of locusts (Peled and Tietz, 1974).

Brown and Johnston (1963) observed that when labeled 2-monoacylglycerols were incubated with homogenates of hamster intestine the label was incorporated into triacylglycerols. The radioactivity was retained in the 2-monoacylglycerols which were produced by pancreatic hydrolysis of the synthetic product. It had been established previously (Clark and Hübscher, 1961) that *rac*-1-monoacylglycerol stimulated the incorporation of acyl CoA into triacylglycerols of intestinal homogenates of rabbit. Homogenates of rat intestinal mucosa were also shown to incorporate labeled *rac*-1-monoacylglycerol into higher glycerides. Further work with doubly labeled 2- and *rac*-1-monoacylglycerols in homogenates (Johnston and Brown, 1962; Brown and Johnston, 1964*a*,*b*) or slices (Johnson and Borgström, 1964) of hamster intestinal mucosa indicated that intact monoacylglycerols were esterified to triacylglycerols. The 1-isomer, however, is not the major product of pancreatic lipase hydrolysis in the lumen. Kinetic data (Ailhaud *et al.*, 1964; Brown and Johnston, 1964) indicated that the 2-monoacylglycerol is the preferred substrate for acylation. In hamster intestinal homogenates or microsomes, both *rac*-1- and 2-monoacylglycerols have been claimed to be converted to X-1,3-diacyl- and X-1,2-diacylglycerols, both of which were converted to triacylglycerols (Brown and Johnston, 1964*a*, *b*; Johnston *et al.*, 1965). With similar preparations of rat intestine, the *rac*-1-monoacylglcyerols were converted to X-1,3-diacylglycerols with no further esterification, while the 2-monoacylglycerols were esterified to triacylglycerols through X-1,2-diacylglycerol intermediates (Ailhaud *et al.*, 1964; Johnston *et al.*, 1965). Work on the more stable glycerol monoethers and everted sacs or

Table III. Comparison of the Acylation of 1- and 2-Acyl- and -Alkylglycerols by Microsomes of Hamster and Rat Intestinal Mucosa[a]

Substrate	Product		
	X-1,2-Diacylglycerol	X-1,3-Diacylglycerol	Triacylglycerol
		(nmoles/mg protein/hr)	
Hamster			
2-Hexadecanoylglycerol	352	4	800
2-Hexadecylglycerol	612	38	672
rac-1-Hexadecanoylglycerol	0	202	646
rac-1-Hexadecylglycerol	0	116	396
rac-1-Octadecenoylglycerol	52	328	256
1-Octadecenyl-*sn*-glycerol	0	300	188
Rat			
2-Hexadecanoylglycerol	84	2	16
2-Hexadecylglycerol	20	2	0
rac-1-Hexadecanoylglycerol	18	82	2
rac-1-Hexadecylglycerol	2	52	0
rac-1-Octadecenoylglycerol	16	28	0
1-Octadecenyl-*sn*-glycerol	0	52	0

[a] From Paltauf and Johnston (1971).

homogenates of rat (Gallo *et al.*, 1968; Sherr and Treadwell, 1965) and hamster (Kern and Borgström, 1965) intestine confirmed the above differences. The specificity of the utilization of a series of monoalkylglycerols as substrates for triacylglycerol synthesis *in vitro* has been investigated by Paltauf and Johnston (1971). These authors found that while the *sn*-1- and 2-alkylglycerols were satisfactory substrates, the *sn*-3 derivatives were not acylated by intestinal mucosal microsomes. The rat preferentially utilized the 2-acylglycerols while the mucosal microsomes of the hamster utilized either 1- or 2-alkyl or -acylglycerols for the synthesis of triacylglycerols or their ether analogues (Table III). In all these studies, the ethers were utilized less readily for complete acylation than the corresponding monoacylglycerols, as evidenced by the proportionally high accumulation of the intermediates in the incubation mixtures. Johnston *et al.* (1970) have demonstrated that diacylglycerols synthesized via the monoacylglycerol pathway are mainly used for triacylglycerol synthesis whereas diacylglycerols synthesized via the phosphatidic acid pathway are mainly used in the synthesis of phosphatidylcholine.

In intestinal mucosa, monoacylglycerol acyltransferase activity is found predominantly in the villus cells (Negrel and Ailhaud, 1975). Monoacylglycerol acyltransferase activity has been extensively studied in preparations of isolated villus cells under a variety of physiological conditions (O'Doherty and Kuksis, 1975*a*, *b*; O'Doherty *et al.*, 1972*b*,*c*, 1973, 1975). In isolated villus cells (O'Doherty, 1974; O'Doherty and Kuksis, 1975*b*), everted sacs (Breckenridge and Kuksis, 1975), and microsomes (O'Doherty, 1974; O'Doherty and Kuksis, 1974*b*) of rat intestine, the yield of diacylglycerol synthesized appeared to depend on the nature of both the monoacylglycerol and fatty acid supplied. Monoacylglycerol acyltransferase activity occurs predominantly in the microsomal fraction of all species tested (Senior and

Isselbacher, 1962; Brindley and Hübscher, 1965; Brown and Johnston, 1964*a*, *b*), although some activity has been reported with the brush border fraction (Forstner *et al.*, 1965; Gallo and Treadwell, 1970). Using histochemical techniques, Higgins and Barrnett (1971) showed that the enzymes of the monoacylglycerol pathway are mainly located on the inner surface of the smooth endoplasmic reticulum in rat intestinal mucosa, while the α-glycerophosphate pathway acyltransferases are mainly localized in the rough endoplasmic reticulum. The monoacylglycerol acyltransferase of rat intestine has been shown to require phosphatidylcholine for optimal activity, with unsaturated species of phosphatidylcholine more effective than saturated species (O'Doherty *et al.*, 1974*b*).

Several studies have been reported on the specificity of monoacylglycerol acyltransferase with respect to both the monoacylglycerol and fatty acid substrates. Early experiments with rat intestinal preparations indicated a preference for long-chain acyl CoAs (Ailhaud *et al.*, 1964). Brindley and Hübscher (1965), using homogenates of cat intestinal mucosa, found a preferential uptake of acids $C_{16:0}$ to $C_{18:1}$ as compared with $C_{6:0}$, $C_{8:0}$, and $C_{18:0}$ from an equimolar mixture of fatty acids ranging from $C_{6:0}$ to $C_{18:1}$. In microsomes of rat intestinal mucosa, rat liver, and Ehrlich ascites cells (O'Doherty, 1974; O'Doherty and Kuksis, 1974*b*), it has been found that the V_{max} of the enzyme decreased with different acyl acceptors in the order 2-monooleoylglycerol > 2-monopalmitoylglycerol > 2-hexadecylglycerol > 2-monostearoylglycerol > 2-monomyristoylglycerol > *rac*-1-monooleoylglycerol ≥ *rac*-1-monopalmitoylglycerol. Using an assay system in which palmitoyl CoA was generated from palmitoyl-(−)carnitine with microsomes of guinea pig intestinal mucosa, Short *et al.* (1974) showed that the V_{max} of monoacylglycerol acyltransferase decreased with different acyl acceptors in the order 2-monopalmitoylglycerol > 2-hexadecylglycerol > *rac*-1-monopalmitoylglycerol. In addition, it has been observed in microsomes of rat intestine, rat liver, and Ehrlich ascites tumor cells (O'Doherty, 1974; O'Doherty and Kuksis, 1974*b*) that the degree of utilization of each monoacylglycerol depends on the nature of the fatty acid supplied (Table IV). Similar results have been obtained with isolated intestinal epithelial cells (O'Doherty and Kuksis, 1975*b*) and everted sacs (Breckenridge and Kuksis, 1975), indicating that the activity of monoacylglycerol acyltransferase depends on both the nature of the monoacylglycerol and fatty acid supplied.

The stereospecificity of monoacylglycerol acyltransferase has been studied both *in vivo* and *in vitro*. Johnston *et al.* (1970) examined the stereospecificity of monoacylglycerol acylation in hamster mucosal microsomes, using 2-monopalmitoylglycerol and palmitic acid as substrates, and found that 1,2-diacyl-*sn*-glycerols are almost exclusively formed in this system. A similar finding has been reported by Sundler and Akesson (1970) for pig liver microsomes. Using microsomes of rat intestine, rat liver, and Ehrlich ascites tumor cells, O'Doherty and Kuksis (1974*b*) showed that although 1,2-diacyl-*sn*-glycerol was the main product (73–85%) of 2-monoacylglycerol acylation, some 2,3-diacyl-*sn*-glycerols (6–26%) were also formed (Table V). Studies on the stereospecificity of rat intestinal monoacylglycerol acyltransferase in isolated epithelial cells (O'Doherty and Kuksis, 1975*b*), in everted sacs (Breckenridge and Kuksis, 1975), and *in vivo* (Breckenridge *et al.*, 1976*b*) are at variance with the results obtained with microsomes. These studies indicated that although 1,2-diacyl-*sn*-glycerols were the major intermediates (55–65%) formed,

Table IV. Comparison of Monoacylglycerol Acyltransferase Activity in Microsomes of Rat Intestine, Liver, and Ehrlich Ascites Cells[a]

Substrates	Source of microsomes: Intestine	Liver	Ehrlich ascites cells
	(nmoles/mg protein/hr)		
2-Monooleoyl-*sn*-glycerol			
[1-^{14}C]Myristate	211	20	49
[9,10-^{3}H]Palmitate	221	24	56
[1-^{14}C]Stearate	206	17	51
[1-^{14}C]Oleate	228	27	63
[1-^{14}C]Linoleate	234	28	69
[1-^{14}C]Arachidonate	230	26	67
2-Monostearoyl-*sn*-glycerol			
[1-^{14}C]Myristate	98	9	25
[9,10-^{3}H]Palmitate	103	12	33
[1-^{14}C]Stearate	92	9	24
[1-^{14}C]Oleate	113	15	41
[1-^{14}C]Linoleate	118	17	44
[1-^{14}C]Arachidonate	129	16	40
2-*O*-Hexadecyl[2-^{3}H]glycerol			
Myristate	108	12	35
Palmitate	133	17	39
Stearate	120	14	34
Oleate	141	18	46
Linoleate	156	20	48
Arachidonate	148	18	45
rac-1-Monooleoylglycerol			
[9,10-^{3}H]Palmitate	43	4	11
[1-^{14}C]Oleate	47	5	12

[a] From O'Doherty and Kuksis (1974*b*).

significant amounts (35–45%) of 2,3-diacyl-*sn*-glycerols were also formed. It is possible either that a species difference is exhibited in the acylation of monoacylglycerols in the hamster and rat or that the acylation of 2-monoacylglycerols may take place via two different enzymes which exhibit varying lability toward cellular disruption. Relevant to this is the study of Gallo and Treadwell (1970). These investigators examined the structure of the diacylglycerol synthesized in rat intestine by determining the specificity of the diacylglycerol formed for diacylglycerol kinase. Since only part of the available diacylglycerol was utilized as a substrate by the enzyme, it was suggested that some 2,3-diacyl-*sn*-glycerols were also synthesized. Using smooth endoplasmic reticulum of rat intestinal cells, it was shown that the amounts of 1,2- and 2,3-diacyl-*sn*-glycerols synthesized may depend on the concentrations of monoacylglycerol and fatty acyl CoA used in the assay (O'Doherty, unpublished observations). The symmetrical acylation of 2-monoacylglycerols in the intestine would be of benefit to the animal as it would allow a much more efficient absorption and resynthesis of dietary fat than would be possible if the

Table V. Distribution of Radioactivity in 1,2- and 2,3-Diacyl-sn-glycerols Synthesized by Incubating 2-Mono-oleoyl-sn-glycerol and Labeled Fatty Acids in Microsomes of Rat Intestine, Liver, and Ehrlich Ascites Cells[a]

Source of microsomes	Diacylglycerol isomer	Lipid precursors (% total X-1,2 isomers)					
		[9,10-^{3}H]Palmitate + MO[b]		[1-^{14}C]Stearate + MO		[1-^{14}C]Oleate + MO	
		A[c]	B[c]	A	B	A	B
Intestine	1,2	83	85	84	82	81	82
	2,3	17	15	16	18	19	18
Liver	1,2	90	92	94	91	88	90
	2,3	10	8	6	9	12	10
Ehrlich ascites cells	1,2	82	85	81	83	74	73
	2,3	18	15	19	17	26	27

[a] From O'Doherty and Kuksis (1974*b*).
[b] Abbreviation: MO, 2-monooleoyl-*sn*-glycerol.
[c] A and B represent separate incubations.

conversion to triacylglycerols were confined to only one of two possible isomers. Further work is needed to unravel the stereospecificity of monoacylglycerol acylation.

Waite and Sisson (1976) have described the transacylation of monoacylglycerols to diacylglycerols by means of a monoacylglycerol acyltransferase present in the liver. This enzyme occurs in the plasmalemma and has been previously known as phospholipase A_1. It utilizes 1-acylglycerols and 1,2- and 1,3-diacylglycerols, as well as 1-acyl- and 1,2-diacylglycerophosphorylethanolamine as acyl donors and 1- and 2-acylglycerols as acyl acceptors. The 2-acylglycerol, triacylglycerol, and 2-acylglycerophosphorylethanolamine did not serve as acyl donors, and diacylglycerol and 1-acyl- and 2-acylglycerophosphorylethanolamine did not serve as acyl acceptors.

6.4.2. *Synthesis of Diacylglycerols from Phosphatidic Acid*

Diacylglycerols can also be generated within the cell by the action of phosphatidate phosphohydrolase, an enzyme which catalyzes the hydrolysis of phosphatidic acid to diacylglycerols and inorganic phosphate. This enzyme occupies an important branch point in lipid metabolism since the 1,2-diacyl-*sn*-glycerols that result from the dephosphorylation of phosphatidic acid can be used as precursors of phosphatidylcholine, phosphatidylethanolamine, and triacylglycerols. Surprisingly, however, phosphatidate phosphohydrolase is one of the poorly understood lipolytic enzymes, perhaps in part because of the inherent difficulties encountered in purifying the particle-bound enzyme. Phosphatidate phosphohydrolase was first discovered in plants by Kates (1955) and was subsequently identified in animal tissues (Weiss *et al.*, 1956; Smith *et al.*, 1967). Numerous investigations have been carried out to study its properties in the mammalian liver (Pieringer and Hokin, 1962; Wilgram and Kennedy, 1963; Sedgwick and Hübscher, 1965, 1967; Mitchell *et al.*, 1971), brain (Rossiter and Strickland, 1958; Hokin and Hokin, 1959; Agranoff, 1962; Strickland *et al.*, 1963; McCaman *et al.*, 1965), kidney (Bartlett *et al.*, 1962; Coleman and Hübscher, 1962, 1963), intestinal mucosa (Johnston and Bearden, 1962; Coleman and Hübscher, 1962; Brindley and Hübscher, 1965; Mitchell *et al.*, 1971), adipose tissue (Steinberg *et al.*, 1961; Vaughan, 1961), erythrocyte membranes (Hokin and Hokin, 1961*a*, *b*; Hokin *et al.*, 1963), and avian salt glands (Hokin and Hokin, 1961*b*).

The intracellular distribution of phosphatidate phosphohydrolase has been studied in several tissues. Coleman and Hübscher (1962) demonstrated that 63% of the enzyme activity of porcine kidney homogenates was recovered in the microsomal fraction, while Johnston and Bearden (1962) claimed that in the intestinal mucosa of the hamster, 80% of the enzyme was equally distributed in the mitochondrial and microsomal fractions. The rat liver enzyme, however, was reported to be primarily localized in the lysosomal fraction (Wilgram and Kennedy, 1963). Sedgwick and Hübscher (1967) suggested that in rat liver, phosphatidate phosphohydrolase was a true constituent not only of lysosomes but also of microsomes and mitochondria. They found, in agreement with the other studies, that only a maximum of 10% of the enzymatic acitivity was in the cytosol. Johnston *et al.* (1967), however, claimed that phosphatidate, bound to microsomes of intestinal mucosa, was chiefly converted into 1,2-diacyl-*sn*-glycerol by the phosphatidate phosphohydrolase of the cytosol rather than by the microsomal enzyme. These workers showed that the addition of cytosol resulted in a severalfold increase in triacylglycerol synthesis in microsomes. Similar findings were reported by Smith *et al.* (1967), who studied the conversion of mitochondrial-bound phosphatidate from rat liver and microsomal-bound phosphatidate from cat intestinal mucosa. These observations led to the conclusion that the function of the cytosol is to supply phosphatidate phosphohydrolase for the conversion of microsomal phosphatidic acid into 1,2-diacyl-*sn*-glycerols, which can then be used as a substrate for triacylglycerol synthesis. Phosphatidylcholine synthesis can also be slightly stimulated in rat liver microsomes by the addition of cytosol. Since present evidence suggests that it is primarily the phosphatidate phosphohydrolase of the cytosol that is involved in microsomal lipid synthesis, it remains unclear which function can be attributed to the microsomal and mitochondrial phosphatidate phosphohydrolase.

Several studies have been reported on the substrate specificity of phosphatidate phosphohydrolase. Early studies showed that brain phosphatidate phosphohydrolase could utilize hexadecylphosphate as well as phosphatidate as a substrate (Agranoff, 1962), while the hamster intestinal mucosa enzyme could degrade both phosphatidate and lysophosphatidate at equal rates (Johnston and Bearden, 1962). A fifteenfold purification of rat liver mitochondrial phosphatidate phosphohydrolase has been shown to catalyze the hydrolysis of phosphatidate and hexadecylphosphate (Sedgwick and Hübscher, 1967). Several laboratories have tested the possibility of whether or not the fatty acid composition of phosphatidic acid influences the activity of the enzyme. McCaman *et al.* (1965) found that phosphatides derived from soybean, yeast, and egg phosphatidylcholine were hydrolyzed at about equal rates by the phosphatidate phosphohydrolase from homogenates of rat brain, suggesting that the specific fatty acid distribution of phosphatidic acid is

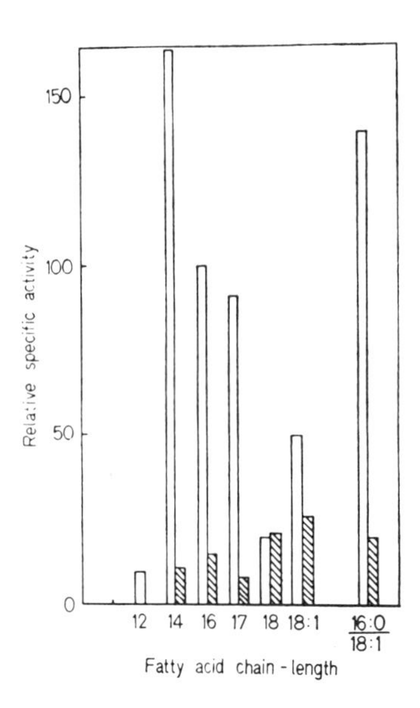

Fig. 4. Relative activities of the mitochondrial and soluble phosphatidate phosphohydrolase with phosphatidate bound to mitochondrial membranes (Mitchell *et al.*, 1971). □, Soluble; ▧, mitochondrial.

preserved during its conversion into 1,2-diacyl-*sn*-glycerol. Mitchell *et al.* (1971) assessed the substrate specificities of soluble and particulate phosphatidate phosphohydrolases with preparations from rat liver by using membrane-bound phosphatidates of various fatty acid composition as substrates. The soluble enzyme had the greatest activity with dimyristoyl and palmitoyl-oleoyl phosphatidates, while dilauroyl and distearoyl phosphatidates were poor substrates (Fig. 4). The particulate enzyme had in general much lower activities which were highest with distearoyl and dioleoyl phosphatidates. Work by Akesson *et al.* (1970*a*, *b*) indicates a nonselectivity of phosphatidate phosphohydrolase with respect to the fatty acid composition of its substrate. These investigators followed the distribution of glycerol incorporated into the various molecular species of phosphatidic acid and diacylglycerols of rat liver at short time intervals after the administration of [2-^{3}H]glycerol. They found that the isotope distribution among the various diacylglycerol fractions is very similar to that among the different species of phosphatidic acid, supporting the concept that phosphatidate phosphohydrolase does not exhibit a selectivity with respect to the fatty acid composition of phosphatidic acid.

6.4.3. *Synthesis of Other Diacylglycerols*

Aside from glycerol phosphate and the various neutral acylglycerols, other glycerolipids may also serve as acceptors of fatty acids yielding acylglycerolipids. Since these acylations have been demonstrated to exhibit marked fatty acid specificity, they have been summarized here, although a detailed consideration of glycerophospholipid metabolism is not a part of this discussion.

The acyl transfer from acyl CoA to 1-acylglycerophosphorylcholine was first demonstrated by Lands (1960). This enzyme is located in the microsomal fraction. A selective reaction with linoleate rather than with oleate was observed with 1-acylglycerophosphorylcholine prepared either by phospholipase A_2 digestion of egg phosphatidylcholine or by an acyl rearrangement of a 2-acylglycerophosphorylcholine. A detailed comparison of the specificities of the enzyme activity toward different fatty acids has shown a definite effect of chain length and molecular shape of the CoA thiol esters. The geometric isomer of 9,12-octadecadienoyl CoA showed that the *trans* configuration at the 12 position decreased the rate of reaction much more than a *trans* configuration at the 9 position (Lands *et al.*, 1966*a*). The monoene most commonly found esterified at the 2 position of the glycerol molecule (*cis*-9) was the preferred substrate for the freshly prepared enzyme, but the activity was lost rapidly on storage. Table VI summarizes the findings for the rat and pig liver microsomes. Acyl transfer to 1-acylglycerophosphorylethanolamine resembled that for the choline derivatives, with linoleate reacting more rapidly than stearate (Merkl and Lands, 1963). Van den Bosch *et al.* (1967) have shown that mixtures of saturated and unsaturated acylglycerophosphorylcholines are esterified at about the same rate, leading to no clear-cut selection of specific molecular species, in agreement with previous findings of Lands and Merkl (1963).

The selective esterification of octadecenoate isomers (and their analogues) to the 2 position of 1-acylglycerol-3-phosphorylcholine is shown in Fig. 5. These data suggest that the synthesizing enzymes favor 18-carbon acyl chains that have π-bonds located at positions 5, 9, and 12 along the chain. Since this pattern of selectivity

Table VI. Rates of Esterification of the sn-2 Position of 1-Acyl-sn-glycerol-3-phosphorylcholine

Acyl group[a]	Rat[b]	Pig[c]	Acyl group[a]	Rat[d]	Pig[d]	Acyl group[a]	Pig[e]
			(nmoles/mg protein/min)				
18:0	2	3	18:2 (2,5)	1	1	18:1 (2)	0
16:0	4	5	18:2 (3,6)	0	0	18:1 (3)	0
14:0	4	6	18:2 (4,7)	2	1	18:1 (4)	0
12:0	7	9	18:2 (5,8)	18	11	18:1 (5)	4
16:1 (c9)	6	3	18:2 (6,9)	6	7	18:1 (6)	1
18:1 (c9)	12	12	18:2 (7,10)	1	1	18:1 (7)	0
18:1 (t9)	10	12	18:2 (8,11)	7	5	18:1 (8)	3
18:2 (c9,c12)	23	17	18:2 (9,12)	40	22	18:1 (9)	7
18:2 (c9,t12)	9	13	18:2 (10,13)	14	13	18:1 (10)	6
18:2 (t9,c12)	14	12	18:2 (11,14)	31	21	18:1 (11)	5
18:2 (t9,t12)	6	10	18:2 (12,15)	21	16	18:1 (12)	22
18:3 (all c)	15	15	18:2 (13,16)	76	46	18:1 (13)	11
20:4 (all c)	20	14	18:2 (14,17)	17	14	18:1 (14)	6
						18:1 (15)	6
						18:1 (16)	7
						18:1 (17)	3

[a] The numbers in parentheses denote the positions of the double bonds.
[b] Lands *et al.* (1966*b*).
[c] Jezyk and Lands (1968).
[d] Reitz *et al.* (1968).
[e] Reitz *et al.* (1969).

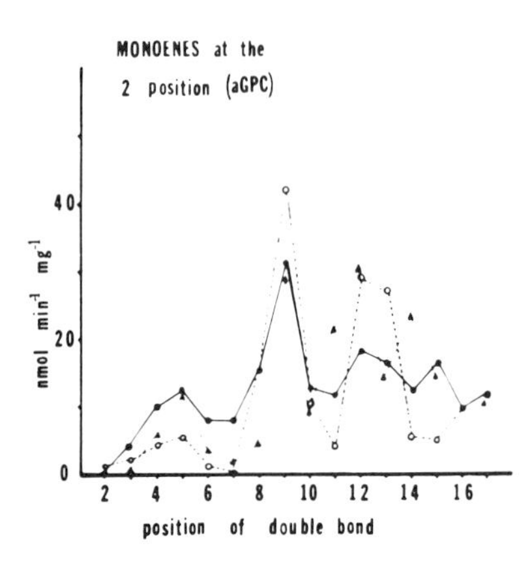

Fig. 5. Selective esterification of octadecenoate isomers (and their analogues) to the *sn*-2-position of 1-acylglycerol-3-phosphorylcholine (Lands, 1976). O, *cis*; ●, *trans*; ▲, yne.

occurs with *cis*, *trans*, or acetylenic analogues, the configuration and/or melting point is not directly influencing the biosynthetic events (Lands, 1976).

An acyl transfer from acyl CoA to a 2-acylglycerol-3-phosphorylcholine by a liver microsomal system was first demonstrated by Lands and Merkl (1963). Stearate and palmitate were incorporated more readily than linoleate and oleate. More detailed studies revealed that the enzyme in rat liver microsomes catalyzing the acylation of 2-acylglycerophosphorylcholine was particularly inactive with unsaturated acyl CoA derivatives containing a *cis* double bond at the 9 position. In contrast to these, the *trans*-ethylenic configuration appeared to be slightly preferred even to the naturally occurring saturated chains (Lands *et al.*, 1966*a*). These results indicate that the enzyme was extremely sensitive to configurational differences. An optimum chain length of 16 carbons was found for saturated acids (Jezyk and Lands, 1968). The results with configurational isomers of oleate and linoleate suggested that the more linear, extended hydrocarbon chains with the higher melting points effected an optimal rate of transfer, in contrast to the *cis* isomers of unsaturated fatty acids. Acids containing *cis*-ethylenic bonds at a distance of ten or more carbon atoms from the carboxyl end were rapidly utilized. Similar results were obtained with enzyme preparations from livers of rat, cow, pig, and pigeon (Reitz *et al.*, 1968). Table VII compares the rates of esterification at the *sn*-1 position of 2-acylglycerophosphorylcholine. The ethanolamine analogue, 2-acylglycerophosphorylethanolamine, was also shown to react with stearate more readily than with linoleate (Merkl and Lands, 1963), and later studies demonstrated specificities similar to those for 2-acylglycerophosphorylcholine (Lands and Hart, 1965).

There are striking differences between certain positional isomers of *cis*- and *trans*-octadecenoates (Fig. 6). Whereas the positional isomers of the *cis*-ethylenic acids had a selectivity nearly identical to that of the *cis*-methylene (cyclopropene) acids, the *trans* isomers differed markedly from both (Okuyama *et al.*, 1969). Thus the biosynthesizing activity that usually esterifies saturated fatty acids appears to be sensitive to configuration of the acyl chain in such a way as to exclude 18-carbon acids that contain a *cis*-9 or *cis*-11 configuration (Lands, 1976).

Table VII. Rates of Esterification of the sn-1-Position in 2-Acyl-sn-glycerol-3-phosphorylcholine

Acyl group[a]	Rat[b]	Pig[c]	Acyl group[a]	Rat[d]	Acyl group[a]	Pig[e]
			(nmoles/mg protein/min)			
20:0	—	5	18:2 (2,5)	0	18:1 (2)	0
18:0	20	28	18:2 (3,6)	2	18:1 (3)	1
16:0	20	38	18:2 (4,7)	4	18:1 (4)	3
14:0	8	24	18:2 (5,8)	3	18:1 (5)	14
12:0	7	12	18:2 (6,9)	3	18:1 (6)	5
16:1 (c9)	0	—	18:2 (7,10)	1	18:1 (7)	18
18:1 (c9)	4	—	18:2 (8,11)	27	18:1 (8)	49
18:1 (t9)	23	34	18:2 (9,12)	2	18:1 (9)	11
18:2 (c9,12)	2	—	18:2 (10,13)	26	18:1 (10)	22
18:2 (c9,t12)	2	—	18:2 (11,14)	31	18:1 (11)	35
18:2 (t9,c12)	14	—	18:2 (12,15)	28	18:1 (12)	45
18:2 (t9,t12)	28	43	18:2 (13,16)	33	18:1 (13)	28
18:3 (all c)	2	—	18:2 (14,17)	27	18:1 (14)	35
20:4 (all c)	2	—			18:1 (15)	21
					18:1 (16)	16
					18:1 (17)	12

[a] The numbers in parentheses denote the positions of the double bonds.
[b] Lands *et al.* (1966*b*).
[c] Jezyk and Lands (1968).
[d] Reitz *et al.* (1968).
[e] Reitz *et al.* (1969).

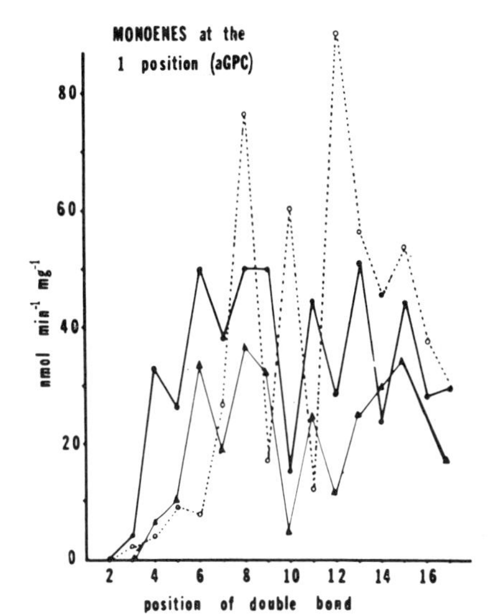

Fig. 6. Selective esterification of octadecenoate isomers (and their analogues) to the *sn*-1 position of 2-acylglycerol-3-phosphorylcholine (Lands, 1976). ○, *cis*; ●, *trans*; ▲, yne.

6.5. Biosynthesis of Triacylglycerols

6.5.1. Synthesis of Triacylglycerols from Diacylglycerols

The synthesis of triacylglycerols from diacylglycerols is catalyzed by the enzyme diacylglycerol acyltransferase (EC 2.3.1.20). It is pertinent to point out that the diacylglycerols utilized as substrates in this reaction can originate from both the monoacylglycerol and phosphatidic acid pathways. Since it is not yet known whether one or two distinct enzymes are involved in this step, their known properties will be considered together.

Diacylglycerol acyltransferase was first described in particulate fractions of chicken liver by Weiss and Kennedy (1956). The acylation of diacylglycerols has now also been observed in cell-free systems of rat and pig liver (Stein and Shapiro, 1958; Akesson, 1969; De Kruyff *et al.*, 1970; O'Doherty *et al.*, 1972*a*), chicken adipose tissue (Goldman and Vagelos, 1961), rat, rabbit, and hamster intestinal mucosa (Clark and Hübscher, 1960, 1961; Ailhaud *et al.*, 1964; O'Doherty *et al.*, 1972*a*), mammary glands of lactating guinea pigs and goats (McBride and Korn, 1964; Pynadath and Kumar, 1964), and Ehrlich ascites tumor cells (O'Doherty and Kuksis, 1974*b*). Diacylglycerol acyltransferase is found predominantly in the microsomal fraction of liver (Wilgram and Kennedy, 1963; Sarzala *et al.*, 1970). In the intestinal mucosa, some of this enzyme is reported to be localized in the brush border (Forstner *et al.*, 1965; Gallo and Treadwell, 1970). The diacylglycerol acyltransferase of rat intestinal mucosa has been shown to require phosphatidylcholine for optimal activity, with unsaturated species of phosphatidylcholine more effective than saturated species (O'Doherty *et al.*, 1974*b*). Some recent evidence (O'Doherty and Kuksis, 1975*c*) suggests that the Z protein may be involved in the acylation of diacylglycerols to triacylglycerols, as addition of Z protein to microsomes of rat liver and intestinal mucosa resulted in a fifteenfold stimulation of diacylglycerol acyltransferase activity. Since the Z protein of liver cytosol is known to have different affinities for various fatty acyl CoAs (Ockner *et al.*, 1972; Mishkin *et al.*, 1972) it is possible that it and other cytosolic proteins play a role in fatty acid specificity in triacylglycerol biosynthesis.

The stereospecificity of diacylglycerol acyltransferase has been studied in a number of tissues. Using chicken liver microsomes, Weiss *et al.* (1960) showed that 1,2-diacyl-*sn*-glycerol was about twice as effective a substrate as 2,3-diacyl-*sn*-glycerol in accepting palmitic acid. With chicken adipose tissue, Goldman and Vagelos (1961) found that racemic, mixed diacylglycerols containing palmitate tended to be better substrates than the analogous 1,2 isomers containing stearate. In these studies, however, no attempt was made to demonstrate that direct acylation of the diacylglycerols occurred. Using 1,2-diacyl-*sn*-glycerols and 2,3-diacyl-*sn*-glycerols of different degrees of unsaturation, it was shown that direct acylation of both diacylglycerol isomers can take place in rat liver, rat intestinal mucosa, and Ehrlich ascites tumor cells (O'Doherty *et al.*, 1972*a*; O'Doherty and Kuksis, 1974*b*). It was found by these investigators that when an equimolar mixture of the isomeric diacylglycerols was used as substrate, liver microsomes acylated less of the 2,3-diacyl-*sn*-glycerols than did microsomes of intestinal mucosa and Ehrlich ascites cells. Furthermore, the total yield of triacylglycerol in each tissue depended on the nature of the fatty acid used in the incubation mixture (Table VIII).

Table VIII. Triacylglycerol Synthesis from a Mixture of 1,2- and 2,3-Dioleoyl-sn-glycerols in Microsomes of Rat Liver and Ehrlich Ascites Cells

Source of microsomes	Fatty acid incubated	Triacylglycerol synthesized (nmoles/hr/mg protein)	Percent distribution of label in phenols	
			Phosphatidylphenols[a]	Lysophosphatidylphenols[b]
Liver[c]	[9,10-^{3}H]Palmitate	109	78	22
	[1-^{14}C]Oleate	113	81	19.
	[1-^{14}C]Linoleate	119	87	13
	[1-^{14}C]Arachidonate	124	85	15
Ehrlich ascites cells[c]	[9,10-^{3}H]Palmitate	88	58	42
	[1-^{14}C]Oleate	93	65	35
	[1-^{14}C]Linoleate	99	59	41
	[1-^{14}C]Arachidonate	112	62	38
Intestine[d]	[1-^{14}C]Myristate	135	63	37
	[9,10-^{3}H]Palmitate	160	69	31
	[1-^{14}C]Stearate	130	24	76
	[1-^{14}C]Oleate	140	68	32
	[1-^{14}C]Linoleate	145	89	11

[a] In these phenols the labeled fatty acid is in position 3; thus this represents the utilization of 1,2-diacyl-*sn*-glycerols.
[b] In the lysophosphatidylphenols the label is in position 1, representing the utilization of 2,3-diacyl-*sn*-glycerols.
[c] O'Doherty and Kuksis (1974*b*).
[d] O'Doherty *et al.*(1972*a*).

Table IX. Acylation of 1,2-Diacyl-sn-glycerol and 2,3-Diacyl-sn-glycerol in Microsomes of Rat Liver and Ehrlich Ascites Cells Using a Variety of Fatty Acids[a]

Substrates	Source of microsomes	
	Liver	Ehrlich ascites cells
	(nmoles/mg protein/hr)	
1-Stearoyl-2-linolenoyl-*sn*-glycerol		
[1-^{14}C]Myristate	120	108
[9,10-^{3}H]Palmitate	131	124
[1-^{14}C]Stearate	121	118
[1-^{14}C]Oleate	144	131
[1-^{14}C]Linoleate	152	140
[1-^{14}C]Arachidonate	163	147
2-Palmitoyl 3-oleoyl-*sn*-glycerol		
[1-^{14}C]Myristate	32	58
[9,10-^{3}H]Palmitate	41	76
[1-^{14}C]Stearate	35	67
[1-^{14}C]Oleate	47	88
[1-^{14}C]Linoleate	52	102
[1-^{14}C]Arachidonate	64	111

[a] From O'Doherty and Kuksis (1974*b*).

Using rat liver slices, Hill *et al.* (1968) demonstrated that diacylglycerol acyltransferase exhibited no specificity with respect to the fatty acid composition of the 1,2-diacyl-*sn*-glycerol substrate. Similar results were obtained with microsomes of rat liver (De Kruyff *et al.*, 1970; O'Doherty and Kuksis, 1974*b*), rat intestine, and Ehrlich ascites cells (O'Doherty and Kuksis, 1974*b*). In addition, it has been shown that diacylglycerol acyltransferase exhibits no specificity with respect to the fatty acid composition of the 2,3-diacyl-*sn*-glycerol substrate in microsomes of rat liver, rat intestine, and Ehrlich ascites cells (O'Doherty, 1974; O'Doherty and Kuksis, 1974*b*). Table IX shows the rates of acylation of selected *sn*-1,2- and *sn*-2,3-diacylglycerols with a variety of long-chain fatty acids in microsomes of rat liver and Ehrlich ascites cells.

6.5.2. Triacylglycerol Synthetase

With the exception of phosphatidate phosphohydrolase, the enzymes involved in triacylglycerol synthesis are located in subcellular membranes to which they are tightly bound. When considering these enzymes in terms of a multienzyme complex, it is likely that the spatial orientation of the individual enzymes to one another within the subcellular membranes is of special importance because the substrates or products have mainly hydrophobic properties (Hübscher, 1970). If the overall rate of triacylglycerol synthesis is to be governed by the kinetic properties of the participating enzymes rather than by the rate of transport of the substrates or products from one enzyme to the next, it is probable that this transport could be effected by a very close spatial orientation of the individual enzymes of a multienzyme complex.

Rao and Johnston (1966) have provided evidence for the existence of a tightly linked multienzyme complex in the mucosal triacylglycerol synthetase. They described an approximately seventyfold purification of the enzyme from hamster small intestine which contained acyl CoA synthetase, monoacylglycerol acyltransferase, and diacylglycerol acyltransferase. The increase in specific activity achieved in each of the four purification steps was of the same order for all three enzymes, pointing to the purification of a multienzyme complex rather than of three separate enzymes. This contention was substantiated further by the sedimentation properties of the partially purified synthetase in sucrose density gradients. It was also mentioned that the partially purified synthetase contained phospholipid. The nature of this phospholipid was not determined, but it has been suggested (Hübscher, 1970) that the multienzyme may have required a lipid-rich medium for optimum activity. O'Doherty *et al.* (1974*b*) have demonstrated that both the monoacylglycerol and diacylglycerol acyltransferases of rat intestinal mucosa require phosphatidylcholine for optimal activity, suggesting that phosphatidylcholine could perform this role. Further attempts to purify the multienzyme complex to a greater degree of homogeneity proved unsuccessful (Schiller, 1970).

6.5.3. Relative Contributions of the Major Pathways of Triacylglycerol Synthesis

It was established in the preceding sections that the intestinal mucosa and adipose tissue are capable of synthesizing diacylglycerols and triacylglycerols by at least two independent pathways. Several studies have been carried out with intestinal mucosa to evaluate the relative importance of each pathway. The relative contributions of the two pathways are uncertain, however, since a potential mechanism exists for the degradation of monoacylglycerols. DiNella *et al.* (1960) reported an enzyme hydrolyzing monoacylglycerols that had different properties than pancreatic lipase since mono-, di-, and triacylglycerols were hydrolyzed at identical rates. In other studies (Tidwell and Johnston, 1960), however, an enzyme was isolated which demonstrated only specificity for long-chain monoacylglycerols but had no effect on di- and triacylglycerols. Senior and Isselbacher (1963) showed it to be a microsomal enzyme, and subsequent purification (Pope *et al.*, 1966) required its solubilization with deoxycholate. The relative rate of hydrolysis was the same for long-chain 1- and 2-monoacylglycerols, but it attacked short-chain water-soluble monoacylglycerols at a faster rate. The true metabolic role of this enzyme has not been adequately characterized since most investigators (Brown and Johnston, 1964; Senior and Isselbacher, 1962) have claimed that monoacylglycerols are usually esterified directly.

Studies on the structure of resynthesized triacylglycerols by Savary *et al.* (1961) indicated that 2-monoacylglycerols might be absorbed intact since it was noted that 80% of the fatty acids in position 2 of the ingested glycerides were not exchanged during the absorption. After the feeding of labeled synthetic triacylglycerols of known structure, Mattson and Volpenhein (1964) claimed that in the rat intestinal lumen, triacylglycerol hydrolysis yielded 72 parts 2-monoacylglycerol, 6 parts 1-monoacylglycerol, and 22 parts free glycerol. Approximately 75% of the glycerol was absorbed as monoacylglycerol, and this was claimed to be reesterified directly to triacylglycerol, since the labeled fatty acid at position 2 in the fed triacylglycerol was largely retained in the same position of the lymph triacylglycerols. Similar results have also been reported by Reghavan and Ganguly (1969) for the mucosal triacylglycerols. The finding of large proportions of the original fatty acid at position 2, however, does not indicate that the total synthesis was via the monoacylglycerol pathway, since a random acylation of free fatty acid mixtures would give 33% of the acid at position 2. Their calculations assumed that 1-monoacylglycerol was hydrolyzed and that glycerol was not reutilized. In actual fact, glycerol was recovered to a significantly lower extent than the labeled fatty acid.

It has also been shown by other investigators that endogenous dilution of the fatty acid occurs, which may vary depending on the form of the fed fat. Verdino *et al.* (1965) showed that the feeding of tripentadecanoin resulted in the interesterification of the exogenous acids with those of endogenous origin. The resynthesis of samples of lard and corn oil in the same study resulted in a nonrandom distribution of the fatty acids in the different chemical classes of triacylglycerols, but the characteristic composition of position 2 of the fed triacylglycerol was largely retained in the lymph triacylglycerols. It was claimed that when coconut oil, butter oil, corn oil, and lard were fed to dogs and the lymph triacylglycerols analyzed, the final glycerides failed to resemble those of the diet but the molecular weight distributions were in general accord with the concept that the lymph triacylglycerols were largely synthesized from absorbed 2-monoacylglycerols and fatty acids. Endogenous dilution was estimated to be about 12% at the peak of absorption. Paris and Clement (1968) studied the absorption of doubly labeled 2-monopalmitoylglycerol in the presence of oleic and palmitic acids with loops of rat intestine. Their data indicated that 18% of the monopalmitoylglycerol was hydrolyzed. These authors suggested that 85% of the labeled glycerides were synthesized via direct acylation and 15% by esterification of *sn*-glycerol-3-phosphate. These values were only estimates. Breckenridge *et al.* (1976*a*) studied the structure of mucosal triacylglycerols in rat intestinal mucosa *in vivo* during the absorption of a low-molecular-weight fraction of butter oil and of the corresponding free fatty acids. In this study, it was shown that endogenous dilution varied from a minimum of 5% during triacylglycerol biosynthesis from monoacylglycerols to 15% during their synthesis from free fatty acids and was characterized by a preferential placement of the endogenous acids in the *sn*-3 and 2 positions of the triacylglycerol molecules. The phosphatidic acid pathway contributed a minimum of 20% of the total triacylglycerol yield at all times. From these *in vivo* experiments, it has been estimated that the 2-monoacylglycerol pathway accounts for about 80% of the resynthesized products.

The importance of the two pathways to triacylglycerol synthesis has also been assessed *in vitro*. One such study is that of Kern and Borgström (1965), who used incubations of everted sacs of hamster intestine with micelles of oleic acid and 1-monooleoylglycerol. They assumed that for every mole of monoacylglycerol that was incorporated into triacylglycerol, 2 moles of fatty acid would be incorporated. With this assumption, they claimed that the relative incorporation of the two components indicated that the monoacylglycerol pathway accounted for 80–100% of the synthesis. The value of this study remains in doubt because of the arbitrary substitution of the 1-monoacylglycerol for the natural 2-monoacylglycerol. If monoacylglycerol or free fatty acid was added individually to the everted sac incubations, endogenous fatty acids or monoacylglycerol precursors allowed as much synthesis as when both substrates were added, but it was not determined if such contributions were significant when only one precursor was added. Indeed, the fact that 2 moles of fatty acid were taken up for each mole of monoacylglycerol does not allow the conclusion that all synthesis was via the monoacylglycerol pathway when endogenous lipid can contribute. Breckenridge and Kuksis (1975) have examined the molecular specificity of the biosynthesis of triacylglycerols in rat intestine by means of radioactivity and mass tracers, argentation chromatography, and radio-gas chromatography. Bile salt micelles of alternatively labeled monoacylglycerols and free fatty acids were incubated with everted sacs of intestine for various periods of time. Analysis of the molecular species of the triacylglycerols labeled from monoacylglycerols showed that the 2-monoacylglycerol pathway was responsible for the biosynthesis of a maximum of 90% and the X-1-monoacylglycerol pathway for about 10% of the total radioactive triacylglycerols. Detailed analyses of the molecular species of triacylglycerols labeled from free fatty acids showed that the phosphatidic acid pathway contributed a minimum of 20–30% of the total labeled triacylglycerol formed. There was a preferential utilization in triacylglycerol biosynthesis of the more unsaturated diacylglycerols arising from the monoacylglycerol pathway and of the more saturated diacylglycerols originating from the phosphatidic acid pathway.

Another aspect that has received some attention of late is whether or not one of the major pathways of triacylglycerol synthesis can regulate the functioning and activity of the other. Rodgers (1970) noted that the activity of monoacylglycerol acyltransferase was greater in fed than in fasting animals, suggesting that the small intestine responds, at least to some degree, to the lipid load of the diet presented for absorption. This observation was further confirmed by a later study (Singh *et al.*, 1972) indicating that an increase in the fat content of the diet was associated with a further increase in this mucosal enzymatic activity. The effect of this lipid feeding on the enzymes of the phosphatidic acid pathway, however, was not examined. Polheim *et al.* (1973) have claimed that the synthesis of phosphatidic acid and di- and triacylglycerols via the *sn*-glycerol-3-phosphate pathway is markedly inhibited by 2-monooleoyl ether in microsomal and whole cell preparations obtained from intestinal and adipose tissue. Monoacylglycerols were also inhibitors under conditions in which their hydrolysis was minimized. It was suggested by these authors that the intracellular concentration of monoacylglycerol may regulate the activity of the *sn*-glycerol-3-phosphate pathway. A similar study by Brindley (1973), using microsomes of guinea pig intestinal mucosa, showed that although the addition of *rac*-1-monopalmitoylglycerol resulted in a slight decrease in the rate of incorporation of *sn*-glycerol-3-phosphate into higher glycerides, it did not decrease the rate of incorporation of *sn*-glycerol-3-phosphate into phospholipids. In a detailed study of diacylglycerol biosynthesis in everted sacs of rat intestinal mucosa, Breckenridge and Kuksis (1975) found that the operation of the phosphatidic acid pathway was not inhibited by the presence of monoacylglycerols. *In vivo* studies (Breckenridge *et al.*, 1976*a*, *b*) also indicated that the phosphatidic acid pathway was not noticeably inhibited by the absorption of 2-monoacylglycerols. These latter studies are consistent with the earlier work of Gurr *et al.* (1963), who observed that following an oral administration of olive oil and an injection of ^{32}P there was no increase in the specific activity of the phosphatidic acid fraction above that noted for the feeding of glucose, although the labeling in some of the other phospholipids increased. This would imply that there was a basic level of phosphatidic acid regardless of the extent of fat absorption. This apparent lack of inhibition of phosphatidic acid synthesis is consistent with the concept that a continuous synthesis of phosphatidylcholine is required for chylomicron and lipoprotein synthesis and secretion in intestinal mucosa (O'Doherty *et al.*, 1973, 1974*b*, 1975).

6.6. *Factors Influencing the Biosynthesis of Acylglycerols*

Before discussing some particular examples of factors influencing acylglycerol biosynthesis, a few generalizations can be made. Since higher glycerides can be synthesized from glycerol phosphate and dihydroxyacetone phosphate, factors that control the levels of these two metabolites can directly control higher glyceride synthesis. The NADH formed in the reaction catalyzed by glyceraldehyde phosphate dehydrogenase may be utilized by lactate dehydrogenase, malate dehydrogenase, glycerol phosphate dehydrogenase, or peroxisomes, disregarding other biosynthetic or hydroxylation reactions which might directly utilize the NADH produced in the cell sap. In connection with balance studies of the utilization of glucose carbon via the glycolytic or pentose phosphate pathways, it has been shown (Flatt and Ball, 1964; Katz *et al.*, 1966) that the amounts of NADH formed slightly exceeded those required for the reduction of dihydroxyacetone phosphate to glycerol phosphate, or of pyruvate to lactate. While sufficient NADH is thus available to satisfy both reactions, the relative amounts of glycerol phosphate and lactate formed vary, depending on the type of experiment and on the tissue examined. In experiments using the cytosol from either liver (Lea and Walker, 1965) or intestinal mucosa (Srivastava and Hübscher, 1966), about equal amounts of glycerol phosphate and lactate were formed from glucose. In *in vivo* experiments, however, relatively more lactate than glycerol phosphate was found in a number of mammalian tissues (Hohorst *et al.*, 1959; Wichert, 1962). However, factors controlling glycolysis, gluconeogenesis, and lipogenesis probably exert the main influence (Hübscher, 1970). The levels of a few key enzymes in these pathways can change markedly in the livers of animals starved or refed after a period of fasting.

6.6.1. Nutritional Factors

The tissue levels of glycerol phosphate have been shown to change with the nutritional state of the animal (Rapoport *et al.*, 1943; Bortz and Lynen, 1963). Tzur *et al.* (1964) carried out detailed studies and observed that after 48 hr of starvation the level of glycerol phosphate in rat liver fell to 34% of its original level; within as little as 2 hr after refeeding, the level rose to 80% of the fed control value. Since the enzyme system synthesizing glycerides has a low affinity for glycerol phosphate, it was suggested that the tissue level of glycerol phosphate may be the controlling factor in glyceride synthesis (Tzur *et al.*, 1964). The importance of the glycerol phosphate level in the biosynthesis of glycerides is further supported when the tissue levels of ATP, long-chain acyl CoA, and long-chain acylcarnitine are taken into account. Changes in the tissue levels of these compounds could equally influence glyceride biosynthesis. In any case, there is good evidence, as discussed above, that the tissue concentration of glycerol phosphate is a determining factor in the rate of glyceride biosynthesis.

Recent studies have indicated that the acylation of both monoacylglycerols and diacylglycerols is also affected by the nutritional status of the animal. Powell and McElveen (1974) reported that rats subjected to a 3-day fast showed marked decreases in intestinal monoacylglycerol acyltransferase activity and suggested that this decrease in enzyme activity was due to decreased cell proliferation in starvation. Singh *et al.* (1972) provided evidence that both fatty acid CoA ligase and monoacylglycerol acyltransferase increased their specific activities in intestinal microsomes of rats in response to increased dietary lipid loads. In these studies, the animals were fed for a period of 3 weeks, clearly longer than the intestinal mucosal cell turnover time (Lipkin *et al.*, 1963). Using hamster intestine, Mansbach (1975) showed that fat supplementation resulted in increased activity of diacylglycerol acyltransferase in the distal portion of the intestine. Increased activity of cholinephosphotransferase and lysophosphatidylcholine acyltransferase was also observed in this study. This is consistent with the concept that intestinal diacylglycerol acyltransferase can be stimulated by phosphatidylcholine (O'Doherty *et al.*, 1974*b*). Fallon *et al.* (1975) have demonstrated that in rats fed a high fructose diet for 11 days there was an increased rate of neutral lipid formation from *sn*-glycerol-3-phosphate by liver microsomal preparations. This was attributed to a reported increase in liver phosphatidate phosphatase activity. The significance of this increase was supported by the finding of a fall in microsomal phosphatidate content and a doubling in microsomal diacylglycerol. In addition, diacylglycerol acyltransferase, measured with microsomal-bound diacylglycerol, was increased twofold.

6.6.2. Hormonal Factors

To date, only a few studies have been reported on the effects of hormones on acylglycerol synthesis. Tzur *et al.* (1964) found that the injection of epinephrine to rats reduced the liver levels of glycerol phosphate. These workers reasoned that the reduced levels of glycerol phosphate in liver may have been due to the fact that epinephrine treatment accelerates fatty acid mobilization from the depots. The influx of larger amounts of fatty acids to the liver would accelerate the expenditure of α-glycerophosphate for their assimilation. Leal and Greenbaum (1961) found that administration of growth hormone to rats decreased the rate of incorporation of [1-^{14}C]acetate into liver di- and triacylglycerols. It was suggested that this resulted from a slowing down of the conversion of acetate into palmitoyl CoA.

Diacylglycerols occupy a branch point in glycerolipid biosynthesis, and virtually nothing is known about the factors that control whether diacylglycerols are acylated to triacylglycerols or are converted to phosphoglycerides via the cytidine nucleotide pathways. It was reported by Young and Lynen (1969) that the two biosynthetic enzymes utilizing diacylglycerols as substrates were altered in alloxan-diabetic rats and rats treated with triiodothyronine and propylthiouracil. Diacylglycerol acyltransferase was induced in both the ketotic diabetic and the triiodothyronine-treated animals but was repressed in the propylthiouracil-treated group. Cholinephosphotransferase activity, on the other hand, was unchanged by propylthiouracil treatment. In the ketotic diabetic and triiodothyronine-treated animals, the magnitude of changes in cholinephosphotransferase activity differed from that of diacylglycerol acyltransferase. These results lend support to the concept that the activity of these enzymes is important in regulating the relative amounts of diacylglycerol incorporated into triacylglycerol and phosphatidylcholine. Roncari and Murthy (1975) showed that the administration of thyroid hormone to rats resulted in an increased synthesis of triacylglycerols in liver and heart and a decreased synthesis of triacylglycerols in adipose tissue, suggesting that there may be some tissue specificity in the effect of hormones on glycerolipid synthesis. There is also some evidence that the monoacylglycerol pathway is under hormonal control. Rodgers *et al.* (1967) claimed that adrenalectomy resulted in decreased activity of the mono- and diacylglycerol acyltransferases in rat intestinal mucosa. It was not determined, however, in this study if the administration of adrenal hormones could increase the activities of the acylglycerol acyltransferases in the adrenalectomized animal.

6.6.3. Pharmacological Factors

The study of the effects of antibiotics and drugs on glycerolipid metabolism has gained much momentum in the last few years. The reduction in concentration of serum lipids as a means of either preventing or arresting the progress of atherosclerosis has been an underlying stimulus for much of the current research effort on the effect of drugs.

Although the effects of actinomycin D, cycloheximide, and puromycin on protein synthesis are well established (Pestka, 1971), their action on lipid biosynthesis is poorly understood. To date, no systematic study has been carried out on the effects of these antibiotics on glycerolipid synthesizing enzymes *in vivo* or *in vitro*. Mangiapane *et al.* (1973) studied the effect of administering actinomycin D on glycerolipid synthesis during liver regeneration after subtotal hepatectomy. The injection of actinomycin D prevented the increase in activity of phosphatidate phosphohydrolase but did not prevent the accumulation of triacylglycerol. Thus actinomycin D did not affect the enzymes of triacylglycerol synthesis, other than phosphatidate phosphohydrolase. Whether or not the increase in triacylglycerol content of regenerating liver may be due to a combination of increased fatty acid concentration in the liver and an altered distribution of fatty acid between triacylglycerol and phospholipid was not determined.

In a study on glycerolipid synthesis in isolated intestinal epithelial cells, it was found that the *in vivo* administration of puromycin did not affect the biosynthesis of triacylglycerols via either the phosphatidic acid or monoacylglycerol pathway (O'Doherty *et al.*, 1972*c*, 1973). In a later study, these workers (O'Doherty and Kuksis, 1974*a*) showed that there was no change in activity of the mono- and diacylglycerol acyltransferases in mucosal microsomes from puromycin-treated rats when compared with controls. In these studies, puromycin was observed to bring about a decrease in phosphatidylcholine synthesis in intestinal mucosa. The differential effect of puromycin on the glyceride- and phosphoglyceride-synthesizing enzymes observed in these studies may be due, in part, to the subcellular localization of these enzymes. Higgins and Barrnett (1971), using cytochemical techniques, found that the acyltransferases of the monoacylglycerol pathway were located on the inner surface of the smooth endoplasmic reticulum in rat intestinal mucosa. The α-glycerophosphate pathway acyltransferases, on the other hand, were mainly localized on the rough endoplasmic reticulum. In morphological studies on the effects of puromycin on rat intestinal mucosa (Friedman and Cardell, 1972; Yousef *et al.*, 1973), it was found that puromycin-treated mucosal cells contained much less rough endoplasmic reticulum. It may therefore be suggested that the observed inhibition of phosphatidylcholine may have been due to the subcellular location of the enzymes. Since puromycin does not appear to affect the smooth endoplasmic reticulum (Friedman and Cardell, 1972; Yousef *et al.*, 1973), it follows that the acyltransferases of the monoacylglycerol pathway should not be affected. It has also been observed that the *in vitro* addition of puromycin to isolated intestinal epithelial cells did not suppress triacylglycerol synthesis from [2-^{3}H]glycerol or 2-hexadecylglycerol (O'Doherty and Kuksis, 1975*a*).

Cycloheximide has also been demonstrated to have little effect on triacylglycerol synthesis. Bar-on *et al.* (1972) have shown that in liver microsomes from cycloheximide-treated rats there was only a slight decrease in the activity of diacylglycerol acyltransferase. From the foregoing discussion, it would therefore appear that neither actinomycin D, puromycin, nor cycloheximide affects triacylglycerol biosynthesis to any great extent.

During the last few years, our understanding of the mechanism of action of the hypolipidemic drugs has increased considerably. Clofibrate (ethyl-L-*p*-choloro-phenoxyisobutyrate) is probably the most widely used hypolipidemic drug, both in this country and in Europe. It was originally developed in England and was first considered to act by displacing androsterone, a steroid hormone with hypolipidemic effects, from albumin so as to increase its biological activity (Havel and Kane, 1973). The intraperitoneal injection of 250 mg/kg of clofibrate doubles hepatic mitochondrial α-glycerophosphate dehydrogenase in 4–6 hr, together with stimulation of mitochondrial protein synthesis; this is associated with reduction by half of hepatic concentration of α-glycerophosphate (Pereira and Holland, 1970). When the livers of animals given such doses of clofibrate are perfused with either labeled palmitate or glucose, the secretion of labeled triacylglycerols is also reduced by about one-half (Pereira and Holland, 1970). These studies suggest that the effect of clofibrate on mitochondrial α-glycerophosphate dehydrogenase is causally related to its hypotriglyceridemic action through inhibition of hepatic triacylglycerol synthesis. Other studies confirmed the decrease in α-glycerophosphate level (Westerfield *et al.*, 1968), and that hepatic triacylglycerol synthesis is reduced in treated rats (Adams *et al.*, 1971). However, feeding 0.25% clofibrate in the diet inhibits the incorporation of labeled glycerol into hepatic triacylglycerols in as little as 6 hr with no change in the level of hepatic α-glycerophosphate and addition of 5 mM clofibrate to liver homogenates inhibits the incorporation of α-glycerophosphate into triacylglycerols (Adams *et al.*, 1971). Additional studies have shown that this is associated with substantial inhibition of acyl CoA-α-glycerophosphate acyltransferase, with no effect on fatty acyl CoA synthetase, phosphatidic acid phosphatase, or diacylglycerol acyltransferase (Lamb and Fallon, 1972). A minor part of this effect could be attributed to displacement of palmitoyl CoA from albumin present in the incubation mixture with increased binding to microsomes, but the remainder appeared to reflect inhibition of the initial reaction of the synthesis of glycerolipids. Other enzymatic changes observed in livers of rats treated with clofibrate suggest that the capacity for fatty acid synthesis may also be decreased. These include reduced activity of acetyl CoA carboxylase, fructose-1,6-diphosphate aldolase, and glucose-6-phosphate dehydrogenase (Zakim *et al.*, 1970; Maragoudakis *et al.*, 1972).

It has recently been shown that two classes of drugs, namely the derivatives of clofibrate [ethyl 12-(4-chlorophenoxy)-2-methylpropionate] and fenfluramine [2-ethylamino-1-(*m*-trifluoromethylphenyl)propane hydrochloride], inhibit glycerolipid synthesis *in vitro*. Fenfluramine and its derivatives inhibit the synthesis of glycerides in preparations of rat and human liver (Marsh and Bizzi, 1972; Bowley *et al.*, 1973), rat intestine (Dannenburg *et al.*, 1973), and human adipose tissue (Wilson and Galton, 1971). Clofenapate [sodium 4-(4′-chlorophenyl)phenoxyisobutyrate] has been demonstrated to be more effective than *p*-chlorophenoxyisobutyrate in decreasing the concentration of serum triacylglycerol (Craig, 1972) and in inhibiting glycerol phosphate esterification in homogenates of human liver and in mitochondrial and microsomal fractions of rat liver (Brindley *et al.*, 1973). In addition, clofenapate was more potent in inhibiting the esterification of dihydroxyacetone phosphate than that of glycerol phosphate in rat liver mitochondrial fractions (Bowley *et al.*, 1973). In a detailed study on the effects of drugs on glycerolipid synthesis, Brindley and Bowley (1975) claimed that the amphiphilic anions clofenapate and 2-(*p*-chlorophenyl)-2-(*m*-trifluoromethylphenoxy)acetate both inhibited glycerol phosphate acyltransferase and diacylglycerol acyltransferase at approximately 1.6 and 0.7 mM, respectively, in cell-free preparations and slices of rat liver. Clofenapate (1 mM) also inhibited the incorporation of glycerol into lipids by rat liver slices without altering the relative proportions of the different lipids synthesized. The amphiphilic amines mepyramine, fenfluramine, norfenfluramine, hydroxyethylnorfenfluramine, *N*-(2-benzoyloxyethyl)norfenfluramine, cinchocaine, chlopromazine, and demethylimipramine inhibited phosphatidate phosphohydrolase by 50% at concentrations between 0.2 and 0.9 mM. The last four compounds inhibited glycerol phosphate acyltransferase by 50% at concentrations between 1 and 2.6 mM. None of the amines examined appeared to be an effective inhibitor of diacylglycerol acyltransferase. Norfenfluramine, hydroxyethylnorfenfluramine, and *N*-(2-benzoyloxyethyl)norfenfluramine produced less inhibition of glycerol incorporation into total lipids than was observed with equimolar clofenapate. The major effect of these amines in liver slices was to inhibit triacylglycerol and phosphatidylcholine synthesis and to produce a marked accumulation of phosphatidate. The ability of the amphiphilic amines to inhibit phosphatidate phosphohydrolase may be of importance in controlling glycerolipid synthesis, since this enzyme is thought to have a regulatory function in this process (Hübscher, 1970; Mangiapane *et al.*, 1973; Lamb and Fallon, 1972; Brindley and Bowley, 1975). These compounds should prove to be useful biochemical probes for future investigations on the parameters that contribute to the control of acylglycerol biosynthesis.

6.7. Methods Employed in Studying the Biosynthesis of Acylglycerols

It is beyond the scope of this chapter to present a detailed account of any of the myriad methods and techniques that were used in collecting much of the information described in the previous sections. Nevertheless, a few general comments can be said about the different levels of cellular organization that have been used to study the metabolism of acylglycerols.

In the intact animal, the constituent enzymes and substrates possess definite geometric and chemical relationships to each other. The rates at which metabolites are delivered to and removed from a given tissue are affected or modified by the activities of other tissues. Many anabolic and catabolic processes are regulated by products discharged into the bloodstream, and the various organs of the animal serve to limit the rates at which substrates or products enter or leave certain compartments. Although the use of the intact animal presents several empirical difficulties, there are several methods available to investigate what is happening. These include the feeding of diets deficient in some normal nutrients, addition by dietary or parenteral route of either an abnormal material or an excess of a normal material, or the introduction into the animal of some metabolite labeled in such a fashion that the subsequent distribution of the label can be studied. Surgical procedures can also be introduced, such as the establishment of fistulas from various segments of the gastrointestinal tract, the biliary tract, or the lymphatic system. The study of the effects of partial or complete removal of an organ can also be useful, although here the magnitude of the derangement is somewhat greater. As discussed in earlier sections, the intact animal has been extensively used to study acylglycerol synthesis in both the intestinal mucosa and the liver.

The next level of cellular organization is the perfused, isolated organ. In such studies, there may be loss of some of the regulatory mechanisms such as hormonal and/or neural which operate on the organ in its normal locus. A methodological advance in the application of this technique to the study of acylglycerol metabolism is represented by the work of Kuksis *et al.* (1975*a*), who studied glycerolipid synthesis in isolated rat livers perfused with deuterated water. These workers were able to recognize distinct subsets of acylglycerols arising during their biosynthesis in the presence of deuterated water by determining the distribution of both labeled fatty acids and glycerol among the molecular species. This approach circumvents the difficulty of introducing the acylglycerol precursors at optimum levels and in the proper chemical and physical form. The use of stable isotopes in acylglycerol biosynthesis may possibly permit a much needed measurement of the pool size and turnover rate of the molecular species of acylglycerols which participate in each metabolic transformation. It may then be possible to assess the rate of lipid synthesis as it relates to other metabolic processes such as protein synthesis and membrane biogenesis in the absence of the obscuring effect of conventional measurements of radioactivity and the confusion arising from metabolic reutilization of the substrates (Kuksis *et al.*, 1975*b*).

The next level of organization is the tissue or organ slice. This approach for the study of acylglycerol biosynthesis has been used by Hill *et al.* (1968) for examining the specificity of diacylglycerol acyltransferase. In this technique, tissue slices approximately 50–100 μm in thickness provide to the bathing fluid a surface sufficient to permit adequate exchange of nutrients and waste products so that viability is maintained for several hours. Although reaction rates within the cell of a sliced tissue may deviate from normal rates, this technique has proven extremely useful because of the simplicity of subsequent experimental operations as there is good control not only of the organ and the previous nutritional status of the animal but also of the composition of the bath fluid and of the gas phase with which it is equilibrated.

Another technique one can employ for the study of acylglycerols is that of using isolated cells. The ability to prepare suspensions of cells from various organs which retain their metabolic and structural integrity offers a productive avenue of approach to the study of acylglycerol metabolism. The ease of manipulation and reproducibility of sampling provide significant advantages over other systems employing more intact tissue preparations. More important, the use of cell suspensions allows a degree of homogeneity in cell populations not possible with classical techniques. Cell suspensions can be made by both enzymatic and nonenzymatic means. Isolated cells have been used to study acylglycerol biosynthesis in rat intestinal mucosa (O'Doherty, 1974; O'Doherty *et al.*, 1973, 1975; O'Doherty and Kuksis, 1975*a*, *b*) and rat liver (Sundler *et al.*, 1974).

An extension of the use of cell suspensions is cell culture. Cell culture is a useful system for biochemical studies because it provides a source of large, homogeneous populations of cells. Experimental conditions can be more easily manipulated to allow quantitative experiments. Moreover, cells are isolated from the hormonal and other superimposed physiological regulations of cells *in situ*. Lipidologists appear to have been slow to use the advantages of animal cell culture systems, in which well-controlled studies can be made with quiescent and growing normal cells, tumor cells, cells in different phases of the growth cycle, virus-transformed and chemically transformed cells, and cells growing in medium containing delipidized serum with added lipids. The advantages of using cultured cells for the study of lipid metabolism have been discussed in a review by Howard and Howard (1974). The increasing availability of differentiated cell types especially from vascular tissues should enable more rapid future progress in our understanding of the parameters that regulate acylglycerol synthesis.

A lower level of organization is the use of subcellular fractions. After rupture of the cell membranes, many relationships that are obtained between parts of the normal cell no longer exist. This procedure has proven particularly useful in studies designed to determine the location of metabolic processes within the cell. By addition of suitable substrates, the presence or absence of a given enzymatic activity in these cellular fractions may be ascertained. In acylglycerol metabolism, this activity is confined almost entirely to the microsomal fraction. Extensive studies on acylglycerol synthesis using microsomes from various organs have been outlined in an earlier section.

A yet lower level of cellular organization is the enzyme in solution. A purified enzyme is a prerequisite to a complete understanding of the reaction which it catalyzes. This permits studies of the kinetic and regulatory properties of the enzyme and of the thermodynamic characteristics of the reaction. To date, none of the enzymes involved in acylglycerol synthesis have been purified to homogeneity.

6.8. *Kinetic Problems Encountered in Enzyme Assays Involving Lipid Substrates*

Conventional enzyme kinetics apply mainly to reactions which take place in aqueous solutions. Enzymatic reactions involving lipids differ, however, in one essential aspect: they are heterogenous reactions because the enzymes are water soluble but the substrates are not. Consequently, the enzyme–substrate interaction must take place at the interface of the aggregated substrate and water. Since rigid kinetics for enzymes involving lipid substrates have yet to be developed, a few of the inherent problems are worthy of note. A basic problem is the dispersion of the lipid in an aqueous solution as a micelle or emulsion. When using emulsions one must bear in mind that this may not be the form of the substrate encountered *in vivo* by the enzyme and that the results obtained may not be readily extrapolated to normal physiological conditions. A clear case now of how misinterpretation and confusion can be introduced by using lipid emulsions is demonstrated by the enzyme phosphatidate phosphohydrolase. Originally this enzyme was shown to be implicated in glycerolipid synthesis, and enzymatic assays using emulsions of phosphatidate were reported (Johnston and Bearden, 1962; Coleman and Hübscher, 1962). The phosphohydrolase activity was shown to be in several cell fractions (Wilgram and Kennedy, 1963; Sedgwick and Hüscher, 1965; Brindley and Hübscher, 1965), and it was assumed that both the mitochondrial and microsomal activities were involved in acylglycerol synthesis. Subsequent work on higher acylglycerol synthesis showed that a cytosolic stimulating factor was in fact a soluble phosphatidate phosphohydrolase (Johnston *et al.*, 1967; Smith *et al.*, 1967). This soluble enzyme hydrolyzed phosphatidic acid which had been synthesized on mitochondrial or microsomal membranes but was quite inactive with phosphatidate emulsions. On the other hand, the particulate phosphatidate phosphohydrolase was inactive toward membrane-bound biosynthetic phosphatidate. The characteristics of the soluble phosphohydrolase indicated that parameters such as the surface charge of the substrate or the necessity of a specific lipoprotein complex may be critical for the demonstration of enzymatic activity. Recent work suggests that the soluble phosphohydrolase may be a different enzyme from that found in particulate fractions and that the former may regulate triacylglycerol synthesis (Hübscher, 1970; Brindley, 1973; Mangiapane *et al.*, 1973). Since the earlier studies implicated the particulate enzyme in glyceride synthesis, the discovery of the soluble enzyme was delayed.

Another kinetic problem commonly encountered when lipids are used as substrates is describing substrate concentrations. For a more detailed discussion, the reader should refer to Brockerhoff and Jensen (1974). As these authors explain, the K_m values of conventional enzymology assume that substrates are dispersed as single molecules in true solutions. Lipids, on the other hand, form micelles and emulsions at concentrations above their critical micellar concentrations. It is important to know whether the enzyme acts on the monomolecular form of the substrate or on the aggregate. When the enzyme acts on single molecules in solution, it is often an advantage to work below saturating substrate concentration and thus avoid inhibition at high substrate concentrations when micelles form (Brockerhoff and Jensen, 1974). If the enzyme under study is active only on lipid particles, then the effective substrate concentration is proportional to the surface area of the emulsion rather than the molar concentration of lipid added. How to measure this surface area presents a basic problem. The distribution of lipid substrates can become quite complex in an incubation medium containing both albumin and cell organelles. An acyl CoA ester in a single incubation medium could be bound to albumin or bound to membranes in micellar form or exist as single molecules. As Lamb and Fallon (1972) have pointed out, changes in this distribution may produce alterations in the kinetic behavior of the enzyme utilizing the acyl CoA, and continual monitoring of this distribution is cumbersome. It should also be pointed out that long-chain acyl CoA esters also exhibit detergent activity and detergents do present difficulties for the lipidologist who wishes to measure reactions under defined conditions. Abou-Issa and Cleland (1969) have shown that the addition of albumin not only relieves the inhibition produced by long-chain acyl CoA esters, but it also increases the gross concentration of acyl CoA esters required for saturating conditions. The reaction rate can be more readily controlled if acyl CoA is generated throughout the period of incubation. This can be achieved by generating acyl CoA with the endogenous acyl CoA synthetase. A useful method for generating acyl CoA is by using an excess of carnitine palmitoyltransferase, acyl carnitine, and CoA (Daae and Bremer, 1970; Sanchez *et al.*, 1973; Short *et al.*, 1974). Even though both acylcarnitine and the salts of free fatty acids are themselves detergents and can inhibit enzymatic activity, they do so less than acyl CoA esters. The development of techniques for assaying enzyme reactions is still in its infancy, and it is to be hoped that the next few years will provide inroads in this direction. One particularly attractive advancement of late is the use of membrane-bound substrates (Mansbach, 1973; Fallon *et al.*, 1975). Clearly much more work has to be done in this area.

6.9. Enzymatic Lipolysis of Acylglycerols

Enzymatic lipolysis is essential for both the absorption and turnover of acylglycerols. In contrast to the phospholipases, which display a highly stereochemical course of action, some of the acylglycerol lipases appear to be unusual in apparently hydrolyzing enantiomeric acylglycerols at comparable rates. The purpose of this section will be to discuss some studies on the specificities of the acylglycerol lipases rather than to give a complete survey of each enzyme. For the latter purpose, the reader should refer to Brockerhoff and Jensen (1974).

6.9.1. Pharyngeal Lipases

Until very recently, it was believed that in most animals dietary lipids pass through to the stomach with little chemical alteration. In 1973, however, Hamosh and Scow (1973) described a potent lipolytic enzyme in the serous glands of rat tongue which hydrolyzed triacylglycerols to mostly diacylglycerols and free fatty acids. A similar activity was also found in the soft palate, anterior oral pharyngeal wall, and lateral oral pharyngeal glands. The studies showed that dietary triacylglycerols are readily hydrolyzed in the stomach to partial glycerides and free fatty acids and that this reaction is catalyzed by the lingual lipase (Hamosh and Scow, 1973). Further studies in man (Hamosh *et al.*, 1975) demonstrated a lipolytic activity in esophageal aspirates which resembled closely that in the stomach and also that in rat tongue (Hamosh and Scow, 1973). The stereospecificity of the lingual lipase has been examined by Paltauf *et al.* (1974), who found that it preferentially releases 1,2-diacyl-*sn*-glycerols. The stereospecificities of the other pharyngeal lipases have yet to be determined. Hamosh and Scow (1973) proposed that the reaction catalyzed by the lingual lipase is the first step in the digestion of dietary lipid. However, its role in lipid digestion in relation to the established secretions of the pancreas remains to be evaluated.

6.9.2. Pancreatic Lipase

Pancreatic lipase is the best known and most investigated of all the lipolytic enzymes. It has been purified from pancreas and pancreatic juice (Sarda *et al.*, 1957, 1958; Wills, 1960; Marchis-Mouren *et al.*, 1959, 1960), the most highly purified preparations being chromatographically and electrophoretically pure (Marchis-Mouren *et al.*, 1960). Pancreatic lipase acts on tri-, di-, and monoacylglycerols, and it appears reasonably certain that the same enzyme is responsible for the hydrolysis of the three types of glycerides (Marchis-Mouren *et al.*, 1959;

Wills, 1960), although the reaction rate is lower with partial glycerides than with triacylglycerols (DiNella *et al.*, 1960; Wills, 1960; Brockerhoff, 1968).

Evidence that pancreatic lipase is not stereospecific was supplied by Tattrie *et al.* (1958) and by Karnovsky and Wolff (1960). Tattrie *et al.* synthesized *sn*-glycerol-1,2-dipalmitate-3-oleate and *sn*-glycerol-1-oleate-2,3-dipalmitate and subjected them to pancreatic lipase. The diacylglycerols formed were in both cases equal mixtures of glycerol palmitate oleate; this proved random hydrolysis at positions 1 and 3. Further studies on the lack of stereospecificity of pancreatic lipase have been reported by Morley *et al.* (1974). These workers examined the stereochemical course of hydrolysis of synthetic *sn*-glycerol-1-palmitate-2-oleate-3-linoleate, *sn*-glycerol-1,2-dipalmitate-3-oleate, and their antipodes by thin-layer and gas chromatography of the diacylglycerol intermediates. In agreement with the study of Tattrie *et al.* (1958), it was observed that the hydrolysis of the synthetic triacylglycerols by pancreatic lipase did not show any significant specificity for either of the primary positions. On the basis of the composition of the diacylglycerol intermediates, however, the enzyme appeared to attack the unsaturated fatty acids in marked preference to the saturated acids in both positions 1 and 3 of the triacylglycerol molecule. Anderson *et al.* (1967) had earlier claimed that the preferential release of unsaturated acids by pancreatic lipase depends on the interfacial conditions of the incubation medium and the overall structure of the substrate. In summary, therefore, the evidence to date strongly suggests that pancreatic lipase lacks stereospecificity.

6.9.3. Lipoprotein Lipase

It is now well established that lipoprotein lipase, often referred to as clearing-factor lipase, is an enzyme whose level rises in blood after an injection of heparin (Fredrickson and Gordon, 1958; Robinson, 1963). Lipolytic activity similar to that in the blood has been demonstrated in a great number of animal tissues (Brockerhoff and Jensen, 1974). Lipoprotein lipase hydrolyzes the triacylglycerols present in chylomicrons, in serum lipoproteins, and in lipoproteins from egg yolk, as well as simple triacylglycerols of plant and animal origin (Brockerhoff and Jensen, 1974).

The question of the positional specificity of lipoprotein lipase has repeatedly found attention. Early investigations (Korn, 1961; Greten *et al.*, 1970; Nilsson-Ehle *et al.*, 1971) on the positional specificity resulted in conflicting conclusions, which appear to have been partly due to a failure to differentiate between the two primary ester groups in the triacylglycerol molecule. The less extensive hydrolysis of the secondary in relation to the primary positions reported by Korn (1961) and Nilsson-Ehle *et al.* (1971) was based on comparisons of the total rate of hydrolysis of the two primary positions. The preferential hydrolysis of the secondary ester bond demonstrated by Greten *et al.* (1970) involved a relation of the rate of release of the acids of the second and third positions of the glyceride molecule only. On the basis of stereospecific analysis of the diacylglycerol intermediates, Morley and Kuksis (1972) claimed that the hydrolysis of trioleoylglycerol by lipoprotein lipase may proceed via a preferential attack on position 1 of triacyl-*sn*-glycerol. This was subsequently confirmed by chromatographic examination of the diacylglycerols released from triacyl-*sn*-glycerols of known structure (Morley *et al.*, 1974). In all instances, the diacylglycerol intermediates contained a preponderance of 2,3-diacyl-*sn*-glycerols. In agreement with these results, Paltauf *et al.* (1974) reported that the intermediate products of digestion of alkyl diacylglycerols by lipoprotein lipase from different sources were predominantly 3-alkyl-2-acyl-*sn*-glycerols. It was pointed out, however, by Morley *et al.* (1974) that the disproportionation of the diacylglycerol intermediates could also have arisen from a preferential further hydrolysis of the 1,2-diacyl-*sn*-glycerols, regardless of the proportion in which the 1,2 and 2,3 isomers may have been released from the original triacylglycerols. To determine whether or not this could have occurred, Morley *et al.* (1975) studied the course of hydrolysis of several sets of enantiomeric diacylglycerols. These investigators showed a preference by the enzyme for lipolysis at position 1 but with less specificity than previously was shown in *sn*-triacylglycerol hydrolysis (Morley and Kuksis, 1972; Morley *et al.*, 1974). These results precluded the possibility that the predominance of 2,3-diacyl-*sn*-glycerol intermediates during triacylglycerol hydrolysis is due solely to a preferential breakdown of the 1,2 isomers and reinforces the concept that lipoprotein lipase is specific for position 1.

Subsequently, Paltauf and Wagner (1976) have reported that lipoprotein lipase fails to discriminate between the *sn*-1- and *sn*-3 positions in alkylacylglycerols in which one of the alkyl groups occupies the 2 position. These findings may account for the observations of Assman *et al.* (1973) that lipases in postheparin plasma of hepatic

Table X. Diacylglycerol Composition of Hydrolysate during Digestion of Synthetic Triacylglycerols by Lipoprotein Lipase[a]

		Argentation chromatography		Gas–liquid chromatography	
Substrate	Emulsifier	1,2-[b] (%)	2,3-[b] (%)	1,2-[c] (%)	2,3-[c] (%)
sn-Glycerol-1-palmitate-2-oleate-3-linoleate	Glycocholate	21.3	78.7	34.9	65.1
		35.5	64.5	29.9	70.1
				31.4[d]	68.6[d]
	Lysolecithin	23.1	76.9	20.3	79.7
		27.2	72.8	23.8	76.2
sn-Glycerol-1-linoleate-2-oleate-3-palmitate	Glycocholate	20.7	79.3	23.9	76.1
		25.2	74.8	30.0	70.0
				35.5[d]	64.5[d]
	Lysolecithin	22.7	77.3	17.5	82.5
		22.0	78.0	24.8	75.2
sn-Glycerol-1-oleate-1,2-dipalmitate	Lysolecithin			24.1	75.9
sn-Glycerol-1,2-dipalmitate-3-oleate	Lysolecithin			34.6	65.4

[a] From Morley *et al.* (1974).
[b] Means of duplicate determinations of diacylglycerol fractions as monoenes and trienes.
[b] Means of duplicate determinations of diacylglycerol fractions as carbon number 32, 34, or 36.
[d] The source of the lipoprotein lipase in these two experiments was rat postheparin plasma. In all other cases, the source was bovine skim milk.

and extrahepatic origin attack *sn*-1 and *sn*-3 positions in dialkylacylglycerols at about the same rates.

The original findings of Morley and Kuksis (1972) and Paltauf *et al.* (1974) have been confirmed and extended by Akesson *et al.* (1976). The latter workers have shown that the stereospecificity for the heparin-releasable hepatic lipase is the same as that previously reported for lipoprotein lipase (Morley and Kuksis, 1972; Paltauf *et al.*, 1974; Morley *et al.*, 1974). Table X indicates the stereochemical composition of diacylglycerol intermediates during the hydrolysis of synthetic triacylglycerols by lipoprotein lipase (Morley *et al.*, 1974). The pseudoenantiomers of acylglycerols were resolved either by argentation thin-layer chromatography or by high-temperature gas–liquid chromatography. In all instances, a greater proportion of the *sn*-2,3- than of the *sn*-1,2-diacylglycerol enantiomers was found.

The above *in vitro* observations have been confirmed by analyses of the diacylglycerol intermediates in the plasma during triacylglycerol clearance *in vivo*. Under a variety of experimental conditions, a preferential accumulation of the *sn*-2,3 isomers was found (Morley *et al.*, 1977). Akesson *et al.* (1976) also found that the hepatic lipase released during liver perfusion had the specificity of lipoprotein lipase. The acid lysosomal lipase of the rat liver, however, showed no specificity and resembled pancreatic lipase in this respect (Akesson *et al.*, 1976). In contrast to pancreatic lipase, the lipoprotein lipase has been shown (Morley and Kuksis, 1977) to exhibit little or no specificity for the nature of the oligo- and polyunsaturated fatty acids in the acylglycerol molecule.

6.9.4. Hormone-Sensitive Lipase

It was first shown by Gordon and Cherkes (1958) that the lipolytic activity of rat epididymal adipose tissue could be increased by incubating the tissue with epinephrine. Subsequently, glucagon, ACTH, and TSH were added to the list of hormones stimulating lipolytic activity (Hollenberg *et al.*, 1961; Vaughan *et al.*, 1964) while the combination of epinephrine and ATP was found to give a greater stimulation than either of them alone (Rizack, 1961). These hormones are now known to exert their effects on adipose tissue by stimulating adenylate cyclase. An increased formation of cyclic AMP brought about by the lipolytic hormones has been demonstrated with rat epididymal fat pads (Butcher *et al.*, 1965) and with isolated fat cells (Jungas, 1966; Butcher and Sutherland, 1967). Furthermore, cyclic AMP has been shown to stimulate lipase activity in cell-free preparations (Rizack, 1964, 1965), although the exact mechanism of this stimulation is as yet unknown. The hormone-sensitive lipase occurs in the epididymal, perirenal, and mesenteric adipose of rats, guinea pigs, and man and in rat heart (Brockerhoff and Jensen, 1974). Its appearance therefore does not appear to be restricted to adipose tissue.

The substrate specificity and stereospecificity of this enzyme are not fully known. Trilinolein, trilaurin, tripalmitin, tristearin, and triolein, listed in order of their reactivity, are all suitable substrates. With tributyrin, the reaction rate is about 10 times that obtained with trilinolein (Rizack, 1961). The enzyme is reported to be specific for the primary positions of triacylglycerols (Björntorp and Furman, 1962) and to bring about the partial hydrolysis of glycerides (Rizack, 1961).

6.9.5. Monoacylglycerol Lipase

The outstanding property of monoacylglycerol lipase which justifies its classification as a lipase is its high reactivity with monoacylglycerols as compared with di- and triacylglycerols. The enzyme has been reported to occur in several tissues, including intestine, liver, and adipose tissue (Brockerhoff and Jensen, 1974). Fractionation of fat-free homogenates of rat adipose tissue showed that 71% of the total activity occurred in the cytosol fraction, while most of the remainder was recovered in the microsomal fraction (Gorin and Shafrir, 1964). In contrast, Senior and Isselbacher (1963) claimed that in rat small intestine the activity was equally distributed in four subcellular fractions.

Kupiecki (1966) reported that the purified preparation from rat adipose tissue hydrolyzed the monoacylglycerols of $C_{12:0}$, $C_{14:0}$, $C_{16:0}$, and $C_{18:0}$ acids at almost equal rates, but monobutyrin was hydrolyzed at an appreciably higher rate, suggesting the presence of a contaminating esterase. Senior and Isselbacher (1963), on the other hand, claimed a markedly different substrate with the enzyme from rat intestinal microsomes. Comparing the hydrolysis of a homologous series of monoacylglycerols ranging from monobutyrin to monostearin, they showed a sharp optimum at monocaprin. Monobutyrin and monostearin were barely hydrolyzed. This discrepancy in the hydrolysis of longer-chain monoacylglycerols may possibly indicate the presence of more than one type of monoacylglycerol lipase. The substrate specificity of monoacylglycerol lipase is still in doubt. Senior and Isselbacher (1963) claimed that the 2 isomer of monoacylglycerols was attacked somewhat faster than the 1 isomer by some intestinal preparations, whereas other preparations indicated no preference for either isomer (Tidwell *et al.*, 1963). The monoacylglycerol lipase from adipose tissue was found to be more active with 2-monoacylglycerols than with 1-monoacylglycerols.

6.10. Summary and Conclusions

This chapter contains a summary of our present knowledge of the synthesis and degradation of acylglycerols in mammalian tissues. Within the last decade or so, the development of powerful chromatographic techniques coupled with the availability of labeled substrates has resulted in the elucidation of many of the pathways of acylglycerol metabolism. Much remains to be learned, however. Research in the next few years should provide us with answers to some of the unresolved problems in acylglycerol metabolism.These include the relative contributions of each of the biosynthetic pathways, how they vary from organ to organ and how they may further be influenced by the nutritional state of the animal and by hormonal factors; the possible role of peroxisomes in mammalian acylglycerol metabolism; and the mechanism of fatty acid esterification and positional specificity on the glycerol molecule. Moreover, we need an understanding of the action of the enzymes involved in acylglycerol metabolism. This will necessitate determination of the correlation between the flow of substrate and enzyme activity, elucidation of the concentrations of substrate, product, and pH within the microenvironment of the membrane, and estimation of the rate of flow of product out of the membrane into the surrounding medium. Clearly there is plenty to do, with no shortage of important problems.

ACKNOWLEDGMENT

This chapter was completed at the University of Wisconsin—Madison during the tenure of a fellowship from the Muscular Dystrophy Association of Canada.

6.11. References

Aas, M., 1971, Organ and subcellular distribution of fatty acid activating enzymes in the rat, *Biochim. Biophys. Acta* **231**:32.

About-Issa, H. M., and Cleland, W. W., 1969, Studies on the microsomal acylation of L-glycerol-3-phosphate. II. The specificity and properties of the rat liver enzyme, *Biochim. Biophys. Acta* **187**:692.

Adams, L. L., Webb, W. W., and Fallon, H. J., 1971, Inhibition of hepatic triglyceride formation by clofibrate, *J. Clin. Invest.* **50**:2339.

Agranoff, B. W., 1962, Hydrolysis of long chain alkyl phosphates and phosphatidic acid by an enzyme purified from pig brain, *J. Lipid Res.* **3**:190.

Agranoff, B. W., and Hajra, A. K., 1971, The acyl dihydroxyacetone phosphate pathway for glycerolipid biosynthesis in mouse liver and Ehrlich ascites tumor cells, *Proc. Natl. Acad. Sci. (USA)* **68**:411.

Ailhaud, G. P., and Vagelos, P. R., 1966, Palmityl-acyl carrier protein as acyl donor for complex lipid biosynthesis in *Escherichia coli*, *J. Biol. Chem.* **241**:3866.

Ailhaud, G., Samuel, D., and Desnuelle, P., 1963, Localisation subcellulaire de l'acyl-CoA synthetase de la muqueuse intestinale, *Biochim. Biophys. Acta* **67**:150.

Ailhaud, G., Samuel, D., Lazdunski, M., and Desnuelle, P., 1964, Quelques observations sur le mode d'action de la monoglyceride transacylase de la diglyceride transacylase de la muqueuse intestinale, *Biochim. Biophys. Acta* **84**: 643.

Akesson, B., 1969, The acylation of diacylglycerols in pig liver, *Eur. J. Biochem.* **9**:406.

Akesson, B., 1970, Initial esterification and conversion of intraportally injected [1-^{14}C]linoleic acid in rat liver, *Biochim. Biophys. Acta* **218**:57.

Akesson, B., Elovson, J., and Arvidson, G., 1970*a*, Initial incorporation into rat liver glycerolipids of intraportally injected [9,10-^{3}H]palmitic acid, *Biochim. Biophys. Acta* **218**:44.

Akesson, B., Elovson, J., and Arvidson, G., 1970*b*, Initial incorporation into rat liver glycerolipids of intraportally injected [^{3}H]glycerol, *Biochim. Biophys. Acta* **210**:15.

Akesson, B., Gronowitz, S., and Herslof, B., 1976, Stereospecificity of hepatic lipases, *FEBS Lett.* **71**:241.

Anderson, R. E., Bottino, N. R., and Reiser, R., 1967, Pancreatic lipase hydrolysis as a source of diglycerides for the stereospecific analysis of triglycerides, *Lipids* **2**:440.

Assmann, G., Kraus, R. M., Fredrickson, D. S., and Levy, R. I., 1973, Positional specificity of triglyceride lipases in postheparin plasma, *J. Biol. Chem.* **248**:7184.

Bailey, J. M., Howard, B. V., and Tillman, S. F., 1973, Lipid metabolism in cultured cells. XI. Utilization of serum triglycerides, *J. Biol. Chem.* **248**:1240.

Barden, R. E., and Cleland, W. W., 1969, 1-Acylglycerol 3-phosphate acyltransferase from rat liver, *J. Biol. Chem.* **244**:3677.

Bar-on, H., Stein, O., and Stein, Y., 1972, Multiple effects of cycloheximide on the metabolism of triglycerides in the liver of male and female rats, *Biochim. Biophys. Acta* **270**:444.

Barrnett, R. J., and Rostgaard, J., 1965, Absorption of particulate lipid by intestinal microvilli, *Ann. N.Y. Acad. Sci.* **131**:13.

Belfrage, P., 1964, Acylation of glycerol ethers and hydroxylated hydrocarbons in epididymal adipose tissue, *Biochem. J.* **92**:41p.

Bickerstaffe, A., and Annison, E. F., 1969, Triglyceride synthesis by the small intestinal epithelium of the pig, sheep and chicken, *Biochem. J.* **111**:419.

Bjorntorp, P., and Furman, R. H., 1962, Lipolytic activity in rat heart, *Am. J. Physiol.* **203**:323.

Bortz, W. M., and Lynen, F., 1963, Elevation of long chain acyl CoA derivatives in livers of fasted rats, *Biochem. Z.* **339**:77.

Bowley, M., Manning, R., and Brindley, D. N., 1973, The tritium isotope effect of *sn*-glycerol 3-phosphate oxidase and the effects of clofenapate and *N*-(2-benzoyloxyethyl) norfenfluramine on the esterification of glycerol phosphate and dihydroxyacetone phosphate by rat liver mitochondria, *Biochem. J.* **136:**421.

Breckenridge, W. C., and Kuksis, A., 1975, Diacylglycerol biosynthesis in everted sacs of rat intestinal mucosa, *Can. J. Biochem.* **53:**1170.

Breckenridge, W. C., Yeung, S. K. F., and Kuksis, A., 1976*a*, Biosynthesis of triacylglycerols by rat intestinal mucosa *in vivo, Can. J. Biochem.* **54:**145.

Breckenridge, W. C., Yeung, S. K. F., Kuksis, A., Myher, J. J., and Chan, M., 1976*b*, Biosynthesis of diacylglycerols by rat intestinal mucosa *in vivo, Can. J. Biochem.* **54:**137.

Brindley, D. N., 1973, The relationship between palmitoyl-coenzyme A synthetase activity and esterification of *sn*-glycerol 3-phosphate by the microsomal fraction of guinea pig intestinal mucosa, *Biochem. J.* **132:**707.

Brindley, D. N., and Bowley, M., 1975, Drugs affecting the synthesis of glycerides and phospholipids in rat liver, *Biochem. J.* **148:**461.

Brindley, D. N., and Hübscher, G., 1965, The intracellular distribution of the enzymes catalyzing the biosynthesis of glycerides in the intestinal mucosa, *Biochim. Biophys. Acta* **106:**495.

Brindley, D. N., Bowley, M., Brooks, R. J., and Malik, S. P., 1973, The effect of *p*-chlorophenoxy-isobutyrate, clofenapate, fenfluramine, the methane sulphanate of *m*-trifluoromethylphenyl-1[β(benzoyoxy)ethyl]amino-2-propane and amphetamine on the esterification of *sn*-glycerol 3-phosphate in human and rat liver, *Abstr. 9th Int. Congr. Biochem. Stockholm,* p. 416

Brockerhoff, H., 1968, Substrate specificity of pancreatic lipase, *Biochim. Biophys. Acta* **159:**296.

Brockerhoff, H., and Jensen, R. G., 1974, in: *Lipolytic Enzymes*, pp. 1–330, Academic Press, New York.

Brown, J. L., and Johnston, J. M., 1963, Distribution of fatty acids in triglycerides synthesized from monoglycerides, *Biochim. Biophys. Acta* **70:**603.

Brown, J. L., and Johnston, J. M., 1964*a*, The mechanism of intestinal utilization of monoglycerides, *Biochim. Biophys. Acta* **84:**264.

Brown, J. L., and Johnston, J. M., 1964*b*, The utilization of 1- and 2-monoglycerides for intestinal triglyceride biosynthesis, *Biochim. Biophys. Acta* **84:**448.

Bublitz, C., and Kennedy, E. P., 1954, Synthesis of phosphatides in isolated mitochondria. III. The enzymatic phosphorylation of glycerol, *J. Biol. Chem.* **211:**951.

Butcher, R. W., and Sutherland, E. W., 1967, The effects of the catecholamines, adrenergic blocking agents, prostaglandin E1 and insulin on cyclic AMP levels in the rat epididymal fat pad *in vitro, Ann. N.Y. Acad. Sci.* **139:**849.

Butcher, R. W., Ho, R. J., Meng, H., and Sutherland, E. W., 1965, Adenosine 3′,5′-monophosphate in tissues and the role of the cyclic nucleotide in the lipolytic response of fat to epinephrine, *J. Biol. Chem.* **240:**4515.

Cheniae, G. M., 1965, Phosphatidic acid and glyceride synthesis by particles from spinach leaves, *Plant Physiol.* **40:**235.

Clark, B., and Hübscher, G., 1960, Biosynthesis of glycerides in the mucosa of the small intestine, *Nature* (*London*) **185:**35.

Clark, B., and Hübscher, G., 1961, Biosynthesis of glycerides in subcellular fractions of intestinal mucosa, *Biochim. Biophys. Acta* **46:**479.

Coleman, R., and Hübscher, G., 1962, Metabolism of phospholipids. V. Studies of phosphatidic acid phosphatase, *Biochim. Biophys. Acta* **56:**479.

Coleman, R., and Hübscher, G., 1963, Metabolism of phospholipids. VII. On the lipid requirement for phosphatidic acid phosphatase activity, *Biochim. Biophys. Acta* **73:**257.

Coniglio, J. G., and Cate, D. L., 1959, Intestinal absorption of C^{14}-palmitic acid and C^{14}-tripalmitin in the rat, *Am. J. Clin. Nutr.* **7:**646.

Craig, G. M., 1972, A comparison of clofibrate and its derivative methyl clofibrate, *Atherosclerosis* **15:**265.

Daae, L. N. W., 1972, The acylation of glycerophosphate in rat liver mitochondria and microsomes as a function of fatty acid chain length, *FEBS Lett.* **27:**46.

Daae, L. N. W., 1973, The acylation of glycerol 3-phosphate in different rat organs and in the liver of different species including man, *Biochim. Biophys. Acta* **306:**186.

Daae, L. N. W., and Bremer, J., 1970, The acylation of glycerophosphate in rat liver: A new assay procedure for glycerophosphate acylation. Studies on its subcellular and submitochondrial localization and determination of the reaction products, *Biochim. Biophys. Acta* **210:**92.

Dannenburg, W. N., Kardian, B. C., and Morrell, L. Y., 1973, Fenfluramine and triglyceride synthesis by microsomes of the intestinal mucosa in the rat, *Arch. Int. Pharmacodyn. Ther.* **201:**115.

DeKruyff, B., Van Golde, L. M. G., and Van Deenen, L. L. M., 1970, Utilization of diacylglycerol species by cholinephosphotransferase, ethanolaminephosphotransferase and diacylglycerol acyltransferase in rat liver microsomes, *Biochim. Biophys. Acta* **210:**425.

Dils, R., and Clark, B., 1962, Fatty acid esterification in lactating rat mammary gland, *Biochem. J.* **84:**19p.

Dimick, P., McCarthy, R. D., and Patton, S., 1965, Paths of palmitic acid incorporation into milk fat triglycerides, *Biochim. Biophys. Acta* **116:**159.

DiNella, R. R., Meng, H. C., and Park, C. R., 1960, Properties of intestinal lipase, *J. Biol. Chem.* **235:**3076.

Fallon, H. J., Barwick, J., Lamb, R. G., and Van den Bosch, H., 1975, Studies of rat liver microsomal diglyceride acyltransferase and cholinephosphotransferase using microsomal-bound substrate: Effects of high fructose intake, *J. Lipid Res.* **16:**107.

Flatt, J. P., and Ball, E. G., 1964, Studies on the metabolism of adipose tissue. XV. An evaluation of the major pathways of glucose catabolism as influenced by insulin and epinephrine, *J. Biol. Chem.* **239:**675.

Forstner, G. G., Riley, E. M., Daniels, S. J., and Isselbacher, K. J., 1965, Demonstration of glyceride synthesis by brush borders of intestinal epithelial cells, *Biochem. Biophys. Res. Commun.* **21:**83.

Fredrickson, D. S., and Gordon, R. S., 1958, Transport of fatty acids, *Physiol. Rev.* **38:**585.

Friedmann, H. I., and Cardell, R. R., Jr., 1972, Effects of puromycin on the structure of rat intestinal epithelial cells during fat absorption, *J. Cell Biol.* **52:**15.

Gallo, L., and Treadwell, C. R., 1970, Localization of the monoglyceride pathway in subcellular fractions of rat intestinal mucosa, *Arch. Biochem. Biophys.* **141:**614.

Gallo, L., Vahouny, G. V., and Treadwell, C. R., 1968, The 1- and 2-octadecyl glyceryl ethers as model compounds for study of triglyceride resynthesis in cell fractions of intestinal mucosa, *Proc. Soc. Exp. Biol. Med.* **127:**156.

Garfinkel, D., and Hess, B., 1964, Metabolic control mechanisms. VII. A detailed computer model of the glycolytic pathway in ascites cells, *J. Biol. Chem.* **239:**971.

Goldfine, H., 1966, Acylation of glycerol 3-phosphate in bacterial extracts, *J. Biol. Chem.* **241:**3864.

Goldman, P., and Vagelos, P. R., 1961, The specificity of triglyceride synthesis from diglyceride in chicken adipose tissue, *J. Biol. Chem.* **236:**2620.

Gordon, R. S., and Cherkes, A., 1958, Production of unesterified fatty acids from isolated rat adipose tissue incubated *in vitro, Proc. Soc. Exp. Biol. Med.* **97:**150.

Gorin, E., and Shafrir, E. 1964, Lipolytic activity in adipose tissue homogenate toward tri-, di- and monoglyceride substrates, *Biochim. Biophys. Acta* **84:**24.

Greenbaum, A. L., Gumaa, K. A., and McLean, P., 1971, The distribution of hepatic metabolites and the control of the pathways of carbohydrate metabolism in animals of different dietary and hormonal status, *Arch. Biochem. Biophys.* **143:**617.

Greten, H., Levy, R. I., Fales, H., and Fredrickson, D. S., 1970, Hydrolysis of diglyceride and glyceryl monoester diethers with lipoprotein lipase, *Biochim. Biophys. Acta* **210:**39.

Gurr, M. I., Pover, W. F. R., Hawthorne, J. N., and Frazer, A. C., 1963, The phospholipid composition and turnover in rat intestinal mucosa during fat absorption, in: *Biochemical Problems of Lipids*, Vol. 1 (A. C. Frazer, ed.), pp. 236–243, Elsevier, Amsterdam.

Hajra, A. K., 1968, Biosynthesis of acyl dihydroxyacetone phosphate in guinea pig liver mitochondria, *J. Biol. Chem.* **243:**3458.

Hajra, A. K., and Agranoff, B. W., 1968*a*, Acyl dihydroxyacetone phosphate: Characterization of a ^{32}P-labeled lipid from guinea pig liver mitochondria, *J. Biol. Chem.* **243:**1617.

Hajra, A. K., and Agranoff, B. W., 1968*b*, Reduction of palmitoyl dihydroxyacetone phosphate by mitochondria, *J. Biol. Chem.* **243:**3542.

Hajra, A. K., Seguin, E. B., and Agranoff, B. W., 1968, Rapid labeling of mitochondrial lipids by labeled orthophosphate and adenosine triphosphate, *J. Biol. Chem.* **243:**1609.

Hamosh, M., and Scow, R. O., 1973, Lingual lipase and its role in the digestion of dietary fat, *J. Clin. Invest.* **52**:88.

Hamosh, M., Klaeveman, H. L., Wolf, R. O., and Scow, R. O., 1975, Pharyngeal lipase and digestion of dietary triglyceride in man, *J. Clin. Invest.* **55**:908.

Havel, R. J., and Kane, J. P., 1973, Drugs and lipid metabolism, *Ann. Rev. Pharm.* **13**:287.

Hayashi, S., and Lin, E. C., 1967, Purification and properties of glycerol kinase from *Escherichia coli*, *J. Biol. Chem.* **242**:1030.

Higgins, J. A., and Barrnett, R. J., 1971, Fine structural localization of acyltransferases: The monoglyceride and α-glycerophosphate pathways in intestinal absorptive cells, *J. Cell Biol.* **50**:102.

Hill, E. E., and Lands, W. E. M., 1968, Incorporation of long-chain and polyunsaturated acids into phosphatidate and phosphatidylcholine, *Biochim. Biophys. Acta* **152**:645.

Hill, E. E., Husbands, D. R., and Lands, W. E. M., 1968, The selective incorporation of ^{14}C-glycerol into different species of phosphatidic acid, phosphatidylethanolamine and phosphatidylcholine, *J. Biol. Chem.* **243**:4440.

Hohorst, H. L., Kreutz, F. H., and Bücher, T., 1959, Über Metabolitgehalte and Metabolitkonzentrationen in der Leber der Ratte, *Biochem. Z.* **332**:18.

Hokin, L. E., and Hokin, M. R., 1959, The synthesis of phosphatidic acid from diglyceride and adenosine triphosphate in extracts of brain microsomes, *J. Biol. Chem.* **234**:1381.

Hokin, L. E., and Hokin, M. R., 1961*a*, Diglyceride kinase and phosphatidic acid phosphatase in erythrocyte membranes, *Nature* (*London*) **189**:836.

Hokin, L. E., and Hokin, M. R., 1961*b*, Studies on the carrier function of phosphatidic acid in sodium transport. I. The turnover of phosphatidic acid and phosphoinositide in the avian salt gland on stimulation of secretion, *J. Gen. Physiol.* **44**:61.

Hokin, L. E., and Hokin, M. R., 1963, Diglyceride kinase and other pathways for phosphatidic acid synthesis in the erythrocyte membrane, *Biochim. Biophys. Acta* **67**:470.

Hokin, M. R., and Hokin, L. E., 1964, The synthesis of phosphatidic acid and protein bound phosphorylserine in salt glands homogenates, *J. Biol. Chem.* **239**: 2116.

Hollenberg, C. H., Raben, M. S., and Astwood, E. B., 1961, The lipolytic response to corticotropin, *Endocrinology* **68**:589.

Howard, B. V., and Howard, W. J., 1974, Lipid metabolism in cultured cells, *Adv. Lipid Res.* **12**:51.

Hübscher, G., 1961, Esterification of monoglycerides by soluble preparations from mammalian tissues, *Biochim. Biophys. Acta* **52**:582.

Hübscher, G., 1970, Glyceride metabolism, in: *Lipid Metabolism*, (S. J. Wakil, ed.), pp. 279–370, Academic Press, New York.

Husbands, D. R., and Lands, W. E. M., 1970, Phosphatidate synthesis by *sn*-glycerol-3-phosphate acyltransferase in pigeon liver particles, *Biochim. Biophys. Acta* **202**:129.

Jezyk, P., and Lands, W. E. M., 1968, Specificity of acyl-CoA: Phospholipid acyltransferases: Solvent and temperature effects, *J. Lipid Res.* **9**:525.

Johnston, J. M., 1959, The absorption of fatty acids by the isolated intestine, *J. Biol. Chem.* **234**:1065.

Johnston, J. M., and Bearden, J. H., 1962, Intestinal phosphatidate phosphatase, *Biochim. Biophys. Acta* **56**:365.

Johnston, J. M., and Borgström, B., 1964, The intestinal absorption and metabolism of micellar solutions of lipids, *Biochim. Biophys. Acta* **84**:412.

Johnston, J. M., and Brown, J. L., 1962, The intestinal utilization of doubly labeled α-monopalmitin, *Biochim. Biophys. Acta* **59**:500.

Johnston, J. M., and Paltauf, F., 1970, Lipid metabolism in inositol-deficient yeast, *Saccharomycetes carlsbergensis*. II. Incorporation of labeled precursors into lipids by whole cells and activities of some enzymes involved in lipid formation, *Biochim. Biophys. Acta* **218**:431.

Johnston, J. M., Rao, G. A., and Reistad, R., 1965, Species difference in the synthesis of triglycerides from monoglycerides, *Biochim. Biophys. Acta* **98**:432.

Johnston, J. M., Rao, G. A., Lowe, P. A., and Schwartz, B. E., 1967, The nature of the stimulatory role of the supernatant fraction on triglyceride synthesis by the α-glycerophosphate pathway, *Lipids* **2**:14.

Johnston, J. M., Paltauf, F., Schiller, C. M., and Schultz, L. D., 1970, The utilization of the α-glycerophosphate and monoglyceride pathways for phosphatidylcholine biosynthesis in the intestine, *Biochim. Biophys. Acta* **218**:124.

Jungas, R. L., 1966, Role of cyclic-3′,5′-AMP in the response of adipose tissue to insulin, *Proc. Natl. Acad. Sci. (USA)* **56:**757.
Karnovsky, M. L., and Wolff, D., 1960, Studies on the stereospecificity of lipases, in: *Biochemistry of Lipids* (G. Popjak, ed.), pp. 53–59, Pergamon Press, New York.
Kates, M., 1955, Hydrolysis of lecithin by plant plastid enzymes, *Can. J. Biochem. Physiol.* **33:**575.
Katz, J., Landau, B. R., and Bartsch, G. E., 1966, The pentose cycle, triose phosphate isomerization and lipogenesis in rat adipose tissue, *J. Biol. Chem.* **241:**727.
Kern, F., and Borgström, B., 1965, Quantitative study of the pathways of triglyceride synthesis by hamster intestinal mucosa, *Biochim. Biophys. Acta* **98:**520.
Korn, E. D., 1961, The fatty acid and positional specificities of lipoprotein lipase, *J. Biol. Chem.* **236:**1638.
Kornberg, A., and Pricer, W. E., 1953, Enzymatic esterification of L-glycerophosphate by long chain fatty acids, *J. Biol. Chem.* **204:**345.
Krebs, H. A., and Veech, R. L., 1970, Regulation of the redox state of the pyridine nucleotides in rat liver, in: *Pyridine Nucleotide-Dependent Dehydrogenases* (H. Sund, ed.), pp. 413–434, Springer-Verlag, New York.
Kuhn, N. J., and Lynen, F., 1965, Phosphatidic acid synthesis in yeast, *Biochem. J.* **94:**240.
Kuksis, A., Myher, J. J., Marai, L., Yeung, S. K. F., Steiman, I., and Mookerjea, S., 1975*a*, Distribution of newly formed fatty acids among glycerolipids of isolated perfused rat liver, *Can. J. Biochem.* **53:**509.
Kuksis, A., Myher, J. J., Marai, L., Yeung, S. K. F., Steiman, I., and Mookerjea, S., 1975*b*, Distribution of newly formed palmitate and stearate among molecular species of choline and ethanolamine phosphatides, *Can. J. Biochem.* **53:**519.
Kupiecki, F. P., 1966, Partial purification of monoglyceride lipase from adipose tissue, *J. Lipid Res.* **7:**230.
La Belle, E. F., Jr., and Hajra, A. K., 1972, Enzymatic reduction of alkyl and acyl derivatives of dihydroxyacetone phosphate by reduced pyridine nucleotides, *J. Biol. Chem.* **247:**5825.
Lamb, R. G., and Fallon, H. J., 1970, The formation of monoacylglycerophosphate from *sn*-glycerol 3-phosphate by a rat liver particulate preparation, *J. Biol. Chem.* **245:**3075.
Lamb, R. G., and Fallon, H. J., 1972, Inhibition of monoacylglycerophosphate formation by chlorophenoxyisobutyrate and *β*-benzalbutyrate, *J. Biol. Chem.* **247:**1281.
Lands, W. E. M., 1960, Metabolism of Glycerolipids. II. The enzymatic acylation of lysolecithin, *J. Biol. Chem.* **235:**2233.
Lands, W. E. M., 1976, Selectivity of microsomal acyltransferases, in: *The Essential Fatty Acids: Miles Symposium 1975* (W. W. Hawkins, ed.), Nutrition Society of Canada and Miles Laboratories, Toronto.
Lands, W. E. M., and Hart, P., 1964, Metabolism of glycerolipids. V. Metabolism of phosphatidic acid, *J. Lipid Res.* **5:**81.
Lands, W. E. M., and Hart, P., 1965, Metabolism of glycerolipids. VI. Specificities of acyl coenzyme A: phospholipid acyltransferases, *J. Biol. Chem.* **240:**1905.
Lands, W. E. M., Blank, M. L., Nutter, L. J., and Privett, O. S., 1966*a*, A comparison of acyltransferase activities *in vitro* with the distribution of fatty acids in lecithins and triglycerides *in vivo, Lipids* **1:**224.
Lands, W. E. M., Pieringer, R. A., Slakey, S. P. M., and Zschocke, A., 1966*b*, A micromethod for the stereospecific determination of triglyceride structure, *Lipids* **1:**444.
Lapetina, E. G., and Hawthorne, J. N., 1971, The diglyceride kinase of rat cerebral cortex, *Biochem. J.* **122:**171.
Lea, M. A., and Walker, D. G., 1965, Factors affecting hepatic glycolysis and some changes that occur during development, *Biochem. J.* **94:**655.
Leal, R. S., and Greenbaum, A. L., 1961, The effect of pituitary growth hormone on phospholipid synthesis, *Biochem. J.* **80:**27.
Leat, W. M. F., and Cunningham, H. M., 1968, Pathways of lipid synthesis in the sheep intestine, *Biochem. J.* **109:**38p.
Lipkin, M., Sherlock, P., and Bell, B., 1963, Cell proliferation kinetics in the gastrointestinal tract of man. II. Cell renewal in stomach, ileum, colon and rectum, *Gastroenterology* **45:**721.
Lowry, O. H., Passonneau, J. V., Hasselberger, F. X., and Schultz, D. W., 1964, Effect of ischemia on known substrates and cofactors of the glycolytic pathway in brain, *J. Biol. Chem.* **239:**18.

Lynch, R. D., and Geyer, R. P., 1972, Uptake of *rac*-glycerol 1-oleate and its utilization for glycerolipid synthesis by strain L fibroblasts, *Biochim. Biophys. Acta* **260:**547.

Mangiapane, E. H., Lloyd-Davis, K. A., and Brindley, D. N., 1973, A study of some enzymes of glycerolipid biosynthesis in rat liver after subtotal hepatectomy, *Biochem. J.* **134:**103.

Manning, R., and Brindley, D. N., 1972, Tritium isotope effects in the measurement of the glycerol phosphate and dihydroxyacetone phosphate pathways of glycerolipid biosynthesis in rat liver, *Biochem. J.* **130:**1003.

Mansbach, C. M., II, 1973, Complex lipid synthesis in hamster intestine, *Biochim. Biophys. Acta* **296:**386.

Mansbach, C. M., II, 1975, Effect of acute dietary alteration upon intestinal lipid synthesis, *Lipids* **10:**318.

Maragoudakis, M. E., Hankin, H., and Wasvary, J. M., 1972, On the mode of action of lipid-lowering agents, *J. Biol. Chem.* **247:**342.

Marchis-Mouren, G., Sarda, L., and Desnuelle, P., 1959, Purification of hog pancreatic lipase, *Arch. Biochem. Biophys.* **83:**309.

Marchis-Mouren, G., Sarda, L., and Desnuelle, P., 1960, Purification de la lipase a portir du sac pancreatique de porc, *Biochim. Biophys. Acta* **41:**358.

Marsh, J. B., and Bizzi, A., 1972, Effects of amphetamine and fenfluramine on the net release of triglycerides of very low density lipoproteins by slices of rat liver, *Biochem. Pharmacol.* **21:**1143.

Martensson, E., and Kanfer, J., 1968, The conversion of α-glycerol-^{14}C-3-phosphate into phosphatidic acid by solubilized preparation from rat brain, *J. Biol. Chem.* **243:**497.

Mattson, F. H., and Volpenhein, R. A., 1962, Rearrangement of glyceride fatty acids during digestion and absorption, *J. Biol. Chem.* **237:**53.

Mattson, F. H., and Volpenhein, R. A., 1964, The digestion and absorption of triglycerides, *J. Biol. Chem.* **239:**2772.

McBride, O. W., and Korn, E. D., 1964, Acceptors of fatty acid for glyceride synthesis in guinea pig mammary gland, *J. Lipid Res.* **5:**448.

McCaman, R. E., Smith, M., and Cook, K., 1965, Intermediary metabolism of phospholipids in brain tissue. II. Phosphatidic acid phosphatase, *J. Biol. Chem.* **240:**3513.

McMurray, W. C., Strickland, K. P., Berry, J. F., and Rossiter, R. J., 1957, Incorporation of ^{32}P-labeled intermediates into the phospholipids of cell-free preparations of rat brain, *Biochem. J.* **66:**634.

Merkl, I., and Lands, W. E. M., 1963, Metabolism of glycerolipids. IV. Synthesis of phosphatidylethanolamine, *J. Biol. Chem.* **238:**905.

Mishkin, S., and Turcotte, R., 1974, Stimulation of monoacylglycerophosphate formation by Z protein, *Biochem. Biophys. Res. Commun.* **60:**376.

Mishkin, S., Stein, L., Gatmaitan, Z., and Arias, I. M., 1972, The binding of fatty acids to cytoplasmic proteins: Binding to Z protein in liver and other tissues of the rat, *Biochem. Biophys. Res. Commun.* **47:**997.

Mitchell, M. P., Brindley, D. N., and Hübscher, G., 1971, Properties of phosphatidate phosphohydrolase, *Eur. J. Biochem.* **18:**214.

Monroy, G., Rola, F. H., and Pullman, M. E., 1972, A substrate- and position-specific acylation of *sn*-glycerol 3-phosphate by rat liver mitochondria, *J. Biol. Chem.* **247:**6884.

Monroy, G., Chroboczek-Kelker, H., and Pullman, M. E., 1973, Partial purification and properties of an acyl coenzyme A: *sn*-glycerol 3-phosphate acyltransferase from rat liver mitochondria, *J. Biol. Chem.* **248:**2845.

Morley, N., and Kuksis, A., 1972, Positional specificity of lipoprotein lipase, *J. Biol. Chem.* **247:**6389.

Morley, N., and Kuksis, A., 1977, Lack of fatty acid specificity in the lipolysis of oligo- and polyunsaturated triacylglycerols by milk lipoprotein lipase, *Biochim. Biophys. Acta* **487:**332.

Morley, N. H., Kuksis, A., and Buchnea, D., 1974, Hydrolysis of synthetic triacylglycerols by pancreatic and lipoprotein lipase, *Lipids* **9:**481.

Morley, N. H., Kuksis, A., Buchnea, D., and Myher, J. J., 1975, Hydrolysis of diacylglycerols by lipoprotein lipase, *J. Biol. Chem.* **250:**3414.

Morley, N., Kuksis, A., Hoffman, G. A. D., and Kakis, G., 1977, Preferential *in vivo* accumulation of *sn*-2,3-diacylglycerols in postheparin plasma of rats, *Can. J. Biochem.* **55:**1075.

Negrel, R., and Ailhaud, G., 1975, Localization of the monoglyceride pathway enzymes in the villus tips of intestinal cells and their absence from the brush-border, *FEBS Lett.* **54:**183.

Neptune, E. M., Sudduth, H. C., Brigance, W. H., and Brown, J. D., 1963, Lipid glyceride synthesis by rat skeletal muscle, *Am. J. Physiol.* **204:**933.

Nilsson-Ehle, P., Belfrage, P., and Borgström, B., 1971, Purified human lipoprotein lipase: Positional specificity, *Biochim. Biophys. Acta* **248:**114.

Numa, S., and Yamashita, S., 1974, Regulation of lipogenesis in animal tissues, in: *Current Topics in Cellular Regulation*, Vol. 8 (B. L. Horecker and E. R. Stadtman, eds.), p. 197, Academic Press, New York.

Ockner, R. K., Manning, J. A., Poppenhausen, R. B., and Ho, W. K. L., 1972, A binding protein for fatty acids in cytosol of intestinal mucosa, liver, myocardium and other tissues, *Science* **177:**56.

O'Doherty, P. J. A., 1974, Studies on the control of triacylglycerol synthesis and release by rat intestine, thesis, University of Toronto, pp. 1–261.

O'Doherty, P. J. A., and Kuksis, A., 1974*a*, Differential effect of puromycin on triacylglycerol and phosphatidylcholine synthesis in rat mucosal microsomes, *Can. J. Biochem.* **52:**170.

O'Doherty, P. J. A., and Kuksis, A., 1974*b*, Microsomal synthesis of di- and tri-acylglycerols in rat liver and Ehrlich ascites cells, *Can. J. Biochem.* **52:**514.

O'Doherty, P. J. A., and Kuksis, A., 1975*a*, Effect of puromycin *in vitro* on protein and glycerolipid biosynthesis in isolated epithelial cells of rat intestine, *Int. J. Biochem.* **6:**435.

O'Doherty, P. J. A., and Kuksis, A., 1975*b*, Glycerolipid biosynthesis in isolated rat intestinal epithelial cells, *Can. J. Biochem.* **53:**1010.

O'Doherty, P. J. A., and Kuksis, A., 1975*c*, Stimulation of triacylglycerol synthesis by Z protein in rat liver and intestinal mucosa, *FEBS Lett.* **60:**256.

O'Doherty, P. J. A., Kuksis, A., and Buchnea, D., 1972*a*, Enantiomeric diglycerides as stereospecific probes in triglyceride synthesis *in vitro*, *Can. J. Biochem.* **50:**881.

O'Doherty, P. J. A., Yousef, I. M., and Kuksis, A., 1972*b*, Glyceride metabolism in isolated mucosal cells, *J. Am. Oil Chem. Soc.* **49:**306A.

O'Doherty, P. J. A., Yousef, I. M., and Kuksis, A., 1972*c*, Differential effect of puromycin on triglyceride and phospholipid biosynthesis in isolated mucosal cells, *Fed. Proc. Fed. Am. Soc. Exp. Biol.* **31:**2739.

O'Doherty, P. J. A., Yousef, I. M., and Kuksis, A., 1973, Effect of puromycin on protein and glycerolipid biosynthesis in isolated mucosal cells, *Arch. Biochem. Biophys.* **156:**586.

O'Doherty, P. J. A., Kuksis, A., and Buchnea, D., 1974*a*, Utilization of acyl-*sn*-glycerol cyclic phosphodiesters in glycerolipid synthesis, *J. Am. Oil Chem. Soc.* **51:**504A.

O'Doherty, P. J. A., Yousef, I. M., and Kuksis, A., 1974*b*, Effect of phosphatidylcholine on triacylglycerol synthesis in rat intestinal mucosa, *Can. J. Biochem.* **52:**726.

O'Doherty, P. J. A., Yousef, I. M., Kakis, G., and Kuksis, A., 1975, Protein and glycerolipid biosynthesis in isolated intestinal epithelial cells of normal and bile fistula rats, *Arch. Biochem. Biophys.* **169:**252.

Okuyama, H., and Lands, W. E. M., 1970, A test for the dihydroxyacetone phosphate pathway, *Biochim. Biophys. Acta* **218:**376.

Okuyama, H., and Lands, W. E. M., 1972, Variable selectivities of acyl coenzyme A: monoacylglycerophate acyltransferase in rat liver, *J. Biol. Chem.* **247:**1414.

Okuyama, H., Lands, W. E. M., Christie, W. W., and Gunstone, F. D., 1969, Selective transfer of cyclopropane acids by acyl coenzyme A: phospholipid acyltransferase, *J. Biol. Chem.* **244:**6514.

Paltauf, F., and Johnston, J. M., 1971, The metabolism *in vitro* of enantiomeric 1-*O*-alkyl glycerols and 1,2- and 1,3-alkyl glycerols in the intestinal mucosa, *Biochim. Biophys. Acta* **239:**47.

Paltauf, F., and Wagner, E., 1976, Stereospecificity of lipases, enzymatic hydrolysis of enantiomeric alkyldiacyl- and dialkylacylglycerols by lipoprotein lipase, *Biochim. Biophys. Acta* **431:**359.

Paltauf, F., Esfandi, F., and Holasek, A., 1974, Stereospecificity of lipases: Enzymic hydrolysis of enantiomeric alkyl diacylglycerols by lipoprotein lipase, lingual lipase and pancreatic lipase, *FEBS Lett.* **40:**119.

Paris, R., and Clement, G., 1965, Differences de comportement des acides oleique et palmitique au cours de la synthese de triglycerides a partir de 1-monopalmitine par la muqueuse intestinale de rat, *Biochim. Biophys. Acta* **106:**634.

Paris, R., and Clement, G., 1968, Biosynthese de triglycerides a partir de 2-monopalmitine doublement marquee dans la muqueuse intestinale de rat, *Biochim. Biophys. Acta* **152:**63.

Paris, R., and Clement, G., 1969, Biosynthesis of lysophosphatidic acid from ATP and 1-mono-olein by subcellular particles of intestinal mucosa, *Proc. Soc. Exp. Biol. Med.* **131:**363.

Peled, Y., and Tietz, A., 1974, Acylation of monoglycerides by locus fat-body microsomes, *FEBS Lett.* **41**:65.

Pereira, J. N., and Holland, G. F., 1970, Studies of the mechanism of action of *p*-chlorophenoxy-isobutyrate, in: *Atherosclerosis: Proceedings of the Second International Symposium*, pp. 549–554, Springer-Verlag, New York.

Pestka, S., 1971, Inhibitors of ribosome functions, *Ann. Rev. Biochem.* **40**:697.

Pieringer, R. A., and Hokin, L. E., 1962, Biosynthesis of lysophosphatidic acid from monoglyceride and adenosine triphosphate, *J. Biol. Chem.* **237**:653.

Plackett, P., and Rodwell, A. W., 1970, Glycerolipid biosynthesis by *Mycoplasma* strain Y, *Biochim. Biophys. Acta* **210**:230.

Polheim, D., David, J. S. K., Schultz, F. M., Wylie, M. B., and Johnston, J. M., 1973, Regulation of triglyceride biosynthesis in adipose and intestinal tissue, *J. Lipid Res.* **14**:415.

Pollock, R. J., Hajra, A. K., and Agranoff, B. W., 1975*a*, The relative utilization of the acyl dihydroxy-acetone phosphate and glycerol phosphate pathways for synthesis of glycerolipids in various tumors and normal tissues, *Biochim. Biophys. Acta* **380**:421.

Pollock, R. J., Hajra, A. K., Folk, W. R., and Agranoff, B. W., 1975*b*, Use of [1 or 3-^{3}H, U-^{14}C]glucose to estimate the synthesis of glycerolipids via acyl dihydroxyacetone phosphate, *Biochem. Biophys. Res. Commun.* **65**:658.

Pope, J. L., McPherson, J. C., and Tidwell, H. C., 1966, A study of a monoglyceride-hydrolyzing enzyme of intestinal mucosa, *J. Biol. Chem.* **241**:2306.

Possmayer, F., Scherphof, G. L., Dubbelman, T. M. A. R., Van Golde, L. M. G., and Van Deenen, L. L. M., 1969, Positional specificity of saturated and unsaturated fatty acids in phosphatidic acid from rat liver, *Biochim. Biophys. Acta* **176**:95.

Powell, G. K., and McElveen, M. A., 1974, Effect of prolonged fasting on fatty acid re-esterification in rat intestinal mucosa, *Biochim. Biophys. Acta* **369**:8.

Prottey, C., and Hawthorne, J. N., 1967, The biosynthesis of phosphatidic acid and phosphatidyl-inositol in mammalian pancreas, *Biochem. J.* **105**:379.

Pynadath, T. I., and Kumar, S., 1964, Incorporation of short- and long-chain fatty acids into glycerides by lactating goat mammary tissue, *Biochim. Biophys. Acta* **84**:251.

Raghavan, S. S., and Ganguly, J., 1969, Studies on the positional integrity of glyceride fatty acids during digestion and absorption in rats, *Biochem. J.* **113**:81.

Rao, G. A., and Abraham, S., 1973, α-Glycerolphosphate dehydrogenase activity and levels of glyceride-glycerol precursors in mouse mammary tissues, *Lipids* **8**:232.

Rao, G. A., and Johnston, J. M., 1966, Purification and properties of triglycerides synthetase from the intestinal mucosa, *Biochim. Biophys. Acta* **125**:465.

Rapoport, S., Leva, E., and Guest, G. M., 1943, The distribution of acid-soluble phosphorus in the livers of rats, fed and fasting, *J. Biol. Chem.* **149**:57.

Reitz, R. C., Lands, W. E. M., Christie, W. W., and Holman, R. T., 1968, Effects of ethylenic bond position upon acyltransferase activity with isomeric *cis,cis*-octadecadienoyl coenzyme A thiol-esters, *J. Bol. Chem.* **243**:2241.

Reitz, R. C., El-Shiekh, M., Lands, W. E. M., Ismail, I. A., and Gunstone, F. D., 1969, Effects of ethylenic bond position upon acyltransferase activity with isomeric *cis*-octadecenoyl coenzyme A thiol esters, *Biochim. Biophys. Acta* **176**:480.

Rizack, M. A., 1961, An epinephrine-sensitive lipolytic activity in adipose tissue, *J. Biol. Chem.* **236**:657.

Rizack, M. A., 1964, Activation of an epinephrine-sensitive lipolytic activity from adipose tissue by adenosine 3′,5′-phosphate, *J. Biol. Chem.* **239**:392.

Rizack, M. A., 1965, Mechanism of hormonal control of adipose tissue lipase, *Ann. N.Y. Acad. Sci.* **131**:250.

Robinson, D. S., 1963, The clearing factor lipase and its action in the transport of fatty acids between the blood and the tissues, *Adv. Lipid Res.* **1**:133.

Rodgers, J. B., 1970, Lipid absorption and lipid-reesterifying enzyme activity in small bowel of the protein-deficient rat, *Am. J. Clin. Nutr.* **23**:1331.

Rodgers, J. B., Riley, E. M., Drummey, G. D., and Isselbacher, K. J., 1967, Lipid absorption in adrenalectomized rats: The role of altered enzyme activity in the intestinal mucosa, *Gastro-enterology* **53**:547.

Rognstad, R., Clark, D. G., and Katz, J., 1974, Pathways of glyceride glycerol synthesis, *Biochem. J.* **140**:249.

Roncari, D. A. K., and Hollenberg, C. H., 1967, Esterification of free fatty acids by subcellular preparations of rat adipose tissue, *Biochim. Biophys. Acta* **137**:446.

Roncari, D. A. K., and Murthy, V. K., 1975, Effects of thyroid hormones on enzymes involved in fatty acid and glycerolipid synthesis, *J. Biol. Chem.* **250**:4134.

Rossiter, R. J., and Strickland, K. P., 1958, Biogenesis of phosphatides and triglycerides, *Ann. N.Y. Acad. Sci.* **72**:790.

Sanchez de Jimenez, E., and Cleland, W. W., 1969, Studies on the microsomal acylation of L-glycerol-3-phosphate. I. The specificity of the rat brain enzyme, *Biochim. Biophys. Acta* **176**:685.

Sarda, L., Marchis-Mouren, G., and Desnuelle, P., 1957, Sur les interactions de la lipase pancréatique avec triglycerides, *Biochim. Biophys. Acta* **24**:425.

Sarda, L., Marchis-Mouren, G., and Desnuelle, P., 1958, Nouveaux essais de purification de la lipase pancréatique de porc, *Biochim. Biophys. Acta* **30**:224.

Sarzala, M. G., Van Golde, L. M. G., De Kruyff, B., and Van Deenen, L. L. M., 1970, The intramitochondrial distribution of some enzymes involved in the biosynthesis of rat liver phospholipids, *Biochim. Biophys. Acta* **202**:106.

Sastry, P. S., and Hokin, L. E., 1966, Studies on the role of phospholipids in phagocytosis, *J. Biol. Chem.* **241**:3354.

Sastry, P. S., and Kates, M., 1966, Biosynthesis of lipids in plants. II. Incorporation of glycerophosphate-^{32}P into phosphatides by cell-free preparations from spinach leaves, *Can. J. Biochem.* **44**:459.

Savary, P., Constantin, M. J., and Desnuelle, P., 1961, Sur la structure des triglycerides des chylomicrons lymphatiques du rat, *Biochim. Biophys. Acta* **48**:562.

Schiller, C. M., 1970, Purification and properties of triglyceride synthetase, thesis, University of Texas, pp. 1–122.

Schultz, F. M., and Johnston, J. M., 1971, The synthesis of higher glycerides via the monoglyceride pathway in hamster adipose tissues, *J. Lipid Res.* **12**:132.

Sedgwick, B., and Hübscher, G., 1965, Metabolism of phospholipids. IX. Phosphatidate phosphohydrolase in rat liver, *Biochim. Biophys. Acta* **106**:63.

Sedgwick, B., and Hübscher, G., 1967, Metabolism of phospholipids. X. Partial purification and properties of a soluble phosphatidate phosphohydrolase from rat liver, *Biochim. Biophys. Acta* **144**:397.

Senior, J. R., and Isselbacher, K. J., 1962, Direct esterification of monoglycerides with palmityl coenzyme A by intestinal epithelial subcellular fractions, *J. Biol. Chem.* **237**:1454.

Senior, J. R., and Isselbacher, K. J., 1963, Demonstration of an intestinal monoglyceride lipase: An enzyme with a possible role in the intracellular completion of fat absorption, *J. Clin. Invest.* **42**:187.

Shapiro, B., Statter, M., and Rose, G., 1960, Pathways of triglyceride formation in adipose tissue, *Biochim. Biophys. Acta* **44**:373.

Sherr, S. A., and Treadwell, C. R., 1965, Triglyceride biosynthesis from monoglycerides in isolated segments of intestinal mucosa: Utilization of an ether analogue of 2-monostearin, *Biochim. Biophys. Acta* **98**:539.

Short, V. J., Brindley, D. N., and Dils, R., 1974, A new assay procedure for monoglyceride acyltransferase, *Biochem. J.* **141**:407.

Singh, A., Balint, J. A., Edmonds, R. G., and Rodgers, J. B., 1972, Adaptive changes of the rat small intestine in response to a high fat diet, *Biochim. Biophys. Acta* **260**:708.

Skipski, V. P., Morehouse, M. G., and Deuel, H. J., Jr., 1959, The absorption in the rat of a 1,3-dioleoyl-2-deuteriostearyl glyceride-C^{14} and a 1-monodeutiostearyl glyceride-C^{14}, *Arch. Biochem. Biophys.* **81**:93.

Smith, M. E., Sedgwick, B., Brindley, D. N., and Hübscher, G., 1967, The role of phosphatidate phosphohydrolase in glyceride biosynthesis, *Eur. J. Biochem.* **3**:70.

Snyder, F. L., 1972, *Ether Lipids: Chemistry and Biology*, Academic Press, New York.

Snyder, F., Piantodosi, C., and Malone, B., 1970, The participation of 1- and 2-isomers of *O*-alkylglycerols as acyl acceptors in cell-free systems, *Biochim. Biophys. Acta* **202**:244.

Srivastava, L. M., and Hübscher, G., 1966, Glucose metabolism in the mucosa of the small intestine, *Biochem. J.* **100:**458.

Stein, Y., and Shapiro, B., 1958, Glyceride synthesis by microsome fractions of rat liver, *Biochim. Biophys. Acta* **30:**271.

Stein, Y., Stein, O., and Shapiro, B., 1963, Enzymic pathways of glyceride and phospholipid synthesis in aortic homogenates, *Biochim. Biophys. Acta* **70:**33.

Steinberg, D., Vaughan, M., and Margolis, S., 1961, Studies of triglyceride biosynthesis in homogenates of adipose tissue, *J. Biol. Chem.* **236:**1631.

Stoffel, W., Schiefer, H. G., and Wolf, G. D., 1966, Untersuchunger über die Biosynthese von Membranphospholipoiden: Acylierung des Lysolecithins und der Lysophosphatidsäure durch Polyenfettsäuren, *Z. Physiol. Chem.* **347:**102.

Stoffel, W., Tomas, M. E. D., and Schiefer, H. G., 1967, Die enzymatische Acylierung von Lysophosphatidsäure, gesättigtem und ungesättigtem Lysolecithin, *Z. Physiol. Chem.* **348:**882.

Strickland, K. P., Subrahmanyam, D., Pritchard, E. T., Thompson, W., and Rossiter, R. J., 1963, Biosynthesis of lecithin in brain: Participation of cytidine diphosphate choline and phosphatidic acid, *Biochem. J.* **87:**128.

Sundler, R., and Akesson, B., 1970, The acylation of monoacylglycerol isomers by pig liver microsomes, *Biochim. Biophys. Acta* **218:**89.

Sundler, R., Akesson, B., and Nilsson, A., 1974, Effect of different fatty acids on glycerolipid synthesis in isolated rat hepatocytes, *J. Biol. Chem.* **249:**5102.

Tamai, Y., Lands, W. E. M., Barve, J. A., and Gunstone, F. D., 1973, Selective transfer of acetylenic acids to form lecithins, *Biochim. Biophys. Acta* **296:**563.

Tattrie, N. H., Bailey, R. A., and Kates, M., 1958, The action of pancreatic lipase on stereoisomeric triglycerides, *Arch. Biochem. Biophys.* **78:**319.

Terner, C., and Korsh, G., 1962, The biosynthesis of C^{14}-labeled lipids by isolated bull spermatozoa, *Biochemistry* **1:**367.

Tidwell, H. C., and Johnston, J. M., 1960, An *in vitro* study of glyceride absorption, *Arch. Biochem. Biophys.* **89:**79.

Tidwell, H. C., Pope, J. L., Askins, R. E., and McPherson, J. C., 1963, Specificity of a lipase of the intestinal mucosa, in: *Biochemical Problems of Lipids*, Vol. 1 (A. C. Frazer, ed.), pp. 217–222, Elsevier, Amsterdam.

Tzur, R., Tal, E., and Shapiro, B , 1964, α-Glycerophosphate as regulatory factor in fatty acid esterification, *Biochim. Biophys. Acta* **84:**18.

Van den Bosch, H., Van Golde, L. M. G., Eibl, H., and Van Deenen, L. L. M., 1967, The acylation of acylglycero-3-phosphorylcholines by rat liver microsomes, *Biochim. Biophys. Acta* **144:**613.

Van den Bosch, H., Van Golde, L. M. G., and Van Deenen, L. L. M., 1972, Dynamics of phosphoglycerides, *Rev. Physiol.* **66:**13.

Vaughan, M., 1961, Effect of hormones on glucose metabolism in adipose tissue, *J. Biol. Chem.* **236:**2196.

Vaughan, M., Berger, J. E., and Steinberg, D., 1964, Hormone-sensitive lipase and monoglyceride lipase activities in adipose tissue, *J. Biol. Chem.* **239:**401.

Verdino, B., Blank, M. L., and Privett, O. S., 1965, Endogenous lipid composition of the intestinal lymph of rats raised on a fat-free lard or corn oil diet. *J. Lipid Res.* **6:**356.

Waite, M., and Sisson, P., 1976, Mode of action of the plasmalemma phospholipase from rat liver, in: *Lipids*, Vol. 1 (R. Paoletti, G. Porcellati, and G. Jacini, eds.), pp. 127–139, Raven Press, New York.

Weiss, S. B., and Kennedy, E. P., 1956, The enzymatic synthesis of triglycerides, *J. Am. Chem. Soc.* **78:**3550.

Weiss, S. B., Smith, S. W., and Kennedy, E. P., 1956, Net synthesis of lecithin in an isolated enzyme system, *Nature (London)* **178:**594.

Weiss, S. B., Kennedy, E. P., and Kiyasu, J. Y., 1960, The enzymatic synthesis of triglycerides, *J. Biol. Chem.* **235:**40.

Westerfeld, W. W., Richert, D. A., and Ruegamer, W. R., 1968, The role of the thyroid hormone in the effect of *p*-chlorophenoxyisobutyrate in rats, *Biochem. Pharmacol.* **17:**1003.

Wichert, P. V., 1962, Enzymatische Bestimmung von α-2-Glycerophosphat in normalen und belasteten Wormblüterorganën, *Biochem. Z.* **336:**49.

Wieland, O., and Suyter, M., 1958, Glycerokinase; Isolierung und eigenschaften des Enzymes, *Biochem. Z.* **329:**320.

Wilgram, G. F., and Kennedy, E. P., 1963, Intracellular distribution of some enzymes catalyzing reactions in the biosynthesis of complex lipids, *J. Biol. Chem.* **238:**2615.

Wills, E. D., 1960, The relation of metals and -SH groups to the activity of pancreatic lipase, *Biochim. Biophys. Acta* **40:**481.

Wilson, J. S. D., and Galton, D. J., 1971, The effect of drugs on lipogenesis from glucose and palmitate in human adipose tissue, *Horm. Metab. Res.* **3:**262.

Yamashita, S., and Numa, S., 1972, Partial purification and properties of glycerophosphate acyltransferase from rat liver, *Eur. J. Biochem.* **31:**565.

Yamashita, S., Hosaka, K., and Numa, S., 1972, Resolution and reconstitution of the phosphatidate-synthesizing system of rat liver microsomes, *Proc. Natl. Acad. Sci.* (*USA*) **69:**3490.

Yamashita, S., Hosaka, K., and Numa, S., 1973, Acyl-donor specificities of partially purified 1-acylglycerophosphate acyltransferase, 2-acylglycerophosphate acyltransferase and 1-acylglycerophosphoryl-choline acyltransferase from rat liver microsomes, *Eur. J. Biochem.* **38:**25.

Yamashita, S., Nakaya, N., Miki, Y., and Numa, S., 1975, Separation of 1-acylglycerolphosphate acyltransferase and 1-acylglycerolphosphorylcholine acyltransferase of rat liver microsomes, *Proc. Natl. Acad. Sci.* (*USA*) **72:**600.

Young, D. L., and Lynen, F., 1969, Enzymatic regulation of 3-*sn*-phosphatidylcholine and triacylglycerol synthesis in states of altered lipid metabolism, *J. Biol. Chem.* **244:**377.

Yousef, I. M., O'Doherty, P. J. A., and Kuksis, A., 1973, Ribosome profiles of mucosal cells of normal, bile fistula and puromycin treated rats, *J. Nutr.* **103:**27.

Zahler, W. L., and Cleland, W. W., 1969, Studies on the microsomal acylation of L-glycerol 3-phosphate. III. Time course of the reaction, *Biochim. Biophys. Acta* **176:**699.

Zakim, D., Paradini, R. S., and Herman, R. H., 1970, Effect of clofibrate (ethyl-chlorophenoxyisobutyrate) feeding on glycolytic and lipogenic enzymes and hepatic glycogen synthesis in the rat, *Biochem. Pharmacol.* **19:**305.

Chapter 7

Composition of Selected Dietary Fats, Oils, Margarines, and Butter

Alan J. Sheppard, John L. Iverson, and John L. Weihrauch

7.1. Introduction

Fats and oils provide the most concentrated source of energy in the diet, about 9 kcal/g. Animal tissues, dairy products, and oils extracted from certain seeds contribute most of the fat in the diet (see Table I). The source, chemical nature, and processes that fats and oils are subjected to before they reach the market shelf are important aspects of the study of lipids. A fat is distinguished from an oil by its solid state at a particular temperature, usually room temperature. Some oils are winterized to remove triglycerides that would cause the oil to solidify in the refrigerator; other oils are lightly hydrogenated to convert highly unsaturated fatty acids to more saturated acids to prevent the development of off-flavors and rancidity.

Fats and oils consist primarily of the three-carbon glycerol molecule with three straight-chain fatty acids attached as esters. This large molecule is commonly called a triglyceride and represents about 95% of the weight of most oils from seeds. However, minor components are also present, such as free fatty acids, sterols, sterol esters, phosphatides, glycolipids, hydrocarbons, glyceryl ethers, fat-soluble vitamins, and monoglycerides and diglycerides. The term "lipid" is more appropriate than the term "fat," since lipid includes not only triglycerides but these additional chemical substances that are present in varying amounts depending on which food fat is being discussed.

Fatty acids are most often named in the scientific literature, especially in older articles, by common names rather than by their scientific, systematic nomenclature. Unfortunately, common names may vary from one part of the world to another, which leads to confusion for readers. It is highly desirable to use the scientific names of the International Union of Chemistry (IUC). The fatty acids normally encountered are composed of 4 to 24 carbon atoms. The fatty acids with 4 to 12 carbons generally are saturated. A number of monounsaturated fatty acids

Alan J. Sheppard and John L. Iverson • Division of Nutrition, Food and Drug Administration, Washington, D.C. 20204. ***John L. Weihrauch*** • Consumer and Food Economics Institute, Agricultural Research Service, U.S. Department of Agriculture, Hyattsville, Maryland 20782.

Table I. Percentage of Fat Contributed by Major Food Groups (1972)[a,b]

Food	Percent
Fats and oils	38.7
Meat	32.0
Dairy products except butter	12.6
Dry beans, nuts, and soy products	3.6
Butter	3.1
Eggs	3.1
Poultry	2.3
Flour and cereal products	1.3
Fish	1.0
Fruits and vegetables	0.9
Miscellaneous	1.3

[a] Fat consumption for 1972 was 158 g fat/day per capita.
[b] Adapted from *Food Consumption, Prices, and Expenditures*, U.S. Department of Agriculture (1968, 1972).

containing one double bond may be present, including 14:1, 16:1, 18:1, 20:1, 22:1, and 24:1 (number of carbon atoms:number of double bonds). Oleic acid is the most common monounsaturated acid.

The numbering procedure generally used assigns the number 1 to the carbon of the carboxyl group and progressively increases toward the noncarboxyl end of the carbon chain. The Greek letters α, β, etc., are sometimes used. The α-carbon is the carbon adjacent or next in sequence to the carboxyl carbon, and therefore is carbon-2 in the numerical system. Carbon-3 is β and so on. The suffix "-anoic" denotes a saturated fatty acid carbon chain. The suffix "-enoic" denotes some degree of unsaturation in the fatty acid carbon chain.

Polyunsaturated fatty acids (PUFAs) have two or more double bonds. Linoleic acid has 18 carbon atoms, is lacking four hydrogen atoms, and has two double bonds, at the 9–10 and 12–13 positions, separated by a methylene group in the 11 position. The fatty acids of oils as removed from the seed have the double bonds in the *cis* configuration rather than the *trans* configuration. Thus the ICU name for linoleic acid is *cis*-9,*cis*-12-octadecadienoic acid. The natural form of unsaturation is the *cis* form, with hydrogen atoms at the point of unsaturation on the same side of the carbon chain with the molecule bent back on itself. When the hydrogen atoms at the point of unsaturation are on opposite sides of the carbon chain, the unsaturation is *trans* and the molecule is less bent. The geometric isomeric configuration has metabolic implications relating to enzyme specificity. This minor structural difference also imparts large physical differences which are useful in food technology. Oleic acid melts at 4°C, while elaidic acid melts at 44°C. This melting point difference is most useful in modifying oils for particular properties in foods since *trans* acids are more solid at room temperature than *cis* acids. Nutritionally, the *cis* configuration is of prime importance since the essential fatty acids are all of the *cis* form.

The lipids from lean meats of domesticated animals contain relatively low amounts of arachidonic acid, which is recognized as an essential fatty acid. It is either absent in the fatty tissues or present in trace amounts. Arachidonic acid has 20 carbon atoms with 8 hydrogen atoms missing and 4 double bonds of the *cis* configuration, with the double bonds separated by a methylene unit. Its more scientific name (ICU) is *cis*-5,*cis*-8,*cis*-11,*cis*-14-eicosatetraenoic acid. Marine oils contain large amounts of 20- and 22-carbon-chain acids obtained by chain elongation and desaturation of linolenic acid. The 22-carbon-atom unsaturated acids with 10 and 12 hydrogen atoms missing have 5 and 6 double bonds, respectively. The nutritional importance of these long-chain polyunsaturated acids has not been determined, even though they are present in trace amounts in butter oils and animal fats.

Two or more isomers of unsaturated acids are possible. In lipid chemistry, the two types of isomers are positional and geometric. Three positional isomers of the 18-carbon monounsaturated fatty acids are found in nature: the most common one is oleic acid with the double bond in the 9 position; petroselenic acid has the double bond in the 6 position; vaccenic acid has the double bond in the 11 position and *trans* configuration.

The fatty acids in natural fats are attached to the 3 position available on the glycerol moiety ($CH_2OHCHOHCH_2OH$). In nature, the more unsaturated fatty acids tend to occupy the center position on the molecule. When only two different fatty acids are present in a triglyceride, eight structural isomers (combinations of fatty acids esterified to glycerol) are possible, with markedly different physical properties. A single isomer (a set combination of fatty acids forming the triglyceride) will have a sharp melting point, while a mixture of isomers (triglycerides) will have a wider melting point. Cocoa butter contains palmitic, stearic, and oleic acids. However, the three isomers of 2-oleodipalmitin, 2-oleopalmitostearin, and 2-oleodistearin represent approximately 80% of the total triglyceride content (Dutton *et al.*, 1961; Sampugna and Jensen, 1969); this gives cocoa butter its sharp melting point so desirable in milk chocolate.

Table II. Fatty Acid Composition of "Soy Lecithin"[a]

Fatty acid	g/100 g lecithin
12:0	0.1
14:0	0.1
16:0	11.6
18:0	3.2
20:0	0.1
Total saturated	15.2
16:1	0.4
18:1	14.3
18:2	43.2
18:3	5.6
Total unsaturated	63.4

[a] Commercial "soy lecithin" is a mixture of about 29% lecithin, 31% cephalin, and 40% inositol phosphatide; the commercially available product is 70% by weight "soy lecithin" in soy oil.

Monoglycerides and diglycerides, which are used as stabilizers, consist of the glycerol moiety with one or two fatty acids attached. They are prepared by mixing free glycerin and triglycerides with sodium hydroxide as a catalyst and heating to 100–250°C in a procedure known as glycerolysis. The product contains various amounts of monoglycerides and diglycerides depending on the specific conditions used. Lecithin is a phospholipid containing glycerol, which is esterified to two fatty acids and phosphoric acid. The latter in turn is attached to the alcohol group of the organic base choline. Another description of lecithin is to say that it is a mixture of the diglycerides of palmitic, stearic, and oleic acids linked to the choline ester of phosphoric acid. Frequently, at least one of the fatty acids is unsaturated (55% in soybean lecithin). Lecithin is used as an emulsifier. The fatty acid composition of lecithin from soybeans is presented in detail in Table II.

7.2. Methodology

The methodology used for lipid or fat extraction, lipid subfractionation, and esterification is not a simple matter and is heavily dependent on the end-product or data requirements. For example, if only a total fat value is needed, the methodology requirement is not so stringent as for a total lipid measurement. Conversely, if further chemical analyses are to be performed on the lipid extract, such as determining structures and bond configurations, the methodology requirements become even more demanding. In the latter case, the extraction must be quantitative for all lipid classes yet gentle enough not to destroy or alter the lipid and fatty acid structure from that existing initially. Therefore, the methodology selected is governed to a large degree by the end-product or data requirements.

7.2.1. Extraction Techniques

7.2.1.1. Total Fat

The term "total fat" is widely used in the analysis of foods and feeds. It is a classical term which by definition means ether extract. Ether extracts primarily the neutral lipids and does not quantitatively remove the more polar lipids such as cholesterol, cholesterol esters, and the phospholipids. Thus the term "total fat" must not be confused with the term "lipid." Depending on the type of product being extracted, the values may be identical or may differ markedly. The Association of Official Analytical Chemists (AOAC) has collaboratively studied the procedure for total fat and it is now well standardized (AOAC, 1975*a*). There are slight variations in the methodology depending on the material to be analyzed, e.g., bread, cheese, or meat. Basically, the methodology is quite simple. The sample is first prepared for analysis by dicing, grinding, homogenizing, etc., followed by weighing 3–4 g by difference into a thimble containing a small amount of sand or asbestos. The sample size may vary somewhat based on the expected fat content. The thimble contents are mixed using a glass rod, and the thimble and rod are placed in a 50-ml beaker and dried in an oven 6 hr at 100–102°C or 1.5 hr at 125°C. Then the thimble is placed in a Bailey-Walker or a Goldfisch apparatus and extracted with anhydrous ether or petroleum ether. The extraction period may vary from 4 hr at a condensation rate of 5–6 drops/sec to 16 hr at 2–3 drops/sec. Following extraction, the extract is dried 30 min at 100°C, cooled, and weighed. The AOAC extract is undesirable as a total lipid extract on a quantitative basis and for subsequent fatty acid analysis since it has definitely been demonstrated that PUFAs are damaged by the procedure (Sheppard, 1963; Sheppard *et al.*, 1974*a*; Hubbard *et al.*, 1976).

7.2.1.2. Lipid

Lipid is the material extractable from plant and animal tissues by the so-called fat solvents such as alcohol, ether, and chloroform. The material extracted is a complex mixture, depending on the tissue and solvent(s) used, and usually contains representatives of several or all of the classes of organic compounds known as fats (neutral lipids, e.g., glycerides), waxes, phospholipids, glycolipids, and sterol esters, as well as the hydrolytic products of some of these compounds. These various substance types have in common one very important characteristic, namely, they all yield fatty acids upon hydrolysis.

The distribution of the various lipid classes in an extract is quite dependent on the solvent or combination of solvents used in the extraction process. For example, ether will remove the neutral lipids quantitatively, but sterols and phospholipids are only partially extracted. The problem is mainly related to solubilities and polarities. If a more polar solvent such as chloroform is used, the more polar lipids are solubilized and extracted. Another important consideration is that the lipid extracted must be unaltered, e.g., in the original state as it existed in the starting material. For example, if the goal of the analysis is to measure the various PUFAs, the method must be capable of extracting them without forming polymers, isomerizing, shifting double bonds, destroying double bonds, or fragmenting the carbon chain. Unfortunately, more often than not, too little consideration is given to the extraction procedure; thus subsequent analysis reflects the extraction rather than the true total lipid or lipid components. Some comparison studies have been reported in recent years (Sheppard, 1963; Sheppard *et al.*, 1974*a*; Engler and Bowers, 1975; Hubbard *et al.*, 1976). The most satisfactory general-purpose method for extraction of lipids from foods resulting from these studies is the use of a mixture of chloroform and methanol as the extracting reagent. The procedure is that of Folch *et al.* (1957) or a modification of it (Bligh and Dyer, 1959). Our experience with these two procedures has resulted in our using the Folch *et al.* procedure because it exhibits less variation than the Bligh and Dyer procedure. However, we have never found a statistical difference in the mean values obtained. The adaptation of the Folch *et al.* method for food extraction has been incorporated into methodology for CFR 1.18 (Code of Federal Regulations, 1975; fatty acid cholesterol regulation) (Sheppard *et al.*, 1974*b*). The procedure is widely employed in the food field for both research and routine analyses. However, it may be necessary to use highly tailored extraction techniques for unusual or specialized analyses.

The method is outlined as follows:

Choose a sample size to contain approximately 1 g fat. Quantitatively transfer this sample to a homogenization vessel (e.g., Vertis, Waring) and homogenize with a volume of chloroform–methanol (2 : 1, v/v) that is twentyfold that of the original sample size (i.e., for 10 g sample, use 200 ml reagent). Homogenize for 2 min unless it is visually obvious that a longer time is required because of some unusual characteristics of the sample. Quantitatively transfer and filter the homogenized sample through a qualitative grade (rapid) filter paper into a separatory funnel of appropriate size. Rinse blender (bowl, blade, lid, etc.) with at least two chloroform–methanol rinses (0.2 vol homogenate) and filter these washes as previously done with homogenate to ensure quantitative transfer. Wash the homogenate residue trapped in the filter paper twice with two chloroform–methanol rinses (0.2 vol homogenate). Pass all rinses through the filter paper into the separatory funnel. Wash the extract by thoroughly mixing it with 0.2 its volume of water (i.e., 200 ml extract, use 40 ml water). Let the mixture stand until the layers completely separate. Drain most of the lower-phase chloroform into an appropriate-size round-bottom flask with a 24/40 fitting. Wash the upper phase (water–methanol) three times with chloroform–methanol (2 : 1). The volume of these washes is equal to the volume of the water–methanol remaining in the separatory funnel. Drain the washes into the round-bottom flask to form a composite lipid extract. Attach the round-bottom flask to a flash evaporator with a 50°C water bath. Remove the chloroform, using vacuum provided by a water aspirator or other vacuum source. Quantitatively transfer the dried residue to a tared receiver, using ethyl ether as the transferring solvent. Remove all traces of ether from the composite extract under a stream of nitrogen. Then transfer the tared receiver to a vacuum oven. Dry the extract at 60°C and under 26 inches of vacuum. When all water has been removed and a constant weight has been achieved, transfer the tared receiver to a nitrogen-flushed desiccator and allow the sample and receiver to temperature equilibrate. Following cooling, weigh the ether extract and record the weight. Take up the dried lipid extract in redistilled petroleum ether and quantitatively transfer it to a 100-ml volumetric flask. This procedure establishes volume weight. Throughout the entire operation, use good analytical techniques, such as using tongs rather than fingers to move tared receivers and keeping the sample extractions and esters under nitrogen whenever possible.

Cooking and salad oils generally do not require extraction and can be utilized directly when preparing esters of the fatty acids.

7.2.2. *Esterification Techniques for Fatty Acids*

The sulfuric acid–methanol method (AOAC, 1965), hydrochloric acid–methanol method (Stoffel *et al.*, 1959), and boron trifluoride method (AOAC, 1975*b*) as modified by Solomon *et al.* (1974) perform equally well for preparing the methyl esters of the fatty acids in foods other than dairy products such as butter that contain substantial quantities of the short-chain volatile fatty acids. The substitution of *n*-butanol for methanol in the boron trifluoride method prepares the butyl esters of the fatty acids (Iverson and Sheppard, 1977). The losses associated with recoveries of the short-chain fatty acids are avoided when the butyl esters are used. The butyl esters are more soluble in the fat solvents, less volatile, and markedly less water soluble than are the comparable short-chain fatty acid methyl esters.

The boron trifluoride–methanol procedure of the AOAC as modified by Solomon *et al.* (1974) is a good general-purpose procedure that has wide application and is routinely used in the senior author's laboratory for food analysis. The method is outlined as follows:

As described under Section 7.2.1.2, transfer to a 50-ml reaction flask an aliquot of the chloroform–methanol extract containing a known weight of total lipid of not less than 350 mg but not more than 500 mg, or directly weigh a sample of cooking or salad oil into a round-bottom or Erlenmeyer flask. The choice of type of flask depends solely on whether subsequent heating of the flask is with a hot plate or a heating mantle. Add 6 ml of 0.5 N methanolic KOH and a boiling chip. Attach a condenser and heat the mixture under reflux until the fat globules dissolve (usually 5–10 min). Then add 7 ml of 12.5–14% boron trifluoride–methanol reagent via the condenser and boil 2 min; add 2–5 ml heptane through the condenser and boil 1 min longer. Remove the flask from the heat source, cool to room temperature, remove the condenser, and add sufficient saturated aqueous sodium chloride solution to float the heptane solution of fatty acid methyl esters. To recover dry esters, quantitatively transfer the mixture to a separatory funnel (usually 125 ml) using both distilled water and redistilled petroleum (30–60°C) ether to complete the transfer. Extract twice with 50-ml portions of redistilled petroleum ether. Combine the ether extracts in a second separatory funnel and wash the extract four times with petroleum ether to remove any residual acid and the boron trifluoride breakdown products. Quantitatively transfer the washed esters to a tared receiver, evaporate the petroleum ether under a stream of nitrogen, and weigh. Transfer to a 25-ml volumetric flask, using *n*-hexane as the carrier solvent.

When free fatty acids are to be esterified, bypass the saponification and follow the above procedure beginning with the introduction of the boron trifluoride–methanol reagent.

7.2.3. Gas Chromatographic Analysis of Fatty Acid Methyl Esters

For gas chromatographic analysis of fatty acid methyl esters, use a 2 × 4 mm inside-diameter column (preferably glass) packed with 12–15% (w/w) ethylene glycol succinate liquid phase coated on 100/120 mesh Gas-Chrom P (Applied Science Laboratories, Inc., P.O. Box 440, State College, Pennsylvania 16801) or its equivalent. Maintain the column operating temperature in the range of 170–190°C and obtain the desired retention times by adjusting the carrier gas flow rate. Methyl stearate should appear 12–15 min after injection, and methyl stearate and methyl oleate should be completely resolved from each other. Use either hydrogen flame or β-argon ionization detection systems with the specified column. Thermal conductivity detectors are excellent but lack the sensitivity of the ionization detectors. Obtain a weight response plot for each component of interest, using an equal-weight mixture of the fatty acid methyl esters. Inject three different amounts. Graphically plot the response in area vs. amount for each component of interest. Be certain that the response plot for individual fatty acid methyl esters brackets the peak area of that ester in the sample.

Obtain a chromatogram of the sample and obtain the peak area of each component of interest either electronically or by actual physical measurement. Then determine the amount of compound present from the appropriate standard response plot. Calculate the amount (weight) of each fatty acid methyl ester in the total GLC analysis and then obtain a sum total weight of the measured fatty acid methyl esters. Compare this total weight to the weight of methyl esters obtained from the esterification procedure, relating the GLC sample size injected to the volume of the esters from the sample. Such a comparison serves as a check to ensure that the GLC analysis reasonably accounts for the total fatty acid content.

7.2.4. Sterol Derivatization

The preparation of the butyrate derivative of the sterol content of the sample is initiated using a portion of the *n*-hexane solution of the fatty acid methyl ester preparation.

7.2.4.1. *When Cholesterol or Sterol Content Is Unknown*

Take a 5- or 10-ml aliquot of the fatty acid methyl ester preparation and quantitatively transfer it to a 50-ml round-bottom flask fitted with a 24/40 joint. Experience with a product will dictate the size of aliquot to use on a routine basis.

7.2.4.2. *When Approximate Cholesterol or Sterol Content Is Known*

Based on label claim, laboratory experience with the product, etc., transfer a sufficient amount of the *n*-hexane solution of methyl esters to achieve a final concentration of 1–2 mg cholesterol or other sterol/ml after completing the derivatization process.

7.2.4.3. *General Procedure for Cholesterol and Sterols*

Evaporate the *n*-hexane completely, using a stream of nitrogen. Add 10 ml benzene–absolute ethanol (5 : 1, v/v). Evaporate to dryness, with a stream of nitrogen, on a steam bath. If traces of water are still visible in the flask, repeat the addition and evaporation of the benzene–absolute ethanol. Pipette 4 ml of a solution consisting of butyric anhydride–pyridine (2 : 1, v/v) into the residue in the round-bottom flask. (WARNING: This reagent must be prepared fresh daily.) Attach the round-bottom flask containing the residue and reagent to a condenser, reflux 10 min, and then cool the flask under a stream of tap water. After cooling the flask, evaporate the solution under a stream of nitrogen just to dryness. Quantitatively transfer the semidried residue to a 5-ml volumetric flask, using *n*-hexane. Dilute to volume with *n*-hexane. With some samples it may be necessary to either dilute or concentrate the preparation. The GLC response will determine whether an adjustment in concentration is necessary. Be certain to record the volume change, because this information will be needed in subsequent calculations.

7.2.4.4. *Cholesterol and Sterol Standards*

The preparation of the standard follows the procedure outlined above for the cholesterol or sterol in a food extract, with minor differences which are listed in detail. The final concentration should be 1 mg/ml. Difficulty will be encountered in getting any amount in excess of 1 mg/ml into solution. Accurately weigh 500 mg cholesterol or sterol. Quantitatively transfer it to a 500-ml volumetric flask with *n*-hexane and dilute to volume with *n*-hexane. This constitutes a *stock solution*. Make certain that all of the cholesterol is in solution. Repeated shaking may be necessary. Take a 10-ml aliquot of the cholesterol stock solution and quantitatively transfer to the 500-ml round-bottom flask as outlined for sample. Proceed as with the sample above. The final concentration in terms of the starting cholesterol is 2 mg/ml or 2 μg/μl. This constitutes the *operating standard*. With some samples, it may be necessary to dilute the standard; if this is the case, be certain to determine the new concentration (μg/μl).

7.2.5. *Gas Chromatographic Analysis of Sterols*

The gas chromatograph must be capable of an electrometer output of 10×10^{-10} A, maintenance of column oven temperatures of 275°C, maintenance of injector and detector temperatures of 325°C, and accommodation of an all-glass column and injection port. Metal columns and metal injection ports tend to lead to the destruction of the sterols. On-column injection is highly desirable.

The gas chromatographic operating parameters are as follows: Column: 2 m × 4 mm inside-diameter glass. Column packing: 1% SE-30 (by weight) coated on 100/120 mesh Gas-Chrom Q (Applied Sciences Laboratories, Inc., P.O. Box 440, State College, Pennsylvania 16801) or its equivalent. Electrometer: output equivalent to a system equipped with an electrometer giving an output of 10×10^{-10} A with a 2-mV recorder, full-scale response with the attenuator at maximum sensitivity. Detector: hydrogen flame ionization or other ionization detectors if flame is not available. Operating temperatures: column 250–265°C, detector 300–315°C, and injection area 300–315°C. Cholesteryl butyrate peak time is approximately 15 min, adjusted by varying the carrier gas flow rate and/or column temperature (the latter within 250–265°C specifications). The approximate relative retention times are cholesteryl butyrate, 1.00; campesteryl butyrate, 1.30; stigmasteryl butyrate, 1.38; and sitosteryl butyrate, 1.59.

The gas chromatograph should be calibrated for each individual sterol that is to be measured, using its butyl derivative as prepared under Section 7.2.4. The following discussion refers to cholesterol, but the approach is identical for each individual sterol. The calibration procedure is as follows: Obtain a weight response plot by injecting a minimum of three different amounts of the prepared standard. Each point should be obtained in duplicate. Graphically plot the average response in area (y axis) vs. the amount for each quantity of cholesterol (x axis). Be certain that the response plot brackets the peak area of the cholesteryl butyrate found in the sample. Read from the graphic plot the quantity of cholesterol (x axis) in the sample that corresponds to the sample peak area (y axis).

Remember that the peak shown on the gas chromatogram represents a given amount of cholesterol. The amount of cholesterol reacted and placed in a given volume must be kept in mind since the cholesteryl butyrate is merely a vehicle for transforming the cholesterol into a derivative suitable for chromatographing. For analytical purposes, the derivative as such is not of interest for calculating purposes, but rather the cholesteryl butyrate peak represents a specific amount of cholesterol.

This eliminates the problem of making mathematical conversions from the derivative to cholesterol.

The butyrate derivatives were selected because they gave a resolution that was superior to that of the unreacted sterols or the acetate and propionate derivatives. An additional advantage is that the solvent, fatty acid methyl esters, and vitamin E butyrate clear the column before the appearance of the sterol butyrate series.

7.2.6. Lipoxidase:*cis,cis*-Polyunsaturated Fatty Acid Analysis

The following method is applicable to the enzymatic determination of the *cis,cis*-methylene-interrupted PUFAs (triglycerides) in fats and oils. The fatty acids in the sample are saponified to form their potassium salts. The salts of the *cis,cis*-methylene-interrupted PUFAs are then conjugated and oxidized to hydroperoxides by atmospheric oxygen in the presence of the enzyme lipoxidase. The weight in grams of total *cis,cis*-methylene-interrupted PUFAs/100 g sample is calculated from the absorbance of the conjugated diene hydroperoxide measured at 234 nm.

7.2.6.1. Reagents

Potassium Borate Buffer, 1.0 M, pH 9.0. Dissolve 61.9 g H_3BO_3 and 25.0 g KOH in approximately 800 ml distilled water by stirring and heating. Cool to room temperature and adjust pH to 9.0 by adding 1.0 N HCl or 1.0 N KOH as required. Dilute to 1 liter with distilled water and mix.

Potassium Borate Buffer, 0.2 M, pH 9.0. Dilute 200 ml of 1.0 M potassium borate buffer to 1 liter with distilled water and mix.

Lipoxidase Stock Solution. Dissolve 20 mg lipoxidase (ICN Nutritional Biochemicals, Cleveland, Ohio 44128) or its equivalent in 10 ml ice-cold 0.2 M potassium borate buffer.

Dilute Lipoxidase Solution. Mix 2 ml stock solution with 8 ml ice-cold 0.20 M buffer. If a large number of analyses are to be performed, 5 ml stock solution mixed with 20 ml ice-cold 0.2 M buffer, giving 25 ml dilute lipoxidase solution, is more desirable.

Boiled Dilute Lipoxidase Solution. Transfer 4 ml dilute lipoxidase solution to a 10-ml volumetric flask and hold in boiling water for 5 min.

Alcoholic KOH, 0.5 N. Dissolve 1.40 g KOH in 95% ethanol and dilute to 50 ml with 95% ethanol. Prepare fresh daily.

7.2.6.2. Procedure

a. Vegetable Oils. Accurately weigh 100 mg sample and transfer with *n*-hexane into a 100-ml volumetric flask which has just been flushed with nitrogen. Dilute to volume with *n*-hexane. Transfer 1.00 ml *n*-hexane solution to a 100-ml volumetric flask and completely evaporate the solvent under a stream of nitrogen.

b. Margarines. Quantitatively transfer a 20-ml aliquot of the petroleum ether–total lipid extract previously prepared per the instructions given above under Section 7.2.1.2 to a 100-ml volumetric flask. Experience with a specific product may indicate that a different-size aliquot is more desirable. Evaporate the petroleum ether (must be completely removed) and take up the residue in *n*-hexane; further dilute with pure-grade *n*-hexane to reach a desired final concentration. Transfer 1.00 ml of the *n*-hexane solution to a 100-ml volumetric flask and completely evaporate the solvent under a stream of nitrogen.

c. Shortenings, Meats, Solid Fats, and Other Foodstuffs (*moderate to high PUFA content*). Follow the instructions provided for margarines under (b) above.

d. Saponification of Samples. Add 2 ml 0.5 N alcoholic KOH solution to the solvent-free sample in the 100-ml volumetric flask, mix, and let stand 5 hr in the dark or overnight. After saponification is complete, add 20 ml of 1.0 M potassium borate buffer solution and 50 ml distilled water, mix, add 2 ml 0.5 N HCl, and dilute to volume with distilled water.

e. Determination of the PUFA Content. Pipette 3.00 ml saponified fatty acid solution into each of four test tubes (13 × 100 nm). To the first two tubes (blanks) add 0.10 ml of the boiled (inactive) dilute enzyme solution and mix well. To the third and fourth test tubes (duplicate samples) add 0.10 ml of the (active) dilute enzyme solution, mix, and let the test tubes stand exposed to air at room temperature for 30 min. Transfer the contents of the tubes to matched 1-cm quartz cells and place in the spectrophotometer. Adjust the instrument with the blank samples and measure the absorbance of the samples at 234 nm.

7.2.6.3. Calculations

Standard Curve. Weigh 100 mg linoleic acid (99% pure, Nu Chek Prep, P.O. Box 172, Elysian, Minnesota) or its equivalent and transfer to a 100-ml volumetric flask with *n*-hexane. Quantitatively transfer a 10-ml aliquot to a 100-ml volumetric flask and dilute to volume with *n*-hexane. Quantitatively transfer 1-, 3-, 6-, and 9-ml aliquots to separate 100-ml volumetric flasks. Reduce the aliquot in each volumetric flask to dryness with a stream of nitrogen. Pipette 2 ml absolute ethanol into each fatty acid-containing volumetric flask; then add 20 ml of 1.0 M potassium borate solution, 50 ml of distilled water, and 2 ml of 0.5 N HCl. Dilute to final volume with distilled water. Pipette 3.00 ml of the prepared fatty acid solution from each of the standard samples into four separate test tubes. To the first and second tubes add 0.1 ml of the boiled enzyme and mix well. To the third and fourth tubes add 0.1 ml of the unboiled (active) enzyme, mix well, and let stand at room temperature for 30 min. Repeat this procedure for each of the four standard levels. Read and record the absorbance at each level. Draw a standard curve (Lambert–Beer plot), using absorbance vs. μg/ml for each level of standard.

Sample. Obtain the absorbance of the sample. Using the standard curve, determine the amount of *cis,cis*-PUFAs in the sample. The following relationship is applicable: $\mu g \times$ dilution factor $\times 10^{-4}$ g = g/100 g product.

7.2.6.4. Special Problem Samples

Special problems are encountered in extracting the lipids from starchy foods such as cereal products and baked goods. In these products, the major portion of polar lipids (consisting primarily of phospholipids and glycolipids) occurs bound to protein and starch (Mecham, 1971). Monoglycerides are increasingly being used in baked goods. They tend to bind with amylose and are not extracted with chloroform, methanol, or mixtures of these solvents. To overcome these difficulties, acid hydrolysis has been employed to extract all lipid material. This procedure is unsatisfactory because the original lipid is destroyed and only the fatty acid portion of the lipid is extracted.

Recently, cereal chemists have been employing water-saturated *n*-butanol according to Mecham and Mohammad (1955) to obtain maximum extraction of intact lipids from cereals. At room temperature, nearly quantitative extraction of total lipid is attained. Acker and Schmitz (1967) extracted starch lipids nearly quantitatively at 70°C, while Morrison *et al.* (1975) obtained quantitative extraction of total starch lipids with water-saturated *n*-butanol at 90–100°C without damage to the lipid.

The reader is referred to Sheppard *et al.* (1974*a*) and Hubbard *et al.* (1976) for procedures applicable to samples presenting unusual analytical problems.

7.3. *Basic Processing*

7.3.1. *General Discussion*

Specific processes are used for separating the oil from the oilseed and converting the crude oil into finished consumer products. These processes will not be considered in great depth unless the products have nutritional significance.

Originally, most of the oil was extracted from the seed by cold or hot pressing methods, which left large amounts of oil remaining in the press cake. These press methods have been almost totally replaced or combined with solvent extraction methods which are considerably more efficient. The seeds may be ground, cooked, or rolled to break down the cell structure and heated to achieve maximum extraction of oil from the seed (Hutchins, 1968; Galloway, 1976). Solvent-extracted oil contains fewer impurities than pressed oils. Residues of the extraction solvents, e.g., *n*-hexane, are removed during processing. Soybean and cottonseed are flaked to expose a large surface to the solvent and cooked to denature the protein to facilitate oil extraction (Hutchins, 1976). The protein-rich meal contains less than 1% oil. The *n*-hexane is recovered by evaporation and condensation with little loss. The solvent-free crude oil is then ready to undergo a series of operations designed to produce a finished product for specific uses.

Crude oil contains fatlike substances other than triglycerides. These other substances include monoglycerides and diglycerides, free fatty acids, sterols, proteinaceous and mucilaginous materials, and phosphatides. They are largely removed by treating the oil with alkali to form water-soluble compounds that are removed by washing with hot water, filtering, and centrifuging (Sullivan, 1968; Cowan, 1976; Carr, 1976). The oil is dried under vacuum to remove moisture, other volatile material, and distillable pigments during the deodorizing procedure (Zehnder and McMichael, 1967; Evans *et al.*, 1964*b*), resulting in a bland finished product. The process of deodorization calls for blowing steam through hot oil at 210–274°C and vacuum of 1–6 mm Hg (Zehnder, 1976). The steam strips off the odoriferous materials. The final step in deodorization is the addition of citric acid to prevent color reversion. The maximum amount of citric acid that can be added is 0.01% in the United States.

Winterization is a cooling procedure used to remove the more saturated high-melting fractions from an oil so that it will not solidify or become cloudy during refrigeration. To produce a salad oil, originally cottonseed oil was held in tanks outdoors in the winter and the solid and liquid portions separated. Today, refrigeration and solvent winterization techniques are used with increased yields of salad oil.

The most pronounced change in physical and chemical properties takes place when an oil is hydrogenated. Hydrogenation is the addition of hydrogen atoms at the points of unsaturation in the fatty acids. In the hydrogenation process, hot oil is whipped under pressure with hydrogen gas in the presence of a catalyst. The hydrogenation process is easily controlled and the nickel catalyst generally used is completely removed (Coenen, 1976; Johnson *et al.*, 1962; Lefebur and Baltes, 1975; Beal *et al.*, 1969).

The value of hydrogenation can be illustrated by applying it to soybean oil. Soybean oil contains about 7% linolenic acid, which is unstable and responsible for off-odors and flavors, and in the extreme may become rancid. The hydrogenation process can be selectively controlled and stopped after most of the linolenate has been converted to a more saturated molecule while the oil still remains liquid. More extensive hydrogenation would produce soft, semisoft, and solid oils used in margarines, shortenings, and bakery formulations. Complete hydrogenation would result in a hard, brittle solid composed of fully saturated triglycerides at normal temperatures.

The hydrogenation conditions can be varied to produce a finished product with particular chemical and physical properties. In nonselective hydrogenation, the formation of high-melting glycerides is favored, since the catalyst once attached to a molecule in a random manner proceeds to hydrogenate the double bond. In selective hydrogenation, hydrogen is preferentially attached to a molecule to favor one type of structure over another. Thus soybean oil could be preferentially hydrogenated to convert linolenic acid to a less saturated acid with a minimum reduction in linoleic acid, an essential fatty acid, and minimum production of highly saturated triglycerides (Evans *et al.*, 1964*a*; Scholfield *et al.*, 1964; Allen, 1962). The oil is then winterized to remove the saturated triglycerides, and salad oil or margarine base oils are produced. The same approach could be followed using other starting oils.

Unfortunately, from a nutritional viewpoint, selective hydrogenation can also be used to maximize the production of *trans* double bonds. Fatty acids with *trans* configuration do not function as essential fatty acids. The geometric configuration of the double bond has a pronounced effect on the melting point of the fatty acid. Technically, in the production of household shortening the manufacturer can maximize the production of *trans* isomers if he so desires.

Molecular rearrangement is a process used to alter the physical characteristics of an oil by redistributing the fatty acids on the glycerol molecule. Normally, fatty acids are attached selectively on the glycerol molecule, with the more unsaturated molecule taking the center position. Sodium methoxide is the usual catalyst, and temperatures of approximately 100°C are used to redistribute the fatty acids in a random manner on the glycerol molecule (Weiss, 1970). In the process, the fatty acids are transferred from one position to another without any other change. In the natural state, lard has a very narrow temperature range where it has good consistency for home use. However, interesterified lards have satisfactory consistency over a wide temperature range. This is due to the changed crystalline structure of the solidified interesterified lard.

Interesterification is more often applied to a mixture of two different oils, for example, coconut oil with large amounts of lauric acid and another vegetable oil containing predominantly palmitic, stearic, and oleic acids. These oil mixtures can be further altered by partial hydrogenation, which produces a fat that has the physical characteristics for a specific need. If the two oils are coconut and soybean, the resulting oil is neither coconut nor soybean but a new base oil that can be used for margarines, shortenings, or special hard butters.

Bleaching is the process used to remove undesirable color from an oil. Bleaching of the crude oil is achieved by using bleaching clays, also known as fuller's earth, or activated carbon, which is more expensive (Lantondress, 1967; Rich, 1970). The chemistry by which the pigments and other colored matter are removed is not clearly understood. Some oils such as cottonseed contain unbleachable pigments, and poor-quality oils may be unbleachable. A final bleaching is sometimes used to remove color that has developed during processing.

In countries where rapeseed oil is used, the oil is sometimes completely hydrogenated to produce a saturated long-chain fatty acid glyceride with a high melting point. This product can be used in a much smaller amount (1–2%) than the shorter-chain saturated glycerides to impart the same degree of firmness to plastic products (Seiden, 1967). Some margarines contain a large amount of *trans* isomers to impart firmness. Special procedures are available to promote the formation of *trans* isomers (Kurucz-Lusztig, *et al.*, 1974), with the resulting product containing approximately 30% *trans* isomers (Mijanders and Lincklaen, 1971). Techniques are available to prepare shortenings with high all-*cis* configuration to replace the commonly high *trans*-content shortening. High polyunsaturated margarines can be prepared by blending a highly polyunsaturated oil with a hydrogenated vegetable oil to produce a margarine with 25–40% linoleic acid with a polyunsaturated/saturated ratio of approximately 2 (Gooding, 1963). Highly polyunsaturated margarines can be prepared by using safflower oil and hardened peanut oil. Very highly polyunsaturated soft margarines can be prepared by transesterification of 90 parts safflower oil with a high stearine oil (McNaught, 1973). Thus the techniques are available for preparing highly nutritious margarines, fats, and oils that are tailored for technical purposes (Babayan, 1975; Potter, 1968).

7.3.2. Additives

Emulsifiers are usually added to margarines and shortenings to facilitate their blending with other ingredients in food preparation (Lauridsen, 1976). Monoglycerides and diglycerides are most useful because the hydroxyl group is water soluble and the fatty acid group is lipid soluble, thus producing a stable oil-and-water suspension. Lecithin is a phospholipid usually obtained from soybean oil; it is used as an emulsifier. Propylene glycol esters, lactated monoglycerides, and polysorbate esters are also used as emulsifiers, particularly in shortening for commercial use. Petrowski (1975) has compiled a list of food-grade emulsifiers.

Stabilizers, when used in fats and oils, are added to prevent or delay the development of rancidity associated with oxidation. Propyl gallate, butylated hydroxyanisole, and butylated hydroxytoluene are antioxidants that are widely used. The antioxidants have a synergistic effect, and a combination of antioxidants is more effective than a single antioxidant present at the same level (Weiss, 1970; Sherwin, 1972). A new antioxidant, 6-hydroxy-2,5,7,8-tetramethylchroman-2-carboxylic acid, is under development but has not been approved for food use (Cort *et al.*, 1975*a*, *b*).

The silicones are antifoam agents which cause the bubbles of moisture formed during the frying of moist foods to break and thus prevent splattering and boiling over. The silicones are most useful in continuous operations where an inhibitor such as lecithin, oxystearin, or polyglycerol esters of fatty acids is added to increase the length of time before crystals appear in a salad oil at refrigeration temperatures. Citric or phosphoric acid is frequently used as a chelating agent during processing to remove trace amounts of metals that would catalyze oxidation of the finished product. Preservatives such as salt, vinegar, spices, sodium benzoate, benzoic acid, and potassium sorbate added to a food product inhibit bacterial spoilage. Finally, pigments and flavors are added to give products such as margarines their characteristic color and taste.

7.3.3. Mayonnaise and Salad Dressing

Mayonnaise and salad dressing are emulsified semisolid foods prepared from vegetable oils, with mayonnaise containing not less than 65% by weight of vegetable oil (CFR 25.1) and salad dressing not less than 30% (CFR 25.3) (Code of Federal Regulations, 1975). Mayonnaise must contain egg yolk and egg white, for they are the major emulsifying compounds in this food. A typical mayonnaise contains 77–82% salad oil, 5.3–5.8% fluid egg yolk, and 2.8–4.5% vinegar (Weiss, 1970). Salad dressing must contain not less than 30% by weight of vegetable oil and not less than 4% by weight of liquid egg yolks. Typically, a salad dressing contains 35–50% salad oil, 4–4.5% liquid egg yolks, 9–12.0% vinegar (100 grain), and 9–12.5% sugar. Proposed amendments to CFR 25.1 and 25.3 (Code of Federal Regulations, 1975) would require label declaration of ingredients. The proposed changes would allow the use of functional classes of safe and suitable ingredients that would not modify the fundamental characteristic properties of mayonnaise and salad dressing. These changes would allow more flexibility in mayonnaise and salad dressing formulation, but the prohibition against the use of any spice, except saffron or turmeric, which imparts color simulating that provided by egg yolk would still be maintained. Mustard and paprika are permitted, but they can be used only in small amounts because of flavor or color limitations.

French dressing is the separable liquid or emulsified viscous food prepared from vegetable oils; it contains not less than 30% by weight of vegetable oil. The standard of identity for French dressing (CFR 25.2) (Code of Federal Regulations, 1975) has already been amended and requires label declarations of the ingredients used. A good French dressing contains 55–65% oil.

A large variety of nonstandard dressings such as Thousand Island, Russian, Blue Cheese, and Caesar are on the market.

7.4. Source Oils

Vegetable oils are the usual source of oil for margarines, cooking oils, shortenings, salad oils, special products such as cheese spreads, and potato chip-like products. The data are tabulated in the subsequent tables according to the fatty acid composition in g/100 g fat. As part of its responsibility for compiling United States tables of food composition, the Nutrient Data Research Group of the U.S. Department of Agriculture conducted an exhaustive search of the literature spanning the years from 1960 to 1975 for data on fatty acids in foods. Data were also collected from private industry, university, and government laboratories. Ramifications of the search and criteria for comprehensive evaluation of data for purposes of developing fatty acid tables were discussed by Kinsella *et al.* (1975).

Tables of fatty acid composition in this chapter were adapted from more extensive tabulations of the fatty acid content of foods that were developed from the new body of data and which are being published by a number of authors in a series of articles entitled "Comprehensive Evaluation of Fatty Acids in Foods" (Posati *et al.*, 1975*a*, *b*; Fristrom *et al.*, 1975; Brignoli *et al.*, 1976; Exler *et al.*, 1975; Weihrauch *et al.*, 1977). In some of the tables involving the fatty acid content of various commercial products, unpublished analytical data from the laboratory of two of the authors (A. J. S. and J. L. I.) are included. The tabulations include a one-standard-deviation range for each fatty acid when sufficient data were available to calculate a valid value.

Table IIIA,B is a compilation of the fatty acid composition of seed oils that are used at a significant level in North America, with the emphasis on the oils of United States commerce. There is considerable natural variation in oils; thus a one-standard-deviation range for each fatty acid is provided whenever sufficient data were available. The one-standard-deviation range theoretically should encompass two-thirds of the oils one would encounter. This is a newer approach to tabulating nutrient composition data and is necessary to emphasize the natural variation that actually occurs. In addition to the natural variation, analytical variation is a portion of the one-standard-deviation figure; it is virtually impossible to separate these two values in the literature data available.

7.4.1. Soybean

The soybean (*Glycine max*) is a native of Eastern Asia and has been grown in North America in quantity only since 1930 (Wolf and Cowan, 1971). Its use has expanded rapidly in the last 35 years so that it has become the major oil seed in North America and also in the world. Soybeans now supply more than half of the "visible fats" (Wolf and Cowan, 1971) consumed in the United States.

Plant geneticists and processors have worked together to change the composition of soybean oil to overcome problems associated with its high linolenic acid content (Wolf and Cowan, 1971; Lundberg, 1972; Howell, 1972). The principal problems of undesirable odor and flavor and development of rancidity through autoxidation have been overcome. This is accomplished by hydrogenating, bleaching, and deodorizing soybean oil and markedly changing its original fatty acid composition (Lundberg, 1972). Linolenic acid (18 : 3) is changed from 6–10% to less than 1%, linoleic acid is changed from 45–55% to approximately 35%, and oleic acid is changed from 20–30% to approximately 50%. In mild hydrogenation, there is little change in the stearic acid content. However, varying amounts of *trans* isomers formed during hydrogenation are present in the unsaturated acids. Lundberg (1972) is not overly concerned with the presence of these unnatural fatty acids if they are kept at low levels.

7.4.2. Cottonseed

Cottonseed (*Gossypium*) was the major vegetable oil until recent years when it was supplanted by soybean oil. Currently, the availability of cottonseed for cottonseed oil production is limited by the market for cotton fiber. Cottonseed oil contains the cyclopropenoid acids (malvalic and sterculic), which are undesirable in the diet of laying hens since they may produce eggs with pink whites and other abnormalities (Evans *et al.*, 1967). Special procedures have been developed to eliminate these undesirable responses during processing (Rayner *et al.*, 1966; Carter, 1972). New glandless cottonseed varieties have been developed (Watts, 1965). Thus the inherent difficulties with cottonseed oils have been solved.

7.4.3. Peanut

The peanut (*Arachis hypogaea*) is a legume, closely related to beans and peas, that flowers above ground; the resulting pod becomes buried in the soil where the seed develops. It grows in warm and tropical countries, with Africa, India, and the People's Republic of China accounting for 75% of the world production (Wilson, 1972). In these countries, the peanut is grown primarily for its oil content, although the protein is used for animal feed or fertilizer. In the United States, peanuts are grown for consumption both as the nut and as peanut butter, and for use in confectionery products. Low-grade nuts are used for oil production. Peanut oil is difficult to winterize and therefore is not used in salad dressings. Peanut oil is used largely in margarines, cooking oils, and commercial frying oils. It is difficult to attach significance of fatty acid compositional differences to the different varieties of peanut because specific genotypes and geographic and seasonal variations are also important (Young *et al.*, 1974; Holaday and Pearson, 1974).

7.4.4. Corn

Corn (*Zea mays*) oil is a by-product in the production of cornstarch. Corn oil is high in polyunsaturates and in good demand for use in margarines and cooking oil. Since corn is not grown primarily for its oil content, the corn oil supply is fairly constant and cannot be readily expanded. Although there are various genotypes of corn (Jellum, 1970; Thompson *et al.*, 1973), the fatty acid composition of commercial corn oil is remarkably constant (Beadle *et al.*, 1965). Since the major asset of corn oil is its high linoleic acid content, the product is usually not hydrogenated but may be mixed with other oils in margarines, mayonnaise, and cooking oils.

Table IIIA. *Fatty Acid Composition (g/100 g fat) of Selected Vegetable Oils*[a,b]

Fatty acid	Soybean ($n = 29$)	Cottonseed ($n = 31$)	Corn ($n = 17$)	Safflower ($n = 37$)	Safflower (high oleic) ($n = 6$)	Sunflower: Northern ($n = 16$)	Sunflower: Southern ($n = 17$)
10:0		0.5					
12:0	0.1	0.4					
14:0	0.2 ± 0.1	0.8 ± 0.2		0.1		0.1	
16:0	10.7 ± 1.2	22.0 ± 2.9	10.7 ± 0.8	6.4 ± 1.0	4.9 ± 0.5	5.8 ± 0.6	6.3 ± 1.3
18:0	3.9 ± 0.6	2.2 ± 0.7	1.7 ± 0.2	2.5 ± 1.0	1.5 ± 0.3	4.1 ± 1.3	3.9 ± 0.8
20:0	0.2 ± 0.1	0.2	0.3 ± 0.2	0.5 ± 0.3		0.3	0.7 ± 0.8
Total saturated	15.1	26.1	12.7	9.5	6.4	10.3	10.9
16:1	0.3	0.8 ± 0.5	0.1 ± 0.1	0.6		0.1	0.2
18:1	22.8 ± 2.7	18.1 ± 2.6	24.6 ± 1.7	11.9 ± 1.9	76.9 ± 2.6	21.7 ± 6.1	34.0 ± 8.9
18:2	50.8 ± 2.1	50.3 ± 3.6	57.3 ± 2.5	73.3 ± 2.2	12.3 ± 2.4	66.4 ± 5.7	49.4 ± 8.0
18:3	6.8 ± 1.6	0.4 ± 0.5	0.8 ± 0.3	0.5 ± 0.3		0.3 ± 0.2	0.3 ± 0.2
Total unsaturated	80.7	69.6	82.8	86.3	89.2	88.5	83.9

[a] Minor fatty acids of less than 0.1 g were not included.
[b] Mean and standard deviation of n analyses.

Table IIIB. Fatty Acid Composition (g/100 g fat) of Selected Vegetable Oils[a,b]

Fatty acid	Sesame ($n = 42$)	Olive ($n = 30$)	Peanut ($n = 15$)	Rapeseed: Zero erucic acid ($n = 43$)	Rapeseed: Low erucic acid ($n = 25$)	Rapeseed: Medium erucic acid ($n = 35$)	Rapeseed: High erucic acid ($n = 71$)
12:0	0.3 ± 0.1						
14:0	0.1 ± 0.1		0.1 ± 0.1		0.1		0.1
16:0	9.4 ± 2.9	11.5 ± 1.8	9.5 ± 0.9	4.8 ± 0.7	3.3 ± 0.6	2.9 ± 0.6	2.5 ± 0.5
18:0	4.8 ± 1.6	2.3 ± 0.5	2.3 ± 0.7	1.5 ± 0.6	1.3 ± 0.3	1.1 ± 0.2	0.9 ± 0.2
20:0	0.6 ± 0.4	0.4 ± 0.2	1.4 ± 0.7	0.6 ± 0.4	0.7 ± 0.2	0.7 ± 0.1	0.7 ± 1.0
Total saturated	15.2	14.2	17.3[c]	6.9	5.4	4.7	4.2
16:1	0.3 ± 0.2	1.0 ± 0.4		0.5 ± 0.3	0.3 ± 0.2	0.3 ± 0.1	0.4 ± 0.3
18:1	39.1 ± 3.5	71.5 ± 3.4	45.6 ± 6.0	53.2 ± 7.8	30.1 ± 6.4	17.0 ± 3.9	11.2 ± 2.0
18:2	40.0 ± 4.1	8.2 ± 1.9	31.0 ± 5.9	22.2 ± 6.0	17.5 ± 3.0	14.1 ± 1.5	12.8 ± 1.2
18:3	0.5 ± 0.3	0.7 ± 0.4		11.0 ± 2.0	9.5 ± 1.8	8.8 ± 1.2	8.6 ± 1.3
20:1	0.2 ± 0.1		1.2 ± 0.4	1.0 ± 0.6	10.7 ± 2.5	10.7 ± 1.4	7.3 ± 1.4
22:1	0.4 ± 0.2			0.2 ± 0.3	19.2 ± 6.6	36.8 ± 4.4	48.1 ± 3.0
Total unsaturated	80.5	81.4	77.8	88.1	87.3	87.7	88.4

[a] Minor fatty acids of less than 0.1 g were not included.
[b] Mean and standard deviation of n analyses.
[c] Includes about 2.7 g 22:0 and 1.3 g 24:0.

7.4.5. Safflower

Safflower (*Carthamus tictorius*) oil has come into use in the last 20 years as a dry-land crop which benefits from irrigation. Plant scientists have altered the basic nature of the plant through genetic manipulation to markedly increase the oil content of the seed and drastically reduce the amount of hull (Kneeland, 1966). The most important aspect of safflower oil is its high linoleic acid content (approximately 80%). Plant geneticists have observed that the proportions of linoleic and oleic acids are due to a single gene and that the proportions of these two acids can be reversed from 79% linoleic and 12% oleic to 15% linoleic and 80% oleic (Purdy and Campbell, 1967). There are now two commercial safflower oils, one high in linoleic and the other high in oleic acid content (Applewhite, 1966). Geneticists are now developing intermediate linoleic and high stearic acid genotypes (Knowles, 1972).

7.4.6. Sunflower

The sunflower (*Helianthus annuus*) is a leading world oilseed crop, and the oil is economically more important than its protein (Robertson, 1972; Trotter and Givan, 1971; Gandy, 1972). Sunflower oil is commercially more important in Canada and Russia than in the United States since it is drought resistant and grows well in cool climates. It is now making the transition from use as bird feed and a specialty item for human consumption to use as food oil. The fatty acid composition of sunflower oil depends on variety, planting conditions, and climatic conditions during the growing season (Robertson, 1972). The oil of the crop grown in the northern United States contains 65–75% linoleic acid, whereas that of the crop grown in the South has a 30–60% linoleic acid content.

7.4.7. Rapeseed

Rapeseed grows well in cooler climates and is a major crop in Eastern and Northern Europe and Canada, Argentina, and Chile. The two main crops are *Brassica napus*, commonly known as rape or colz, and *Brassica campestris*, which is also known as turnip rape (Downey, 1971). Agronomists and plant geneticists have succeeded in increasing the yield per acre by 50% and in altering the fatty acid composition of the oil. Rapeseed oil differs from other vegetable oils in having erucic acid, a C_{22} monounsaturated acid, present up to about 50%. Nutritional research has firmly established that when erucate oils are fed as the only oil there is reduced food consumption and retarded animal growth (Sheppard *et al.*, 1971; Rocquelin *et al.*, 1971). Rakow and McGregor (1973) discovered that the erucic acid content is controlled by two genes with no genetic dominance. Thus it is possible to develop zero-erucic-acid rapeseed oils. The Canadians have developed a zero-erucate rapeseed oil, and their commercial production is now shifting to these low-erucate oils. Additional genetic effort is now continuing in an attempt to modify the levels of the C_{18} unsaturated fatty acids in rapeseed oil.

7.4.8. *Other Cruciferae*

Utilization of Cruciferae seed oils is not limited to rapeseed oils. Appelqvist (1968, 1970, 1971) in Sweden has characterized white mustard (*Sinapis alba*), brown mustard (*Brassica nigra*), and false flax (*Camelina sativa*), which are oilseeds of some current commercial importance or of potential usage. There is worldwide interest in more extensive utilization of high-erucate oils for potential industrial uses (Nieschlag and Wolff, 1971; Mikolajczak *et al.*, 1961).

7.4.9. *Olive*

Olive (*Olea erupea*) oil is primarily an oil of the Mediterranean countries. The traditional virgin olive oil is pressed from the fleshy part of the fruit and not from the pit. High pressures are not used, and the oil is not deodorized or further refined. This highly flavored and prized olive oil is difficult to obtain since more efficient procedures for extracting olive oil are now in use, including solvent extraction procedures. The consumer now receives "pure" olive oil extracted from the lower-grade olives and olive pits. Although rather extensive studies using modern instrumental techniques have been made to establish the composition of olive oil, these studies have not led to satisfactory criteria for distinguishing virgin from "pure" olive oil (Cucurachi, 1967; Eisner *et al.*, 1965; Colakoglu, 1969; Tiscornia and Bertini, 1972; Sutton *et al.*, 1966).

7.4.10. *Sesame*

The sesame seed (*Sesamum indicum*) is used in baked goods and confectionery products. The seed oil is a good salad oil requiring no winterization (Lyon, 1972; Yermanos *et al.*, 1972). Sesame is an important crop in many countries. It is a minor crop in this country, largely because of difficulties in harvesting; importation of the seed and oil is rapidly increasing. The oil contains natural antioxidants that make it more stable than other oils. The minor components are sesamin (0.4–1.1%), sesamolin (0.3–0.6%), and traces of sesamol. Sesamol is a very effective antioxidant.

7.4.11. *Cocoa*

Cocoa butter is a vegetable butter which is hard at room temperature and melts rapidly at temperatures slightly below body temperature. Cocoa butter is unique in possessing a pleasing characteristic odor and a sharp melting point. Traditionally and legally in the United States, cocoa butter is the fat obtained from roasting sound *Theobroma cacao* beans, removing the shell, breaking the nibs, and pressing the fractured nibs to remove a portion of the oil (Code of Federal Regulations, 1975). In some other countries, the whole bean including the shell can be used, and solvent extraction procedures are allowed. To facilitate international trade, it is hoped that these differences can be resolved. Cocoa butter is a unique fat in that it contains palmitic acid, stearic acid, and oleic acid in nearly equal amounts, with oleic acid occupying the center position on the triglyceride molecule (Woidich *et al.*, 1964; Doro and Remoli, 1965; Wijngaarden *et al.*, 1968; Iverson *et al.*, 1969). Substitutes for cocoa butter are illipe butter and replacement fats such as coberine obtained by fractionation of other fats (Landman *et al.*, 1961; Feuge *et al.*, 1973).

7.4.12. Palm and Palm Kernel

The use of palm oil and palm kernel oil is rapidly increasing with increased demand for oil and improved agricultural practices. Palm oil is obtained from the fruit of the palm tree, *Elaeis guiniensis*, a species commonly divided into three varieties depending on the amount of shell, pulp, and kernel. The *dura* has a thick shell, the *tenera* has a thin shell, and the *pisifera* has no shell. Two distinctly different oils are obtained from the fruit (*Congopalm*, 1970; Guam, 1972; Litchfield, 1970), depending on the portion used. Palm oil contains palmitic acid and is obtained from the fleshy portion of the fruit after removal of the shell and kernels. Palm oil is used in margarines and shortenings, while palm kernel oil is used as a hard butter or special confectionery fat (Pritchard, 1969). Palm kernel oil is obtained from the nut or kernel and contains predominantly lauric acid. To produce palm kernel oil, the shell and the fleshy part are removed, and the oil is extracted from the kernel.

7.4.13. Coconut

The coconut palm (*Cocos nucifera*) originated in Melanesia and is now grown in most of the tropics. Fresh coconut is an important item in tropical diets. The nut is split and dried, and the dried meat, copra, is a major export commodity containing 57–75% oil (Cornelius, 1973). The oil is extracted by crude stamp presses in undeveloped areas and by expellers and solvent extraction procedures in modern facilities. Coconut oil contains 44–51% lauric acid. The babasu palm (*Orbignya martin*) in Brazil and the cohune palm (*Orbignya cohune*) in Central America contain similar levels of lauric acid. Coconut oil is used as a cooking oil in tropical countries, where it is liquid. In colder areas where the oil is solid, it is used as a shortening and in margarine manufacture. When the oil is cooled slowly, it separates into liquid and solid fractions. When the solid fraction is partially hydrogenated, it is used as a cocoa butter substitute or for confectionery coatings and as a shortening in bakery products. About 1 billion pounds of lauric acid oils were imported into the United States during 1968, with about half of the poundage being used for nonfood items, primarily in soaps and in lauric acid derivatives such as synthetic alcohol used in manufacturing detergents (Sonntag, 1969). Although large amounts of lauric acid oils are imported, it is difficult to determine precisely how much is used in foods and in what foods it appears.

7.5. Confectionery Products

Confectionery fats are highly specialized products possessing specific melting and organoleptic qualities that the average fat or oil cannot fulfill (Paulicka, 1976). The fatty acid compositions of the more commercially important confectionery fats are summarized in Table IV.

Table IV. Fatty Acid Composition (g/100 g fat) of Confectionery Fats[a,b]

Fatty acid	Cocoa butter ($n = 158$)	Coconut ($n = 24$)	Palm kernel ($n = 15$)	Palm ($n = 20$)	Illipe ($n = 4$)	Shea nut ($n = 4$)	Coberine ($n = 1$)
6:0		0.7 ± 0.9	0.2 ± 0.1				
8:0		7.5 ± 1.6	3.3 ± 1.6		0.2	0.2	
10:0		5.9 ± 0.9	3.8 ± 1.1		0.2	0.2	
12:0		43.7 ± 3.1	46.4 ± 3.8	0.2 ± 0.1	1.0	1.4	1.2
14:0	0.1 ± 0.1	16.4 ± 1.7	16.2 ± 1.8	1.1 ± 0.4	0.3	0.3	0.9
16:0	25.4 ± 1.5	8.2 ± 1.8	7.8 ± 1.3	42.0 ± 4.8	16.3 ± 1.6	4.4 ± 0.9	36.3
18:0	33.2 ± 1.2	3.0 ± 1.8	2.7 ± 1.3	4.3 ± 1.5	43.4 ± 2.7	39.0 ± 3.2	26.3
20:0	0.9 ± 0.2	0.9 ± 0.2	1.0	0.3 ± 0.1	1.6 ± 0.2	1.2 ± 0.1	1.0
Total saturated	59.6	86.3	81.4	47.9	63.0	46.7	65.7
16:1	0.4 ± 0.3	0.4 ± 0.3		0.5 ± 0.5	0.2	0.1	
18:1	32.6 ± 1.5	5.7 ± 1.7	11.4 ± 3.4	37.9 ± 6.9	30.9 ± 1.4	43.7 ± 2.7	29.5
18:2	2.8 ± 0.6	1.8 ± 0.9	1.5 ± 0.6	9.0 ± 2.3	1.5 ± 1.2	4.9 ± 1.1	0.5
18:3	0.2 ± 0.1			0.3 ± 0.2	0.1	0.3	
Total unsaturated	36.0	7.9	12.9	47.7	32.7	49.0	30.0

[a] Minor fatty acids of less than 0.1 g were not included.
[b] Mean and standard deviation of n analyses.

Cocoa butter is used in the manufacture of confectionery products, for it imparts pleasing melting characteristics that enhance the organoleptic properties of candies. It is a light yellow fat with a sharp melting point of approximately 30°C, and it is not further refined as most oils are before use. The dominant triglyceride of cocoa butter contains oleic acid in the center (β) position and the saturated acids stearic and palmitic in the α and α' positions (ends) of the glycerol molecule. There are four polymorphic forms of crystalline cocoa butter; the most stable form melts at 34–35°C (Minifie, 1970). Few other fats can be mixed with cocoa butter and yield a properly crystallized product.

Considerable effort has been spent on the development of cocoa butter substitutes. Illipe (*Madhuca longifolia*) butter is similar to cocoa butter in fatty acid composition (see Table IV). However, the triglyceride composition of illipe butter is different, and when it is used in appreciable quantities the crystalline characteristics of the product are altered. The lauric acid oils such as coconut and palm kernel can be hydrogenated for a soft type of replacement. Fractions obtained from winterizing and fractionating other oils such as cottonseed oils (Feuge *et al.*, 1973) and edible tallow (Luddy *et al.*, 1973, 1975) or fractions obtained from palm oil have been proposed as replacements (Padley *et al.*, 1972). However, most fats sold as replacements for cocoa butter can be used only in limited quantities, for they change the physical properties of the product and cause it to deteriorate more rapidly than the traditional product. Lauric acid oils tend to develop off-flavors and odors when stored.

The triglyceride molecule can be rearranged or interesterified and then

Table V. The Fatty Acid Composition

Fatty acid	Almond ($n = 15$)	Beechnut ($n = 1$)	Brazil ($n = 7$)	Cashew ($n = 8$)	Chestnut, European ($n = 2$)	Filbert ($n = 19$)
12:0				1.9 ± 0.0		
14:0	0.1 ± 0.0		0.2 ± 0.2	0.8 ± 0.0		0.3 ± 0.1
16:0	6.1 ± 1.0	8.6	14.9 ± 1.0	10.3 ± 0.6	15.5	4.9 ± 1.1
18:0	1.7 ± 0.5	1.5	10.4 ± 1.1	7.0 ± 1.2	0.9	1.8 ± 1.1
20:0	0.2 ± 0.2	3.4		0.7 ± 0.3	0.5	0.2 ± 0.0
Total saturated	8.1	13.5	25.5	20.2[c]	16.9	7.2
16:1	0.6 ± 0.4		0.4 ± 0.2	0.5 ± 0.3	0.7	0.3 ± 0.1
18:1	67.7 ± 2.6	39.1	32.5 ± 3.2	55.8 ± 2.7	35.8	76.8 ± 3.4
18:2	18.2 ± 2.0	40.1	37.3 ± 4.0	17.0 ± 1.7	35.1	10.2 ± 3.5
18:3	0.5 ± 0.4			0.4 ± 0.1	3.9	0.2 ± 0.1
Total unsaturated	87.0	79.2	70.2	73.8	75.5	87.6
Total fat, g/100 g nut	53.9	50.3	68.2	45.6	2.7	64.7

[a] Minor fatty acids of less than 0.1 g were not included.
[b] Mean and standard deviation of n analyses.
[c] Contains about 0.3 g each of 6:0 and 8:0.
[d] Contains 0.2 g 20:4.

hydrogenated to a particular point to produce a product with a more desirable melting characteristic for confectionery use than the original oil. The trend now is to produce hard butters for particular uses such as chocolate-type coatings or for use in candy products that are used by themselves and are not substitutes for the traditional products. Compound coatings made with hard butter as the only source of fat yield more uniform products. The problems of compatibility with cocoa butter do not arise, and the special tempering conditions required for cocoa butter are not necessary.

7.6. Nuts

Nuts are an important item in some diets because they are high in lipid content and are a significant source of oil. The oils removed from various nuts, e.g., peanut oil, are discussed in Section 7.4. This discussion is limited to nuts *per se*. The best data available on lipid content and fatty acid composition of nuts, compiled by Fristrom *et al.* (1975), are presented in Table V. The common varieties used in North America contain 46–76 g fat/100 g edible kernel. Compositional data on coconut, palm kernel, and other confectionery fats are given in Table IV. The saturated fatty acids predominate in these tree nuts, ranging from 47% to 86%. Conversely, the almond, Brazil nut, pecan, black walnut, and English walnut are rich in polyunsaturates.

(g/100 g fat) of Selected Nuts[a,b]

Macadamia ($n = 5$)	Pecan ($n = 15$)	Pilinut ($n = 1$)	Pine ($n = 6$)	Pistachio ($n = 25$)	Walnut, black ($n = 4$)	Walnut, English ($n = 19$)
				0.1		
1.1 ± 0.5			0.1	0.2		1.1 ± 1.1
9.1 ± 1.6	5.9 ± 1.0	27.7	7.9 ± 2.5	11.7 ± 2.1	5.7 ± 3.3	7.1 ± 2.1
2.1 ± 0.9	2.6 ± 0.6	11.5	3.3 ± 0.4	1.3 ± 0.7	2.9 ± 1.6	2.1 ± 0.5
2.0 ± 0.2			0.8 ± 0.4	0.5 ± 0.3		0.7 ± 0.7
14.3	8.5	39.2	12.1	13.8	8.6	11.0
18.6 ± 1.4	0.3 ± 0.1		0.3 ± 0.1	0.5 ± 0.3	0.2	0.2 ± 0.1
57.1 ± 2.3	60.1 ± 7.2	46.8	37.3 ± 2.2	67.2 ± 5.4	17.9 ± 5.9	15.3 ± 2.3
1.5 ± 0.2	23.7 ± 6.6	9.6	43.5 ± 2.4	12.6 ± 4.7	61.7 ± 10.5	55.0 ± 4.2
1.4 ± 0.0	1.2 ± 0.3		0.7 ± 0.1	0.5 ± 0.2	6.7 ± 3.1	10.9 ± 3.8
78.6	85.5[d]	56.4	81.8	80.9	86.5	81.4
75.7	71.4	63.0	51.0	53.6	59.6	63.4

Table VI. Fatty Acid Composition (g/100 g fat) of Selected Household and Commercial Cooking and Salad Oils[a,b]

Fatty acid	Household: Soybean ($n = 38$)	Household: Soy–cottonseed blend (80%:20%) ($n = 6$)	Commercial: Soybean ($n = 3$)	Commercial: Soy–cottonseed blend ($n = 7$)	Commercial: Sunflower ($n = 4$)
16:0	9.8 ± 1.1	13.7 ± 2.6	8.9 ± 0.6	10.5 ± 1.0	7.1 ± 0.4
18:0	5.0 ± 1.8	4.1 ± 0.9	5.8 ± 2.6	11.7 ± 4.1	5.5 ± 0.6
Total saturated	14.8	17.8	14.7	22.2	12.6
16:1	0.3 ± 0.4	0.3 ± 0.1			
18:1	43.5 ± 2.5	29.1 ± 4.4	59.6 ± 1.4	32.2 ± 8.0	46.0 ± 0.2
18:2	33.6 ± 3.1	44.9 ± 2.8	20.4 ± 1.9	37.3 ± 6.1	35.3 ± 0.7
18:3	2.5 ± 0.9	2.8 ± 0.6	0.9	4.0 ± 1.6	0.9 ± 0.2
Total unsaturated	79.9	77.1	80.9	73.5	82.2

[a] Minor fatty acids of less than 0.1 g were not included.
[b] Mean and standard deviation of n analyses.

7.7. Cooking Oils

Cooking oils may be processed to enhance their appearance, taste, shelf life, and stability during use. This is accomplished, as previously described, by bleaching, winterizing, deodorizing, and alkali refining. The enumerated processing techniques have little effect on the fatty acid distribution, and therefore the data shown in Tables IIIA and IIIB apply to unhydrogenated cooking oils.

When oils are hydrogenated and/or blended, the fatty acid distribution is quite different from that of the source oils (Table VI). The content of linoleic and linolenic acids is lower and the content of oleic and stearic acids is frequently higher than that of unhydrogenated oils. Products developed for the retail market are more polyunsaturated than corresponding industrial products. Blending combines desirable features from two or more oils. The most common blends contain hydrogenated soybean oil and cottonseed oil.

7.8. Shortenings

At present, there are no requirements for labeling of source oils for shortenings; however, steps are presently being taken by the Food and Drug Administration to rectify this matter. Shortenings are usually designed for specific purposes, as illustrated in Table VII. Manufacturers of edible oils still rely on characteristics such as iodine number, solid-fat index, melting range, and crystal structure to monitor the

Table VII. Fatty Acid Composition (g/100 g fat) of Selected Household and Commercial Shortenings[a,b]

Fatty acid	Household		Commercial			
	National brand[c] (n = 7)	Store brand[c] (n = 16)	Lauric oil from whipped toppings, nondairy products, coffee creamers (n = 3)	Animal–vegetable shortening for heavy-duty frying (n = 8)	Animal–vegetable shortening for pie crust (n = 3)	Soy/palm shortening for cake mix (n = 4)
16:0	13.8 ± 1.5	13.2 ± 2.3	10.7 ± 1.0	22.2 ± 1.6	19.2 ± 2.1	40.6 ± 3.6
18:0	10.8 ± 0.7	11.1 ± 1.6	21.9 ± 9.3	20.8 ± 2.4	14.6 ± 0.4	6.9 ± 1.5
Total saturated	24.6	24.3	91.0[d]	45.9[e]	35.0	47.5
16:1				4.3	1.4 ± 0.1	
18:1	41.7 ± 2.3	56.9 ± 5.2	2.2 ± 0.6	35.9 ± 1.9	38.0 ± 3.4	40.6 ± 3.4
18:2	26.6 ± 1.8	11.0 ± 5.5	1.0	5.7 ± 3.6	18.1 ± 1.2	7.5 ± 4.4
18:3	2.2 ± 0.8	0.6 ± 0.6		1.0 ± 0.5	1.0 ± 0.5	
Total unsaturated	70.5	68.5	3.2	47.8[f]	59.1	48.1

[a] Minor fatty acids of less than 0.1 g were not included.
[b] Mean and standard deviation of *n* analyses.
[c] All-purpose, soy–cottonseed (80% : 20%).
[d] Contains 8:0 = 4.2 ± 1.1; 10:0 = 4.3 ± 1.1; 12:0 = 35.8 ± 5.7; 14:0 = 14.4 ± 2.5.
[e] Contains 14:0 = 2.9 ± 0.7.
[f] Contains 14:1 = 1.4.

quality of their products. This is especially true of industrial fats, which are essentially unaffected by the current consumer demand for relatively unsaturated shortenings. The great number of fat combinations and processing techniques presently available may yield final products with similar desirable physical characteristics but with wide variations in fatty acid composition.

Generally, household shortenings are made from all-vegetable oils and are more unsaturated than commercial shortenings. The latter are frequently vegetable–animal fat blends. The presence of animal fat is indicated by relatively high levels of the 16:1 fatty acid (Table VII). Heavy-duty frying oils are hydrogenated to enhance resistance to oxidation and to impart other desirable characteristics such as higher smoke and flash points.

As the amount of palmitic acid in an oil is relatively unaffected by hydrogenation, the amount of this acid in a blend of hydrogenated fats may provide a clue to the identity of the source oils. On the average, soybean oil, cottonseed oil, and palm oil contain, respectively, 10.7, 22.0, and 46.4 g palmitic acid/100 g oil. The presence of unhydrogenated soybean oil is indicated by relatively high values for linolenic acid; a high amount of lauric acid usually indicates the presence of coconut oil or a palm kernel oil.

7.9. Margarines

Margarines are, in essence, water-in-fat emulsions exhibiting the characteristics of the fat phase. There are presently three major categories of margarines on the market that comply with the Standards of Identity (Code of Federal Regulations, 1975) promulgated by the Food and Drug Administration. The standards require that stick, tub, and liquid margarines contain a minimum of 80% total fat by weight. Two relatively new additions to the margarine family have been introduced, called diet imitation margarines and low-calorie spreads, which contain 40% and 60% fat, respectively. These latter types of product currently are not included under a Standard of Identity, but such a requirement may be stipulated in the future.

Examples of popular stick, tub, liquid, and diet margarines are shown in Tables VIII and IX. Generally, tub margarines are more polyunsaturated than stick margarines made from the same source oils. Data obtained by the lipoxidase method for *cis,cis*-methylene-interrupted polyunsaturation range from 50% to 100% of PUFAs as measured by GLC. Meager data on *trans* acids reveal that *trans*-monoenoic acids contribute by far the greatest portion of total *trans* acids; the rest come from *cis-trans* isomers. Only small amounts of *trans,trans* isomers have been reported by Carpenter and Slover (1973). These investigators also reported appreciable amounts of positional isomers in margarines. The levels of the non-all-*cis* isomers in predominantly fat-type products such as margarines and oils are of considerable interest to nutritional biochemists. This interest stems from the fact that the *trans* isomers do not function in the essential fatty acid metabolic pathways; if the *trans* isomer content should rise to a high level in dietary fat, there is always the long-range worry about essential fatty acid deficiency.

Table VIII. Fatty Acid Composition (g/100 g margarine) of Selected Stick and Tub Margarines[a]

	Source of oil[b]					
	Stick			Tub		
Fatty acid	CpHC ($n = 13$)	pHSB, pHCS ($n = 35$)	CN + (P)[c] ($n = 3$)	CpHC ($n = 12$)	pHSB, pHCS ($n = 6$)	S, pHCS, pHPN ($n = 3$)
16:0	8.9 ± 1.2	8.7 ± 0.8	7.2 ± 0.2	9.2 ± 0.6	8.7 ± 0.9	8.1 ± 0.8
18:0	4.7 ± 0.8	5.9 ± 1.0	3.7 ± 0.9	4.9 ± 1.3	5.5 ± 0.9	4.9 ± 1.8
Total saturated	14.1	15.1	56.5[d]	14.5	14.7	13.4
18:1	38.2 ± 2.0	45.8 ± 5.5	8.1 ± 2.1	29.8 ± 2.8	36.9 ± 2.0	16.1 ± 3.3
18:2	22.4 ± 2.1	13.4 ± 4.3	11.2 ± 1.7	29.7 ± 3.1	22.6 ± 1.3	48.4 ± 4.8
18:3	0.4 ± 0.2	0.9 ± 0.7	0.3 ± 0.4	1.3 ± 1.0	1.4 ± 0.7	
Total unsaturated	62.4	61.4	20.0	62.4	61.9	64.7
Fat content, %	80.1	80.1	80.0	80.4	80.1	81.7

[a] Mean and standard deviation of n analyses.
[b] C, Corn; pHC, partially hydrogenated corn oil; pHSB, partially hydrogenated soybean oil; pHCS, partially hydrogenated cottonseed oil; CN, coconut oil; S, safflower oil; pHPN, partially hydrogenated peanut oil.
[c] (P) may include sunflower seed oil, hydrogenated coconut oil, hydrogenated peanut oil, hydrogenated palm oil, and liquid safflower oil.
[d] 6:0 = 0.3 g; 8:0 = 4.7 g; 10:0 = 3.2 g; 12:0 = 28.2 g; 14:0 = 7.8 g.

Table IX. Fatty Acid Composition (g/100 g product) of Liquid and Diet Margarine, Mayonnaise, and Similar Products[a]

Fatty acid	Liquid margarine[b] ($n = 5$)	Diet (tub) margarine[c] ($n = 16$)	Mayonnaise ($n = 109$)	Dietetic mayonnaise ($n = 1$)	Salad dressing (French) ($n = 33$)	Sandwich spread ($n = 8$)
16:0	8.3 ± 0.1	4.3 ± 0.1	8.0 ± 1.1	1.5	3.6 ± 1.0	4.6 ± 0.3
18:0	4.4 ± 1.1	1.8 ± 0.1	3.0 ± 0.4	0.4	1.3 ± 0.3	1.4 ± 0.2
Total saturated	12.7	6.1	11.0	1.9	4.9	6.0
16:1			2.0 ± 0.2		0.2 ± 0.1	
18:1	22.0 ± 4.7	14.3 ± 0.2	21.6 ± 2.5	2.3	8.6 ± 1.9	8.7 ± 0.4
18:2	35.9 ± 4.3	16.1 ± 0.3	36.5 ± 2.0	3.8	15.4 ± 2.6	21.3 ± 0.5
18:3	4.8 ± 1.6	0.2 ± 0.1	4.3 ± 1.1	0.8	1.7 ± 0.5	2.2 ± 0.6
Total unsaturated	62.7	30.6	64.4	6.9	25.9	32.2
Fat content, %	80.1	38.7	79.1	9.2	33.4	40.0

[a] Mean and standard deviation of n analyses.
[b] Made from partially hydrogenated soybean oil, soybean oil, and cottonseed oil.
[c] Made from corn oil and partially hydrogenated corn oil.

7.10. Dairy Products and Creamers

Milk fat is considered to be the most complex lipid mixture known. It is primarily triglyceride, containing about 500 fatty acids. Posati *et al.* (1975*a*) have critically evaluated the data from the recent literature. In considering the variables that can affect fatty acid composition, they reported that breed had little effect on composition; feeding practices affect primarily the fat content, and those that depress the content of milk fat usually increase unsaturation. Effects identified with seasonal variation cause milk fat to contain, in the summer, on the average about 25.4% of 16:0 and 30.7% of 18:1 fatty acids, while winter milk contains on the average 34.4% and 22.4%, respectively, of the two acids. They further report that the processing of milk into various products primarily affects the fat content but does not affect the fatty acid composition of the milk fat. Three cheeses were selected to demonstrate this observation (Table X). In spite of greatly differing production techniques, the fatty acid pattern remains relatively unaltered.

Analysis of milk-fat fatty acids is made difficult by the presence of large amounts of butyric and caproic acids and the relative volatility of their methyl esters, and the solubility in water of the free acids. For this reason, short-chain acids are frequently analyzed separately as the less volatile propyl or butyl esters. The composition of butterfat is presented in Table XI.

A host of non-milk-fat products that are available on the market may contain hydrogenated coconut or palm kernel oil and/or palm oil and hydrogenated soybean

Table X. Fatty Acid Composition (g/100 g fat) of Selected Cheeses[a]

Fatty acid	Blue ($n = 10$)	Cheddar ($n = 18$)	Cottage, creamed ($n = 7$)
4:0	2.3 ± 0.2	3.2 ± 0.7	3.2
6:0	1.3 ± 0.6	1.6 ± 0.6	0.7
8:0	0.9 ± 0.2	0.9 ± 0.3	0.8 ± 0.3
10:0	2.1 ± 0.4	1.9 ± 0.4	1.8 ± 0.2
12:0	1.7 ± 0.8	1.7 ± 0.8	1.6 ± 0.8
14:0	11.5 ± 1.1	10.1 ± 0.7	10.5 ± 0.8
16:0	31.9 ± 3.1	29.4 ± 3.9	30.2 ± 1.3
18:0	11.3 ± 1.7	12.2 ± 2.4	11.4 ± 1.3
Total saturated	65.0	64.6	63.3
14:1	1.2 ± 0.3	1.5 ± 0.4	1.5 ± 0.2
16:1	2.8 ± 1.2	3.1 ± 0.6	3.6 ± 0.8
18:1	23.0 ± 1.7	22.8 ± 6.1	23.4 ± 2.2
18:2	1.9 ± 0.6	1.8 ± 0.6	2.2 ± 0.5
18:3	0.9 ± 0.3	1.1 ± 0.6	0.9 ± 0.3
Total unsaturated	29.8	30.3	31.6
Total fat	29.6	32.8	4.0

[a] Mean and standard deviation of n analyses.

Table XI. Fatty Acid Composition (g/100 g fat) of Butter, Modified Milk, and Nondairy Creamers[a]

Fatty acid	Butter ($n = 88$)	Filled milk[b] ($n = 29$)	Coffee creamer[b] (powder) ($n = 8$)	Coffee creamer[b] (liquid) ($n = 18$)	Coffee creamer[c] (liquid) ($n = 7$)
4:0	3.2 ± 0.4				
6:0	1.9 ± 0.4				
8:0	1.1 ± 0.3	5.1 ± 1.5	3.8 ± 2.3	3.5 ± 1.9	0.1
10:0	2.5 ± 0.5	3.8 ± 1.3	4.1 ± 1.5	2.8 ± 1.6	0.1 ± 0.2
12:0	2.8 ± 0.6	60.2 ± 11.5	38.3 ± 3.0	66.3 ± 11.0	0.6 ± 0.8
14:0	10.1 ± 1.2	11.7 ± 4.0	16.9 ± 2.6	10.0 ± 3.7	0.2 ± 0.2
16:0	26.3 ± 3.1	5.6 ± 2.8	10.6 ± 2.1	4.3 ± 2.2	8.9 ± 2.7
18:0	12.1 ± 1.5	4.3 ± 1.9	17.9 ± 7.4	6.0 ± 3.8	9.8 ± 1.3
Total saturated	62.3	91.2	91.4	93.3	19.4
14:1	1.5 ± 0.3				
16:1	2.3 ± 0.6				
18:1	25.1 ± 2.9	2.9 ± 2.2	2.7 ± 1.3	1.1 ± 1.5	75.8 ± 3.3
18:2	2.3 ± 0.7	0.2 ± 0.5			0.2 ± 0.2
18:3	1.4 ± 0.6				
Total unsaturated	32.6	3.2	2.8	1.1	76.0
Total fat	80.1	3.4	35.6	9.2	11.2

[a] Mean and standard deviations of n analyses.
[b] Made with lauric oil.
[c] Made with hydrogenated vegetable oil.

and cottonseed oil. Examples in Table XI were taken from Posati *et al.* (1975*a*). The predominant fatty acid in creamers based on the lauric oils from palm kernel and coconut is lauric acid.

7.11. Animal Fats

By far the greatest source of fat in the human diet is of animal origin, in spite of a considerable decline in the total production of animal fats in recent years.

The fatty acid composition of lard, beef tallow, mutton tallow, and chicken fat is shown in Table XII. These fats are generally more saturated than most vegetable oils. A distinguishing feature of these animal fats is their relatively high palmitoleic acid content. Beef and mutton tallow are the least polyunsaturated fats, containing on the average 3.7 and 4.8 g linoleic acid, respectively, per 100 g fat. Lard and chicken fat are somewhat more polyunsaturated, containing an average of 10 and 16.5 g of linoleic acid, respectively.

7.12. Marine Products

The use of fish as a source of fat becomes very complex (Gruger *et al.*, 1964; Sidwell *et al.*, 1974). The fat content of the common species of fish varies extensively from less than 1% to over 20% in the edible portion. The lipid content of fat fish such

Table XII. Fatty Acid Composition (g/100 g fat) of Animal Fats[a]

Fatty acid	Lard ($n = 26$)	Beef tallow ($n = 15$)	Mutton tallow ($n = 17$)	Chicken fat ($n = 20$)
10:0	0.1 ± 0.1	0.1		0.2
12:0	0.5 ± 0.7	0.9 ± 1.2		1.0 ± 0.1
14:0	1.4 ± 0.7	3.7 ± 1.7	2.7 ± 0.5	1.2 ± 0.5
16:0	23.7 ± 1.7	24.8 ± 3.5	20.8 ± 2.3	23.8 ± 3.0
18:0	13.0 ± 2.9	18.7 ± 3.7	25.0 ± 4.5	6.4 ± 2.7
20:0	0.8 ± 0.1			
Total saturated	39.5	48.2	48.5	32.6
14:1		1.6 ± 0.3		
16:1	2.6 ± 1.9	4.7 ± 2.4	1.7 ± 1.0	5.8 ± 2.9
18:1	40.9 ± 2.5	36.0 ± 4.9	33.8 ± 4.0	39.7 ± 6.6
18:2	10.0 ± 1.6	3.7 ± 2.0	4.8 ± 2.5	16.5 ± 6.7
18:3	1.4	0.6 ± 0.6	2.7 ± 1.2	1.1 ± 0.5
Total unsaturated	54.9	46.6	43.0	63.1

[a] Mean and standard deviation of n analyses.

as whole Atlantic herring may vary from over 20% in December to as low as 3.2% in April (Stoddard, 1968), while the lipid content of whole Pacific herring varies from less than 2% after spawning to 35% in September (Iverson, 1958). However, fish as consumed can be classified by oil content as high, over 15%; medium, 5–15%; and low, under 5%. The fat content varies in different parts of the fish, with the lowest

Table XIII. Fatty Acid Composition (g/100 g fat) of Selected Marine Oils[a]

Fatty acid	Atlantic herring ($n = 9$)	Menhaden ($n = 5$)	Cod liver ($n = 35$)
14:0	7.2 ± 1.8	9.0 ± 3.3	3.4 ± 0.9
16:0	12.0 ± 2.5	19.0 ± 4.0	10.7 ± 1.3
18:0	1.1 ± 0.4	3.8 ± 1.3	2.3 ± 0.5
Total saturated	21.5	33.3	17.3
16:1	8.3 ± 1.3	13.3 ± 3.0	9.4 ± 2.5
18:1	12.0 ± 1.8	15.5 ± 6.1	22.6 ± 5.4
18:2	1.7 ± 0.5	2.0 ± 1.1	1.8 ± 1.2
18:3	0.6 ± 0.4	1.0 ± 0.4	0.7 ± 0.3
18:4	2.0 ± 1.4	2.4 ± 1.5	1.6 ± 0.7
20:1	13.9 ± 2.2	1.7 ± 0.6	10.2 ± 2.7
20:4	0.4 ± 0.3	1.0 ± 0.5	1.2 ± 0.7
20:5	4.4 ± 3.2	12.5 ± 0.5	8.9 ± 2.1
22:1	21.5 ± 3.4	0.7 ± 0.3	5.8 ± 2.4
22:5	1.0 ± 0.5	1.7 ± 0.4	1.6 ± 1.5
22:6	3.8 ± 1.6	7.9 ± 5.4	9.3 ± 3.3
Total unsaturated	73.9	62.3	77.1

[a] Mean and standard deviation of n analyses.

values in the tail portion and the highest values in the head area (Stansby, 1973; Lambertsen, 1973). High-oil-content fish such as anchovy, herring, mackerel, and salmon contribute appreciably to oil in some diets.

Fish lipids contain long-chain C_{20} and C_{22} polyunsaturates with four, five, and six double bonds. There are also large amounts of C_{16}–C_{22} monounsaturated acids. Fish flesh contains about 0.5% phospholipids; these phospholipids make a large contribution to the lipid content of lean fish. The fatty acid composition of selected commercial marine oils (Table XIII) illustrates this complexity. These commercial oils, normally not used for human food purposes in the United States, are obtained from whole herring and menhaden and extracted from the livers of cod. Extensive data on the fatty acid composition of fish have been published by Exler and Weihrauch (1976).

7.13. Summary and Trends

Fats and oils provide the most concentrated source of energy per unit weight of any dietary constituent. They are primarily esters of fatty acids and glycerol and frequently contain other chemical combinations in varying amounts. Recent interest has focused on the fatty acids to elucidate their physiological effects relative to general health. The necessity of using scientific nomenclature in describing fatty acids can not be overemphasized for clarity of scientific communication. Animal tissues, dairy products, and oil extracted from soybeans as a group contribute most of the fat in the United States diet.

Fats and oils technology is currently undergoing a dynamic change reflecting consumer demands for more highly polyunsaturated fatty acid and lower cholesterol content in the diet. This demand places a serious stress on quality control and research laboratories to develop new analytical methodology to reflect the changes occurring in food manufacturing and processing. The presently used lipoxidase method for measuring *cis,cis*-methylene polyunsaturated fatty acids needs to be replaced by new methodology that is more rapid, is more fatty acid chain-length specific, and exhibits a smaller coefficient of variation. The same general statement can legitimately be made regarding the gas chromatographic methods for fatty acids and sterols, especially cholesterol. The high-pressure liquid chromatograph is probably the instrumentation that will dominate the next generation of methods that will gradually evolve.

The fatty acid patterns of food fats and oils must be continually monitored. The selection of new strains of existing plant varieties, the introduction of new oils from plant sources not now enjoying wide usage, and the development of technology to drastically alter the composition of currently used oils may influence the total fat, sterol, and fatty acid composition and content of the American food supply. The replacement of animal fats with processed vegetable fats has resulted in a significant increase in the content of the *trans* geometric isomers in the diet. In the future, the level of the *trans* isomers may become a subject of major nutritional concern if the level becomes too high. Thus the vast quantity of fats and oils compositional data provided in this chapter are accurate at this writing but may be an inaccurate estimate of the fatty acid content in 5–10 years. Data for nuts and other relatively unprocessed foods will not be undergoing such dynamic technological evolution.

7.14. References

Acker, L., and Schmitz, H. T., 1967, Über die Lipide der Weizenstärke, *Die Stärke* **19:**233.

Allen, R. R., 1962, Practical aspects of hydrogenation, *J. Am. Oil Chem. Soc.* **39:**457.

Appelqvist, L.-A., 1968, Fatty acid composition in seeds of some Svalof varieties and strains of rape, turnip rape, white mustard and false flax, *Acta Agr. Scand.* **18:**1.

Appelqvist, L.-A., 1970, Lipids in Cruciferae. VI. The fatty acid composition of seeds of some cultivated *Brassica* species and of *Sinapis alba L.*, *Fette Seifen Anstrichm.* **72:**783.

Appelqvist, L.-A., 1971, Lipids in Cruciferae. VIII. The fatty acid composition of some wild or partially domesticated species, *J. Am. Oil Chem. Soc.* **48:**740.

Applewhite, T. H., 1966, The composition of safflower seed, *J. Am. Chem. Soc.* **43:**406.

Association of Official Analytical Chemists, 1965, Preparation of methyl esters, in: *Official Methods of Analysis*, p. 429, AOAC, Washington, D.C.

Association of Official Analytical Chemists, 1975*a*, Crude fat or ether extract, in: *Official Methods of Analysis*, p. 135, AOAC, Washington, D.C.

Association of Official Analytical Chemists, 1975*b*, Boron trifluoride method, in: *Official Methods of Analysis*, pp. 497–498, AOAC, Washington, D.C.

Babayan, V. K., 1975, Tailoring fats for nutrients in processed foods, in: *Fats and Carbohydrates* (P. L. White and D. C. Fletcher, eds.), pp. 21–37, Publishing Sciences Group, Acton, Mass.

Beadle, J. B., Just, D. E., Moser, R. E., and Reiners, R. A., 1965, Composition of corn oil, *J. Am. Oil Chem. Soc.* **42:**90.

Beal, R. E., Moulton, K. J., Moser, H. A., and Black, T. J., 1969, Removal of copper from hydrogenated soybean oil, *J. Am. Oil Chem. Soc.* **46:**498.

Bligh, E. G., and Dyer, W. J., 1959, A rapid method of total lipid extraction and purification, *Can. J. Biochem. Physiol.* **3:**911.

Brignoli, C. A., Kinsella, J. E., and Weihrauch, J. L., 1976, Comprehensive evaluation of fatty acids in foods. V. Unhydrogenated fats and oils, *J. Am. Diet. Assoc.* **68:**224.

Carpenter, D. L., and Slover, H. T., 1973, Lipid composition of selected margarines, *J. Am. Oil Chem. Soc.* **50:**372.

Carr, R. A., 1976, Degumming and refining practices in the U.S., *J. Am. Oil Chem. Soc.* **53:**397.

Carter, C. M., 1972, Projection and prospects for cottonseed in the 1970's, *J. Am. Oil Chem. Soc.* **49:**515A.

Code of Federal Regulations, 1975, Title 21.

Coenen, J. W. E., 1976, Hydrogenation of edible oils, *J. Am. Oil Chem. Soc.* **53:**382.

Colakoglu, M., 1969, Analysis of Turkish olive oils produced in 1966–1967, *Ege Univ. Ziraat Fak. Yayin.* **138:**1.

Congopalm, October 1970, *Palm Oil, A Major Tropical Product*, Societé Cooperative Avenue des Aviateurs 538 Kinshasa, selling office: Rue Belliard 35 B1040 Bruxelles, Belgium.

Cornelius, J. A., 1973, Coconuts: A review, *Trop. Sci.* **15:**15.

Cort, W. M., Scott, J. W., Araujo, M., Mergens, W. J., Cannalonga, M. A., Osadca, M., Harley, H., Parrish, D. R., and Pool, W. R., 1975*a*, Antioxidant activity and stability of 6-hydroxy-2,5,7,8-tetramethylchroman-2-carboxylic acid, *J. Am. Oil Chem. Soc.* **52:**174.

Cort, W. M., Scott, J. W., and Harley, J. H., 1975*b*, Proposed antioxidant exhibits useful properties, *Food Technol.* **29(6):**46.

Cowan, J. C., 1976, Degumming, refining, bleaching, and deodorization theory, *J. Am. Oil Chem. Soc.* **53:**344.

Cucurachi, A., 1967, On the chemico-physical characteristics and acidic composition of olive oil, *Riv. Ital. Sostanze Grasse* **44:**172.

Doro, B., and Remoli, S., 1965, Gas chromatographic researches on cocoa butter, *Riv. Ital. Sostanze Grasse* **42:**108.

Downey, R. K., 1971, Agricultural and genetic potentials of cruciferous oil-seed crops, *J. Am. Oil Chem. Soc.* **48:**718.

Dutton, H. J., Scholfield, C. R., and Mounts, T. L., 1961, Glyceride structure of vegetable oils by counter-current distribution. V. Comparison of natural, interesterified, and synthetic cocoa butter, *J. Am. Oil Chem. Soc.* **38:**96.

Eisner, J., Iverson, J. L., Moazingo, A. K., and Firestone, D., 1965, Gas chromatography of unsaponifiable matter. III. Identification of hydrocarbons, aliphatic alcohols, tocopherols, triterpenoid alcohols and sterols present in olive oils, *J. Assoc. Off. Agr. Chem.* **48**:417.

Engler, P. P., and Bowers, J. A., 1975, Methods for analyzing percentage lipid of ground beef and beef-soy blends, *J. Agr. Food Chem.* **23**:588.

Evans, C. D., Beal, R. E., McConnell, D. G., Black, L. T., and Cowan, J. C., 1964*a*, Partial hydrogenation and winterization of soybean oil, *J. Am. Chem. Soc.* **41**:260.

Evans, C. D., Oswald, J. T., and Cowan, J. C., 1964*b*, Soybean unsaponifiables: Hydrocarbons from deodorized condensates, *J. Am. Oil Chem. Soc.* **41**:406.

Evans, R. J., Bandemer, S. L., and Davidson, J. A., 1967, Compounds in cottonseed oil that cause pink-white discoloration in stored eggs, *Poult. Sci.* **46**:345.

Exler, J., and Weihrauch, J. L., 1976, Comprehensive evaluation of fatty acids in foods. VIII. Finfish, *J. Am. Diet. Assoc.* **69**:243.

Exler, J., Kinsella, J. E., and Watt, B. K., 1975, Lipids and fatty acids of important finfish: New data for nutrient tables, *J. Am. Oil Chem. Soc.* **52**:154.

Feuge, R. O., Gajee, B. B., and Lovegren, N. V., 1973, Cocoa butter-like fats from fractionated oil. 1. Preparation, *J. Am. Oil Chem. Soc.* **50**:50.

Folch, J., Lees, M., and Sloane-Stanley, G. H., 1957, A simple method for the isolation and purification of total lipides from animal tissues, *J. Biol. Chem.* **266**:497.

Food Consumption, Prices, and Expenditures, 1968 and 1972 supplement Agricultural Economic Report 138, Economic Research Service, U.S. Department of Agriculture, Washington, D.C.

Fristrom, G. A., Stewart, B. C., Weihrauch, J. L., and Posati, L. P., 1975, Comprehensive evaluation of fatty acids in foods. IV. Nuts, peanuts and soups, *J. Am. Diet. Assoc.* **67**:351.

Galloway, J. P., 1976, Cleaning, cracking, dehulling, decorticating and flaking of oil bearing materials, *J. Am. Oil Chem. Soc.* **53**:271.

Gandy, D. E., 1972, Projection and prospects for sunflower seed, *J. Am. Oil Chem. Soc.* **49**:518A.

Gooding, C. M., July 30, 1963, U.S. Patent 3,099,564, assigned to Corn Products Company.

Gruger, E. H., Jr., Nelson, R. W., and Stansby, M. E., 1964, Fatty acid composition of oils from 21 species of marine fish, freshwater fish and shellfish, *J. Am. Oil Chem. Soc.* **41**:662.

Guam, H. S., 1972, Malaysian palm oil, *J. Am. Oil Chem. Soc.* **49**:322A.

Holaday, C. E., and Pearson, J. L., 1974, Effects of genotype and production area on the fatty acid composition, total oil and total protein in peanuts, *Food Sci.* **39**:1206.

Howell, R. W., 1972, The plant geneticist's contribution toward changing lipid and amino acid composition of soybeans, *J. Am. Oil Chem. Soc.* **49**:30.

Hubbard, W. D., Sheppard, A. J., Newkirk, D. R., Prosser, R., and Osgood, T., 1977, Comparison of various methods for the extraction of total lipids, fatty acids, cholesterol and other sterols from food products, *J. Am. Oil Chem. Soc.* **54**:81.

Hutchins, R. P., 1968, Processing control of crude oil production from oil-seeds, *J. Am. Oil Chem. Soc.* **45**:624A.

Hutchins, R. P., 1976, Continuous solvent extraction of soybeans and cottonseed, *J. Am. Oil Chem. Soc.* **53**:279.

Iverson, J. L., 1958, unpublished data.

Iverson, J. L., and Sheppard, A. J., 1977, Butyl ester preparation for gas–liquid chromatographic determination of fatty acids in butter, *J. Assoc. Off. Anal. Chem.* **60**:284.

Iverson, J. L., Eisner, J., and Firestone, D., 1965, Fatty acid composition of olive oil by urea fractionation and gas–liquid chromatography, *J. Assoc. Off. Anal. Chem.* **48**:1191.

Iverson, J. L., Harrill, P. G., and Weik, R. W., 1969, Fatty acid composition of cocoa butter oil by urea fractionation and programmed temperature gas chromatography, *J. Assoc. Off. Anal. Chem.* **52**:685.

Jellum, M. D., 1970, Plant introductions of maize as a source of oil with unusual fatty acid composition, *J. Agr. Food Chem.* **18**:365.

Johnson, A. E., MacMillan, D., Dutton, H. H., and Cowan, J. C., 1962, Hydrogenation of linolenate. VI. Survey of commercial catalysts, *J. Am. Oil Chem. Soc.* **39**:273.

Kinsella, J. E., Posati, L., Weihrauch, J., and Anderson, B., 1975, Lipids in foods: Problems and procedures in collating data, *Crit. Rev. Food Technol.* **5**:299.

Kneeland, J. A., 1966, The status of safflower, *J. Am. Oil Chem. Soc.* **43**:403.

Knowles, P. F., 1972, The plant geneticist's contribution toward changing lipid and amino acid composition of safflower, *J. Am. Oil Chem. Soc.* **49:**27.

Kurucz-Lusztig, É., Lukács-Hágony, P., Prépostaffy-Jánoshegyi, M., Jeranek-Knapecz, M., and Biacs, P., 1974, Studies on vegetable oil hardening technologies by means of the analytical parameters of fats with special reference to the formation of *trans*-isomer fatty acids, *Acta Aliment. Acad. Sci. Hung.* **3(4):**357.

Lambertsen, G., 1973, Lipids in marine fish, *Wiss. Veroeff. Dtsch. Ges. Ernaehr.* **24:**4.

Landman, W., Lovegren, N. V., and Feuge, R. O., 1961, Confectionery fats. II. Characterization of products prepared by interesterification and fractionation, *J. Am. Oil Chem. Soc.* **38:**466.

Lantondress, E. G., 1967, Hydrogenation and bleaching control procedures, *J. Am. Oil Chem. Soc.* **44:**154A.

Lauridsen, J. B., 1976, Food emulsifiers: Surface activity, edibility, manufacture, composition, and application, *J. Am. Oil Chem. Soc.* **53:**400.

Lefebur, Von J., and Baltes, J., 1975, Nickel-silver catalysts and their application in selective hydrogenation of fats, *Fette Seifen Anstrichm.* **77:**128.

Litchfield, C., 1970, Taxonomic patterns in the fat content, fatty acid composition and triglyceride composition of palm seeds, *Chem. Phys. Lipids* **4:**96.

Luddy, F. E., Hampson, J. W., Herb, S. F., and Rothbart, H. L., 1973, Development of edible tallow fractions for specialty fat uses, *J. Am. Oil Chem. Soc.* **50:**240.

Luddy, F. E., Hampson, J. W., Tunik, M. H., and Rothbart, H. L., September 29, 1975, A simplified and improved tallow fractionation, American Oil Chemists Society, 49th annual fall meeting, Cincinnati, Ohio, abst. 5.

Lundberg, W. O., 1972, Chemical, biochemical and nutritional aspects of soybean oil, *Fette Seifen Anstrichm.* **74:**557.

Lyon, C. K., 1972, Sesame: Current knowledge of composition and use, *J. Am. Oil Chem. Soc.* **49:**245.

McNaught, J. P., July 17, 1973, U.S. Patent 3,746,551, assigned to Lever Brothers Company.

Mecham, D. K., 1971, Lipids, in: *Wheat Chemistry and Technology* (Y. Pomeranz, ed.), p. 393, Monograph Series III (revised), American Association of Cereal Chemists, St. Paul, Minn.

Mecham, D. K., and Mohammad, A., 1955, Extraction of lipids from wheat products, *Cereal Chem.* **32:**405.

Mijanders, A., and Lincklaen, H. W., August 3, 1971, U.S. Patent 3,597,229, assigned to Lever Brothers Company.

Mikolajczak, K. L., Miwa, T. K., Earl, F. R., and Wolff, I. A., 1961, Search for new industrial oils. V. Oils of Cruciferae, *J. Am. Oil Chem. Soc.* **38:**678.

Minifie, B. W., 1970, *Chocolate, Cocoa and Confectionery: Science and Technology*, pp. 56–58, Avi, Westport, Conn.

Morrison, W. R., Mann, D. L., Soon, W., and Coventry, A. M., 1975, Selective extraction and quantitative analysis of non-starch and starch lipids from wheat flour, *J. Sci. Food Agr.* **26:**507.

Nieschlag, H. J., and Wolff, J. A., 1971, Industrial uses of high erucate oils, *J. Am. oil Chem. Soc.* **48:**723.

Padley, F. B., Paulussen, C. N., Soeters, C. J., and Tresser, D., 1972, The improvement of chocolate using mono-unsaturated triglycerides SOS and POS, *Rev. Int. Choc.* **27:**226.

Paulicka, F. R., 1976, Specialty fats, *J. Am. Oil Chem. Soc.* **53:**421.

Petrowski, G. E., 1975, Food-grade emulsifiers, *Food Technol.* **29(7):**52.

Posati, L., Kinsella, J. E., and Watt, B. K., 1975*a*, Comprehensive evaluation of fatty acids in foods. I. Dairy products, *J. Am. Diet. Assoc.* **66:**482.

Posati, L., Kinsella, J. E., and Watt, B. K., 1975*b*, Comprehensive evaluation of fatty acids in foods. III. Eggs and egg products, *J. Am. Diet. Assoc.* **67:**111.

Potter, N. N., 1968, *Food Science*, pp. 419–439, Avi, Westport, Conn.

Pritchard, J. L. R., 1969, Quality of palm oil and palm kernels—User requirements, *Trop. Sci.* **11(2):**103.

Purdy, R. H., and Campbell, B. J., 1967, High oleic acid safflower oil, *Food Technol.* **21:**31A.

Rakow, G., and McGregor, D. I., 1973, Opportunities and problems in modification of levels of rapeseed C_{18} unsaturated fatty acids, *J. Am. Oil Chem. Soc.* **50:**400.

Rayner, E. T., Brown, L. E., and Dupuy, H. P., 1966, A simplified process for the elimination of the halphen-test response in cottonseed oils, *J. Am. Oil Chem. Soc.* **45:**293.

Rich, A. D., 1970, Some fundamental aspects of bleaching, *J. Am. Oil Chem. Soc.* **47:**564A.

Robertson, J. A., 1972, Sunflower: America's neglected crop, *J. Am Oil Chem. Soc.* **49:**239.

Rocquelin, G., Sergiel, J. P., Martin, B., Leclerc, J., and Cluzan, R., 1971, The nutritive value of refined rapeseed oils: A review, *J. Am. Oil Chem. Soc.* **48:**728.

Sampugna, J., and Jensen, R. G., 1969, Stereo specific analysis of the major triglyceride species in the monounsaturated fraction of cocoa butter, *Lipids* **4:**444.

Scholfield, C. R., Butterfield, R. O., Davidson, V. L., and Jones, E. P., 1964, Hydrogenation of linolenate. X. Comparison of products formed with platinum and nickel catalysts, *J. Am. Oil. Chem. Soc.* **41:**615.

Seiden, P., January 17, 1967, U.S. Patent 3,298,837, assigned to The Procter and Gamble Company.

Sheppard, A. J., 1963, Suitability of lipid extraction procedure for gas–liquid chromatography, *J. Am. Oil Chem. Soc.* **40:**545.

Sheppard, A. J., Fritz, J. C., Hooper, W. H., Roberts, T., Hubbard, W. D., Prosser, A. R., and Boehne, J. W., 1971, Crambe and rapeseed oils and energy sources for rats and chicks and some ancillary data on organ weights and body cavity fat composition, *Poult. Sci.* **50:**79.

Sheppard, A. J., Hubbard, W. D., and Prosser, A. R., 1974*a*, Evaluation of eight extraction methods and their effects upon total fat and gas–liquid chromatographic fatty acid composition analyses of food products, *J. Am. Oil Chem. Soc.* **51:**416.

Sheppard, A. J., Hubbard, W. D., and Prosser, A. R., June 11, 1974*b* and as modified by letter on April 16, 1975, Interim Methodology Instructions No. 2 for Implementing Requirements of Section 1.18 of Title 21, Chapter 1. Sub-chapter A, Part 1 ("Labeling of Foods in Relation to Fat, Fatty Acids, and Cholesterol Content"), Division of Nutrition, Bureau of Foods, Food and Drug Administration, Washington, D.C.

Sherwin, E. R., 1972, Antioxidants for food fats and oils, *J. Am. Oil Chem. Soc.* **49:**468.

Sidwell, V. D., Foncannon, P. R., Moore, N. S., and Bonnet, J. C., 1974, Composition of the edible portion of raw (fresh or frozen) crustaceans, finfish, and mollusks. I. Protein, fat, moisture, ash, carbohydrate, energy value and cholesterol, *Mar. Fish. Rev.* **36(3):**21.

Solomon, H. L., Hubbard, W. D., Prosser, A. R., and Sheppard, A. J., 1974, Sample size influence on boron trifluoride-methanol procedure for preparing fatty acid methyl esters, *J. Am. Oil Chem. Soc.* **51:**424.

Sonntag, N. O. V., 1969, Panel discussion and symposium: Nonfood uses of coconut oil: Where are we headed? *J. Am. Oil Chem. Soc.* **46:**632A.

Stansby, M. E., 1973, Polyunsaturates and fat in fish flesh, *J. Am. Diet. Assoc.* **63:**625.

Stoddard, J. H., 1968, *Fat Contents of Canadian Atlantic Herring*, Technical Report No. 79, Fisheries Research Board of Canada, St. Andrews Biology Station, New Brunswick, Canada.

Stoffel, W., Chu, F., and Ahrens, E. H., Jr., 1959, Analysis of long chain fatty acids by gas–liquid chromatography: Micromethod for methyl esters, *Anal. Chem.* **31:**307.

Sullivan, F. E., 1968, Refining of oils and fats, *J. Am. Oil Chem. Soc.* **45:**564A.

Sutton, G., Bertoni, M. H., Cattaneo, P., Abitbol, J., and Denett, J. M., 1966, Olive oil. VI. Effect of the botanical variety on the chemical compositions, *Rev. Argent. Grasas Aceites* **8:**3.

Thompson, D. L., Jellum, M. D., and Young, C. T., 1973, Effect of controlled temperature environments on oil content and fatty acid composition of corn oil, *J. Am. Oil Chem. Soc.* **50:**54.

Tiscornia, E., and Bertini, G. C., 1972, Recent analytical data on the chemical composition and structure of olive oil, *Riv. Ital. Sostanze Grasse* **49:**3.

Trotter, W. K., and Givan, W. D., 1971, Economics of sunflower oil production and use in the United States, *J. Am. Oil Chem. Soc.* **48:**443.

Watts, B. A., 1965, Properties of glandless cottonseed, U.S. Department of Agriculture (ARS 72-38).

Weihrauch, J. L., Brignoli, C. A., Reeves, J. B., and Iverson, J. L., 1977, Fatty acid composition of margarines, processed fats, and oils; A new compilation of data for tables of food composition, *Food Technol.* **31:**80.

Weiss, T. J., 1970, *Food Oils and Their Uses*, Avi, Westport, Conn.

Wijngaarden, D., Thyssen, L. A., and Osinga, T. D., 1968, Zur Gas Chromatographisch Bestimmten Fettsaurenzusammensetsung der Kakaobutter, *Z. Lebensm. Unters. Forsch.* **37(3):**171.

Wilson, C. T., 1972, The present and future role of peanuts in meeting the world's need for food, *J. Am. Oil Chem. Soc.* **49:**343A.

Woidich, H., Gnauer, H., Riedl, O., and Galinovsky, E., 1964, Uber die Zusammensetzung der Kakaobutter, *Z. Lebensm. Unters. Forsch.* **125:**91.

Wolf, W. J., and Cowan, J. C., 1971, Soybeans as a food source, *Crit. Rev. Food Technol.* **2:**81.
Yermanos, D. M., Hemstreet, S., Saleeb, W., and Huszar, C. K., 1972, Oil content and composition of the seed in the world collection of sesame introductions, *J. Am. Oil Chem. Soc.* **49:**20.
Young, C. T., Worthington, R. E., Hammonds, R. O., Matlock, R. S., Walker, G. R., and Morrison, R. D., 1974, Fatty acid composition of Spanish peanut oils as influenced by planting location, soil moisture conditions, variety, and season, *J. Am. Oil Chem. Soc.* **1:**312.
Zehnder, C. T., 1976, Deodorization, *J. Am. Oil Chem. Soc.* **53:**364.
Zehnder, C. T., and McMichael, C. E., 1967, Deodorization, principles and practices, *J. Am. Oil Chem. Soc.* **44:**508.

Chapter 8

Fatty Acid Composition of Glycerolipids of Animal Tissues

Arnis Kuksis

8.1. Introduction

The most complete analyses of fatty acids have been obtained on total lipid extracts, although a prior separation into neutral and polar lipids has also been common. In many instances, the total fatty acid esters have been subjected to a preliminary segregation into subclasses based on unsaturation, molecular weight, and the presence or absence of functional groups. The final composition of the fatty acids in the neutral and polar lipids is then derived by normalization of the data. Detailed analyses of the total fatty acid composition of a lipid have proved essential for obtaining a complete account of the molecular species of various glycerolipids which otherwise might have been overlooked because of losses at intermediate stages of fractionation. Truly meaningful comparisons of the fatty acid composition of tissue lipids are obtained only by the examination of individual lipid classes and molecular species in specific cellular fractions. There has been a steady increase in these data, and those applying to the neutral acylglycerols have been extensively documented in other chapters of this volume.

This chapter contains the results of analyses of both total lipids and individual lipid classes as an indication of the variety of fatty acids that are found in the tissues of man and experimental animals, and as a reference source and guide to future analyses. In anticipation of possible discrepancies between the present and future analyses, the nature of the dietary fats consumed has been noted whenever this information was available, as well as the exact site of sampling and the age and sex of the animals. These factors have proved to have marked influence on the composition of the fatty acids. The compiled data have been taken from the most complete analyses available, but references to selected other determinations have also been included. There remain variations in results which may reflect differences in methods of analysis including extraction and isolation.

Arnis Kuksis • Banting and Best Department of Medical Research, University of Toronto, Toronto, Ontario, M5G 1L6, Canada.

8.2. Methods of Analysis

A valid analysis of the fatty acid composition of an animal tissue requires complex methodology. In addition to selection of a representative sample, there is a need for complete extraction of the lipid with a minimum of decomposition. The total lipid extracts must be freed of nonlipid material and fractionated into the component lipid classes before a meaningful analysis of the fatty acids can be made. There exist a variety of chromatographic techniques which can be routinely employed for this purpose (Nelson, 1975; Kuksis, 1977). In specific instances, however, other methods must be employed to estimate fatty acids which cannot be resolved by chromatographic means. The methods of isolation and identification of natural fatty acids have been discussed in detail in Chapter 1 of this book. The following is a brief summary of the more successful methods of analysis used in the compilation of the data included in this chapter.

8.2.1. Isolation and Preparation of Fatty Acids

The selection of the analytical sample constitutes the first step in the isolation of the lipids and the preparation of fatty acids. The most efficient and carefully executed analyses cannot retrieve the information lost by poor sample selection and by non-representative fatty acid isolation and derivatization.

8.2.1.1. Isolation

There are two basic routines currently in general use which yield essentially quantitative extractions of the major lipid classes when applied to homogenates of whole animal tissue or tissue subfractions. The most popular is the method of Folch *et al.* (1957) which employs chloroform–methanol 2 : 1 in a solvent–tissue ratio of 20 : 1. The other one is that proposed by Bligh and Dyer (1959), which effects a single-phase solubilization of the lipids using chloroform–methanol 1 : 1 in a solvent–tissue ratio of 4 : 1. However, the subsequent partition of the extracts between chloroform and aqueous salt solutions, or chloroform and methanol–water 10 : 9, results in losses of the more polar acidic phospholipids and lysophospholipids (Nelson, 1972*a*, 1975). The crude lipid extract obtained as the final product of the evaporation of the solvent from a chloroform–methanol extract can be effectively freed from all nonlipid material by column chromatography on dextran gels (Wells and Dittmer, 1963; Siakotos and Rouser, 1965; Nelson, 1972*a*; Rouser *et al.*, 1976). The lipid sample is recovered as a mixture of neutral and phospholipids by elution with chloroform–methanol 19 : 1, saturated with water. The purified lipid mixture is then resolved into neutral and phospholipids by thin-layer chromatography (TLC), which also is capable of an effective resolution of individual lipid classes (Skipski and Barclay, 1969; Rouser *et al.*, 1976). Preparative one-dimensional TLC provides ample amounts of pure lipid classes which can be transmethylated to provide the fatty acid methyl esters for subsequent gas–liquid chromatography (GLC) (Kuksis, 1966, 1972).

In those instances where sample size is limited, sufficient amounts of material for fatty acid analysis of individual lipid classes have been obtained by two-dimensional TLC. Rouser *et al.* (1969*a*, 1976) have discussed the two-dimensional TLC systems that separate both acidic and neutral phospholipid classes. The TLC systems suitable for routine resolution of the neutral lipids and phospholipids have been recently reviewed (Nelson, 1975; Kuksis, 1977).

8.2.1.2. Preparation of Fatty Acids

The fatty acids of each lipid class usually have been prepared in the form of fatty acid methyl esters. The methods employed for this purpose have been described in great detail in Chapter 1 of this book. The alkaline transmethylation routine described by Glass (1971) and the methanolic boron trifluoride method of Morrison and Smith (1964) have been found to be best suited for dealing with the more labile fatty acids. However, other acid catalysts may also be employed effectively with most natural fatty acid esters (Christie, 1972; Sheppard and Iverson, 1975). Frequently, the fatty acid methyl esters have been prepared in the presence of silica gel from the scrapings of the TLC plate. For this purpose, the scrapings are covered with about 5 ml of the methylating reagent in a 15-ml-capacity vial equipped with a teflon-lined screw cap. The mixture is heated at 90°C for a minimum of 2 hr. The fatty acid methyl esters are extracted from the reaction mixtures with several washes of hexane, which may contain 0.01% butylated hydroxytoluene (BHT) (Nelson, 1972*a*, 1975). Samples containing material other than fatty acid methyl esters require purification, which may be accomplished on small columns of silicic acid, using diethyl ether in hexane to elute the methyl esters and the antioxidant (Nelson, 1972*a*). Alternatively, the methyl esters may be purified by TLC using pure benzene as the developing solvent, which also allows the recovery of the BHT as a separate band. The purified fatty acid methyl esters are concentrated to a small volume under dry nitrogen, further dried by azeotropic distillation with petroleum ether, and saved for GLC analysis.

8.2.2. Quantitative Determination of Fatty Acids

Two types of quantitation of fatty acids have been commonly performed by GLC. One provides the percent composition of the fatty acid mixture which may be expressed on a mole or weight basis. For this purpose, it is important to obtain complete separation and recovery of all components. The most effective separations of the fatty acid methyl esters are obtained by capillary columns, but recoveries are poor for the higher-molecular-weight components (Ackman, 1972). Despite considerable progress in recent years (Lin *et al.*, 1975; Jaeger *et al.*, 1976), the capillary column methods are not yet available for routine quantitation of fatty acids in natural mixtures. Conventional packed columns have provided the best results and may continue to do so in the future. The development of polar siloxane liquid phases, which allow the resolution of *cis* and *trans* isomers of fatty acids, has somewhat reduced the need for capillary columns (Ottenstein *et al.*, 1976; Golovnya *et al.*, 1976). When calibrated with known mixtures of common standard fatty acids, essentially complete recoveries of all components may be demonstrated with both glass and metal columns and many packing types (Kuksis, 1971). More appropriate secondary standards such as those derived from fish oils (Ackman and Burgher,

1965) or animal testes (Holman and Hofstetter, 1965) have also been employed in the calibration of GLC systems.

The other type of quantitation concerns estimation of the fatty acid content of a lipid fraction. This is best made by means of dilution with suitable internal standards, such as pentadecanoic, heptadecanoic, and eicosanoic acids, which usually occur in natural fatty acid mixtures only in trace amounts (Kuksis, 1966; Privett *et al.*, 1971). For this purpose, the internal standard is added in a ratio of 10–30% depending on the complexity of the mixture of the fatty acids. It is essential that the individual fatty acids be recovered in their true proportions, or at least that the proportions in which they are recovered be known. The linearity of recovery of the fatty acids over the working range must also be established. The method of quantitation of fatty acids by means of internal standard is most useful for the proportionation of the fatty acid subfractions derived by argentation TLC of the total mixture of fatty acid methyl esters.

The GLC quantitation of fatty acids is greatly facilitated by electronic integration of the peak areas, provided that the peaks are well resolved and the baseline is stable. In most instances in the past, however, the fatty acid peaks have been quantitated by triangulation, by disk integration, or by cutting out and weighing the chart paper. Some of the best results of fatty acid quantitation have been realized with the polyester and the Apiezon L columns (Ackman, 1972).

8.2.3. *Quality Control*

Whenever possible, the validity of the GLC analyses of the fatty acid methyl esters must be independently assessed. This can be done in a variety of ways, some of which constitute a part of a sound GLC procedure (Kuksis, 1971). The first indication of the quality of the GLC data is obtained by analyses of appropriate standards under the working conditions. This is not sufficient, however, for analyses of samples containing unknowns. A check on the homogeneity of the peaks and a confirmation of the chain-length distribution may be obtained by comparing the results obtained with polar and nonpolar columns. Alternatively, the chain-length distribution of the fatty acids may be determined by rechromatography following hydrogenation of the sample. In many instances, the saturated and unsaturated fatty acid peaks have been distinguished by bromination, which subtracts the unsaturated fatty acids form the elution profile. An effective segregation of the unsaturated and polyunsaturated fatty acids is obtained by argentation TLC of the fatty acid mixture. Upon rechromatography by GLC, it is possible to assign the number, location, and configuration of the double bonds with much greater confidence. Furthermore, the argentation TLC serves to enrich the minor components and allows a more accurate estimate of their proportion as well as their identification. Recently, a combination of gas chromatography and mass spectrometry has been employed as the method of identification of many unknown fatty acids in the gas chromatograms (see Chapter 1).

In most instances, the data for the tables in this chapter have been selected from those analyses where one or more of the above quality checks had been used.

8.3. Fatty Acids of Different Animal Species and Man

There is considerable evidence which suggests that animal species differ in the fatty acid composition of corresponding tissues (Hilditch and Williams, 1964). Much of this variation is due to differences in the dietary fat, which is known to have a pronounced effect on tissue fatty acids (Carroll, 1965). However, certain variations have been noted also among animals receiving identical diets. The latter discrepancies appear to vary with the type of tissue and have been traced to the level of corresponding cell types, subcellular fractions, and lipoprotein classes, and appear to be genetically determined. In most instances, published analyses of animal tissue glycerolipids have been performed on samples obtained from animals which had consumed fat, often of unknown composition. The following compilation of analytical results serves to illustrate the types and amounts of fatty acids that are found in lipid extracts of animal tissues. The data have been organized by homologous tissues and subcellular components so that any similarities or discrepancies may be seen regardless of their true origin.

8.3.1. Differences among Homologous Tissues

Detailed analyses from a large variety of animal species have shown that among vertebrates there is little or no species variability of phospholipid class distribution of organs or tissues, including subcellular particulates (Rouser *et al.*, 1969*b*). This has been attributed to common functional requirements of the component cell membranes. In order to determine any inherent differences in the fatty acid composition of corresponding tissues of different animal species, the dietary intake of the fatty acids must be known and controlled. Likewise, for valid comparisons, the animals must be selected from the same sex, age, and metabolic state.

8.3.1.1. Adipose Tissue and Liver

It has been known for several decades that dietary fats are deposited in the adipose tissue as triacylglycerols, from which the fatty acids become released for transport to other tissues and for incorporation into other lipid classes (for review, see Carroll, 1965). As a result, the triacylglycerols of the adipose tissue are subject to the greatest and most rapid dietary alteration.

Table I compares the fatty acid compositions of the triacylglycerols of the adipose tissue of swine, rat, and chicken on a fat-free diet (Anderson *et al.*, 1970*a*, *b*). The values for the triacylglycerols of the liver serve as controls. The endogenous adipose tissue triacylglycerols of rat and chicken differ markedly in composition from those of swine, although all three contain the same major fatty acids. The main difference, however, is that the swine triacylglycerols have saturated fatty acids in the 2 position, whereas in rat and chicken that position is preferentially occupied by unsaturated fatty acids (Bottino *et al.*, 1970).

Since the animals in Table I were reared (2–5 months) on fat-free diets, eicosatrienoic acid, which is characteristic of essential fatty acid deficiency, is present in the triacylglycerols of all the livers (Anderson *et al.*, 1970*b*). There are some differences between swine and rat fatty acids. Palmitic acid is much lower and

Table I. Fatty Acid Composition of Triacylglycerols of the Liver and Adipose Tissue of Swine, Rat, and Chicken on Fat-Free Diets

Fatty acids	Liver[a]			Perirenal adipose tissue		
	Swine	Rat	Chicken	Swine[b]	Rat[c]	Chicken[c]
			(mole %)			
14:0	1.4	1.1	0.9	1.4	1.7	0.9
14:1			0.2			
15:0	0.6	—		—	0.4	0.2
16:0	17.7	27.7	25.0	30.0	25.8	25.1
16:1	7.1	9.8	5.8	3.5	11.7	6.4
17:0	2.5	—	—	0.6	—	—
17:1	3.5	0.4	tr	0.8	0.4	—
18:0	5.0	2.3	8.6	16.6	3.8	8.4
18:1	56.6	56.0	57.9	45.7	55.4	58.0
18:2	0.7	0.5	tr	0.3	0.1	—
20:0	—	0.5		0.3	0.1	
18:3	1.7					1.2
20:1	0.2	0.6	1.0	0.8	0.3	
20:2			0.2			
20:3	2.2	1.5	0.5			
20:4	0.2					
20:5	0.1					

[a] Anderson *et al.* (1970*b*).
[b] Anderson *et al.* (1970*a*).
[c] Bottino *et al.* (1970).

linolenic acid and eisosatrienoic are much higher. Also, the swine liver contains significant amounts of heptadecanoic and heptadecenoic acids, which are absent or are present only in trace amounts in the other two species.

Table II compares the acyl and alk-1-enyl chain compositions of adipose tissue phospholipids of pig, ox, and rat (Grigor *et al.*, 1972). The data show that the ethanolamine phospholipids consist of 54%, 53% and 34% plasmalogen for the pig, ox, and rat adipose tissue samples, respectively. The ethanolamine phospholipids contain less palmitate relative to stearate and a higher proportion of polyunsaturated fatty acids than the choline phospholipid fractions in all three samples. The minor differences in the fatty acid composition of these phospholipids in the samples from the different animal species are probably due to dietary differences and therefore may attest to further tissue specificity.

8.3.1.2. Lung

The lung is known to produce large amounts of a surfactant which consists largely of dipalmitoylphosphatidylcholine, and its content would be expected to be reflected in a high palmitic acid content of the glycerophospholipids of this tissue of all animal species. Table III compares the fatty acid compositions of the glycerolipids of the lung of 11 species of animals. All contain high proportions of palmitic acid in the phospholipids (Clements, 1971). In general, the neutral lipids tend to be less saturated than the phospholipids, and both types of lipids tend to be less

Table II. Composition of Acyl and Alk-1-enyl Moieties from Adipose Tissue Phospholipids of Selected Animal Species[a]

	Pig		Ox		Rat	
Component	EP[b]	CP[b]	EP	CP	EP	CP
			(mole %)			
Methyl esters						
16:0	3.0	14.1	3.0	17.4	6.9	21.6
16:1	0.5	1.9	0.4	1.7	1.0	1.4
17:0	0.6	0.9	0.3	1.2	0.7	0.5
17:1	0.4	0.4		0.5	0.5	0.3
18:0	20.4	29.3	13.2	24.1	19.0	25.5
18:1	13.9	22.7	7.3	17.4	16.6	18.5
18:2	18.1	20.2	3.4	7.8	11.5	18.3
18:3	0.2	0.2	0.5	1.3	0.5	0.1
20:1	—		0.2			
20:2	—	0.2	0.7			0.2
20:3	0.8	0.6	5.4	6.6		4.0
20:4	18.6	6.9	21.6	8.4	21.2	9.6
	0.4		2.1	2.6		
22:2	2.3	0.2	8.5	3.7	1.3	0.5
22:3	0.2		1.5	0.5	0.7	0.4
24:1	—	0.6	0.2	0.1	0.2	
22:5	1.1	0.3	11.0	3.0	2.7	0.7
22:6	0.4	0.3	1.9	0.6	5.5	1.4
Dimethylacetals						
16:0	4.6	0.7	4.9	1.2	4.7	0.3
16:1	0.6	0.1	0.9	0.4	0.4	0.2
17:0	0.6		0.6	0.2	0.4	0.1
17:1	—		0.1			
18:0	9.6	0.3	11.0	0.9	3.3	0.1
18:1	3.6		1.0		2.4	
18:2	0.3		0.2		0.4	

[a] From Grigor *et al.* (1972).
[b] Abbreviations: EP, ethanolamine phospholipid fraction; CP, choline phospholipid fraction.

saturated in the poikilotherms than in the homeotherms. The strong effect of a diet rich in polyunsaturated fats shows in the very high level of unsaturation and abundance of longer-chain acyl groups in neutral fats of the sea lion lung, but the acyl groups in the phospholipids in this seagoing mammal do not seem to reflect the dietary influence. An even closer relationship is seen for the purified phosphatidylcholines of the lung of the various animal species (Clements, 1971).

The phosphatidylcholines of the brain of the rat, rabbit, pig, cow, and sheep also have shown close similarities in the composition of the molecular species and fatty acids (Montfort *et al.*, 1971). On the other hand, the phosphatidylcholines of the kidney and liver of the rat, rabbit, pig, sheep, and cow have shown greater dietary and environmental dependence in the composition of the molecular species and fatty acids (Montfort *et al.*, 1971). Differences among animal species may also be seen in the fatty acid composition of the glycerophospholipids of skeletal muscle (Marai and Kuksis, 1973).

Table III. Fatty Acid Composition of Neutral Lipids and Phospholipids of the Lung of Selected Animal Species[a]

Species	14:0		16:0		16:1		18:0		18:1		18:2		18:3		20:4	
	NL[b]	PL[b]	NL	PL	NL	PL	NL	PL	NL	PL	NL	PL	NL	PL	NL	PL
	(mole %)															
Mouse	1.4	1.7	34.1	65.3	6.4	5.4	5.6	14.6	29.6	8.0	22.3	2.5			0.7	4.2
Guinea pig	3.1	2.9	37.4	44.5	3.0	2.3	10.1	11.1	25.1	16.6	17.2	14.8	2.0	0.9	2.1	7.1
Rat	1.9	2.4	31.5	44.6	4.1	4.7	7.9	15.4	22.7	11.3	25.5	8.0	0.7		5.6	13.6
Rabbit	3.0	1.4	45.3	44.3	7.0	4.0	8.2	9.5	23.5	21.2	9.3	9.8	0.3		3.1	9.9
Sea lion	3.3	4.8	23.8	49.0	5.4	6.5	9.5	10.1	32.3	19.4	2.5	0.4	11.2	1.5	12.1	8.5
Chicken	1.2	0.4	29.9	59.2	4.5	0.4	10.8	18.8	43.6	13.3	8.0	1.9	1.2		0.8	6.0
Dog	2.4	2.7	28.9	54.1	5.3	3.1	14.3	21.7	31.9	10.9	7.3	2.3	2.2	0.7	7.7	1.0
Man	4.6	1.9	34.5	52.5	11.4	3.1	10.1	22.1	25.2	17.7	14.1	2.8				
Cow	1.6	1.6	28.0	43.5	2.8	4.1	19.3	18.8	38.3	24.7	7.6	3.2			1.8	4.1
Turtle	2.5	3.2	21.9	29.3	6.6	5.5	10.3	15.2	40.3	25.0	11.3	6.3	3.1	0.4	3.8	14.9
Frog	2.4	1.3	26.2	33.1	7.2	11.5	12.9	12.9	31.0	20.1	10.6	9.3	1.8	0.9	7.7	10.9

[a] Abbreviated from Clements (1971).
[b] Abbreviations: NL, neutral lipid; PL, phospholipid.

8.3.1.3. Testes and Ovaries

The testis and ovary, which produce reproductive cells, contain lipids that are especially rich in polyunsaturated fatty acids. Table IV compares the fatty acid compositions of the various glycerolipid classes isolated from the testes of ox and pig (Holman and Hofstetter, 1965), mouse (Coniglio *et al.*, 1975), and rat (Nakamura *et al.*, 1968). The fatty acid compositions of the corresponding lipid classes do not yield identical patterns. Some polyunsaturated acids exist as traces in the testes of one species and as major components in another species. The polyunsaturated acids of these lipids are predominantly of the linoleate family. The occurrence of many higher metabolites of the oleate family indicates that these acids are normal constituents of tissues of normal animals and are not formed only in essential fatty acid deficiency (Holman and Hofstetter, 1965). The presence of diacylglycerols along with free fatty acids in the testes of ox and pig might suggest lipolysis prior to extraction. However, there is no consistent relationship between the fatty acid composition of phospholipids or triacylglycerols and their possible products of hydrolysis, the diacylglycerols and free fatty acids. Usually, glycerophospholipids are rich in polyunsaturated fatty acids and triacylglycerols are poor in polyunsaturated fatty acids. The testes and the ovaries appear to be an exception, and in these tissues the triacylglycerols may serve a role other than simple energy storage. Moreover, it appears that diacylglycerols, triacylglycerols, and perhaps free fatty acids should not be considered a single pool in these tissues (Nakamura *et al.*, 1968).

Grogan *et al.* (1973) have confirmed the identity of the polyenoic acids of human testicular tissue by chemical means. The following relative average amounts were obtained for these acids (in percent of total fatty acids): 20:3 $\omega6$ (5.9%), 22:4 $\omega4$ (1.3%), 22:5 $\omega6$ (0.5%), 22:5 $\omega3$ (0.6%), and 22:6 $\omega3$ (8.0%). The polyenes (20:4 $\omega6$ and 18:2 $\omega6$) were not characterized because of their known ubiquity. The quite small relative amounts of 22:5 $\omega3$ and large relative amounts of 22:6 $\omega3$ present in human testes are in strong contrast to the large relative amounts of 22:5 $\omega3$ and small relative amounts of 22:6 $\omega3$ found by Ahluwalia and Holman (1969) in human sperm.

Unlike the rat testis (Davis *et al.*, 1966), the human testis retains very little of its $\omega6$ fatty acid as the 22-carbon polyenoic acids. Also, unlike the other species, the human testis shows preference for the less abundant $\omega3$ family when accumulating 22-carbon polyenes. It appears that those variations are not due to dietary differences but are characteristic of the species. It will be of interest to find the factors that are responsible for these differences.

The fatty acid pattern of ram testis (Scott and Setchell, 1968) is essentially similar to that of the bull testis. Comparable differences among the animal species are seen in the fatty acid composition of the triacylglycerols, and the diacylglycerols and free fatty acids, where they have been isolated.

Close similarities in the qualitative and to a lesser extent in the quantitative composition of the fatty acids of the glycerolipids are also seen between the Graafian follicles and residual ovarian tissue of cow and pig (Holman and Hofstetter, 1965). Table V gives the fatty acid compositions of the phospholipids, free fatty acids, and triacylglycerols of the cow and pig ovarian tissues. Unlike the testes, which were rich in fatty acids of the linoleate family, the various lipid classes of the ovarian tissues

Table IV. *Fatty Acid Composition of the Glycerolipids from Different Animal Testes*

Fatty acid	Ox[a]				Pig[a]				Mouse[b]		Rat[c]		
	PL[d]	DG[d]	FFA[d]	TG[d]	PL	DG	FFA	TG	PL	TG	TGE[d]	TG	PL
							(weight %)						
10:0	tr	tr	tr	tr									
12:0	tr	tr	tr	tr	0	0	0	0.1	tr	1.1			
14:0	0.2	2.6	1.0	1.4	1.2	5.8	2.6	10.5	0.8	1.3			
14:0	5.3	10.9	tr	0.8	0.6	0	tr	0.9					
16:0	25.4	20.2	32.3	39.2	25.3	17.5	25.7	19.4	29.2	16.7	20.3	24.7	34.9
16:1 ω7	0.7	2.9	3.3	2.5	0.4	1.5	3.9	4.1	1.3	5.1	tr	0.9	0.5
16:2 ω4	0.7	tr	1.1	0.8	tr	tr	1.9	tr					
18:0	8.7	4.5	7.4	6.9	12.9	10.8	15.0	9.2	8.0	2.7	4.1	2.7	7.3
18:1 ω9	14.3	6.7	17.0	21.2	13.1	7.0	18.1	9.1	12.2	25.0	5.4	9.7	10.5
18:2 ω6	6.0	3.6	6.0	6.6	6.8	11.0	5.4	5.1	2.7	22.3	1.9	4.5	4.5
18:3 ω6	tr	tr	tr	tr	tr	1.7	tr	tr		0.3	0.4	tr	tr
18:3 ω3													
20:0	0.2	1.9	1.4	0.8	2.4	0.8	1.1	1.0					
20:1 ω9	0.3	tr	1.1	0.9									
20:2 ω9	0.0					1.7		1.6					
20:2 ω6	0.4	tr	1.4	1.3	0.8		1.4	tr					
20:3 ω9											1.1	1.5	1.0
20:3 ω6	1.9	1.0	4.1	3.8	3.8	2.8	2.8	3.4	1.4	0.7			
20:4 ω6	20.3	5.2	8.4	3.3	11.3	4.2	13.4	2.1	14.2	1.7	3.3	4.1	15.6
20:5 ω3	tr	tr	tr	0.9	4.6	3.0	0	2.5					
22:3 ω9						0.7		1.1					
22:3 ω6						2.2		2.1					
22:4 ω9						6.6		6.4	1.4	1.6	6.7	6.4	1.5
22:4 ω6	1.0	1.1	2.5	1.5	4.1	2.7	2.5	2.3					
22:5 ω6	4.1	4.5	3.9	4.3	5.6	9.6	5.4	9.9	12.3	5.9	23.9	33.3	18.0
22:5 ω3	8.0	1.7	1.4	1.1		6.2		3.3	0.5	1.1			
22:6 ω3	0.4	8.6	6.9	1.9	7.1	4.3		6.0	8.5	2.9			
24:4 ω6		22.3	0.9						1.1	0.8	11.7	3.4	0.8
A 24:5									1.5	1.3	13.7	6.7	0.6

[a] Holman and Hofstetter (1965).
[b] Coniglio *et al.* (1975).
[c] Nakamura *et al.* (1968).
[d] Abbreviations: PL, phospholipid; DG, diacylglycerol; TG, triacylglycerol; FFA, free fatty acid; TGE, glycerylether diesters.

Table V. Fatty Acid Composition of the Principal Lipids from Cow and Pig Ovarian Tissue[a,b]

Fatty acid	Ovarian tissue						Graafian follicles					
	Cow			Pig			Cow			Pig		
	PL	FFA	TG	PL	FFA	TG	PL	FFA	TG	PL	FFA	TG
	(weight %)											
12:0	tr	tr	tr	tr	0	0						
14:0	1.2	1.9	1.8	0.3	0.8	0.4	2.6	1.3	1.4	0.4	2.9	1.3
14:0	0.5	1.5	0.8	tr	tr	0	17.7	0.6	tr	0.1	tr	9.7
16:0	14.8	11.1	10.1	17.8	13.4	15.7	29.3	10.4	9.9	17.5	27.2	24.1
16:1 ω7	1.9	5.7	6.0	1.3	1.3	1.6	8.2	3.0	3.0	1.3	4.4	6.6
16:2 ω4	tr	tr	tr	tr	1.5	tr	tr	1.5	tr	tr	tr	tr
18:0	15.3	13.7	9.4	16.0	13.8	13.1	9.6	12.0	9.5	18.0	17.8	8.2
18:1 ω9	10.3	14.5	15.3	10.9	15.7	20.3	8.7	11.5	11.9	13.0	21.4	28.2
18:2 ω6	12.1	13.9	17.3	10.2	12.5	14.9	4.9	14.6	14.6	7.9	6.4	8.2
18:3 ω6	0.5	0.8	1.9	0.5	0.6	tr	0.8	0.4	0.4	0.7	0.4	0.1
18:3 ω3 } 20:0	1.8	2.7	2.4	1.0	1.0	1.4	1.3	1.1	1.4	2.8	1.2	1.3
20:2 ω9	tr	0	0	tr	0	0	0	0	0	0	tr	tr
20:2 ω6	2.4	6.1	6.9	0.8	1.7	1.6	1.3	6.3	7.1	1.0	2.2	0.6
20:3 ω9	tr	0	0	0.9	tr	tr	0	0	0	0.5	0.6	0.3
20:3 ω6	3.1	4.9	5.7	1.8	2.0	1.7	2.7	5.7	6.2	2.7	1.0	0.5
20:4 ω6	13.0	11.1	7.8	22.0	19.7	12.8	2.7	13.2	9.8	13.2	2.8	4.6
20:5 ω3	3.9	0.4	tr	1.5	tr	0	2.2	tr	0.3	6.2	0.4	0.1
22:3 ω9	0	tr	tr	0	0	0	0	0	0	0	0.1	0.3
22:3 ω6	0	tr	tr	0	0	0	tr	0.7	0.6	tr	0.5	0.2
22:4 ω9	1.5	tr	0	1.7	0	0						
22:4 ω6	5.3	6.8	9.9	6.4	13.3	15.0	4.4	12.7	17.3	5.6	1.5	3.1
22:5 ω6	0.8	0.3	tr	0.9	0.5	tr	tr	0.6	0.8	0.9	1.0	0.3
22:5 ω3	8.0	4.1	4.0	2.3	1.4	1.4	3.5	3.5	3.1	1.7	0.2	0.7
22:6 ω6	0	tr	0	3.6	0.8	tr	tr	0	tr	3.9	1.1	0.4
22:4 ω6	0	0.4	0.8	0	0	0	tr	1.0	tr	0	6.0	0.9

[a] From Holman and Hofstetter (1965).
[b] Abbreviations as in Table IV.

contain high proportions of the members of the linolenate family. Like the testes, the ovaries also contain high amounts of fatty acids of the oleate family. The overall distribution of fatty acids in neutral lipids and phospholipids from bovine corpora lutea (Scott *et al.*, 1968) is very similar to that reported for the fatty acid composition of the ovarian lipids.

8.3.1.4. Other Tissues

Other complex fatty acid mixtures have been isolated from the glycerolipids of the minor tissues of various animal species. Table VI compares the fatty acid compositions of the thyroid and adhering adipose tissue of ox (Lagrou *et al.*, 1974) and rabbit (Lipshaw and Foa, 1974). Palmitic, stearic, and oleic acids are the major components in the triacylglycerol fraction, myristic, palmitoleic, and linoleic acids being present in much smaller amounts. The free fatty acid pattern is dramatically different, showing the presence of polyunsaturated fatty acids, arachidonic and docosahexaenoic. The total glycerophospholipid fraction shows a predominance of palmitic, stearic, oleic, linoleic, arachidonic, and docosahexaenoic acids. The total fatty acid patterns of the glycerophospholipids of the thyroid tissue are similar to those found for human thyroid by other investigations (Levis *et al.*, 1972). Significant differences, however, could be seen between the phosphatidylcholine and

Table VI. Fatty Acid Composition of Glycerolipids of Thyroid and Adhering Adipose Tissue

	Ox[a]						Rabbit[b]	
	Triacylglycerol		Phospholipid		Free fatty acid			
Fatty acid	Thyroid	Adipose	Thyroid	Adipose	Thyroid	Adipose	Triacyl-glycerol	Phospho-lipid
	(weight %)							
14:0	2.2	3.7			1.6	3.4	4.3	0.9
14:1			2.3				0.9	0.3
16:0	21.2	30.9	22.0	23.9	32.1	25.5	34.8	23.2
16:1	1.9	1.6		1.2	1.4	2.6	10.4	3.8
16:2			1.5					1.7
18:0	17.5	28.1	13.1	27.4	18.2	18.3	7.3	8.2
18:1	55.6	34.6	28.2	30.9	30.4	48.1	27.0	29.5
18:2	1.6	1.2	10.6	2.6	5.2	2.2	11.4	13.4
18:3					1.6		2.2	0.5
20:0			1.7	1.5				0.4
20:1								
20:3								
20:4			11.1	8.0	6.0			10.8
22:0			1.7	2.0				1.9
22:1								
22:6			7.8	2.4	3.7			
23:0								
24:0								1.3
Other								2.5

[a] Lagrou *et al.* (1974).
[b] Lipshaw and Foa (1974).

phosphatidylethanolamine fractions. Much of the phosphatidylethanolamine fraction in human, pig, and dog thyroids was found to be plasmalogen (Levis *et al.*, 1972). Surprisingly, the fatty acid patterns in the different lipid classes (neutral and polar) were the same for both normal and hypertrophic bovine tissues (Lagrou *et al.*, 1974).

Levis *et al.* (1972) have observed that the total phospholipids in human and pig thyroids vary according to sex. Female pig thyroids have a distinctly higher phospholipid content than male glands. The authors point out that similar data found for human thyroids need further confirmation because of the wide range of ages of human thyroids used in the study. Analyses of the fatty acid composition from the normal paranodular tissue and the nodular tissue of one male and one female human thyroid, however, have failed to show any differences. No sex differences were seen in the analyses of fatty acids of pig thyroids.

Analyses of the phospholipid fatty acids reported by other investigators have suggested the presence of nervonic acid in lecithin from human and bovine thyroid (Vilkki and Jaakonmaki, 1966). The data presented by Levis *et al.* (1972) exclude the presence of this fatty acid in glycerophospholipids from human and pig thyroids. These workers have suggested that the previous reports of the presence of nervonic acid in the glycerolipids of the thyroid could have been due to contamination with the glycolipids which overlap with lecithins in certain separation systems.

Prottey and Hawthorne (1966) have described the lipid composition of the pancreas of ox and guinea pig, while Lough and Garton (1968) have determined the lipids of human pancreas. Table VII gives the fatty acid composition of the major lipid classes of human pancreas and of the triacylglcyerols of the pancreas of ox and guinea pig. The triacylglycerols constitute by far the greatest proportion of the total lipids. Regardless of the method of extraction used, the same amount of total lipid containing a similar small proportion of methyl esters (about 1%) is obtained. Similar small amounts of the fatty acid methyl esters have been observed by Meldolesi *et al.* (1971) in the lipids of the guinea pig pancreas. In common with the

Table VII. Fatty Acid Composition of Pancreatic Lipids of Several Animal Species

Fatty acid	Man[a]					Ox[b]	Guinea pig[b]
	TG	FFA	PL	CE[c]	FAME[c]	TG	TG
				(weight %)			
14:0	3.9	3.2	5.5	2.6	4.0	3.6	2.7
14:1	0.9	—	—	0.8	0.5	1.8	—
16:0	22.0	40.9	26.0	24.0	25.2	31.5	43.0
16:1	6.9	4.6	8.3	6.8	9.2	3.3	—
18:0	5.3	6.4	6.9	5.0	6.9	29.4	23.8
18:1	54.0	37.6	42.8	47.7	32.4	29.4	23.8
18:2	6.4	4.0	9.3	8.6	7.4	1.0	6.7
18:3	—	2.7	1.2	3.6	11.6		
Others[d]	0.6	0.6		0.9	2.8		

[a] Lough and Garton (1968).
[b] Prottey and Hawthorne (1966).
[c] Abbreviations: CE, cholesteryl ester; FAME, fatty acid methyl ester; other abbreviations as in Table IV.
[d] Traces of 12:0, together with acids of chain length greater than 18:3.

lipids of human adipose tissue, the major components of all the human pancreatic lipids are palmitic and oleic acids (Lough and Garton, 1968). In general, each class of lipid has a similar overall fatty acid composition. Other preparations of human pancreatic lipids have yielded much higher proportions of fatty acid methyl esters (Leikola *et al.*, 1965) which may have arisen from the use of chloroform–methanol as the extracting solvent (Lough *et al.*, 1962).

8.3.2. Differences among Homologous Cell Types

The differences noted in the fatty acid composition of the glycerolipids of corresponding tissues or organs of different animal species are also seen in homologous cell types from these species. In numerous instances, however, the advantage gained by the isolation of individual cell types of a given tissue has been lost by increasing contamination of the isolated cells with the incubation or culture medium, which contains lipids or lipoproteins.

8.3.2.1. Erythrocytes

Numerous studies have estimated the fatty acid composition of the lipids of red blood cells of man and experimental animals. A thorough knowledge of animal red cell lipids has been considered necessary to study and understand the composition

Table VIII. Fatty Acid Composition of Total Erythrocyte Lipids from Man and Various Animal Species[a]

Fatty acid	Animal species: Man	Guinea pig	Rabbit	Monkey	Dog	Pig	Ox	Hamster	Cat	Mink
					(weight %)					
12:0	—	—	0.4	—	—	—	—	—	—	0.1
14:0	—	—	—	—	0.8	—	—	—	—	0.4
15:0	—	—	—	—	—	—	—	—	—	0.2
16:0	19.9	9.9	23.5	23.8	19.1	21.4	12.1	19.0	20.1	20.5
16:1	—	0.7	1.8	2.2	1.6	2.4	2.7	4.9	2.6	0.8
17:0	—	—	—	—	—	—	—	—	—	0.5
18:0	15.6	37.3	11.7	15.3	27.4	10.4	14.1	14.7	17.8	20.0
18:1	12.9	9.8	17.9	13.4	14.5	32.1	34.5	22.6	11.0	15.8
18:2	9.0	9.1	25.5	14.5	10.7	23.2	21.1	18.1	21.5	11.3
18:3	—	—	1.7	—	—	—	—	1.1	—	0.8
20:0	—	—	—	—	—	—	—	—	—	0.2
20:2	—	—	0.7	—	—	—	—	—	—	0.1
20:3	3.6	1.5	1.7	1.8	—	—	—	—	—	0.8
20:4	14.9	19.3	7.6	25.0	22.8	6.4	4.8	19.0	18.5	22.2
22:4	—	—	1.0	—	—	—	—	—	—	1.5
22:5	2.5	2.5	0.5	—	—	—	—	—	—	0.8
24:0	4.2	1.6	2.4	—	—	—	—	—	—	1.7
24:1	7.6	—	0.5	—	—	—	—	—	—	0.4
Others	5.0	—	1.4	3.4	—	—	—	—	—	2.6

[a] Modified from Nelson (1972*b*).

and dynamic behavior of lipids in abnormal erythrocytes and to achieve insight into the role of lipids in cell membranes. Table VIII compares the total fatty acid compositions reported for the erythrocytes of man and nine experimental animal species (Nelson, 1972*b*). Although not all analyses have been completed in the same detail, there is nevertheless an impressive similarity in the fatty acid composition in all instances. The palmitic acid content shows relatively little variation, while somewhat greater variation is seen for stearic, oleic, and lauric acids. The greatest fluctuations are seen in the content of arachidonic acid. Only minor amounts of other acids are found in the red blood cells of any of the animal species.

8.3.2.2. Lymphocytes

Normal lymphocytes from different animal species exhibit a very similar composition in their phospholipid fatty acids, which differ markedly from those of other tissues. This observation favors the notion that there may exist a cell-specific pattern of individual phospholipids rather than a species-specific pattern, which would reflect adaptation to environment and function. Table IX compares the fatty acid compositions of phosphatidylcholine and phosphatidylethanolamine from lymphocytes of pig, rabbit, and calf (Ferber *et al.*, 1975). All resting lymphocytes analyzed contain relatively low amounts of polyenoic fatty acids when compared to rat liver. The content of polyenoic acids increases during *in vitro* and *in vivo* stimulation. Thus the ratio of polyenoic to saturated fatty acids in position 2 of rabbit lymphocyte phosphatidylcholine was found to increase during *in vitro* stimulation from 0.538 to 0.926. During *in vivo* stimulation, this ratio changed from 2.9 to 6.2 in position 2 of phosphatidylethanolamine (Ferber *et al.*, 1975). It was reasoned that the increases in the content of the polyunsaturated fatty acids and in

Table IX. Fatty Acid Composition of Phosphatidylcholine and Phosphatidylethanolamine[a] from Lymphocytes of Different Species[b]

	Fatty acid						
	16:0	16:1	18:0	18:1	18:2	20:4	22:6
				(mole %)			
PC							
Pig lymphocytes	38.6	3.7	16.6	19.3	9.2	12.4	—
Rabbit lymphocytes	42.4	1.9	9.6	17.9	18.1	8.0	—
Rabbit thymus	47.9	5.8	6.9	17.2	14.4	5.1	—
Calf thymus	34.6	3.2	8.0	29.4	20.2	2.3	—
PE							
Pig lymphocytes	18.7	2.5	28.1	14.4	4.8	31.3	—
Rabbit lymphocytes	12.9	2.4	27.0	14.6	9.9	27.9	0.4
Rabbit thymus	16.8	3.4	24.8	14.2	9.5	26.1	—
Calf thymus	16.1	0.5	25.1	25.4	12.8	19.3	—

[a] Including alk-1-enylacyl ethers from plasmalogen.
[b] From Ferber *et al.* (1975).

the ratio polyunsaturated/saturated acids brought about an increased fluidity of the cell membrane with appropriate physiological consequences.

8.3.2.3. Spermatozoa

Another source of well-defined mammalian cells is spermatozoa. The phospholipids of mammalian spermatozoa appear to serve two main functions. As components of membranes they play a significant role in maintaining the structural integrity of the various highly organized membrane systems, while as nutrients they may supply the spermatozoa with a ready source of oxidizable substrate. Furthermore, the polyunsaturated fatty acids may serve as precursors of prostaglandins, which play a significant role in the metabolism of the spermatozoa. Table X compares the fatty acid and aldehyde compositions of the phospholipid-bound fatty acids of spermatozoa of five mammalian species. The principal fatty acid found in all species examined, except the rabbit, is docosahexaenoic acid. Docosapentaenoic acid is quantitatively the most important fatty acid in rabbit spermatozoa, and is also found in large amounts in the boar. All species contain palmitic, oleic, stearic, and linoleic acids, while arachidonic acid is present in all species except the rabbit. The predominant fatty aldehyde is palmitaldehyde, with values ranging from 91.2% in ram to 52.1% in human spermatozoa. Poulos *et al.* (1973) have suggested that the differences in the fatty acid composition between the species could be related to differences in behavior of mammalian spermatozoa in response to sudden alterations in environment, e.g., cold shock and dilution. Jain and Anand (1976) have recently provided detailed analyses of the fatty acids of buffalo sperm lipids.

8.3.2.4. Other Cell Types

The development of enzymatic methods for the release of cells from tissues and for the growing of mammalian cells in cell culture media has permitted the examination of the lipid composition of cell preparations which normally occur as tissue. An example is provided by the recent analyses of the lipids of pig, human, and rat epidermal cells. Table XI gives the fatty acid composition of phospholipid and glycosphingolipid fractions of pig and human epidermal cells (Gray and Yardley, 1975). Palmitic, stearic, oleic, and linoleic acids are major fatty acids in the diacylglycerophospholipids of pig epidermal cells and presumably in those of the human epidermal cells. The fatty acids of the sphingomyelins of pig and man are also reasonably close, as are the fatty acids of the glycosphingolipids of the two animal species. The sphingolipids from both animal species contain significant amounts of C_{26} and C_{28} fatty acids, which are not normally found in glycosphingolipids or phospholipids from mammalian sources, with the exception of brain, which does contain C_{26} fatty acid (White, 1973). It has also been noted that the lipid compositions of cells that were isolated with or without contact with bovine serum albumin (a possible source of contamination by fatty acids) were the same.

8.3.3. Differences among Homologous Subcellular Fractions

As a further illustration of the variation in the composition of natural fatty acid mixtures, a few examples have been included which compare the fatty acids of

Table X. *Fatty Acid and Aldehyde Composition of Phospholipids of Mammalian Spermatozoa*[a,b]

Animal species	Chain length : number of double bonds														
	14:0	15:0 br	15:0	16:0	16:1	18:0	18:1	18:2	20:0	22:0	20:4	22:4	22:5	22:6	Others
	Fatty acids (weight %)														
Bull	2.0			14.1		5.3	5.7	3.9	ND[c]	ND	3.5	ND	ND	61.3	4.2
Boar	3.0			13.0		5.1	2.6	2.1	ND	1.5	3.2	1.6	27.9	37.7	2.3
Ram	5.2			14.7		5.1	3.8	1.7	ND	1.1	4.5	ND	ND	61.4	2.5
Rabbit	1.5			24.2		18.7	9.1	4.8	1.6	1.2	ND	ND	39.0	ND	
Man	ND			24.8		11.3	8.1	4.0	ND	2.9	5.1	ND	ND	35.2	8.6
	Aldehydes (weight %)														
Bull	2.3	2.3	2.1	87.4	2.0	1.4	1.3								1.2
Ram	1.6	1.8	1.3	91.2	1.0	1.7	ND								1.4
Boar	2.7	2.9	2.2	82.7	4.3	2.3	ND								2.9
Rabbit	1.2	ND	ND	66.3	ND	30.2	ND								2.3
Man	1.8	ND	ND	52.1	6.4	15.9	11.2								1.1

[a] Abbreviated from Poulos *et al.* (1973).
[b] Mean values of three to six replicate analyses.
[c] ND, not determined.

Table XI. Fatty Acid Composition of Phospholipids and Glycosphingolipids in Pig and Human Epidermal Cells[a]

	Pig			Human	
Fatty acid	PC[b]	SP[b]	GSL[b]	SP	GSL
			(weight %)		
14:0	14.7	1.7	tr	2.6	5.1
15:0	1.2	1.0	tr	1.1	3.4
16:0	26.4	22.5	9.4	14.6	8.2
16:1	6.4	tr			
17:0	tr	tr	tr	2.0	2.4
18:0	7.5	4.5	2.7	6.4	4.3
18:1	19.1	3.1	24.7	2.8	17.9
18:2	22.9				
18:3	tr				
20:0		24.6	15.8	11.6	7.7
20:1					
20:2					
20:3					
20:4					
21:0		0.8	tr	1.3	1.7
22:0		8.1	9.8	8.9	4.3
23:0		1.0	1.6	1.6	1.7
24:0		14.4	14.5	18.8	10.0
24:1		10.5	2.6	9.5	2.0
25:0		0.8	tr	2.0	5.2
26:0		2.8	9.0	5.8	5.4
28:0			7.8	0.7	5.9
24:0:OH			tr		2.6
26:0:OH			tr		5.6
NI		4.2	2.1	10.3	8.6

[a] From Gray and Yardley (1975).
[b] Abbreviations: SP, sphingomyelin; GSL, glycosphingolipid; PC, phosphatidylcholine; NI, not identified.

homologous subcellular fractions of different animal species. According to the concept of tissue specificity of lipid composition (Veerkamp *et al.*, 1962), it would be expected that further similarities might be found, especially for those subcellular fractions which possess the more specialized organization and function.

8.3.3.1. Mitochondria

There has been a great interest in the lipid and fatty acid composition of the mitochondria because lipids have been demonstrated to be essential for mitochondrial function. The ox tissue mitochondria analyzed originally (Fleischer and Rouser, 1965) failed to show the simple and characteristic mitochondrial lipid patterns noted later (Fleischer *et al.*, 1967). This apparently was due to insufficient purification of the mitochondrial fraction and because lipid decomposition falsified the analytical results. Table XII gives the fatty acid composition of the adrenal mitochondria of the elephant, rat, and pig. The fatty acid composition of the neutral lipids from the

Table XII. Fatty Acid Composition of Mitochondria of Adrenals of Selected Animal Species[a]

	Elephant[b]		Rat[c]		Pig[d]	
Fatty acid	TG	PL	TG	PC	TG	PC
			(weight %)			
14:0	5.3	2.5	1.0	tr	2.6	1.0
14:1	1.5	—	0.5	—	0.7	—
15:0 (14:2)	—	—	—	—	—	0.2
16:0	41.2	11.3	15.9	13.3	20.8	14.8
16:1	6.3	1.7	3.5	1.7	1.8	0.5
16:2	2.3	1.7	—	1.7	0.7	1.0
17:0	1.5	1.7	—	tr	tr	—
18:0	9.2	23.2	10.9	28.3	17.3	29.7
18:1	28.2	17.0	22.9	11.7	30.5	11.0
18:2	1.5	6.8	11.9	8.3	11.5	8.9
18:3 ω6	—	—	1.0	—	2.6	1.5
20:0	0.7	1.4	1.0	—	—	—
20:1 ω9	2.3	—	1.5	—	1.8	1.0
20:2 ω9		1.8	1.0	—	—	1.0
20:2 ω6		—	—	—	1.8	—
20:3 ω6		4.0	—	—	2.2	2.4
20:4 ω6		19.3	10.0	35.0	—	22.7
20:5 ω3		—	1.0	—	2.8	—
22:0		—	0.5	—	—	—
22:2 ω9		—	—	—	—	2.0
22:3 ω6		—	1.0	—	—	—
22:4 ω6		1.3	10.0	—	0.7	—
22:5 ω6		1.8	3.9	—	—	1.0
22:5 ω3		1.3		—	—	—
22:6 ω3		1.8	1.5	—	—	—
24:0		—	1.0	—	—	1.0

[a] Abbreviations as in Tables IV and XI.
[b] Cmelik and Ley (1974).
[c] Cmelik and Fonseca (1974).
[d] Cmelik and Ley (1975).

elephant (Cmelik and Ley, 1974) and pig (Cmelik and Ley, 1975) adrenal mitochondria differs from that of the rat (Cmelik and Fonseca, 1974) in a markedly reduced content of arachidonic and docosatetraenoic acids as well as in augmented quantities of oleic and palmitic acids. The mitochondrial phospholipids of the rat and elephant adrenals also differ in the content of the polyunsaturated fatty acids. The presence of substantial quantities of palmitoleic and palmitolinoleic acids in all lipid fractions of the elephant adrenals suggests the existence of the ω7 pathway in this animal species, although no higher homologues of these fatty acids have been identified (Cmelik and Ley, 1974). In general, the distribution of the fatty acids of glycerophospholipids in the adrenal mitochondria of the elephant, rat, and pig is similar to that seen in the heart mitochondria of the ox (Wheeldon *et al.*, 1965), pig (Fleischer and Rouser, 1965), and sheep (Richardson *et al.*, 1961), except that the level of linoleic acid is higher in the pig heart particles. Similarities are also seen in the fatty acid composition of the choline and ethanolamine phospholipids of the adrenal, heart, and liver mitochondria, but the fatty acid patterns of the cardiolipins and phosphatidylinositols show significant differences (White, 1973).

8.3.3.2. Microsomes

The intracellular membranes other than mitochondria also contain a considerable amount of polyunsaturated fatty acids. The microsomal lipids are the most unsaturated, followed by those of the Golgi apparatus, whereas those of the lysosomes have the least degree of unsaturation (Henning *et al.*, 1970). The lipids of the microsomal membranes (sarcoplasmic reticulum) of the skeletal muscle have been shown (Martonosi, 1972) to serve as activators of the adenosine triphosphatase located there, as well as of the Ca^{2+} transport function of the membrane. Table XIII compares the fatty acid and aldehyde compositions of the major glycerophospholipids in the sarcoplasmic reticulum of the skeletal muscle of rat, chicken, man, and rabbit. The major acid is palmitic, which together with palmityl aldehyde makes up nearly 40% of the total. Other major chain lengths are oleic and linoleic acids, comprising another 40% of the total. Arachidonic acid constitutes 7–10%, while stearic and other polyunsaturated acids make up 3–5% of the total. The proportion of arachidonic acid in the phosphatidylethanolamines of chicken and man is nearly double that in the phosphatidylcholines, and there are nearly double the amounts of other long-chain polyunsaturated acids when compared to phosphatidylcholines. The sarcoplasmic reticulum of the rat skeletal muscle is unusual in possessing an especially high proportion of docosahexaenoic acid in its phosphatidylethanolamine fraction. Despite the marked differences in the fatty acid composition of several phospholipid classes, there are general similarities in the pairing of fatty acids in the molecular species of the lipids of sarcotubular vesicles in the four animal species investigated (Marai and Kuksis, 1973). On the basis of these studies, it would appear that the physicochemical requirements of the sarcoplasmic membranes of different animal species may be met by a variety of lipid mixtures within a defined range of composition.

8.3.3.3. Plasma Membranes

The fatty acids of the plasma membranes have been commonly observed to be much more saturated than those of the intracellular membranes, which has been attributed to their function as a barrier against the outside environment (Ray *et al.*, 1969). Several investigators have examined the fatty acid profiles of various lipid classes of plasma membranes of a variety of normal and neoplastic tissues. Table XIV gives the fatty acid composition of the triacylglycerols and phosphatidylcholine of plasma membranes of male rat and mouse liver (Van Hoeven *et al.*, 1975). The results show considerably larger amounts of arachidonic acid in the rat lipids than in those of the mouse, which is compensated for by a somewhat higher proportion of docosahexaenoic acid in the mouse liver plasma membranes. The data on the normal rat liver plasma membranes closely agree with those of Henning *et al.* (1970) and Wood (1970). The overall fatty acid composition of the liver plasma membrane phospholipids exhibits increased saturation when compared with that of intracellular membranes. Examination of the pertinent data on intracellular membranes (White,

Table XIII. Fatty Acid and Aldehyde Content of Glycerophospholipids of Sarcotubular Vesicles of Skeletal Muscle of Selected Animal Species[a,b]

Fatty acid	PC[c]				PE[c]				PS[c]				CL[c]			
	Rabbit	Rat	Chicken	Man	Rabbit	Rat	Chicken	Man	Rabbit	Rat	Chicken	Man	Rabbit	Rat	Chicken	Man
	(mole %)															
14:0 A[c]	tr	tr	tr	1.4	tr	0.5	3.6	2.5	—	—	—	—	—	—	—	—
14:0	0.2	0.4	0.4	1.6	0.4	1.4	1.2	2.0	—	—	—	—	1.2	2.1	tr	0.8
15:0									tr	tr	—	1.3				
16:0 A	6.2	0.9	13.3	6.6	20.8	11.9	11.1	14.2	—	—	—	3.3	tr	tr	1.4	1.8
16:0	34.1	37.3	30.1	30.9	4.7	7.0	9.2	6.0	8.2	2.1	8.7	7.4	5.5	12.0	4.1	3.3
16:1	0.8	1.0	0.4	0.5	1.6	0.5	1.4	0.5	2.4	0.5	1.1	1.2	0.8	1.3	1.6	1.9
18:0 A	tr	0.4	1.0	1.1	6.9	4.2	4.8	4.0	—	—	—	—	tr	tr	2.1	1.2
18:0	3.3	7.0	5.3	6.6	7.6}	19.9}	14.4	14.5	21.3	46.3	35.6	33.5	1.7	4.0	4.5	5.5
18:1 A					6.4}			3.6	—	—	—	—				
18:1	16.8	7.7	18.6	12.1	13.2	5.1	10.3	5.9	9.6	6.9	7.0	9.4	8.6	10.1	21.8	18.7
18:2 ω6	24.7	22.5	12.9	23.6	6.3	6.9	4.7	10.1	7.5	3.8	10.0	4.1	76.0	66.2	61.0	62.1
19:0									3.0	0.6	tr	1.9				
20:1	0.6	tr	0.1	tr	0.8	1.0	tr	0.4	tr	tr	tr	1.2				
20:2 ω9	0.3	0.3	0.3	0.1	tr	0.5	tr	tr	tr	tr	tr	0.1	4.6	2.6	1.4	2.0
20:3 ω6	1.3	1.0	1.0	1.0	0.3	0.7	0.3	2.0	3.2	3.8	8.1	1.6	0.9	tr	tr	1.9
20:4 ω6	7.5	12.8	12.9	10.1	15.7	12.8	21.6	21.6	9.5	10.0	11.4	8.2	0.5	1.6	2.1	0.8
20:5 ω3	0.1	1.1	tr	0.3	0.7	0.7	tr	0.4	0.2	tr	tr	0.3				
22:2 ω9	0.4	0.2	tr	0.3	0.4	0.3	0.5	0.6	5.4	3.2	1.5	8.5				
22:3 ω9	1.2	0.6	0.9	0.7	4.6	1.4	3.7	3.4	8.0	5.6	5.7	5.2				
22:4 ω6	0.6	tr	0.3	0.5	2.3	0.5	3.5	1.4	12.1	1.6	2.6	4.8				
22:5	1.6	2.1	0.7	1.3	5.5	4.1	2.6	2.7	6.2	4.1	tr	3.9				
22:6	0.3	4.4	1.5	0.3	1.8	20.5	6.8	4.0	3.2	11.5	8.4	4.1				

[a] From Marai and Kuksis (1973).
[b] Means of two or three analyses.
[c] Abbreviations: PC, PE, PS, choline, ethanolamine, and serine phosphatides; CL, cardiolipin; A, aldehyde.

Table XIV. Fatty Acid Composition of Triacylglycerols and Phosphatidylcholines of Plasma Membranes Derived from Normal Livers and Hepatoma of Rat and Mouse[a,b]

Fatty acid	PC				TG		
	Rat		Mouse		Rat	Mouse	
	Liver	Hepatoma	Liver	Hepatoma	Liver	Liver	Hepatoma
	(mole %)						
14:0	—	3.5	—	—	5.0	3.6	1.0
16:0	28.8	33.8	42.2	41.3	27.5	28.6	27.7
16:1	—	2.0	—	1.0	10.0	9.3	6.3
18:0	18.0	13.7	11.2	8.4	4.5	2.8	3.3
18:1	8.4	16.6	13.9	14.7	28.5	40.2	46.5
18:2	19.4	20.8	16.2	19.1	15.5	11.7	9.9
18:3	—	1.0	—	—	5.0	3.3	1.9
20:3	1.0	0.5	2.0	1.6	—	—	—
20:4	17.0	6.1	6.2	6.9	1.0	0.5	1.0
20:5	2.4	1.0	1.0	0.5	—	—	—
22:5	1.5	—	1.0	—	2.0	—	—
22:6	3.5	1.0	6.3	6.5	1.0	—	2.4

[a] From Van Hoeven *et al.* (1975).
[b] Abbreviations as in Tables IV and XI.

1973) reveals that this is due to the relatively high content of saturated sphingomyelins in the plasma membrane.

The fatty acid composition (16:0 and 18:0) of phosphatidylcholine shows small but significant differences according to whether the fatty acids are derived from the liver plasma membranes of male or female mice. The sex difference is very pronounced for sphingomyelin, which contains as much as 20 mole % more 22:0 in the case of male than of female livers, at the expense of 16:0 and 24:1 (Van Hoeven *et al.*, 1975).

The fatty acid patterns of the various lipid classes of rat hepatoma plasma membranes differ from those of the corresponding liver membranes by a pronounced decrease of polyunsaturated acids and an increase in mono- and diunsaturated fatty acids. The fatty acid patterns of the liver and hepatoma membrane lipids show much more mutual resemblance in the case of the mouse than of the rat.

8.3.4. Differences among Homologous Lipoproteins

The concept of tissue specificity in lipid and fatty acid composition may be extended to the homologous classes of lipoproteins of different animal species. However, the similarities are not equally obvious in all instances. Again, dietary fat and environmental conditions affect the fatty acid composition greatly, but certain basic features of molecular structure and physicochemical properties appear to be maintained in most cases.

8.3.4.1. Milk Lipoproteins

The lipoproteins of milk occur largely as constituents of the fat globule membrane, a complex lipid–protein system, which is oriented at the fat/plasma interface and serves to stabilize the fat emulsion. In many respects, these particles resemble chylomicrons and other low-density lipoproteins and may therefore be classified with them. Table XV compares the fatty acid compositions of milk triacylglcyerols from several animal species. Marked differences are seen in the short-chain fatty acid content between ruminant and nonruminant milk fats. Although high-carbohydrate diets lead to the production of increased amounts of short- and medium-chain-length fatty acids also in nonruminant milk fats, the very-short-chain-length fatty acids such as butyric acid do not appear to accumulate (Insull *et al.*, 1959). Feeding of protected unsaturated oils to cows results in a replacement of many of the short-chain-length acids by linoleic acid (Mills *et al.*, 1976). The proportion of the short-chain fatty acids in the milk fat therefore appears to be governed by the need to maintain a certain degree of fluidity of the milk lipoproteins. This is achieved by including either short-chain acids or long-chain unsaturated fatty acids into the largely saturated long-chain triacylglycerols produced by the

Table XV. Fatty Acid Composition of Milk Fat Triacylglycerols[a]

Fatty acid	Dog	Guinea pig	Man	Cow		Goat	Sheep	Horse
				Jersey	Holstein			
				(mole %)				
4:0	—	—	—	9.8	8.5	8.2	10.3	—
6:0	—	—	—	5.0	2.9	6.9	3.4	0.7
8:0	—	—	—	2.4	1.4	5.8	2.3	5.4
10:0	tr	—	0.6	4.8	2.3	7.9	3.4	12.3
12:0	0.4	tr	3.0	4.1	2.1	1.9	1.8	8.5
14:0	3.6	3.1	5.3	11.8	7.5	2.6	5.0	6.9
15:0	1.4	0.7	0.6	1.7	1.2	0.7	0.9	0.2
16:0	23.6	30.9	26.5	36.5	28.0	16.0	20.9	21.3
16:1	5.1	3.3	4.0	1.1	1.6	1.2	1.2	4.5
16:2	1.4	—	—	—	—	—	—	—
17:0	2.6	1.0	1.1	0.8	0.7	2.4	2.9	0.5
18:0	8.5	3.0	7.8	8.6	14.6	14.3	15.5	2.1
18:1	41.5	39.0	37.6	13.0	26.5	30.4	27.2	17.4
18:2	8.9	17.1	10.0	0.4	1.5	1.7	2.9	14.7
18:3	—	—	0.6	—	1.0	—	2.4	5.6
20:0	2.4	2.0	—	—	tr	—	tr	—
20:1	0.6	—	0.6	—	—	—	—	—
20:2	—	—	0.5	—	—	—	—	—
20:3	—	—	0.4	—	—	—	—	—
20:4	—	—	0.8	—	—	—	—	—
22:2	—	—	0.3	—	—	—	—	—
22:5	—	—	0.1	—	—	—	—	—
22:6	—	—	0.3	—	—	—	—	—

[a] From Breckenridge and Kuksis (1967).

mammary gland. More complete analyses of the milk fats from a much larger number of animal species have been compiled by Morrison (1970).

8.3.4.2. Plasma Lipoproteins

The various lipid classes of the plasma lipoproteins possess characteristic fatty acid profiles. Since different density classes of the lipoproteins differ in the lipid class content, they also differ in fatty acid composition. There is evidence, however, that in the various lipoprotein density classes the patterns of fatty acids of individual lipid classes do not change, although further systematic studies are necessary (Skipski, 1972).

Although animals have been widely employed in studies of lipoprotein synthesis and turnover, only fragmentary information is available concerning the lipid composition of the plasma lipoproteins of various animal species. Furthermore, it is difficult to interpret the reported data since most workers have assumed that methods which successfully separate human plasma lipoproteins can be used to fractionate animal plasma. Recent work, however, has shown that the flotation scheme used to fractionate human plasma lipoproteins is not directly applicable to the preparation of low-density lipoproteins from the rat or pig (Margolis, 1969). In general, the low-density lipoproteins in the rabbit, chicken, and dog have compositions similar to those found in man.

Of the plasma lipoproteins, the chylomicrons are subject to the greatest dietary influence, and care must be taken to control the composition of the dietary fat before any comparisons of fatty acid composition are made. Table XVI gives the composition of the major fatty acids in the triacylglycerols of the oil phase and the membrane of the chylomicrons in dog, man, and rat receiving corn oil (Zilversmit, 1965). There are only minor differences in the fatty acid composition of the triacylglycerols of the oil phase among three animal species. Somewhat greater are the differences among the animal species in the triacylglycerol composition of the membranes of chylomicrons. The membranes of chylomicrons of man appear to accommodate proportionally less of the unsaturated fatty acids of the oil phase, yielding a more saturated membrane phase. In all instances, the fatty acid composition of the triacylglycerols of the chylomicron membrane is much more saturated than that of the triacylglycerols of the oil phase. It has been suggested (Zilversmit, 1969) that the difference in the degree of unsaturation between triacylglycerols of membrane and oil phases results from the selective accumulation of saturated triacylglcyerols in the membrane phase during freezing. Chylomicron membranes prepared at room temperature do not appear to exhibit as marked differences as those prepared at lower temperatures. Similar observations have been made with the membranes of the milk fat globule, which have yielded a high-melting triacylglycerol mixture (Huang and Kuksis, 1967).

8.3.4.3. Bile Lipoproteins

Bile from most animal species is known to contain large amounts of phosphatidylcholine which exists in micellar solution in association with bile acids, cholesterol, and an apoprotein (Swell *et al.*, 1968). The function of the phosphatidyl-

Table XVI. Fatty Acid Composition of Triacylglycerols of the Oil and Membrane Phase of Chylomicrons of Selected Animal Species[a,b]

	14:0		16:0		18:0		18:1		18:2		Saturated	
Animal species	Oil	Membrane	Oil	Membrane	Oil	Membrane	Oil	Membrane	Oil	Membrane	Oil	Membrane
					(weight %)							
Dog	0.1	1.8	12	44	1.7	11	29	17	56	25	14	58
	0.3	1.4	12	49	2.0	8.2	29	15	53	22	14	59
Man	3.2	9.0	16	58	3.6	13	29	9.8	40	5.5	27	84
Rat	0.2	1.0	9.5	32	1.3	5.1	28	21	60	39	11	38
	0.6	0.4	12	47	1.6	6.9	31	19	53	25	15	55

[a] Abbreviated from Zilversmit (1965).
[b] Peaks of less than 3% of total area have been excluded (no more than 11% of total area).

cholines is believed to be that of a detergent. The structural requirements of the phosphatidylcholines in this instance could conceivably be very different from those of the same lipid in tissue membranes, which might be reflected in an altered fatty acid makeup. Since these phosphatidylcholines would serve a comparable physicochemical function in all animal species, it is anticipated that they would possess comparable fatty acid compositions. Table XVII compares the fatty acids of the phosphatidylcholines from several animal biles. Linoleic acid is the principal unsaturated fatty acid of pig bile phosphatidylcholine, but that from the ruminants contains appreciable quantities of 9-*cis*,11-*trans*-octadecadienoic acid, 9-*cis*,11-*trans*,15-*cis*-octadecatrienoic acid, and linolenic acid, together with other polyunsaturated fatty acids of the $\omega3$ series. The palmitic, stearic, oleic, and linoleic acids are generally the major components, while arachidonic acid, which is a common constituent of phosphatidylcholine, is present in appreciable quantities only in pig bile. A study of the molecular species of the biliary phosphatidylcholines (Christie, 1973) has shown that there exist considerable specificity and similarity in the manner in which the different unsaturated fatty acids are associated with the C_{16} and C_{18} saturated components, especially when expressed in terms of degree of preference of the unsaturated fatty acids for palmitic acid in each chemical species. It is therefore possible that the principal structural requirement of phosphatidylcholine for its detergent action in bile is that it be liquid, and this requirement may be satisfied by the presence of palmitic acid in position 1 and of an unsaturated fatty acid in position 2 (Christie, 1973).

8.4. Fatty Acids of Individual Animal Species and Man

The fatty acid composition of an animal species is characteristic of its genetic makeup, its diet, and its environment. Individual members of a species, however, may also show genetic differences, when matched for age, sex, and diet (Chin, 1970; Kang *et al*., 1971). Comparisons of fatty acids of glycerolipids from specific tissues must therefore be matched also for genetic homogeneity of the population.

8.4.1. Tissue Differences

It has been well established that animal tissues differ in their fatty acid composition (Hilditch and Williams, 1964). This is due largely to differences in content of various lipid classes which possess characteristic fatty acid profiles. However, certain variation has also been found in the fatty acid composition of the same lipid class when isolated from different tissues. Other differences among tissues and lipid classes arise because of the diet, age, sex, and metabolic state of the animal. In the following discussion, some of these effects have been illustrated by examples from different animal species.

8.4.1.1. Rat

The fatty acid composition of rat tissue lipids has been investigated more extensively than that of any other mammalian tissues, including human. This has

Table XVII. Composition of Fatty Acids of Bile Phosphatidylcholines of Selected Animal Species

Animal species	Fatty acid															
	16:0	16:1	18:0	18:1	18:2 (ω6)	18:2 conj.	18:3 (ω6)	18:3 (ω3)	18:3 conj.	18:4 (ω3)	20:3 (ω6)	20:4 (ω6)	20:4 (ω3)	20:5	22:5	22:6
	(mole %)															
Pig[a]	34.6	1.5	11.7	15.9	23.0	—	0.3	1.0	—	—	1.2	6.1	—	0.9	1.0	1.5
Sheep[a]	36.0	3.0	9.8	27.9	6.6	4.7	—	4.9	1.4	—	—	0.8	—	1.4	0.8	0.8
Cow[a]	33.3	4.5	12.2	25.6	10.2	1.1	0.6	2.9	—	0.4	1.5	1.6	1.0	1.7	1.2	0.5
Dog[b]	22.1	1.9	12.5	14.5	35.3	—	—	tr	—	—	—	13.3	—	—	—	—
Rat[b]	26.9	1.0	11.3	8.0	32.1	—	—	tr	—	—	—	18.3	—	—	—	—
Man[b]	56.7	3.0	3.3	12.5	23.4	—	—	tr	—	—	—	1.0	—	—	—	—

[a] Christie (1973).
[b] Balint *et al.* (1965).

Table XVIII. Fatty Acid Composition of Rat Tissue Lipids on Standard Laboratory Diet[a,b]

Fatty acid	Heart				Kidney				Skeletal muscle				Liver			
	CE	TG	FFA	PL	CE	TG	FFA	PL	CE	TG	FFA	PL	CE	TG	FFA	PL
	(weight %)															
12:0					1.0				1.3							
13:0					0.6											
14:0 br					0.9				0.6							
14:0	1.6	1.2	1.4		3.2	1.1	0.7		2.1	0.6	1.5	0.6	1.0	0.9	0.5	
14:1					2.1				1.5							
15:0	0.6				0.9				0.8		0.7					
15:1	0.4															
A[c]	0.8			1.0	1.6			1.0			0	1.9	0.8			
16:0	26.5	15.5	15.4	14.9	30.2	27.1	15.9	8.2	23.9	16.9	15.5	17.8	24.2	17.1	15.2	16.9
16:1	3.5	5.2	3.0	1.4	6.1	4.2	2.5	1.1	2.9	3.5	5.2	3.1	2.9	3.8	2.7	1.1
17:0 br									0.8							
17:0	1.0	0.9	0.6	0.6	1.2	0.6	0.9		1.3	0.5	0.5		0.8	0.5		0.9
16:2							0.6		0.8						0.6	
B[c]					0.5			1.1				1.0	0.5			
18:0	23.8	15.5	16.6	24.7	19.2	11.3	11.9	18.6	20.0	7.1	9.5	23.1	19.7	14.2	11.4	37.6
18:1	18.9	37.9	24.5	13.1	14.9	33.1	17.9	15.6	19.1	43.4	32.9	14.4	15.2	30.3	14.3	9.6
18:2	5.5	16.1	21.1	30.5	2.3	12.9	13.5	13.7	4.9	24.7	20.4	21.1	16.1	24.2	24.7	15.2
18:3	0.6	0.6	1.3	0.7		0.7	1.8		1.1	0.8	1.0				1.4	
20:0	1.3		1.2		1.6	1.8	1.7		3.4		1.0		0.5			
20:1	0.6	0.7				1.9			5.1		0.7		0.8	0.6		
21:0	3.0	0.5	2.0		2.8	2.4	2.0		2.6		1.2		1.2	0.7	1.1	
20:4	4.9	1.2	10.2	11.3	3.5	1.1	28.0	37.2	2.4	0.6	6.1	15.3	13.7	5.5	24.1	16.7
22:0	2.9				6.5		1.8		2.4		0.9		0.8			

[a] From Connellan and Masters (1965).
[b] Abbreviations as in Tables IV and VII.
[c] Unknown acids.

been so because of the general utility of the rat as an experimental animal and because of the ease of induction of essential fatty acid deficiency in this species.

One of the most complete analyses of the fatty acid composition of both hepatic and extrahepatic tissues of the rat under defined dietary conditions has been provided by Connellan and Masters (1965). Table XVIII gives the fatty acid composition of the heart, kidney, skeletal muscle, and liver lipids of rats fed a standard laboratory diet. The C_{12}–C_{22} fatty acids represent in general more than 90% of all the fatty acids present. Comparisons of individual lipid fractions show that the concentration of the fatty acids in the different cholesterol ester fractions is similar for most of the major components. The linoleic and arachidonic acid content in liver, however, is considerably higher than that in the extrahepatic tissues. With the triacylglycerols, greater variations of the component percentages occur between tissues. The main component of all triacylglycerol fractions is oleic acid. The palmitic acid content of kidney triacylglycerols is markedly higher than that of other tissues, and arachidonic acid is present in liver triacylglycerols to a greater extent than in the extrahepatic tissues. The glycerophospholipid fraction exhibits some differences between liver and extrahepatic tissues. Although all the phospholipid fractions display high arachidonic acid contents, kidney phospholipid is the richest source of this acid. Connellan and Masters (1965), however, failed to obtain an estimate for such polyunsaturated fatty acids as 20:5, 22:4, 22:5, and 22:6, all of which are readily detectable in the phospholipids of most tissues of the rat and other animals.

Table XIX gives the composition of long-chain unsaturated fatty acids of various tissues of the rat during intake of diets containing saturated and unsaturated fat or no fat at all (Egwim and Kummerow, 1972). In accordance with the general observations of other investigators (Carroll, 1965), the present study illustrates the profound influence of diet on the fatty acid composition of rat tissues. It was concluded by Egwim and Kummerow (1972) that the linoleate-deficient nature of the hydrogenated fat, rather than its high content of *trans* acids, explained the high tendency of this fat to induce accumulation of long-chain $\omega 9$ fatty acids in the phospholipids and cholesteryl esters of the tissues studied. Unfortunately, the changes in the C_{16} and C_{18} fatty acids were not reported.

The complete composition of the fatty acids of the cholesteryl esters, total phospholipids, and triacylglycerols of the adrenal lipids of rats given a fat-free diet and a commercial chow diet is given in Table XX (Takayasu *et al.*, 1970). There is a replacement of arachidonic and linoleic acids by palmitoleic, oleic, and eicosatrienoic acids. Dietary deficiency of essential fatty acids results in a fall in the proportion of linoleic acid and a rise in those of palmitoleic, oleic, and eicosatrienoic acids in neutral and phospholipid fractions when compared to the corresponding values for stock-fed animals. Other studies have shown similar effects for other rat tissues (Guarnieri and Johnson, 1970), although the fatty acids have not been analyzed in the same detail as in the present illustrations. Complete analyses of the fatty acids of molecular species of selected glycerophospholipid classes of normal and essential fatty acid deficient rats have been presented by Van Golde *et al.* (1968).

8.4.1.2. Pig and Dog

The tissue lipids of pig and dog have been frequently investigated because these animals yield larger amounts of tissue and because their lipid metabolism has been

Table XIX. Dietary Fat and Long-Chain Fatty Acid Composition of Phospholipids from Rat Tissues[a,b]

Tissue and diet	Fatty acid												
	20:1 (ω9)	20:2 (ω9)	20:2 (ω6)	20:2 (ω6 + ω7)	20:3 (ω9)	20:3 (ω6)	20:4 (ω6)	20:4 (ω6 + ω7)	22:1 (ω6)	22:2 (ω9)	22:3 (ω9)	22:3 (ω6)	22:4 (ω6)
	(weight %)												
Liver													
Corn oil	tr					tr	24.4						0.6
Hydrogenated fat					20.6	2.7		2.7	0.5				
Fat free					15.5	4.0		3.4					
Heart													
Corn oil						4.6	11.6		1.0			2.6	3.8
Hydrogenated fat		0.7		1.4	1.5	1.4		15.0		0.8	0.3		
Fat free				1.3	22.4			11.5					
Adrenals													
Corn oil	2.5		3.0		1.5		8.8						7.3
Hydrogenated fat				5.0	13.0			25.5		1.0	10.0	1.9	
Fat free	1.2			4.4	21.5			13.8	1.0			2.0	3.0

[a] From Egwin and Kummerow (1972)
[b] The hydrogenated soybean fat contained 16:0 (8.1%), 18:0 (13.2%), 18:1 (76.5%) of which 48.4% was the *trans* isomer, and 18:2 (1%) and traces of other minor acids.

Table XX. Fatty Acid Composition of Adrenal Lipids from Rats Given Fat-Free Diet for 10 Weeks and Rats Given Commercial (Solid) Diet Control[a,b]

	Fat-free diet			Control		
Fatty acid	CE	PL	TG	CE	PL	TG
			(weight %)			
14:0	tr	tr	1.8		tr	
16:0	10.5	8.7	25.7	10.5	6.9	25.6
16:1	4.5	0.5	9.3	2.0	0.1	5.2
18:0	2.9	28.9	6.2	2.6	31.9	7.4
18:1	19.7	17.2	54.9	11.9	8.3	37.8
18:2	1.2	0.5	0.4	4.5	4.7	19.1
18:3 ω6	tr			0.2	tr	tr
18:3 ω3	0.1			0.2	tr	0.3
20:0	2.2	0.3	tr	0.9	0.3	0.1
20:1	5.9	0.8	0.5	2.5	0.5	1.2
20:2 ω9	1.2		0.1			
20:2 ω6	tr			0.7	0.3	0.1
20:3 ω9	6.0	3.0	0.3	tr	0.1	tr
20:3 ω6	0.7	0.2		2.2	0.4	0.1
20:4 ω6	6.3	36.7	0.4	16.2	42.6	1.4
20:5 ω3	tr			1.6	tr	0.1
22:0	0.4	0.3		0.8	0.2	
22:1 ω9	2.4			1.4		
22:3 ω9	8.8					
22:4 ω6	20.0	2.1	0.3	21.5	1.8	0.6
22:5 ω6	3.3			1.7	tr	tr
22:5 ω3	0.6			5.0	0.3	0.3
22:6 ω3	1.8			12.3	0.8	0.7
24:0	tr	0.4		tr	0.5	
24:1 ω9	1.5	0.4		1.3	0.3	

[a] From Takayasu *et al.* (1970).
[b] Abbreviations as in Tables IV and VII.

thought to be more like that of man. Table XXI gives the fatty acid composition of the triacylglycerols from various pig tissues. In all instances, the major fatty acids are the C_{16} and C_{18} saturated and the C_{18} mono- and diunsaturated acids. Nevertheless, significant differences are seen between the various tissues, although those of the blood and liver and those of the kidney and adipose tissue are rather similar. Table XXII shows that the fatty acid composition differs little among the triacylglycerols isolated from different sites of the adipose tissue in either the dog or the pig.

8.4.1.3. Sheep and Ox

The ruminant lipids are peculiar in the content of unusually high proportions of stearic acid in depot fats. In addition, their tissue lipids contain branched-chain fatty acids and a complex mixture of unsaturated fatty acids, which includes geometric and positional isomers of oleic, linoleic, and linolenic acids. There is a continuous

Table XXI. Fatty Acid Composition of Triacylglycerols from Various Pig Tissues[a]

Fatty acid	Liver	Kidney	Heart	Adrenal	Blood	Mesentery	Milk
				(mole %)			
14:0	1.1	1.8	1.5	1.8	1.8	2.2	4.3
16:0	25.4	30.7	25.8	30.2	27.8	32.1	31.6
16:1	2.6	2.7	2.7	2.5	3.4	2.7	9.4
18:0	10.1	18.1	15.9	19.4	11.1	19.9	4.5
18:1	34.2	37.0	47.7	39.1	41.5	38.3	38.4
18:2	17.1	7.6	6.8	7.1	11.7	5.0	10.4
18:3	0.9	0.5			0.7		1.4
20:1	0.6	1.0			0.5		
20:3 ω9	0.5	0.1			0.6		
20:3 ω6	0.9	0.3					
20:4	3.5	0.2			0.9		
20:5	0.7						
22:5	1.8						
22:6	0.6						

[a] Calculated from Christie and Moore (1970).

supply from the rumen of short-chain volatile fatty acids, including formic, acetic, propionic, and butyric acids. The lipids of sheep and ox tissues are a major source of unusual dietary fatty acids in man.

Table XXIII gives the fatty acid composition of the neutral lipids of ovine fetal and maternal tissues (Scott *et al.*, 1967). The proportion of oleic acid in the neutral lipids of maternal heart is well below the fetal value and the proportion of stearic acid is greater. The differences in the degree of saturation of fatty acids in the neutral lipid components are less pronounced. The proportion of C_{18} dienoic and trienoic fatty acids in the maternal phospholipids is much higher than that occurring in the phospholipids in the fetal tissue. This probably reflects differences in the levels of fatty acid desaturases and acyltransferases in these tissues.

8.4.1.4. Rabbit

Table XXIV gives the fatty acid composition of the neutral lipids and phospholipids of rabbit tissues (Comai *et al.*, 1975). The discrepancies among the different tissues are comparable to those seen for the corresponding lipid classes in the rat fed laboratory chow. Odutuga *et al.* (1973) have provided detailed fatty acid composition of rabbit brain at different stages of development.

8.4.1.5. Man

The much more limited analyses in man have in general confirmed the differences demonstrated in the fatty acid composition of animal tissues, as well as the variations with age, sex, and diet. In many instances, however, the analyses in man have been confined to autopsy material and may therefore represent partially autolyzed products. Furthermore, much of the more detailed data have come from

Table XXII. Fatty Acid Composition of Triacylglycerols of Various Sites of Adipose Tissues of Dog and Pig

	Dog[a]				Pig[b]				
Fatty acid	Mesenteric	Subcutaneous	Pericardial	Perirenal	Inner back	Outer back	Perinephric	Stomach	Mesenteric
	(mole %)								
14:0	3.0	3.1	3.1	1.6	1.6	1.7	3.0	2.3	2.2
16:0	21.0	21.9	22.8	20.8	29.1	27.4	30.7	34.4	32.1
16:1	5.7	6.4	4.7	4.9	2.5	2.9	2.9	2.7	2.7
18:0	8.6	6.9	10.3	11.0	18.0	13.0	16.7	20.3	19.9
18:1	46.6	45.9	45.4	44.3	40.7	45.8	40.8	35.7	38.3
18:2	13.7	13.9	12.9	15.6	8.1	9.3	7.0	4.6	5.0
18:3	0.9	1.1	0.5	1.3					
20:4	0.2	0.3		0.2					

[a] Gold (1968).
[b] Calculated from Christie and Moore (1970).

Table XXIII. Fatty Acid Composition of Neutral Lipids of Ovine Fetal and Maternal Tissues[a]

Fatty acid	Neutral lipids								Phospholipids			
	Heart		Liver		Kidney		Brain		Heart		Liver	
	Fetal	Adult	Fetal	Adult	Fetal	Adult	Fetal	Adult	Fetal	Adult	Fetal	Adult
	(weight %)											
10:0	0.2	0.1			0.3		0.5					
12:0	0.3	0.1	0.4	0.6	0.7	0.5	0.7	0.4			tr	tr
14:0	2.0	2.7	5.0	2.3	2.4	2.0	2.0	1.0	0.6	0.4	0.6	0.2
14:1	1.3	0.3		1.0	0.5	0.3	0.5	0.3	tr	0.4	tr	0.1
15:0	1.1	0.4	0.3	0.8	0.8	0.3	0.3	0.3	tr	0.6	0.6	0.4
15:1	0.6	0.3		0.5	0.3	0.2	0.3	1.8				
16:0	20.2	20.4	25.1	28.3	25.4	28.9	13.6	9.4	15.9	12.8	19.8	15.7
16:1	3.7	2.6	3.0	5.0	3.9	4.3	2.8	2.1	5.4	5.1	5.7	1.0
17:0	0.8	1.8	0.8	1.5	1.1	0.8	0.6	0.4	1.4	1.3	1.6	1.2
17:1	2.8	1.1		1.3	0.7	0.5	0.5	2.0	1.8	2.6	0.8	0.4
18:0	13.6	28.6	14.0	12.5	11.8	13.7	18.2	30.5	11.3	13.3	19.6	29.5
18:1	48.0	37.0	45.0	37.3	43.9	35.0	18.5	31.4	38.9	14.3	28.3	23.6
18:2	1.3	3.4	5.8	5.3	1.6	6.5	0.9	2.2	3.0	23.6	1.7	9.2
18:3		0.8		0.8	0.8	1.3	0.6	2.0		1.7	tr	1.8
20:4	1.5	0.4	tr	2.1	3.0	2.5	4.1	8.4	9.5	15.6	11.3	9.7
22:0	1.2				1.9				4.4		3.2	
Hydroxy							22.5	7.8				
Unknown	0.8		0.3	0.3	0.7	1.8	5.5		3.4	4.9	3.2	4.9

[a] Modified from Scott *et al.*(1967).

Table XXIV. Fatty Acid Composition of Rabbit Tissue Glycerolipids and Cholesteryl Esters[a,b]

	Heart		Liver			Adrenal			Testes			Depot	Spleen	
Fatty acid	TG	PL	TG	CE	PL	TG	CE	PL	TG	CE	PL	TG	CE	PL
							(mole %)							
14:0	4.0	1.1	4.2	2.8	0.2	2.6	2.6	0.3	2.5	3.0	0.4	3.1	3.8	0.3
16:0	29.3	20.7	37.9	37.3	23.4	21.5	14.7	20.2	23.5	31.5	28.7	25.0	28.6	22.4
16:1	3.5	0.7	4.0	4.2	0.7	3.4	3.0	0.5	9.1	5.0	0.8	5.2	4.7	0.6
18:0	4.5	13.9	3.7	6.9	18.6	5.0	5.5	19.1	3.0	5.8	10.6	4.6	9.2	16.8
18:1 ω9	21.4	10.3	29.6	29.6	13.1	21.1	29.0	18.9	25.7	27.0	14.0	25.4	29.6	12.8
18:2 ω6	30.6	24.8	12.7	12.3	31.6	30.7	15.0	8.9	26.2	24.7	7.8	26.6	18.8	19.3
18:3 ω3	5.5	0.8	3.0	1.8	2.8	10.8	2.3	1.5	6.1	3.0	0.4	7.8	2.3	1.0
20:3 ω6	0.1	0.3	0.3		0.6	0.1	1.7	0.4	0.3	1.2	4.5			1.2
20:4 ω6	0.5	19.7	1.0	0.7	5.9	1.2	8.1	27.5	0.6	2.4	15.0	0.3		17.6
22:4 ω6		0.5			0.2	0.3	12.3	0.9	0.2		3.3			1.7
22:5 ω6		0.4	0.4		0.3		1.6	0.1	0.4		10.8			0.5
22:5 ω3		1.9			0.4		2.7	0.8			0.4			2.0
22:6 ω3							0.2							

[a] From Comai *et al.* (1975).
[b] Abbreviations as in Tables IV and VII.

Table XXV. Fatty Acid Composition of Triacylglycerols of Human Tissues

Fatty acid	Liver[a]	Heart muscle[a]	Adipose tissue[a]	Aortic tissue[a]	Myo-cardium[b]	Myocardium[c] Mito-chondria	Myocardium[c] Microsomes
				(weight %)			
12:0						0.8	1.0
14:0	1.1	4.6	4.6	3.1	4.1	4.9	4.7
14:1		0.6	0.7			0.4	1.1
15:0	0.2	0.3	0.5				
16:0	28.2	26.4	23.9	26.6	29.9	30.3	28.1
16:1	2.5	8.6	7.2	5.4	5.7	10.5	12.7
17:0	0.6	0.4	0.5				
17:1		0.3	0.5				
18:0	7.4	6.1	7.5	6.3	8.4	5.8	5.3
18:1	50.5	45.6	46.2	44.5	45.5	42.5	40.8
18:2	7.1	5.5	6.3	9.7	5.3	5.2	5.0
18:3 (20:0)	0.4	0.7	1.0	0.9			
20:1	0.5	0.9	1.1	1.0			
20:2				1.3			
20:3	0.2			0.6			
20:4	1.3			1.2			
20:5				0.4			

[a] Christie *et al.* (1971).
[b] Fletcher (1972).
[c] Gloster and Harris (1969).

patients with specific diseases or clinical conditions. Table XXV gives the fatty acid composition of triacylglycerols of different tissues of man. Marked similarities are seen in the qualitative and quantitative makeup of the fatty acids regardless of the site of origin of the triacylglycerols. The content of the polyunsaturated fatty acids, however, may be somewhat higher in the liver and aortic tissue triacylglycerols than in other tissues.

Unlike the neutral glycerolipids, the glycerophospholipids show marked differences in the fatty acid composition when different tissues of man are compared. However, analyses from single donors of tissues are notably lacking, as are analyses from tissues of individuals of the same age, sex, and metabolic state. Table XXVI compares the fatty acid compositions in the ethanolamine phosphatides in the gray and white matter of human brain at different ages (Svennerholm, 1968). In the cerebral cortex (gray matter), 18:0 is the major fatty acid at all ages and contributes about 30% of total. In younger brains, the polyunsaturated fatty acids of the linoleate family predominate, and 20:4 $\omega6$ and 22:4 $\omega6$ together contribute about 25%. With increasing age, the contribution of the latter fatty acids diminishes and there is a corresponding increase of fatty acids of the linolenate series from about 20% in the fetal brain to 35% in the older brain. The ethanolamine phosphatide of white matter differs from that of cortex in several respects. The 18:1 is the major fatty acid in child and adult brain and comprises about 40%, while 18:0 occurs only to the extent of about 10%. The percentage of fatty acids of the linoleate series is

Table XXVI. *Fatty Acid Composition of Phosphatidylethanolamine in Gray and White Matter of Human Brain at Various Ages*[a, b]

Fatty acid	12 weeks	35 weeks	38 weeks		1 month		7 months		4 years		26 years		82 years	
	Gray	Gray	Gray	White	Gray	White	Gray	White	Gray	White	Gray	White	Gray	White
	(weight %)													
16:0	10.6	6.9	7.6	7.8	6.4	5.9	6.5	5.0	5.0	4.8	5.9	6.2	6.8	5.4
16:1	1.2	0.6	0.6	0.6	0.7	0.9	0.7	0.9	0.4	0.5	0.4	1.1	1.2	0.8
18:0	30.0	29.1	32.8	28.9	34.6	28.7	27.8	13.8	30.1	10.2	30.4	13.8	27.2	10.7
18:1	13.3	9.8	8.5	7.6	8.7	13.0	10.3	23.6	8.8	35.2	8.7	43.2	9.8	40.4
18:2 ω6	0.2	0.2	0.3	0.2	tr	0.1	0.2	0.4	0.4	0.7	0.5	0.5	0.3	0.3
18:3 ω6	0.1	0.1	tr	tr	tr	tr	tr	tr	tr	0.1	0.1	0.1	tr	tr
18:4 ω3	0.3	0.2	0.2	0.2	0.3	0.7	0.5	1.9	0.4	1.9	0.2	1.3	0.2	1.2
20:1 ω9 + 18:3 ω3	0.8	0.5	0.3	0.3	0.3	1.6	0.6	3.6	0.5	6.4	0.5	6.0	0.6	6.2
20:3 ω9	1.6	0.9	0.8	0.7	0.6	0.7	0.7	1.3	0.5	0.7	0.3	0.4	0.4	0.8
20:3 ω6	0.5	0.8	1.0	1.1	1.2	1.4	1.5	1.7	1.6	1.9	1.1	1.0	0.8	0.9
20:4 ω6	17.3	17.8	14.9	16.2	16.5	15.1	16.4	13.4	16.7	9.5	13.2	7.9	10.3	9.2
22:4 ω6	9.5	10.1	10.6	13.3	11.1	13.3	11.7	19.6	9.9	18.0	8.3	13.4	6.3	12.8
22:5 ω6	2.8	4.1	4.4	4.8	2.7	2.3	5.0	3.5	2.4	1.4	1.5	0.5	0.9	0.6
22:5 ω3	0.5	0.6	0.4	0.5	0.7	0.9	0.5	0.8	0.6	0.7	0.5	0.3	1.0	1.1
22:6 ω3	10.8	18.3	17.1	17.7	16.1	15.2	16.9	8.7	22.3	5.7	28.6	3.0	33.9	8.3
24:4 ω6	0.6	0.2	0.5	0.2	0.3	0.5	0.5	1.6	0.5	2.2	0.1	1.5	0.2	1.1

[a] From Svennerholm (1968).

[b] The autopsies were performed 12–48 hr after death. The brains were frozen and stored at −20°C prior to lipid extraction. Cerebral cortex (gray matter) and white matter were separated from the frontal and precentral gyri after thawing at 2°C. Subjects showed no signs of malnutrition, with normal neurological and mental states prior to death and normal microscopic findings for the brain with respect to age.

about the same as in cortex, but that for 22:4 ω6 is larger than that for 20:4 ω6. There is some decrease in the content of the linolenate series with age. In the cortex, 22:6 ω3 is comparatively sparse in white matter after early childhood and does not increase with age. Svennerholm (1968) has also examined the fatty acid composition of other glycerophospholipids of human brain and has found specific fatty acid patterns for each one. Furthermore, the pronounced differences between the fatty acids of the lipids from cerebral cortex and from the adjacent white matter justify speaking of tissue-specific fatty acid patterns of brain phosphoglycerides.

Table XXVII gives the fatty acid composition of the phosphatidylcholines of human skeletal muscle at various ages (Bruce and Svennerholm, 1971). The fatty acid components vary considerably with age, the linoleic acid concentration increasing from 4% in the beginning of the second trimester of gestation to 40% at the age of 1 year. In adults, linoleic acid constitutes about 40% by weight. The concentration of fatty acids of the linolenic acid series is very low. There are large differences between the fatty acid patterns of the individual glycerophospholipids in adult brain and muscles, which makes direct comparison of the age variations less meaningful. Moreover, the changes in the fatty acid patterns of brain glycerophospholipids are in part related to the myelination, while no such morphological changes occur in muscles during the late prenatal or postnatal development.

Bruce (1974) has determined the fatty acid composition of individual glycerophospholipids of human skeletal muscle in male subjects aged 0–55 years. It was found that after 1 year the variations in the percentage amounts of the different fatty acids in the individual glycerophospholipids were rather small.

Table XXVIII gives the fatty acid composition of the cholesterol ester, triacylglycerol, and total phospholipid fractions of human adrenals (Takayasu *et al.*, 1970). In the cholesterol esters, 24 kinds of fatty acids have been estimated. The percentage of 18:1 is highest, and 20:3 ω6 and 22:4 ω6 represent high percentages in polyenoic acids. The adrenal phospholipid contains about 20% of arachidonate. Many more polyenoic acids were found in triacylglycerols of an adrenal adenoma of

Table XXVII. Fatty Acid Composition of Phosphatidylcholine of Skeletal Muscle of Man During Growth[a]

	Fetal age		Postnatal age					
Fatty acid	20 weeks	30–34 weeks	Full term	1.5 months	2.5 months	10 months	5 years	16 years
			(mole %)					
16:0	38.3	37.2	35.1	30.3	31.0	29.0	25.9	27.6
16:1 ω7	3.9	1.5	2.6	1.1	0.4	1.5	1.4	2.0
18:0	7.8	7.5	5.1	10.0	8.0	9.4	11.3	9.1
18:1 ω9	34.9	26.7	23.2	19.0	16.7	12.5	12.4	12.7
18:2 ω6	4.7	9.0	17.2	26.3	31.9	37.4	39.4	39.3
20:3 ω6	0.1	2.4	2.6	2.2	1.7	1.3	1.3	1.0
20:4 ω6	6.7	12.5	10.3	8.5	7.9	7.9	5.7	5.0
22:6 ω3	0.7	1.0	1.0	1.0	0.1	0.2	0.6	0.7

[a] From Bruce and Svennerholm (1971).

Table XXVIII. Fatty Acid Composition of Cholesterol Esters and Glycerolipids of Human Adrenal Glands[a, b]

Fatty acid	CE		PL		TG		TG
	Normals[c]	Adenoma[d]	Normals[c]	Adenoma[d]	Normals[c]	Adenoma[d]	Adipose Tissue[d]
				(weight %)			
14:0	0.5	0.8	0.5	—	1.6	0.4	0.9
16:0	6.7	7.2	18.1	22.9	19.7	23.4	19.4
16:1	4.2	3.3	2.5	2.4	7.4	4.0	4.5
18:0	2.7	3.1	16.2	24.6	3.4	9.4	6.5
18:1	37.6	40.6	22.3	17.4	43.7	22.5	40.3
18:2	7.0	10.5	7.3	8.7	15.5	10.3	20.5
18:3 ω6	0.3	0.5	0.1	0.1	0.1	0.1	tr
18:3 ω3	0.1	0.2	0.0	0.1	0.3	0.1	1.2
20:0	0.2	0.3	0.1	0.5	tr	0.1	0.3
20:1 ω9	3.0	2.3	0.9	0.8	2.7	1.6	tr
20:2 ω9	1.5	1.3	0.2	—	0.3	0.4	0.2
20:2 ω6	1.1	1.6	0.6	0.8	0.4	1.2	0.2
20:3 ω6	8.3	9.9	2.9	3.1	0.5	8.1	0.3
20:4 ω6	4.7	3.4	20.1	9.8	0.9	6.3	tr
20:5 ω3	0.8	0.6	0.2	0.7	tr	3.2	1.6
22:0	0.1	0.2	0.3	1.2	—	—	—
22:1 ω9	1.6	1.3	0.1	0.6	1.0	0.6	tr
22:3 ω6	1.8	2.0	—	—	—	—	—
22:4 ω6	7.5	3.8	2.3	0.7	0.5	1.8	tr
22:5 ω6	1.2	0.6	0.1	—	tr	2.0	tr
22:5 ω3	2.3	1.5	0.9	0.4	0.6	1.2	1.1
22:6 ω3	4.3	3.3	2.1	0.7	1.4	2.6	—
24:0	0.2	0.2	0.5	1.4	—	0.8	—
24:1 ω9	1.7	1.5	1.5	2.8	—	1.1	—

[a] Takayasu *et al.* (1970).
[b] Abbreviations as in Tables IV and VII.
[c] Average of five men.
[d] From a patient with primary aldosteronism.

primary aldosteronism than in the adjacent adrenal tissue, whereas the fatty acid compositions of phospholipids and cholesterol esters in the adenoma resembled those in the adjacent tissue.

It should be considered that small amounts of periadrenal adipose tissue might be mixed in adrenal lipids, but not with lipids of isolated cells. However, the differences in fatty acid composition of triacylglycerols between adenoma and normal adrenal tissue are too large to be ascribed to the admixture of adipose fat.

8.4.2. *Differences among Cell Types*

The discrepancies in fatty acid composition of different tissues of the same animal species are largely due to the presence of different cell types in each tissue. There is evidence, however, that most tissues contain more than one cell type, which may differ in their lipid and fatty acid composition. Thus the adipose tissue cells are known to be closely associated with and occasionally interspersed among other cells of a given tissue. Since the adipocytes tend to possess rather uniform lipid composition, they would tend to minimize any differences in fatty acid makeup among different tissue types. The development of enzymatic methods for dispersion of the cells of a tissue has allowed the isolation of individual cell types and has made possible a comparison of their fatty acid composition. Seldom, however, has there been more than one cell type purified from any one tissue. Therefore, studies of fatty acid composition must be mainly limited to comparisons between the purified cell type and the total tissue. Likewise, the cell lines propagated in cell culture have usually come from different tissues, although cell lines from the same tissue could be theoretically maintained in culture and their lipid and fatty acid compositions compared.

8.4.2.1. Rat

There have been numerous analyses made of the fatty acid composition of dispersed and cultured cells of normal and neoplastic tissues of the rat, but rarely have comparisons been made between the fatty acid composition of the derived cells and that of the original tissue.

Rogers (1974) has analyzed the fatty acids of adult and newborn rat hearts and of cultured, neonatal rat hearts (Table XXIX). In adult heart, the proportion of linoleic acid is higher and that of palmitic acid is lower than in newborn heart or in cultured cells. The relative amounts of linoleic and arachidonic acids in adult heart are affected by the source and amount of dietary fat. In heart cells, after 3 days in culture, the proportion of arachidonic acid was found to resemble that in the newborn and adult rat hearts but showed a gradual and significant decline with age. The gradual shift in fatty acid composition as the cells aged in culture was attributed to outgrowth of mesenchymal cells (fibroblasts and endothelial cells) characterized by a low relative proportion of arachidonic acid. The absence of a significant change

Table XXIX. Comparison of Fatty Acid Composition of Adult and Newborn Rat Hearts with That of Heart Cells in Culture and Calf Serum[a,b]

Fatty acid	Adult heart		Newborn heart	Heart cells in culture			Calf serum
	Purified diet	Chow diet		Day 3	Day 7	Day 21	
			(weight %)				
16:0	11.2	14.6	18.3	19.8	19.6	20.4	17.4
18:0	21.1	19.8	18.8	22.2	21.8	23.4	13.8
18:1	12.4	9.1	12.7	13.0	16.0	17.1	21.2
18:2	18.2	29.2	7.4	6.4	10.4	12.8	27.6
20:4	23.9	13.9	24.7	23.5	19.7	17.9	4.7
22:5	2.0	1.4	2.1	2.6	2.0	1.8	1.8
22:6	5.0	10.1	8.9	6.2	3.8	2.2	1.9

[a] From Rogers (1974).
[b] Samples contained minor to trace amounts of 16:1, 17:0, 20:2, 20:3, and 22:4. Means of 3–10 determinations.

in the phospholipid composition after continued incubation of the heart cell cultures for periods of up to 3 weeks reflected the major structural role of these lipid components in cell membranes.

Table XXX gives the total fatty acid composition of the crypt and villus cells of rat intestinal mucosa (Hoffman and Kuksis, 1977). The animals were maintained on a laboratory chow diet until 16 hr prior to sacrifice. There are no significant differences in the fatty acid composition of the two cell types despite marked differences in their capacity to absorb and transport dietary lipid. Absent also are the discrepancies anticipated in the fatty acids due to differences in lipid class composition of these cells of different degree of maturity and differentiation. It is possible, however, that differences exist in the fatty acid composition of the individual lipid classes, which were not analyzed.

Extensive comparisons of the fatty acid composition have been made between the Ehrlich ascites cells and whole rat liver cells (Wood, 1973). The data have indicated various abnormalities in the fatty acid composition and lipid structure of the neoplastic cells. The choline, ethanolamine, and serine phosphatides of whole cells exhibit very low levels of C_{20} and C_{22} polyunsaturated fatty acids relative to normal rat liver, while the neutral and phospholipids of Ehrlich ascites cell membrane fraction generally have lower levels of C_{20} and C_{22} polyunsaturated acids and

Table XXX. Fatty Acid Composition of Villus and Crypt Cells of Rat Intestinal Mucosa[a,b]

Fatty acid	Villus cells	Crypt cells
	(mole %)	
14:0	0.83	0.74
15:0/16:ODMA	1.05	1.06
16:0	21.48	21.92
16:1 ω7	2.26	2.34
17:0	1.33	1.39
18:0	17.23	17.50
18:1 ω9	12.16	14.34
18:2 ω6	14.28	14.39
20:0	1.20	0.98
18:3 ω3/20:1	1.99	1.61
20:2 ω6	0.44	0.47
20:3 ω9	0.35	0.55
20:3 ω6/22:0	1.45	0.90
20:3 ω3/22:1 } 20:4 ω6	17.28	16.31
22:2 ω6	0.54	0.07
20:5 ω3	0.88	0.56
24:0	0.71	0.10
22:3/22:4 ω6	1.50	1.43
22:5 ω6	0.52	0.37
22:5 ω3	0.88	0.79
22:6 ω3	2.47	2.33

[a] From Hoffman and Kuksis (1977).
[b] Means of two analyses.

Table XXXI. *Fatty Acid Composition of Sphingolipids of Rat Liver and Ehrlich Ascites Cells*[a]

Lipid class	Fatty acid										
	16:0	18:0	18:1	20:0	22:0	22:1	23:0	23:1	24:0	24:1	24:2
	(weight %)										
	Rat liver										
Sphingomyelin	15	13	1	2	10	1	9	2	25	21	—
	Ehrlich ascites cells										
Ceramides	36	11	1	2	8	—	1	—	10	22	9
Sphingomyelin	37	6	1	2	10	—	1	—	12	22	9
	Rat liver plasma membranes										
Ceramides	14	9	9	5	6	6	10	5	21	14	—
Sphingomyelin	18	13	5	2	8	1	10	3	22	16	—
	Ehrlich ascites cell membrane fraction										
Sphingomyelin	47	16	1	2	7	—	1	—	8	11	3

[a] Adapted from Wood (1973).

corresponding higher levels of saturated acids than whole cells. Table XXXI gives the fatty acid composition of the sphingolipids of Ehrlich ascites cells and rat liver, and of the corresponding plasma membranes. Free ceramide and sphingomyelin proportions are similar and are characterized by a relatively high percentage of 24:0 and 24:1 acids. Ehrlich ascites cell sphingolipids contain twice the level of palmitic, one-half the concentration of 24:0, and reduced levels of 23:0 and 23:1 acids relative to rat liver. The most important difference noted is the presence of a long-chain acid in the sphingolipids of Ehrlich ascites cells, which has been tentatively identified as Δ^{15},Δ^{18}-tetracosadienoic acid with either *cis,cis* or *cis,trans* configuration of the double bonds. Other studies have shown (Wood, 1976) that neoplasms exhibit a loss of lipid class specificity for isomeric octadecenoates (oleate and vaccenate).

8.4.2.2. *Man*

There have been considerably fewer studies with isolated cell lines of man. Furthermore, the fatty acid analyses have been less complete, and comparisons to the original tissue have been superficial or absent.

Wilkinson (1970) has examined the distribution of the even-chain homologues of fatty acids in the lipids isolated from epidermal cells grown in tissue culture and has found a distribution remarkably similar to that reported by Ansari *et al.* (1970). Ansari *et al.* (1970) have recalculated these data for direct comparison to their results and the new values are given in Table XXXII. The data suggest that the cells grown in tissue culture produce lipids of a fatty acid composition which is very similar to that found in living human epidermis.

Table XXXIII compares the fatty acid composition of a human diploid cell line WI-38 and of an established cell line WI-38VA13A, which was derived directly from

Table XXXII. Comparison of the Composition of Fatty Acids from Epidermal Cells Grown in Tissue Culture with Those from the Living Layer of Human Sole Epidermis

	Cultured cells[a]			Living-layer tissue[b]		
Carbon number	Saturates[c]	Monoenes[c]	Dienes[c]	Saturates[c]	Monoenes[c]	Dienes[c]
			(weight %)			
14	2.4	—	—	2.2	0.3	—
16	45.2	10.2	—	51.0	6.9	0.4
18	33.6	84.2	98.1	39.4	88.4	96.8
20	2.6	2.8	1.9	1.3	2.6	2.6
22	4.4	2.8	—	2.4	1.0	0.1
24	5.9			2.7	0.6	0.1
26	5.9			1.0	0.2	

[a] Wilkinson (1970).
[b] Ansari *et al.* (1970).
[c] Saturates, monoenes, and dienes represent fatty acids with none, one, and two double bonds.

WI-38 by transformation with oncogenic virus SV40. The same fatty acids are present in both cells, but there is a significant decrease in the amount of arachidonic acid present in both the neutral lipids and the phospholipids of the transformed cells as compared to that of the WI-38 line. This decrease is compensated for by a parallel increase in the content of oleic acid. The significance of this alteration in the fatty acid composition is not known, and it is difficult to appraise in view of the absence of analyses of fatty acids of individual glycerolipid classes. Similarities between the phospholipids of human cells of normal and tumor origin have been claimed by Tsao and Cornatzer (1967), although no determination of fatty acids was made.

Table XXXIII. Fatty Acid Composition of Neutral Lipids and Phospholipids of a Human Diploid (WI-38) and a Derived Heteroploid (WI-38VAI3A) Cell Line[a, b]

	Neutral lipids		Phospholipids	
Fatty acid	WI-38	WI-38VAI3A	WI-38	WI-38VAI3A
		(weight %)		
14:0	1.3	1.2	2.1	2.3
14:1	tr	tr	tr	tr
16:0	21	25	23	25
16:1	0.4	tr	0.35	tr
18:0	18	14	19	16
18:1	32	42	31	36
18:2	8.6	11	6.9	9.2
18:3	tr	tr	1.9	tr
20:4	19	5.5	18	12

[a] Abbreviated from Howard *et al.* (1973).
[b] Means of 3–11 analyses of each lipid class.

8.4.3. *Variations among Subcellular Fractions*

The exact fatty acid composition of biological membranes and cellular subfractions is a matter of continuing discussion and speculation. Analyses of the molecular composition of membrane preparations, the purity and identity of which has been effectively controlled, should lead to a fuller understanding of their molecular organization and metabolic behavior. Despite much work in many laboratories, considerable uncertainty remains about the significance of any differences observed. This is largely because most methods of preparation of cell membranes and subcellular fractions result in a partial equilibrium of lipids among different membranes as well as in losses of membrane lipids, and in contamination with lipids from extraneous sources. Furthermore, there may occur an active exchange of the membrane lipids via specific carrier proteins. Nevertheless, it would appear that the various glycerolipid classes possess, at least for the plasma membrane, distinct fatty acid composition.

Table XXXIV. Fatty Acid Composition of Triacylglycerols and Phosphatidylcholines of Subcellular Fractions of Bovine Heart Muscle[a]

	Subcellular fractions									
	Nuclear		Heavy mitochondria		Light mitochondria		Microsomes		Cytoplasm	
Fatty acid	TG[b]	PC[c]	TG	PC	TG	PC	TG	PC	TG	PC
					(weight %)					
<14:0			tr		2.2		tr			tr
14:0	0.9	tr	tr	3.2		tr	0.8	0.5	2.0	
14:1		tr	1.9	1.4	2.9		tr			
15:0	0.6		tr	4.2	1.4	tr	0.6	1.8	1.1	
16:0	27.5	42.0	25.6	37.1	50.7	45.6	47.9	37.2	21.3	46.7
16:1	0.8	1.4	tr	9.5		1.5	tr		1.8	2.5
17:0	1.3		1.9				5.8			
17:1	tr		tr							
18:0	15.5	32.4	12.6	21.2	7.1	22.2	10.1	15.7	16.1	19.6
18:1	41.5	14.7	22.0	17.0	2.5	24.0	13.8	21.4	37.8	23.6
18:2	8.4	9.5	15.4	6.4	6.7	6.7	2.2	13.7	16.2	6.5
18:3	0.9		tr					0.6	0.7	1.1
20:0	tr		tr		17.3		5.7			
20:4	1.4		6.3	tr		tr	1.2	0.8	3.0	tr
20:5	0.7		tr		14.5		8.3			
20:6			6.3				3.6			
21:0					0.8		tr			
22:0	0.5		1.9				tr	2.5		tr
22:1	tr									
22:2			6.1							
24:0								2.6		tr
24:1								1.1		tr

[a] Abbreviations as in Tables IV and XI.
[b] Nazir *et al.* (1967*a*).
[c] Nazir *et al.* (1967*b*).

8.4.3.1. *Ox*

Fatty acids constitute the principal source of energy for the metabolism of the myocardium, which has generated much interest in the composition of both free and bound fatty acids in the heart tissue. One of the most extensive investigations of fatty acids in the subcellular particles has been performed on the bovine heart muscle (Nazir *et al.*, 1967*a*, *b*). Table XXXIV gives the fatty acid composition of the triacylglycerols and phosphatidylcholines from the nuclear, heavy and light mitochondrial, microsomal, and cytoplasmic particles. In the triacylglycerols, considerable variation is seen in the relative amounts of monounsaturated fatty acids among the different subcellular fractions, the nuclear and cytoplasmic compartments possessing the highest relative concentrations. With the exception of the heavy mitochondrial preparations, all other fractions show insignificant amounts of long-chain fatty acids. Saturated acids constitute generally more than two-thirds of the total triacylglycerol fatty acids. With the exception of the microsomes, all other subcellular fractions have uniform fatty acid composition in their phosphatidylcholines. The microsomes exhibit relatively higher levels of total polyunsaturated acids than other fractions. These authors (Nazir *et al.*, 1967*b*) have shown that the cardiolipins, sphingomyelins, and phosphatidylethanolamines also possess nearly identical fatty acid compositions when isolated from different subcellular fractions of the bovine heart muscle.

8.4.3.2. *Rat*

Extensive analyses of the fatty acids of subcellular fractions have also been performed on rat liver (Keenan and Morre, 1970; Colbeau *et al.*, 1971). Table XXXV shows the fatty acid composition of the major glycerophospholipids isolated from plasma membrane, endoplasmic reticulum, Golgi apparatus, and mitochondria. It is seen that the fatty acids of the choline, ethanolamine, and inositol phosphatides are fairly similar in mitochondrial and microsomal membranes, the fatty acid patterns being more characteristic of the phospholipid classes than of the membrane (Keenan and Morre, 1970). However, the fatty acid pattern of the choline and ethanolamine phosphatides contained in the plasma membrane differs by its high degree of saturation (67%). These results are in general agreement with those of earlier (Ray *et al.*, 1969) and later (Colbeau *et al.*, 1971) reports.

Stoffel and Schiefer (1968) have claimed that the fatty acids of the choline and ethanolamine phosphatides are essentially the same in different submitochondrial fractions from rat liver, while Huet *et al.* (1968) reported a higher ratio of saturated to unsaturated fatty acids in the phospholipids in outer mitochondrial membranes from rat liver. Comparisons of the fatty acid composition of the glycerolipids of the inner and outer mitochondrial membranes from guinea pig liver (Parkes and Thompson, 1970) have shown no significant difference in the level of saturation of the fatty acids, except for cardiolipin. The concentration of microsomal phospholipids was markedly decreased by fasting, whereas that of mitochondria and whole rat liver increased significantly (Imaizumi and Sugano, 1975). The synaptic membranes, vesicles, and mitochondria of the rat brain have been shown to be rich in docosahexaenoic acid (Breckenridge *et al.*, 1971).

Table XXXV. Fatty Acid Composition of the Phospholipids of Rat Liver Plasma Membrane, Microsomes, Golgi Apparatus, and Mitochondria

	PC[a]				PE[a]				PI[a]				PS[c]			CL[a]
Fatty acid	PM[b]	ER[b]	GA[b,c]	MT[b]	PM	ER	GA[c]	MT	PM	ER	GA[c]	MT	PM	ER	GA	MT
								(weight %)								
14:0	—	0.7	0.9	0.4	—	0.3	0.7	0.3		2.5	3.9		1.5	9.0	7.1	0.2
16:0	32.8	24.8	34.7	27.0	30.6	22.6	33.5	26.6	30.7	19.3	36.3	26.3	38.7	11.1	29.6	7.0
16:1 ω7	2.9	3.3		3.9	1.2	2.3	0.4	3.2	8.4	1.8		5.8		1.0	11.8	7.6
18:0	34.9	21.0	22.5	21.6	31.3	23.4	31.8	27.3	36.6	45.0	19.9	38.4	46.1	4.7	8.2	3.6
18:1 ω9	10.2	12.3	8.7	13.0	10.4	9.8	5.1	12.0	13.2	7.2	21.9	14.0	8.4	21.8	40.3	19.9
18:2 ω6	8.1	17.7	18.1	12.4	6.5	10.3	10.0	5.4	2.9	3.2	1.6	4.2	1.0	52.3	2.9	58.8
20:0				1.3												
20:3 ω9	1.1	1.2			0.9											
20:4 ω6	8.4	15.8	14.5	17.7	16.5	23.1	18.3	22.0	8.0	21.4	10.2	7.6	4.2			1.2
22:0																
22:6 ω3	1.6	2.9		2.9	2.9	7.2		3.2				3.2				1.8

[a] Keenan and Morre (1970), except as indicated.
[b] Abbreviations: PM, Plasma membrane; ER, endoplasmic reticulum or microsomes; GA, Golgi apparatus; MT, mitochondria; other abbreviations as in Table VII.
[c] Colbeau *et al.* (1971).

8.4.3.3. Man

The most detailed examination of the fatty acid composition of subcellular fractions of human tissues has been carried out on the brain. An early study (Grossi *et al.*, 1961) had demonstrated a gross difference in the fatty acid composition between tumor and normal brain tissue. This has been confirmed in subsequent work (Sun and Leung, 1974). Table XXXVI compares the fatty acid composition of the phosphatidylethanolamines of various subcellular fractions of normal human brain obtained at autopsy from 61- to 87-year-old subjects (Sun, 1973). Significant differences are seen in the composition of the fatty acids from myelin, microsomes, and synaptosomes, as well as between microsomes of the gray and white matter of the brain. The phosphatidylethanolamines from tumor microsomes have relatively more oleic, linoleic, and arachidonic acids and less stearic, docosatetraenoic, and docosahexaenoic acids than the adult brain gray matter. There are also more of the alkyl and alkenyl species of the phosphatidylethanolamines in the tumor fractions. The alkyl and acyl groups characteristic of tumor phospholipids are present in all subcellular fractions. There is a rather close resemblance between certain lipids of the tumors and the lipids of fetal brain, but significant differences have also been noted (Sun and Leung, 1974). The results have demonstrated obvious alterations in membrane lipid composition of neoplastic brain cells, which has led to the suggestion that they must deviate from normal neurons and glials in metabolism and function.

8.4.4. Discrepancies among Plasma Lipoprotein Classes

All lipids present in blood plasma are bound to proteins forming various lipoprotein classes, which can be resolved by density centrifugation. The various

Table XXXVI. Fatty Acid Composition of Phosphatidylethanolamine from Subcellular Fractions of Human Brain[a,b]

		Microsomes			
Fatty acid	Myelin	Gray matter	White matter	Tumors[c]	Synaptosomes
			(weight %)		
16:0	2.9	4.6	3.7	6.7	5.0
16:1	0.6			—	
18:0	5.8	24.6	9.5	11.3	26.0
18:1	42.2	11.6	37.7	18.4	11.4
18:2		0.7		3.3	0.7
20:1	8.8	0.8	7.1	0.6	0.7
20:3 ω6	0.7	0.7		1.4	0.6
20:4 ω6	8.6	13.0	9.8	32.6	13.0
22:3	2.1		1.7	—	
22:4 ω6	20.9	12.2	18.9	10.3	11.6
22:5 ω6		1.2		1.7	1.4
22:6 ω3	5.7	31.1	9.7	10.3	29.6
24:4 ω6	1.3	—	—	—	—

[a] From Sun (1973).
[b] Mean of five or more analyses.
[c] Sun and Leung (1974).

lipoproteins differ in the content and composition of the lipid classes and in their structural organization. There is evidence that the glycerophospholipids of the various plasma lipoprotein classes become rapidly equilibrated and that this equilibration may be facilitated by special phospholipid exchange proteins (Illingworth and Portman, 1972; Zilversmit, 1971). As a result, little difference would be expected in the fatty acid composition of the glycerophospholipids of the various lipoprotein density classes. Such an equilibration, however, would not appear to take place among the triacylglycerols and cholesteryl esters of the various plasma lipoprotein classes.

8.4.4.1. Man

The most detailed studies of the composition of fatty acids have been made on plasma lipoproteins of man. Table XXXVII compares the fatty acid compositions of different lipid classes of various plasma lipoproteins of man in fasting and nonfasting states (Skipski, 1972). It is seen that the fatty acid composition of the triacylglycerols and cholesteryl esters in the VLDL fraction isolated from the sera of nonfasting subjects is significantly different from that of the VLDL fraction isolated from the plasma of fasting subjects. Furthermore, in the nonfasting samples, the VLDL triacylglycerols contain a higher percentage of linoleic acid, and the VLDL cholesteryl esters contain relatively higher amounts of palmitic and oleic acids and a lower amount of linoleic acid than do the corresponding lipid classes in whole serum or in the other lipoprotein classes. The effect of diet on the composition of the LDL and HDL fatty acids is less obvious.

8.4.4.2. Rat

Much less extensive studies of the fatty acid composition of the plasma lipoproteins of the rat have yielded results comparable to those obtained for human plasma lipoproteins. Table XXXVIII compares the fatty acid composition of the phosphatidylcholines, cholesteryl esters, and triacylglycerols of various density classes of plasma lipoproteins of choline-supplemented and choline-deficient rats. The data show that essentially identical fatty acid compositions are present in the phosphatidylcholines, which suggests their rapid equilibration among the different lipoprotein classes. In contrast, the fatty acid compositions of the triacylglycerols and cholesteryl esters show significant differences among the chylomicrons, VLDL, LDL, and HDL, which would exclude their ready equilibration *in vivo*. A similar lack of equilibration of the triacylglycerols among different lipoprotein classes is also observed in choline deficiency and points to a definite source of compartmentation of plasma lipids. No significant differences have been seen in the fatty acid composition of the phosphatidylcholines of any of the lipoprotein classes after 10 days on the choline-deficient diet. There remains a need, however, for further systematic studies of the fatty acid composition of the plasma lipoprotein classes in view of the potential physiological significance of a compartmentation of molecular species of glycerolipids in lipoprotein classes.

Table XXXVII. Fatty Acid Composition of Lipids of Various Lipoprotein Classes of Human Plasma[a]

Lipoprotein class	Fatty acid									
	<16:0	16:0	16:1	16:1–18:0	18:0	18:1	18:2	18:2–20:4	20:4	>20:4
	(weight %)									
	Nonfasting subjects									
Acylglycerols										
Total serum	4.2	27.9	3.9	1.1	4.4	40.7	14.8	1.8	1.2	—
VLDL[b]	4.8	23.3	4.0	1.1	3.5	36.5	21.8	3.2	1.0	0.9
LDL[b]	10.1	23.1	4.1	1.9	4.1	35.4	15.6	3.5	1.4	0.9
HDL[b]	8.5	23.5	4.0	1.7	4.0	35.3	15.7	4.7	1.6	1.1
Cholesteryl esters										
Total serum	4.4	10.4	3.4	1.7	1.3	18.8	50.4	3.0	6.1	0.6
VLDL	4.5	21.0	3.0	1.4	5.6	34.9	25.3	2.5	1.8	—
LDL	3.8	10.5	3.4	1.3	1.5	18.2	50.7	3.8	5.6	1.6
HDL	3.6	10.3	3.3	1.0	1.2	18.5	51.0	3.0	6.3	1.9
Phospholipids										
Total serum	3.3	30.5	1.3	1.0	13.9	12.1	20.7	4.3	9.9	3.0
VLDL	3.6	33.2	1.5	1.9	19.0	13.4	16.0	4.1	4.9	2.5
LDL	4.2	30.1	1.4	1.5	15.1	12.1	19.3	4.4	7.5	4.4
HDL	3.9	29.6	1.6	1.8	15.5	12.6	18.9	4.7	9.5	1.9
	Fasting subjects									
Acylglycerols										
Total plasma	1.3	25.6	3.1	0.5	2.9	45.1	16.1	1.3	4.5	—
VLDL	1.0	26.9	3.3	—	2.9	45.4	15.7	1.5	2.5	—
LDL	0.7	23.1	2.8	—	3.3	46.6	15.7	1.5	5.4	—
HDL	0.8	23.4	2.9	—	3.5	43.8	15.9	1.2	7.6	—
Cholesteryl esters										
Total plasma	0.8	11.5	2.1	0.5	1.1	23.0	53.5	1.2	6.8	—
VLDL		11.5	1.8	—	1.4	25.6	51.8	0.8	6.1	—
LDL		11.3	1.9	—	1.0	22.4	54.9	0.7	6.9	—
HDL		11.4	2.1	—	1.1	22.4	54.8	0.6	6.3	—
Phospholipids										
Total plasma	0.8	34.0	0.5	0.5	14.3	11.9	20.4	4.0	14.0	—
VLDL	—	33.8	0.5	—	14.7	12.2	20.3	4.1	13.6	—
LDL	—	36.0	0.5	—	14.3	11.6	18.9	4.6	13.2	—
HDL	—	31.8	0.4	—	14.5	12.3	20.6	3.9	15.7	—

[a] Modified from Skipski (1972).
[b] Abbreviations: VLDL, LDL, and HDL represent very-low-, low-, and high-density lipoproteins of plasma.

8.4.5. Characteristics of Skin Surface and Fecal Lipids

The most complex fatty acid composition has been recognized for the skin surface and fecal lipids. The fatty acid spectra of skin lipids show the presence of a number of unusual acids which include branched-chain acids, hydroxy-substituted fatty acids, unsaturated fatty acids, and cyclic fatty acids, as well as various

Table XXXVIII. Fatty Acid Composition of Lipids of Various Lipoprotein Classes of Plasma of Normal and Choline-Deficient Rats[a, b]

Lipoprotein class	Fatty acid								
	14:0	16:0	16:1	18:0	18:1	18:2	20:3	20:4	>20:4
	(mole %)								
	Normal animals								
Acylglycerols									
Chylomicrons	0.4	14.0	0.4	5.3	44.3	23.2	1.7	8.9	—
VLDL + LDL	2.5	25.2	0.5	2.7	44.0	21.2	1.4	2.6	—
HDL	3.2	22.0	tr	6.4	38.2	20.1	4.5	5.0	—
>1.21	2.3	26.4	tr	16.9	28.8	19.3	tr	6.4	—
Cholesteryl esters									
Chylomicrons	1.0	18.6	1.1	12.3	30.9	16.1	1.1	18.9	—
VLDL + LDL	0.7	16.2	1.1	2.0	20.5	22.6	tr	36.9	—
HDL	tr	10.3	1.6	0.6	12.4	30.1	tr	44.1	—
>1.21	tr	7.2	1.0	1.4	18.1	24.1	tr	48.2	—
Phosphatidylcholine									
Chylomicrons	1.1	19.6	1.3	16.5	22.4	26.9	—	8.9	2.2
VLDL + LDL	—	20.6	1.2	19.3	20.1	26.1	—	9.1	2.3
HDL	—	18.9	1.0	17.9	22.5	27.9	—	8.2	2.3
>1.21	—	20.5	1.2	18.4	19.7	30.2	—	8.0	2.0
	Choline-deficient animals								
Acylglycerols									
Chylomicrons	0.6	14.3	0.4	7.1	42.7	23.8	1.5	9.8	—
VLDL + LDL	1.0	21.6	0.6	3.8	46.9	20.1	1.6	4.4	—
HDL	3.0	23.9	tr	4.0	36.9	19.8	4.4	5.0	—
>1.21	1.8	27.9	tr	18.1	28.1	20.9	tr	5.4	—
Cholesteryl esters									
Chylomicrons	1.1	21.9	1.3	12.9	27.2	14.7	1.2	19.8	—
VLDL + LDL	0.7	16.3	1.1	1.6	17.7	29.3	tr	33.3	—
HDL	tr	11.0	1.9	0.7	15.8	35.3	tr	37.4	—
>1.21	tr	8.5	1.2	1.1	20.1	26.6	tr	42.5	—
Phosphatidylcholine									
Chylomicrons	1.0	18.7	1.4	19.9	20.3	26.0	—	9.0	2.3
VLDL + LDL	—	21.2	1.0	19.2	19.8	28.3	—	6.9	2.3
HDL	—	20.0	0.8	17.9	20.2	28.7	—	9.0	2.4
>1.21	—	19.0	1.0	18.5	20.2	30.1	—	8.3	2.2

[a] Modified from Mookerjea *et al.* (1975).
[b] Abbreviations as in Table XXXVII.

positional isomers of unsaturated fatty acids (Nikkari, 1974). Many of the unusual fatty acids of feces arise from bacterial action upon the fatty acids of the shed mucosal cells and the normal secretions of skin, while others accumulate as a result of preferential absorption of the normal-chain saturated and unsaturated fatty acids from the gut. However, the possibility cannot be excluded that some of the fatty acids are specifically synthesized for the secretion by the skin or excretion into the feces where they serve some special biochemical or physiological function.

8.4.5.1. Man

The most extensive analyses have been carried out on the skin lipids of man, and these studies have served as a guide to the isolation and identification of unusual fatty acids from human feces.

Ansari *et al.* (1970) have made detailed analyses of the fatty acid composition of the living-layer and stratum corneum lipids of human sole skin epidermis. The fatty acids of the living layer and stratum corneum of human sole epidermis comprised saturated, monoenoic, dienoic, trienoic, polyenoic, and α-hydroxy fatty acids. The C_{16} and C_{18} chain lengths were major components for each type of acid. Comparisons of epidermal acids with the acids of sebaceous glands showed that each tissue can synthesize the same kinds of acids but in widely different amounts. The extremely complex mixture of the fatty acid methyl esters isolated from the skin lipids required repeated argentation TLC and GLC fractionation, followed by ozonolysis and a TLC/GLC examination of the degradation products. Table XXXIX gives the composition of the positional isomers of monoenoic (living-layer and stratum corneum) and dienoic (skin surface) acids of the total lipids. The Δ^9 isomer is by far the major component in the 16:1 and 18:1 series, and there is very little difference in the distribution of positional isomers for all chain lengths in the living-layer and stratum corneum lipids. Of the dienoic acids, the 18:2 and 20:2 components constitute over 93% of the total. The $\Delta^{5,8}$ and $\Delta^{9,12}$ isomers constitute over 81% of the 18-carbon dienes. The $\Delta^{7,10}$ isomer is the major component of the 20-carbon dienes. The results of these analyses suggest that human skin contains enzymes capable of inserting a second *cis* double bond between the carboxyl group and the first double bond, separated from the latter by a methylene unit (Nicolaides and Ansari, 1969). Of the total unsubstituted methyl esters, dienes make up 20–30%, monoenes 47–48%, the remainder consisting of saturates (Ansari *et al.*, 1970). The α-hydroxy fatty acids exhibit two clusters of chain lengths, one above and one below C_{20} in both living layer and stratum corneum, but a greater proportion of the acids occur above C_{20} in the stratum corneum (Ansari *et al.*, 1970). Nicolaides *et al.* (1976) have completed an elaborate identification of the branched-chain fatty acids in vernix caseosa lipids using the GC/MS/COM system.

The first detailed study of the fecal fatty acids of man was made by James *et al.* (1961), who were able to demonstrate the presence of most of the known fatty acids of C_{10} to C_{20} carbon atoms. The 18:0, 16:0, and 18:1 acids were the major components, while branched and odd-carbon-number derivatives occurred in smaller concentrations. Isomeric octadecenoic acids with double bonds in positions 4, 5, 6, 7, 8, 9, 10, 11, and 12 were also obtained. About 50% of these acids had the *trans* configuration. The few subjects with steatorrhea have been shown (James *et al.*, 1961) to contain large amounts of 10-hydroxystearic acid. In addition, these feces also contained traces of the 6-, 7-, 8-, and 9-hydroxystearic acids. Furthermore, feces are known to contain volatile fatty acids such as acetic, propionic, and valeric (54%) and butyric (36%), while caproic and higher (11%) acids account for the rest of the volatile acidity.

Table XL gives the composition of the nonsubstituted fatty acids of human feces from subjects on controlled experimental diets (Ali and Kuksis, 1967*a*, *b*). On the fat-free regimen, the lipid classes contained about the same fatty acids but in

Table XXXIX. Positional Isomers of Monoenoic and Dienoic Fatty Acids of Living-Layer and Stratum Corneum Lipids of Human Sole Epidermis

	Living layer[a]		Stratum corneum[a]			Skin surface[b]	
Fatty acid	Total (%)	Isomer (%)	Total (%)	Isomer (%)	Fatty acid	Total (%)	Isomer (%)
				(weight %)			
16:1 (6)		1		1	16:2 (5, 8)		88
16:1 (7)	7.3	4	7.7	5	16:2 (6, 9)	1.8	7
16:1 (8)		1		1	16:2 (7, 10)		5
16:1 (9)		94		93	17:2 (4, 7)		10
18:1 (7)		2		2	17:2 (5, 8)		59
18:1 (8)		1		1	17:2 (6. 9)	1.6	12
18:1 (9)	86.2	92	80.2	93	17:2 (7, 10)		8
18:1 (10)		1		1	17:2 (8, 11)		8
18:1 (11)		4		3	17:2 (9, 12)		3
18:1 (12)		1		tr	18:2 (5, 8)		55
18:1 (13)		tr		tr	18:2 (6, 9)		5
20:1 (9)		20		4	18:2 (7, 10)	70.5	2
20:1 (11)	2.5	70	2.9	93	18:2 (8, 11)		4
20:1 (13)		10		3	18:2 (9, 12)		26
22:1 (11)		2		2	18:2 (10, 13)		8
22:1 (13)		90	1.8	95	19:2 (5, 6)		86
22:1 (15)		8		3	19:2 (6, 9)		5
24:1 (13)				5	19:2 (7, 10)	1.8	5
24:1 (15)	0.5	ND	2.9	85	19:2 (8, 11)		3
24:1 (17)				10	19:2 (9, 12)		1
26:1 (15)	0.1	ND	1.0	5	20:2 (4, 7)		1
26:1 (17)				95	20:2 (5, 8)		2
					20:2 (6, 9)	22.6	1
					20:2 (7, 10)		78
					20:2 (8, 11)		5
					20:2 (9, 12)		4
					20:2 (10, 13)		7
					20:2 (11, 14)		2
					22:2 (7, 10)		16
					22:2 (8, 11)	1.2	5
					22:2 (9, 12)		70
					22:2 (10, 13)		9

[a] Ansari *et al.* (1970).
[b] Nicolaides and Ansari (1969).

somewhat varying proportions. Because of the occurrence of large amounts of odd-carbon-number components, among both branched- and normal-chain fatty acids, which are not normally accounted for by mammalian metabolism, the presence of phospholipids in the feces was attributed to bacterial synthesis. On high-fat diets, the fecal lipids assumed the normal fatty acid composition in both total lipids and in individual lipid classes. A change in the activity or the population of the intestinal flora was suggested as the basis for this observation. The availability of different dietary fatty acids, however, should also be considered.

The composition of the fecal fatty acids is quite similar to that of the sebum or skin lipids of man, although detailed comparative studies have not been made.

Table XL. Fatty Acid Composition of Fecal Lipids of Man on a Fat-Free Diet and on Fat Diets[a]

Fatty acid	Free fatty acids			Triglycerides			Lecithins		
	FF[b]	BF[b]	CO[b]	FF	BF	CO	FF[c]	BF[d]	CO[d]
					(mole %)				
12:0	1.18	1.0	0.4	0.9		0.1			
13:0 br	0.5								
13:0	0.5								
13:1	1.1								
14:0 br				0.2			0.1	0.6	0.4
14:0	3.9	3.7	0.4	1.9	1.6	0.4	0.2	3.0	2.0
15:0 br	12.6	0.4	0.5	2.1	1.2	0.2	4.5	3.5	3.0
15:0	6.4	2.7	0.6	1.4	tr	0.2	1.9	1.7	3.2
16:0 br	0.7	0.1		0.7	tr	0.2	1.5	1.4	—
16:0	17.9	39.8	17.1	18.0	30.6	21.7	17.0	21.6	18.7
16:1	4.1			}2.4		}3.6	}3.3	}4.2	
17:0 br		2.2	0.3						
17:0	1.3	1.7		1.0					
18:0 br	0.6			0.3		0.3	2.1	1.0	3.3
18:0	7.4	36.0	20.6	36.5	33.1	28.5	9.3	11.1	16.4
18:1	10.0	10.8	49.3	16.4	19.1	46.4	11.5	18.9	13.4
18:2	0.3	0.3	1.9	1.0	8.0	1.1	3.1	4.0	7.0
19:0							0.5		
20:0	1.1	1.4	0.5	1.2	0.9	0.4		0.8	2.9
20:1							0.3	2.3	
21:0	0.8			1.2			0.4		
21:1								2.2	
22:0	1.3		2.5	2.5	0.6	0.6	4.5	2.0	5.7
22:1								6.2	
23:0	1.3			3.1			3.9	0.6	
23:1	2.2			2.8					
24:0	1.5		5.3	3.1			10.1	0.9	
24:1							4.3	0.5	
26:0	14.5			2.9			8.4	14.6	19.8
27:0 br	4.1						12.8	tr	tr

[a] Abbreviations: FF, fat-free diet; BF, butterfat diet; CO, corn oil diet.
[b] Ali (1965).
[c] Ali and Kuksis (1967*a*).
[d] Ali and Kuksis (1967*b*).

8.4.5.2. Rat

The chemistry of the skin surface lipids of the rat also has been described in detail (Nikkari, 1965, 1974). Table XLI gives the fatty acid composition of pooled skin surface lipids of male and female rats maintained on laboratory chow (Nikkari, 1965). The fatty acids include both odd- and even-carbon-number normal-chain saturated, normal-chain monounsaturated, *iso*-branched-chain saturated, and odd-carbon-number *anteiso*-branched-chain saturated acids. In addition to the unsubstituted acids, α-hydroxy acids with the same range of carbon chains (C_{14}–C_{26}) are also found. Male rats exhibit a higher total amount of *anteiso* fatty acids than do

Table XLI. Fatty Acid Composition of Pooled Skin-Surface Lipids of Male and Female Rats[a, b]

Fatty acid	Males	Females
	(weight %)	
14:0	1.5	1.3
15:0	2.0	1.7
16:0 *iso*	4.9	4.1
16:0	9.5	8.2
16:1	1.4	1.6
16:2	tr	tr
17:0 *anteiso*	4.8	3.5
17:0	2.3	2.1
18:0 *iso*	5.5	4.8
18:1	7.5	9.8
18:2	2.7	3.1
19:0 *anteiso*	2.3	2.0
19:0	0.4	0.4
20:0 *iso*	7.5	7.4
20:0	1.8	1.7
20:1	19.9	13.9
20:2	0.2	0.2
21:0 *anteiso*	4.0	3.3
21:0	0.4	0.4
22:0 *iso*	3.5	3.3
22:0	2.3	2.4
22:1	2.4	2.2
23:0 *anteiso*	1.8	1.5
23:0	0.9	1.2
24:0 *iso*	2.4	2.5
24:0	5.0	4.6
24:1	2.0	1.9
25:0 *anteiso*	1.3	1.2
25:0	1.1	0.9
26:0 *iso*	1.1	1.2
26:0	1.0	1.2
26:1	1.5	1.4

[a] Abbreviated from Nikkari (1965).
[b] Means of five animals each sex.

female rats, but there are no other significant differences between the sexes. Neither the proportion of *anteiso* acids nor the chain-length distribution of fatty acids is affected by dietary lipids. Statistically significant alterations, however, have been found to occur in the saturated, monounsaturated, and diunsaturated fatty acids as a result of a manipulation in dietary linoleic acid content (Nikkari, 1965).

Many of the fatty acid types recognized in the skin surface lipids have been subsequently identified in the fecal lipids of rats (Grigor *et al.*, 1970). Table XLII gives the fatty acid composition of pooled rat fecal lipids. It is seen that the fecal lipids of rats contain a wide range of branched *iso* and *anteiso* fatty acids which have chain lengths up to C_{26} and make up some 20% of the total fatty acids. The branched acids are trace components only in the lipids of the stomach contents, the small

Table XLII. Fatty Acid Composition of Pooled Rat Fecal Lipids[a]

	Chain modification					
Carbon number	Normal saturated	Branched *iso-*	Branched *anteiso-*	Cyclopropane	Mono-unsaturated	Poly-unsaturated
	(weight %)					
14	1.5	0.5			0.1	
15	6.3	2.4	5.9			
16	17.6	1.4			4.8	
17	2.2	1.3	1.3		0.3	
18	10.5	0.9			12.8	1.1
19	0.3		0.2	0.5		
20	2.0	1.2			3.5	0.2
21	0.1		2.7			
22	1.9	0.7			1.1	0.2
23	0.9		0.4		0.1	
24	3.8	0.8			1.3	
25	0.2		0.5		0.1	
26	5.0	0.4			0.9	

[a] Modified from Grigor *et al.* (1970).

intestinal contents, the bile, and the small and large intestinal epithelial cells. However, they are present in amounts approaching those of the fecal lipids in the lipids of the cecal contents. Treatment of rats with neomycin sulfate, while producing evidence of partial elimination of the intestinal bacteria, produces no decrease in the amounts of fecal branched-chain acids (Grigor *et al.*, 1970; Grigor and Duncley, 1973).

Since fatty acids occur in skin lipids and in feces partly in free and partly in bound form, there arise extremely complex mixtures of molecular species of lipids, which differ greatly from those of any other tissue, cell type, subcellular fraction, or cellular secretion. There is evidence that, in the guinea pig at least, the skin surface lipids may contain large amounts of diacylalkyldiols (Downing and Sharaf, 1976) which occur in other tissues in trace amounts only.

8.5. Summary and Conclusions

The purpose of this chapter was to survey the nature of fatty acids in animal tissues and to give an indication of the variability to be found in quantitative data obtained in response to a few factors under direct experimental control. As may be seen from the large number of tables of limited data, this modest goal has been difficult to attain. Despite some 25 years of steady development in chromatographic techniques, there have been very few critical comparative analyses made either of specific glycerolipid classes or of total lipid extracts from different or homologous animal tissues. Usually no effort has been made to resolve or to recover all fatty acids present. Losses of polyunsaturated and long-chain fatty acids frequently have been complete or largely complete. It appears that the more detailed the analyses of the fatty acids, the more trivial the experimental material. Parallel analyses of control tissues seem to have been neglected in favor of quotations of data from dubious earlier analyses. Where detailed analyses of lipid classes have been made, fatty acid determinations, as a rule, have been omitted. As a result, there is a need for repeating many of the lipid class separations on animal tissues and subcellular fractions, so that adequate fatty acid analyses can be reported.

There also has been much variation in the results of and in the form of reporting fatty acid analyses by different laboratories. Data from the same laboratory are generally the most consistent. These data, however, may have given full expression to any bias that may have existed in a given laboratory in regard to the presumed correct composition. Many laboratories and scientific journals well known for excellence in reporting nonlipid metabolite data have been lax in policing reports on fatty acid and lipid analyses. Each instance of potential similarity or discrepancy in fatty acid composition must therefore be examined on the basis of data much more limited than would have been desirable or anticipated considering the apparent general abundance of fatty acid and lipid analyses in the literature. It is to be hoped that the reader will accept the present collection of largely trivial comparisons as a personal challenge and as an encouragement to collect and tabulate better data in the future, as well as to rededicate himself to more vigilant policing of the reports of colleagues.

Nevertheless, it is possible to show that in most tissues and animal species the major fatty acids are the same and comprise the common C_{16} and C_{18} saturated, and mono- and diunsaturated components. The polyunsaturated fatty acids are found to be distributed somewhat more selectively, and distinct tissue and animal species differences may be demonstrated. In general, each glycerophospholipid class appears to have a characteristic fatty acid composition which varies little from tissue to tissue in the same animal species or among animal species. This can be clearly recognized as glycerolipid class specificity of fatty acid composition. There exists, however, also a certain tissue specificity in the fatty acid composition. This is best demonstrated by comparing different tissues in the same animal. Interanimal and interspecies comparisons are obscured by differences in current or previous dietary regimens, as well as by differences in age, sex, and metabolic state of the animals or animal species. In most instances, the differences in fatty acid composition between tissues are greater than those between animal species, even when environmental factors are not identical. The differences among tissues are retained among the cells which make up these tissues. However, subcellular fractions, with the possible exception of the plasma membrane, appear to have the same fatty acid composition for each glycerolipid class. Clear-cut differences are also demonstrable among fatty acid compositions of the glycerolipids retained by the cell membrane and those released as lipoproteins or lung and biliary surfactants.

However, it would appear that considerable diversity exists in the fatty acid compositions of anatomically and functionally similar tissues or membranes. Although it is not immediately obvious how this heterogeneity in composition relates to membrane structure and function, it is possible that different fatty acid compositions may be physiologically equivalent when considered in terms of total lipid and glycerolipid structure. This would be consistent with the concept that physicochemical attributes of the hydrocarbon chains in the glycerolipids are the important determinants of their physiological action, although unique acyl groups may be required for certain biochemical processes in membranes (Silbert *et al.*, 1974; Silbert, 1975). Considerable variation, therefore, ought to be obtained and expected in the fatty acid composition of animal tissues, cells, and cell membranes.

ACKNOWLEDGMENTS

The studies of the author and his collaborators referred to in this chapter were supported by funds from the Ontario Heart Foundation, Toronto, Ontario, the Medical Research Council of Canada, Ottawa, Ontario, and the Special Dairy Industry Board, Chicago, Illinois.

8.6. References

Ackman, R. G., 1972, The analysis of fatty acids and related materials by gas–liquid chromatography, in: *Progress in the Chemistry of Fats and Other Lipids*, Vol. 12 (R. T. Holman, ed.), pp. 165–284, Pergamon Press, Oxford.

Ackman, R. G., and Burgher, R. D., 1965, Cod liver oil fatty acids as secondary reference standards in the GLC of polyunsaturated fatty acids of animal origin: Analysis of a dermal oil of the Atlantic leatherback turtle, *J. Am. Oil Chem. Soc.* **42:**38.

Ahluwalia, B., and Holman, R. T., 1969, Fatty acid composition of lipids of bull, boar, rabbit and human semen, *J. Reprod. Fertil.* **18:**431.

Ali, S. S., 1965, Fecal lipid excretion in man on controlled experimental diets, Ph.D. thesis, pp. 1–249, Queen's University, Kingston, Canada.

Ali, S. S., and Kuksis, A., 1967*a*, Excretion of phospholipids by man on fat-free diets, *Can. J. Biochem.* **45:**689.

Ali, S. S., and Kuksis, A., 1967*b*, Excretion of phospholipids by men on high fat diets, *Can. J. Biochem.* **45:**703.

Anderson, R. E., Bottino, N. R., and Reiser, R., 1970*a*, Animal endogenous triglycerides. I. Swine adipose tissue, *Lipids* **5:**161.

Anderson, R. E., Bottino, N. R., Cook, L. J., and Reiser, R., 1970*b*, Animal endogenous triglycerides. III. Swine, rat and chicken liver: Comparison with adipose tissue, *Lipids* **5:**171.

Ansari, M. N. A., Nicolaides, N., and Fu, H. C., 1970, Fatty acid composition of the living layer and stratum corneum of human sole skin epidermis, *Lipids* **5:**838.

Balint, J. A., Kyriakides, E. C., Spitzer, H. L., and Morrison, E. S., 1965, Lecithin fatty acid composition in bile and plasma of man, dogs, rats and oxen, *J. Lipid Res.* **6:**96.

Bligh, E. G., and Dyer, W. J., 1959, A rapid method of total lipid extraction and purification, *Can. J. Biochem. Physiol.* **3:**911.

Bottino, N. R., Anderson, R. E., and Reiser, R., 1970, Animal endogenous triglycerides. II. Rat and chicken adipose tissue, *Lipids* **5:**165.

Breckenridge, W. C., and Kuksis, A., 1967, Molecular weight distribution of milk fat triglycerides from seven species, *J. Lipid Res.* **8:**473.

Breckenridge, W. C., Gombos, G., and Morgan, I. G., 1971, The docosahexaenoic acid of the phospholipids of synaptic membranes, vesicles and mitochondria, *Brain Res.* **33:**581.

Bruce, A., 1974, Skeletal muscle lipids. III. Changes in fatty acid composition of individual phosphoglycerides in man from fetal to middle age, *J. Lipid Res.* **15:**109.

Bruce, A., and Svennerholm, L., 1971, Skeletal muscle lipids. I. Changes in fatty acid composition of lecithin in man during growth, *Biochim. Biophys. Acta* **239:**393.

Carroll, K. K., 1965, Dietary fat and the fatty acid composition of tissue lipids, *J. Am. Oil Chem. Soc.* **42:**516.

Chin, H. P., 1970, Sphingomyelin fatty acid composition in human twins: The effects of genotype, age and sex, *Biochim. Biophys. Acta* **218**:407.

Christie, W. W., 1972. The preparation of alkyl esters from fatty acids and lipids, in: *Topics in Lipid Chemistry*, Vol. 3 (F. D. Gunstone, ed.), pp. 171–197, Wiley, New York.

Christie, W. W., 1973, The structures of bile phosphatidylcholines, *Biochim. Biophys. Acta* **316**:204.

Christie, W. W., and Moore, J. H., 1970, A comparison of the structures of triglycerides from various pig tissues, *Biochim. Biophys. Acta* **210**:46.

Christie, W. W., Moore, J. H., Lorimer, A. R., and Lawrie, T. D. V., 1971, The structures of triglycerides from atherosclerotic plaques and other human tissues, *Lipids* **6**:854.

Clements, J., 1971, Comparative lipid chemistry of lungs, *Arch. Intern. Med.* **127**:387.

Cmelik, S., and Fonseca, E., 1974, Composition of mitochondrial lipids from rat adrenal glands, *Hoppe-Seyler's Z. Physiol. Chem.* **355**:19.

Cmelik, S., and Ley, H., 1974, Fatty acid composition of some cellular and subcellular elements of the elephant adrenal gland, *Hoppe-Seyler's Z. Physiol. Chem.* **355**:797.

Cmelik, S. H. W., and Ley, H., 1975, Distribution of cholesteryl esters and other lipids in subcellular fractions of the adrenal gland of the pig, *Lipids* **10**:707.

Colbeau, A., Nachbaur, J., and Vignais, P. M., 1971, Enzymic characterization and lipid composition of rat liver subcellular membranes, *Biochim. Biophys. Acta* **249**:462.

Comai, K., Farber, S. J., and Paulsrud, J. R., 1975, Analysis of renal medullary lipid droplets from normal, hydronephrotic and indomethacin treated rabbits, *Lipids* **10**:555.

Coniglio, J. G., Grogan, W. M., Jr., Harris, D. G., and Fitzhugh, M. L., 1975, Lipid and fatty acid composition of testes of quaking mice, *Lipids* **10**:109.

Connellan, J. M., and Masters, C. J., 1965, Fatty acid components of rat tissue lipids, *Biochem. J.* **94**:81.

Davis, J. T., Bridges, R. B., and Coniglio, J. G., 1966, Changes in lipid composition of the maturing rat testis, *Biochem. J.* **98**:342.

Downing, D. T., and Sharaf, D. M., 1976, Skin surface lipids of the guinea pig, *Biochim. Biophys. Acta* **431**:378.

Egwim, P. O., and Kummerow, F. A., 1972, Influence of dietary fat on the concentration of long-chain unsaturated fatty acid families in rat tissues, *J. Lipid Res.* **13**:500.

Ferber, E., De Pasquale, G. G., and Resch, K., 1975, Phospholipid metabolism of stimulated lymphocytes: Composition of phospholipid fatty acids, *Biochim. Biophys. Acta* **398**:364.

Fleischer, S., and Rouser, G., 1965, Lipids of subcellular particles, *J. Am. Oil Chem. Soc.* **42**:588.

Fleischer, S., Rouser, G., Fleischer, B., Casu, A., and Kritchevsky, G., 1967, Lipid composition of mitochondria from bovine heart, liver and kidney, *J. Lipid Res.* **8**:170.

Fletcher, R. F., 1972, Lipids of human myocardium, *Lipids* **7**:728.

Folch, J., Lees, M., and Stanley, G. H. S., 1957, A simple method for the isolation and purification of total lipids from animal tissues, *J. Biol. Chem.* **266**:497.

Glass, R. L., 1971, Alcoholysis, saponification and preparation of fatty acid methyl esters, *Lipids* **6**:919.

Gloster, J., and Harris, P., 1969, The lipid composition of mitochondrial and microsomal fractions of human myocardial homogenates, *Cardiovasc. Res.* **3**:45.

Gold, M., 1968, Triglyceride subclasses of various dog adipose tissue sites, *Lipids* **3**:539.

Golovnya, R. V., Uralets, V. P., and Kuzmenko, T. E., 1976, Characterization of fatty acid methyl esters by gas chromatography on siloxane liquid phases, *J. Chromatogr.* **121**:118.

Gray, G. M., and Yardley, H. J., 1975, Lipid compositions of cells isolated from pig, human and rat epidermis, *J. Lipid Res.* **16**:434.

Grigor, M. R., and Duncley, G. G., 1973, Origin of the high saturated fatty acid content of rat fecal lipids, *Lipids* **8**:53.

Grigor, M. R., Duncley, G. G., and Purves, H. D., 1970, The branched-chain fatty acids of rat fecal lipids: The contribution of the intestinal microorganisms, *Biochim. Biophys. Acta* **218**:400.

Grigor, M. R., Moehl, A., and Snyder, F., 1972, Occurrence of ethanolamine and choline-containing plasmalogens in adipose tissue, *Lipids* **7**:766.

Grogan, W. M., Jr., Coniglio, J. G., and Rhamy, R. K., 1973, Identification of some polyenoic acids isolated from human testicular tissue, *Lipids* **8**:480.

Grossi, E., Paoletti, P., and Paoletti, R., 1961, A gas–liquid chromatographic analysis of fatty acid compositions of human normal and tumoral nervous tissue, in: *Proceedings of the Fourth International Congress of Neuropathology*, Vol. 1 (H. Jacon, ed.), pp. 29–35, Georg Thieme Verlag, Stuttgart.

Guarnieri, M., and Johnson, R. M., 1970, The essential fatty acids, in: *Advances in Lipid Research*, Vol. 8 (D. Kritchevsky and R. Paoletti, eds.), pp. 115–174, Academic Press, New York.

Henning, R., Kaulen, H. D., and Stoffel, W., 1970, Isolation and chemical composition of the lysosomal and the plasma membrane of the rat liver cell, *Hoppe-Seyler's Z. Physiol. Chem.* **351**:1191.

Hilditch, T. P., and Williams, P. N., 1964, *The Chemical Constitution of Natural Fats*, 4th ed., Chapman and Hall, London.

Hoffman, A. G. D., and Kuksis, A., 1977, Comparison of *de novo* synthesis of fatty acids between villus and crypt cells isolated from rat small intestine, *Proc. Can. Fed. Biol. Soc.* **20**:21.

Holman, R. T., and Hofstetter, H. H., 1965, The fatty acid composition of the lipids from bovine and porcine reproductive tissues, *J. Am. Oil Chem. Soc.* **42**:540.

Howard, B. V., Butler, J. D., and Bailey, J. M., 1973, *Lipid Metabolism in Normal and Tumour Cells in Culture* (R. Wood, ed.), pp. 200–214, American Oil Chemists' Society Press, Champaign, Ill.

Huang, T. C., and Kuksis, A., 1967, A comparative study of the lipids of globule membrane and fat core of the milk serum of cows, *Lipids* **2**:453.

Huet, C., Levy, M., and Pascaud, M., 1968, Spécificité de constitution en acides gras des phospholipides des membranes mitochondriales, *Biochim. Biophys. Acta* **150**:521.

Illingworth, D. R., and Portman, O. W., 1972, Independence of phospholipid and protein exchange between plasma lipoproteins *in vivo* and *in vitro*, *Biochim. Biophys. Acta* **280**:281.

Imaizumi, K., and Sugano, M., 1975, Glycerolipid metabolism in cellular and subcellular fractions of fasted rat liver, *J. Nutr. Sci. Vitaminol.* **21**:223.

Insull, W., Jr., Hirsch, J., James, T., and Ahrens, E. H., Jr., 1959, The fatty acids of human milk. II. Alterations produced by manipulation of caloric balance and exchange of dietary fats, *J. Clin. Invest.* **38**:443.

Jaeger, H., Kloer, H. U., and Ditschuneit, H., 1976, Automated glass capillary gas–liquid chromatography of fatty acid methyl esters with reference to *cis-* and *trans-*isomers, *J. Lipid Res.* **17**:185.

Jain, Y. C., and Anand, S. R., 1976, Fatty acids and fatty aldehydes of buffalo seminal plasma and sperm lipid, *J. Reprod. Fertil.* **47**:261.

James, A. T., Webb, J. P. W., and Kelloch, T. D., 1961, The occurrence of unusual fatty acids in fecal lipids from human beings with normal and abnormal fat absorption, *Biochem. J.* **78**:333.

Kang, K. W., Taylor, G. E., Geves, J. H., Staley, H. L., and Christian, J. C., 1971, Genetic variability of human plasma and erythrocyte lipids, *Lipids* **6**:595.

Keenan, T. W., and Morre, D. J., 1970, Phospholipid class of fatty acid composition of golgi apparatus isolated from rat liver and comparison with other cell fractions, *Biochemistry* **9**:19.

Kuksis, A., 1966, Quantitative lipid analysis by combined thin-layer and gas–liquid chromatographic systems, in: *Chromatographic Reviews*, Vol. 8 (M. Lederer, ed.), pp. 172–207, Elsevier, Amsterdam.

Kuksis, A., 1971, Progress in the analysis of lipids. IX. Gas chromatography, Part 1, *Fette Seifen Anstrichm.* **71**:130.

Kuksis, A., 1972, Newer developments in determination of structure of glycerides and phosphoglycerides, in: *Progress in the Chemistry of Fats and other Lipids*, Vol. 12 (R. T. Holman, ed.), pp. 1–163, Pergamon Press, Oxford.

Kuksis, A., 1977, Routine chromatography of simple lipids and their constituents, *J. Chromatogr. Biomed. Appl.* **143**:3.

Lagrou, A., Kierick, W., Christophe, A., and Verdonk, G., 1974, Lipid composition of normal and hypertrophic bovine thyroids, *Lipids* **9**:870.

Leikola, E., Nieminen, E., and Salomaa, E., 1965, Occurrence of methyl esters in the pancreas, *J. Lipid Res.* **6**:490.

Levis, G. M., Karli, J. N., and Malamos, B., 1972, The phospholipids of the thyroid gland, *Clin. Chim. Acta* **41**:335.

Lin, S. N., Pfaffenberger, C. D., and Horning, E. C., 1975, Thermostable glass open tubular capillary columns with a polar liquid phase for gas chromatography, *J. Chromatogr.* **104:**319.

Lipshaw, L. A., and Foa, P. P., 1974, The composition and possible physiological role of the thyroid lipids, in: *Advances in Lipid Research*, Vol. 12 (D. Kritchevsky and R. Paoletti, eds.), pp. 227–250, Academic Press, New York.

Lough, A. K., and Garton, G. A., 1968, The lipids of human pancreas with special reference to the presence of fatty acid methyl esters, *Lipids* **3:**321.

Lough, A. K., Felinski, L., and Gordon, G. A., 1962, The production of methyl esters of fatty acids as artifacts during the extraction or storage of tissue lipids in the presence of methanol, *J. Lipid Res.* **3:**478.

Marai, L., and Kuksis, A., 1973, Comparative study of molecular species of glycerolipids in sarcotubular membranes of skeletal muscle of rabbit, rat, chicken, and man, *Can. J. Biochem.* **51:**1365.

Margolis, S., 1969, Structure of very low and low density lipoproteins, in: *Structure and Function of Lipoproteins in Living Systems* (E. Tria and A. M. Scanu, eds.), pp. 369–424, Academic Press, New York.

Martonosi, A., 1972, Biochemical and clinical aspects of sarcoplasmic reticulum function, in: *Current Topics in Membranes and Transport*, Vol. 3 (F. Bronner and A. Kleinzeller, eds.), pp. 83–197, Academic Press, New York.

Meldolesi, J., Jamieson, J. D., and Palade, G. E., 1971, Cellular membranes in the pancreas of the guinea pig. II. Lipids, *J. Cell Biol.* **49:**130.

Mills, S. C., Cook, L. J., Scott, T. W., and Nestel, P. J., 1976, Effect of dietary fat supplementation on the composition and positional distribution of fatty acids in ruminant and porcine glycerides, *Lipids* **11:**49.

Montfort, A., Van Golde, L. M. G., and Van Deenen, L. L. M., 1971, Molecular species of lecithins from various animal tissues, *Biochim. Biophys. Acta* **231:**335.

Mookerjea, S., Park, C. E., and Kuksis, A., 1975, Lipid profiles of plasma lipoproteins of fasted and fed normal and choline-deficient rats, *Lipids* **10:**374.

Morrison, W. R., 1970, Milk lipids, in: *Topics in Lipid Chemistry*, Vol. 1 (F. D. Gunstone, ed.), pp. 51–106, Logos Press, London.

Morrison, W. R., and Smith, L. M., 1964, Preparation of fatty acid methyl esters and dimethylacetals from lipids with boron trifluoride–methanol, *J. Lipid Res.* **5:**600.

Nakamura, M., Jensen, B., and Privett, O. S., 1968, Effect of hypohysectomy on the fatty acids and lipid classes of rat testes, *Endocrinology* **82:**137.

Nazir, D. J., Alcaraz, A. P., and Nair, P. P., 1967*a*, Lipids of subcellular particles from bovine heart muscle. I. Fatty acids of neutral lipids, *Can. J. Biochem.* **45:**1725.

Nazir, D. J., Alcaraz, A. P., and Nair, P. P., 1967*b*, Lipids of subcellular particles from bovine heart muscle. II. Fatty acids of phospholipids, *Can. J. Biochem.* **45:**1739.

Nelson, G. J., 1972*a*, Quantitative analysis of blood lipids, in: *Blood Lipids and Lipoproteins: Quantitation, Composition and Metabolism* (G. J. Nelson, ed.), pp. 25–73, Interscience, New York.

Nelson, G. J., 1972*b*, Lipid composition and metabolism of erythrocytes, in: *Blood Lipids and Lipoproteins: Quantitation, Composition and Metabolism* (G. J. Nelson, ed.), pp. 317–386, Interscience, New York.

Nelson, G. J., 1975, Fractionation of phospholipids, in: *Analysis of Lipids and Lipoproteins* (E. G. Perkins, ed.), pp. 63–89, American Oil Chemists' Society, Champaign, Ill.

Nicolaides, N., and Ansari, M. N. A., 1969, The dienoic fatty acids of human skin surface lipid, *Lipids* **4:**79.

Nicolaides, N., Apon, J. M. B., and Wong, D. H., 1976, Further studies of the saturated methyl branched fatty acids of vernix caseosa lipid, *Lipids* **11:**781.

Nikkari, T., 1965, Composition and secretion of the skin surface lipids of the rat: Effects of dietary lipids and hormones, *Scand. J. Clin. Lab. Invest. Suppl.* **85:**1.

Nikkari, T., 1974, Comparative chemistry of sebum, *J. Invest. Dermatol.* **62:**257.

Odutuga, A. A., Carey, E. M., and Prout, R. E. S., 1973, Changes in the lipid and fatty acid composition of developing rabbit brain, *Biochim. Biophys. Acta* **316:**115.

Ottenstein, D. M., Bartley, D. A., and Supina, W. R., 1976, Gas chromatographic separation of *cis-trans* isomers: Methyl oleate/methyl elaidate, *J. Chromatogr.* **119:**401.

Parkes, J. G., and Thompson, W., 1970, The composition of phospholipids in outer and inner mitochondrial membranes from guinea pig liver, *Biochim. Biophys. Acta* **196:**162.

Poulos, A., Dari-Bennett, A., and White, I. G., 1973, The phospholipid-bound fatty acids and aldehydes of mammalian spermatozoa, *Comp. Biochem. Physiol.* **46B:**541.

Privett, O. S., Dougherty, K. A., and Castell, J. D., 1971, Quantitative analysis of lipid classes, *Am. J. Clin. Nutr.* **24:**1265.

Prottey, C., and Hawthorne, J. N., 1966, The lipids of mammalian pancreas, *Biochem. J.* **101:**191.

Ray, T. K., Skipski, V. P., Barclay, M., Essner, E., and Archibald, F. M., 1969, Lipid composition of rat liver plasma membranes, *J. Biol. Chem.* **244:**5528.

Richardson, T., Tappel, A. L., and Gruger, E. R., 1961, Essential fatty acids in mitochondria, *Arch. Biochem. Biophys.* **94:**1.

Rogers, C. G., 1974, Fatty acids and phospholipids of adult and newborn rat hearts and of cultured, beating neonatal rat myocardial cells, *Lipids* **9:**541.

Rouser, G., Kritchevsky, G., Yamamoto, A., Simon, G., Galli, C., and Bauman, A. J., 1969*a*, Diethylaminoethyl and triethylaminoethyl cellulose column chromatographic procedures for phospholipids, glycolipids and pigments, in: *Methods in Enzymology*, Vol. 14 (J. M. Lowestein, ed.), pp. 272–317, Academic Press, New York

Rouser, G., Simon, G., and Kritchevsky, G., 1969*b*, Species variations in phospholipid class distribution of organs. I. Kidney, liver and spleen, *Lipids* **4:**599.

Rouser, G., Kritchevsky, G., and Yamamoto, A., 1976, Column chromatographic and associated procedures for separation and determination of phosphatides and glycolipids, in: *Lipid Chromatographic Analysis*, Vol. 3 (G. V. Marinetti, ed.), 2nd ed., pp. 713–776, Marcel Dekker, New York.

Scott, T. W., and Setchell, B. P., 1968, Lipid metabolism in the testes of the ram, *Biochem. J.* **107:**273.

Scott, T. W., Setchell, B. P., and Bassett, J. M., 1967, Characterization and metabolism of ovine fetal lipids, *Biochem. J.* **104:**1040.

Scott, T. W., Hansel, W., and Donaldson, L. E., 1968, Metabolism of phospholipids and the characterization of fatty acids in bovine corpus luteum, *Biochem. J.* **108:**317.

Sheppard, A. J., and Iverson, J. L., 1975, Esterification of fatty acids for gas–liquid chromatographic analysis, *J. Chromatogr. Sci.* **13:**448.

Siakotos, A. N., and Rouser, G., 1965, Analytical separation of non-lipid water soluble substances and gangliosides from other lipids by dextran gel column chromatography, *J. Am. Oil. Chem. Soc.* **42:**913.

Silbert, D. F., 1975, Genetic modification of membrane lipid, *Annu. Rev. Biochem.* **44:**315.

Silbert, D. F., Cronan, J. E., Jr., Beacham, I. R., and Harder, M. E., 1974, Genetic engineering of membrane lipid, *Fed. Proc.* **33:**1725.

Skipski, V. P., 1972, Lipid composition of lipoproteins in normal and disease states, in: *Blood Lipids and Lipoproteins: Quantitation, Composition and Metabolism* (G. J. Nelson, ed.), pp. 471–584, Interscience, New York.

Skipski, V. P., and Barclay, M., 1969, Thin-layer chromatography of lipids, in: *Methods in Enzymology*, Vol. 14 (J. M. Lowestein, ed.), pp. 530–612, Academic Press, New York.

Stoffel, W., and Schiefer, H. G., 1968, Biosynthesis and composition of phosphatides in outer and inner mitochondrial membranes, *Hoppe-Seyler's Z. Physiol. Chem.* **349:**1017.

Sun, G. Y., 1973, Phospholipids and acyl groups in subcellular fractions from human cerebral cortex, *J. Lipid Res.* **14:**656.

Sun, G. Y., and Leung, B. S., 1974, Phospholipids and acyl groups of subcellular membrane fractions from human intracranial tumors, *J. Lipid Res.* **15:**423.

Svennerholm. L., 1968, Distribution and fatty acid composition of phosphoglycerides in normal human brain, *J. Lipid Res.* **9:**570.

Swell, L., Enteman, C., Leong, G. F., and Holloway, R. J., 1968, Bile acids and lipid metabolism. IV. Influence of bile acids on biliary and liver organelle phospholipids and cholesterol, *Am. J. Physiol.* **215:**1390.

Takayasu, K., Okuda, K., and Yoshikawa, I., 1970, Fatty acid composition of human and rat adrenal lipids: Occurrence of $\omega 6$ docosatrienoic acid in human adrenal cholesterol ester, *Lipids* **5:**743.

Tsao, S. S., and Cornatzer, W. E., 1967, Chemical composition of subcellular particles from cultured cells of human tissue, *Lipids* **2**:41.

Van Golde, L. M. G., Pieterson, W. A., and Van Deenen, L. L. M., 1968, Alterations in the molecular species of rat liver lecithin by corn oil feeding to essential fatty acid deficient rat as a function of time, *Biochim. Biophys. Acta* **157**:84.

Van Hoeven, R. P., Emmelot, P., Krol, J. H., and Oomen-Meulemans, E. P. M., 1975, Studies on plasma membranes. XXII. Fatty acid profiles of lipid classes in plasma membranes of rat and mouse livers and hepatomas, *Biochim. Biophys. Acta* **380**:1.

Veerkamp, J. H., Mulder, I., and Van Deenen, L. L. M., 1962, Comparison of the fatty acid composition of lipids from different animal tissues including some tumours, *Biochim. Biophys. Acta* **57**:299.

Vilkki, P., and Jaakonmaki, I., 1966, Role of fatty acids in iodide complexing lecithin, *Endocrinology* **78**:453.

Wells, M. A., and Dittmer, J. C., 1963, The use of Sephadex for the removal of non-lipid contaminants from lipid extracts, *Biochemistry* **2**:1259.

Wheeldon, L. W., Schumert, Z., and Turner, D. A., 1965, Lipid composition of heart muscle homogenate, *J. Lipid Res.* **6**:481.

White, D. A., 1973, The phospholipid composition of mammalian tissues, in: *Form and Function of Phospholipids* (G. B. Ansell, R. M. C. Dawson, and J. N. Hawthorne, eds.), 2nd ed., pp. 441–482, Elsevier, Amsterdam.

Wilkinson, D. I., 1970, Positional isomers of monoene and diene fatty acids of human skin epidermal cells, *Arch. Biochem. Biophys.* **136**:368.

Wood, R., 1970, Plasma membranes: Structural analyses of neutral lipids and phospholipids of rat liver, *Arch. Biochem. Biophys.* **141**:174.

Wood, R., 1973, Tumour lipids: Structural and metabolism studies of Ehrlich ascites cells, in: *Tumour Lipids: Biochemistry and Metabolism* (R. Wood, ed.), pp. 139–182, American Oil Chemists' Society Press, Champaign, Ill.

Wood, R., 1976, Oleic and vaccenic acid levels in lipid classes of tumours, *Lipids* **11**:578.

Zilversmit, D. B., 1965, The composition and structure of lymph chylomicrons in dog, rat and man, *J. Clin. Invest.* **44**:1610.

Zilversmit, D. B., 1969, Chylomicrons, in: *Structural and Functional Aspects of Lipoproteins in Living Systems* (E. Tria and A. M. Scanu, eds.), pp. 329–368, Academic Press, New York.

Zilversmit, D. B., 1971, Stimulation of phospholipid exchange between mitochondria and artificially prepared phospholipid aggregates by a soluble fraction from liver, *J. Biol. Chem.* **246**:2645.

Index